Distribution Theory and Transform Analysis

An Introduction to Generalized Functions, with Applications

by *A. H. Zemanian*

College of Engineering and Applied Sciences
State University of New York
at Stony Brook

DOVER PUBLICATIONS, INC.
New York

To Edna

Copyright © 1965 by A. H. Zemanian.
All rights reserved under Pan American and International Copyright Conventions.

Published in Canada by General Publishing Company, Ltd., 30 Lesmill Road, Don Mills, Toronto, Ontario.
Published in the United Kingdom by Constable and Company, Ltd.

This Dover edition, first published in 1987, is an unabridged, slightly corrected republication of the work first published by the McGraw-Hill Book Company, New York, 1965, in its "International Series in Pure and Applied Mathematics."

Manufactured in the United States of America
Dover Publications, Inc., 31 East 2nd Street, Mineola, N.Y. 11501

Library of Congress Cataloging-in-Publication Data

Zemanian, A. H. (Armen H.)
 Distribution theory and transform analysis.

 Reprint, slightly corrected. Originally published: New York : McGraw-Hill, c1965. (International series in pure and applied mathematics)
 Bibliography: p.
 Includes index.
 1. Distributions, Theory of (Functional analysis) 2. Transformations (Mathematics) I. Title.
QA324.Z46 1987   515.7'82   87-9116
ISBN 0-486-65479-6 (pbk.)

Preface

L. Schwartz's theory of distributions had two important effects in mathematical analysis. First of all, it provided a rigorous justification for a number of formal manipulations that had become quite common in the technical literature. The second and more important effect was that it opened up a new area of mathematical research, which in turn provided an impetus in the development of a number of mathematical disciplines, such as ordinary and partial differential equations, operational calculus, transformation theory, and functional analysis. However, the subject has remained pretty much in the realm of advanced mathematics, and only a few aspects of it have found their way into the technical literature.

To be sure, a certain type of distribution (in particular, the delta function and its derivatives) had been used in the physical and engineering sciences for quite some time before the advent of distribution theory. Indeed, the delta function dates back to the nineteenth century. A summary of its history is given by Van der Pol and Bremmer (see Van der Pol and Bremmer [1], pp. 62–66, in the bibliography, Appendix D). On the other hand, distribution theory appears to have first been formulated in 1936 by S. L. Soboleff (see Soboleff [1]) and then developed in a systematic and thorough way by L. Schwartz (see Schwartz [1]), whose books appeared in 1950 and 1951. A somewhat different version of this theory was proposed by S. Bochner around 1927 (see Bochner [1], chap. VI), who used it to generalize the Fourier transformation for functions $f(t)$ that grow as some power of t as $|t|$ approaches infinity.

This book, which is based on a graduate course given at the State University of New York at Stony Brook, has two objectives. The first is to provide a comparatively elementary introduction to distribution theory, and the second is to describe the generalized Fourier and Laplace transformations and their applications to integrodifferential equations, difference equations, and passive systems. In recent years an ever-increasing number of textbooks have been devoted to the classical Fourier and Laplace transformations. The corresponding distributional transformations, although they are considerably more powerful tools, have not received the same attention in the current textbooks, nor have they been widely employed by scientists and engineers. It is hoped that this book will help to popularize distributional transform analysis.

Actually, one can introduce the delta function and its derivatives without developing a general theory of distributions, and many current books do so. However, these singular functions comprise but a very small subclass of all the distributions. More importantly, distribution theory provides powerful analytical techniques that cannot be described merely in terms of the delta function and its derivatives. An account of some of these techniques is given in this book.

Another theory of generalized functions is provided by Mikusiński's operational calculus (see Mikusiński [2] and Erdélyi [1]), which is related to the distributional Laplace transformation in roughly the same way as Heaviside's operational calculus is related to the classical Laplace transformation. In many problems one can use either Mikusiński's method or the distributional technique to obtain a solution. It is the latter procedure that is discussed in this book, and one possible justification for it is the following. At the present time the classical Laplace transformation has pretty much superseded Heaviside's operational calculus both in the technical literature and in our college courses. It seems, therefore, more natural to extend the Laplace transformation rather than Heaviside's method and thereby build on the training that the student already has. Admittedly, this is purely a pragmatic reason. Assuming, if you will, that time is of no importance, the best answer to the question "Which theory should be studied?" is "Both." In Sec. 6.4 we very briefly discuss Mikusiński's operational calculus and compare it with distribution theory.

There are a variety of other approaches to the theory of generalized functions that are based, in general, on the facts that generalized functions can be represented as sequences of ordinary functions, which converge in a certain way, and that over a finite interval a generalized function is a finite-order derivative (in an unconventional sense) of a continuous function. (See, for example, Beltrami [1]; Bochner [1], chap. VI; Bremermann and Durand [1]; Courant [1], pp. 766–798;

König [1]; Korevaar [1]; Lighthill [1]; Liverman [1]; Mikusiński [1]; Rehberg [1]; Temple [1].) All these methods can be understood in terms of Schwartz's theory, and most of them are, in fact, encompassed by it. If one wishes to delve at some length into the theory of generalized functions, a knowledge of Schwartz's approach, which conceives of generalized functions as certain continuous linear functionals, has become indispensable, in view of the large and ever-increasing body of literature that uses this point of view. For these reasons, our development will employ Schwartz's functional approach.

This book can be used for a graduate course for engineering and science students and possibly for a senior-level undergraduate course for mathematics majors. It is presumed that the reader has already had a course in advanced calculus and is familiar with the standard theorems on the interchange of limit processes. Some knowledge of functions of a complex variable and of matrix manipulations is also assumed. Finally, at certain places we employ some elements of the theory of Lebesgue integration, although most of the text can be followed without having any knowledge of this subject. In any case, whenever theorems or techniques in any of these subjects are used, the reader is referred to various standard books, where he can seek any additional information that he may need.

In Sec. 10.5 we employ Cauer's representation for a positive-real function. Its proof would take us too far afield, and it has therefore been omitted. Instead, we refer the reader to the mathematical literature (see Loomis and Widder [1]).

An attempt has been made to render all the proofs as elementary as possible. For example, Theorem 3.4-2 can be proved in a very brief way if use is made of the Hahn-Banach theorem. This has not been done; instead, a longer but more elementary argument has been employed.

Since this book is addressed both to mathematics students and to science and engineering students, the problems have been designed to develop the reader's understanding of the theory as well as his facility for using distributions. Thus, some of the problems are exercises of proof; they are concerned either with the arguments given in the text or with the extension of the theory and the development of new results. In contrast to this, other problems develop specific convolution and transform formulas. Still others are exercises for solving distributionally various differential and difference equations; they are intended to enlarge the student's ability to apply distribution theory. It is hoped that a sufficiently broad spectrum of problems has been provided to satisfy the diverse needs of various types of students.

Briefly, the structure of this book is as follows. In Chapter 1 the basic definitions of distributions and the operations that apply to them are discussed. The calculus of distributions and, in particular, limits, differentiation, integration, and the interchange of limiting processes are considered in Chapter 2. Some deeper properties of distributions, such as their local character as derivatives of continuous functions, are given in Chapter 3.

Chapter 4 introduces the distributions of slow growth, which arise naturally in the generalization of the Fourier transformation. Chapters 5 and 6 are concerned with the convolution process and its use in representing differential and difference equations.

The distributional Fourier and Laplace transformations are developed in Chapters 7 and 8, and the latter transformation is applied in Chapter 9 to obtain an operational calculus for the solution of differential and difference equations of the initial-condition type.

Some of the previous theory is applied in Chapter 10 to a discussion of the fundamental properties of certain physical systems, and a concise development of the relationship between the positive-reality of a system function and the passivity of the system is obtained there.

Chapter 11 ends the book with a consideration of periodic distributions. This chapter acts in a crude way as a summary of the book, since it follows the broad outline of the preceding portion of the text.

The appendixes contain a table of formulas for the distributional Laplace transformation, a glossary of symbols, and a bibliography. We particularly direct the reader's attention to Appendix C, which contains the definitions of most of the symbols used in the text.

There is enough material in this book for a two-semester course. A one-semester course may be based upon the following portions of the text: Secs. 1.1 to 1.8, 2.1 to 2.4, 2.6, 2.7, 3.1, 3.3, 3.4, 4.1 to 4.5, 5.1 to 5.6, 6.1 to 6.3, 7.1 to 7.5, 8.1 to 8.5, 9.1 to 9.6. These sections are self-contained; indeed, for those wishing a more rapid introduction to the subject without the more specialized or advanced discussions of the other sections, this outline is a good one to follow. We have designated these sections by using stars and diamonds. Sections denoted by diamonds are those whose conclusions are used in the subsequent development of starred and diamond-marked sections but whose proofs are fairly long and technical. The reader may at first skip these proofs and just read the theorems, examples, and explanatory portions of the diamond-marked sections. This will provide a still briefer introduction to distribution theory without any loss of continuity.

All theorems, examples, and figures are triple-numbered; the first two numbers indicate the sections in which they appear. For example, Theorem 2.2-1 is the first theorem in Sec. 2.2. On the other hand, equations, lemmas, and problems are numbered consecutively, starting with 1 in each new section.

The author wishes to express his thanks to I. Gerst, R. Glasheen, E. O'Brien, J. Sheppard, and his students for their interest and advice. Thanks are also extended to P. Barry, T. Loughlin, and B. Queen, who helped the author check the Laplace transform formulas in Appendix B and compute and plot the graphs of Figs. 1.3-1 to 1.3-3, 1.4-1, and 5.5-1. The author is deeply indebted to J. Korevaar for thoroughly reviewing the manuscript and making numerous criticisms and suggestions. The comments of C. Saltzer on various portions of the book and of C. A. Desoer, who read Chapter 10, were also a valuable aid during the final revisions. The encouragement, cooperation, and patience of the author's wife and her assistance in so many tasks during the preparation of the manuscript are also gratefully noted.

A. H. Zemanian

Contents

PREFACE iii

CHAPTER 1 DISTRIBUTIONS: THEIR DEFINITION AND BASIC PROPERTIES 1
- ★1.1 Introduction *1*
- ★1.2 The Space $\mathcal{D}$ of Testing Functions *2*
- ★1.3 Distributions *6*
- ★1.4 Pseudofunctions, Hadamard's Finite Part, and Cauchy's Principal Value *15*
- ★1.5 Testing Functions and Distributions of Several Variables *21*
- ★1.6 Equality of Distributions over Open Sets *24*
- ★1.7 Some Operations on Distributions *26*
- ♦1.8 Distributions as Local Phenomena *30*

CHAPTER 2 THE CALCULUS OF DISTRIBUTIONS 36
- ★2.1 Introduction *36*
- ♦2.2 Convergence of a Sequence of Distributions (Convergence in the Space $\mathcal{D}'$) *36*
- ★2.3 Some Special Cases of Convergence in $\mathcal{D}'$ *40*
- ★2.4 The Differentiation of Distributions *46*
- 2.5 Hadamard's Finite Part and Some Pseudofunctions Generated by It *57*
- ★2.6 The Primitives of Distributions Defined over $\mathcal{R}^1$ *67*
- ♦2.7 Continuity and Differentiability with Respect to a Parameter upon Which the Testing Functions Depend *72*
- 2.8 Distributions That Depend upon a Parameter and Integration with Respect to That Parameter *75*

CHAPTER 3   FURTHER PROPERTIES OF DISTRIBUTIONS   80

★3.1 Introduction   *80*
3.2 A Characterization of the Delta Functional and Its Derivatives   *81*
★3.3 A Local-boundedness Property of Distributions   *83*
♦3.4 Locally Every Distribution Is a Finite-order Derivative of a Continuous Function   *86*
3.5 Only Finite Linear Combinations of the Delta Functional and Its Derivatives Are Concentrated on a Point   *96*

CHAPTER 4   DISTRIBUTIONS OF SLOW GROWTH   99

★4.1 Introduction   *99*
★4.2 The Space $\mathcal{S}$ of Testing Functions of Rapid Descent   *99*
★4.3 The Space $\mathcal{S}'$ of Distributions of Slow Growth   *102*
♦4.4 A Boundedness Property for Distributions of Slow Growth   *109*
♦4.5 A Differentiability Property for the Application of a Distribution in $\mathcal{S}'_\tau$ to a Testing Function in $\mathcal{S}_{t,\tau}$ *111*

CHAPTER 5   CONVOLUTION   114

★5.1 Introduction   *114*
★5.2 The Direct Product of Distributions   *115*
★5.3 The Support, Commutativity, and Associativity of the Direct Product   *118*
★5.4 The Convolution of Distributions   *122*
★5.5 Some Operations on the Convolution Process   *131*
★5.6 The Continuity of the Convolution Process   *135*
5.7 The Convolution of a Distribution in $\mathcal{S}'$ with a Testing Function in $\mathcal{S}$ *138*
5.8 Convolution Operators   *142*

CHAPTER 6   CONVOLUTION EQUATIONS   148

★6.1 Introduction   *148*
★6.2 Convolution Algebras   *149*
★6.3 An Application to Ordinary Linear Differential Equations with Constant Coefficients   *157*
6.4 Mikusiński's Operational Calculus   *168*

CHAPTER 7   THE FOURIER TRANSFORMATION   171

★7.1 Introduction   *171*
♦7.2 The Ordinary Fourier Transformation   *172*

★7.3 The Fourier Transforms of Testing Functions of Rapid Descent *181*
★7.4 The Fourier Transforms of Distributions of Slow Growth *184*
★7.5 The Fourier Transformation of Convolutions of Distributions Having Bounded Supports *191*
 7.6 The Space Z of Testing Functions Whose Fourier Transforms Are in $\mathcal{D}$ *192*
 7.7 The Space Z' of Ultradistributions *198*
 7.8 The Fourier Transforms of Arbitrary Distributions *202*
 7.9 The Fourier Transformation of the Convolution of Two Distributions One of Which Has a Bounded Support *205*
 7.10 The General Solution of a Homogeneous Linear Differential Equation with Constant Coefficients *208*

CHAPTER 8 THE LAPLACE TRANSFORMATION 212

★8.1 Introduction *212*
◆8.2 The Laplace Transforms of Ordinary Right-sided Functions *213*
★8.3 The Laplace Transforms of Right-sided Distributions *222*
★8.4 The Inversion of the Laplace Transformation for Right-sided Distributions *235*
★8.5 The Laplace Transformation of Convolutions of Right-sided Distributions *240*
 8.6 Some Abelian Theorems of the Initial-value Type *243*
 8.7 Some Abelian Theorems of the Final-value Type *249*
 8.8 The Laplace Transforms of Left-sided Distributions *252*
 8.9 The Laplace Transforms of Distributions Having, in General, Unbounded Supports *254*

CHAPTER 9 THE SOLUTION OF DIFFERENTIAL AND DIFFERENCE EQUATIONS BY TRANSFORM ANALYSIS 260

★9.1 Introduction *260*
★9.2 The Use of the Laplace Transformation in Solving Convolution Equations in the Algebra $\mathcal{D}'_R$ *261*
★9.3 Ordinary Linear Differential Equations with Constant Coefficients *266*
★9.4 Ordinary Linear Integrodifferential Equations with Constant Coefficients *273*
★9.5 Ordinary Linear Difference Equations with Constant Coefficients: The Continuous-variable Case *279*
★9.6 Ordinary Linear Difference Equations with Constant Coefficients: The Discrete-variable Case *286*
 9.7 Ordinary Linear Differential Equations with Polynomial Coefficients *290*

CHAPTER 10   PASSIVE SYSTEMS   293

10.1 Introduction   *293*
10.2 One-ports Having Convolution Representations   *294*
10.3 Causality and Passivity   *300*
10.4 The Positive-reality of the Immittance Function   *305*
10.5 A Representation for the Unit Impulse Response Corresponding to a Positive-real Immittance Function   *307*
10.6 The Realizability of Every Positive-real Function   *310*

CHAPTER 11   PERIODIC DISTRIBUTIONS   313

11.1 Introduction   *313*
11.2 The Space $\mathcal{P}_T$ of Periodic Testing Functions   *314*
11.3 The Space $\mathcal{P}'_T$ of Periodic Distributions   *316*
11.4 T-convolution   *322*
11.5 The T-convolution Algebra $\mathcal{P}'_T$ *327*
11.6 The Fourier Series   *330*
11.7 The Finite Fourier Transformation and the Solution of T-convolution Equations   *335*
11.8 Some Applications to Differential and Difference Equations   *339*

APPENDIX A	**THE AXIOMS FOR A LINEAR SPACE**	**345**
APPENDIX B	**TABLES OF FORMULAS FOR THE RIGHT-SIDED LAPLACE TRANSFORMATION**	**346**
APPENDIX C	**GLOSSARY OF SYMBOLS**	**356**
APPENDIX D	**BIBLIOGRAPHY**	**362**

INDEX   367

1 Distributions: Their Definition and Basic Properties

★1.1 INTRODUCTION

A physical variable is customarily thought of as a function, i.e., a rule which assigns a number to each numerical value of some independent variable. For example, if the independent variable is time t and the physical quantity is a force f, then one would say that the force is known if its value $f(t)$ is specified at every instant of time t. However, it is impossible to observe the instantaneous values of $f(t)$. Any measuring instrument would merely record the effect that f produces on it over some nonvanishing interval of time.

As we shall see, another way of describing a physical variable is to specify it as a functional, i.e., as a rule which assigns a number to each function in a set of so-called "testing functions." We shall be exclusively concerned with functionals of a special type, namely, distributions. It turns out that the distribution concept provides a better mechanism for analyzing certain physical phenomena than does the function concept because, for one reason, various entities, such as the delta function, which arise naturally in several mathematical sciences can be correctly described as distributions but not as functions. Moreover, any physical quantity that can be adequately represented as a function can also be characterized as a distribution and, indeed, there is an advantage in using the latter representation. One cannot assign instantaneous values to a distribution, and consequently the problem of physically interpreting such values does not arise.

Actually, the idea of specifying a function not by its values but by its behavior as a functional on some space of testing functions is a concept that is quite familiar to scientists and engineers through their experience with the classical Fourier and Laplace transformations. When specifying a function $f(t)$ by its Fourier transform

$$\tilde{f}(\omega) = \int_{-\infty}^{\infty} f(t) e^{-i\omega t}\, dt \tag{1}$$

the function is, in fact, being considered as a functional on the set of testing functions consisting of all exponential functions $e^{-i\omega t}$ having imaginary exponents. (Here, ω is a fixed real number for each such testing function.) On the other hand, the testing functions on which distributions are defined cannot, in general, be written down in an explicit form. Nevertheless, the functional concept is a basic one both for Fourier transforms and for distributions.

We shall now take up in this first chapter the definitions of testing functions and distributions, their basic properties, and certain operations that one can apply to them.

★1.2 THE SPACE $\mathfrak{D}$ OF TESTING FUNCTIONS

Before we can describe distributions, we must define the *testing functions*, on which distributions operate. Several different spaces of testing functions will be discussed in this book. The one that is considered in this section is, for our purposes, the most important and is the one we shall employ most often.

Throughout this and the next section, the independent real variable t will be assumed to be one-dimensional. When a function has continuous derivatives of all orders on some set of points, we shall say that the function is *infinitely smooth on that set*. If this is true for all points, we shall merely say that the function is *infinitely smooth*. Moreover, whenever we refer to a "complex number" or a "complex-valued function," it is understood that the number may be real or the function may be real-valued.

The space of testing functions, which is denoted by $\mathfrak{D}$, consists of all complex-valued functions $\phi(t)$ that are infinitely smooth and zero outside some finite interval. It is not required that all elements of $\mathfrak{D}$ be zero outside the same finite interval. The reader can find in Appendix A the axioms that a space must satisfy if it is to be a *linear space*. $\mathfrak{D}$ obviously satisfies them.

An example of a testing function in $\mathfrak{D}$ is

$$\zeta(t) = \begin{cases} 0 & |t| \geq 1 \\ \exp \dfrac{1}{t^2 - 1} & |t| < 1 \end{cases} \tag{1}$$

It can be shown that every derivative of this function exists and is zero at $t = \pm 1$. More generally, then, this function has continuous derivatives of all orders for every t, and they are all equal to zero for $|t| \geq 1$.

It is a fact that *any complex-valued function $f(t)$ that is continuous for all t and zero outside a finite interval can be approximated uniformly by testing functions.* That is, given an $\varepsilon > 0$ there will exist a $\phi(t)$ in $\mathfrak{D}$ such that $|f(t) - \phi(t)| \leq \varepsilon$ for all t.

To show this, we shall construct a set of testing functions $\{\phi_\alpha(t)\}$ that converges uniformly to $f(t)$ as $\alpha \to 0$. Let

$$\gamma_\alpha(t) = \frac{\zeta(t/\alpha)}{\int_{-\infty}^{\infty} \zeta(t/\alpha)\,dt} \qquad \alpha > 0 \tag{2}$$

where $\zeta(t)$ is given by (1). This testing function is zero for $|t| \geq \alpha$, is positive elsewhere, and satisfies the condition

$$\int_{-\infty}^{\infty} \gamma_\alpha(t)\,dt = 1 \tag{3}$$

[$\gamma_\alpha(t)$ is plotted in Fig. 1.3-1 for various values of α.] The desired testing functions are given by

$$\phi_\alpha(t) = \int_{-\infty}^{\infty} f(\tau)\gamma_\alpha(t - \tau)\,d\tau \tag{4}$$

[The right-hand side of (4) is, in fact, the *convolution* of f with γ_α, a process that we shall discuss in Chap. 5.]

That $\phi_\alpha(t)$ is truly in $\mathfrak{D}$ can be shown as follows. Suppose that $f(t)$ is zero outside the interval $a < t < b$. Since $\gamma_\alpha(t)$ is zero for $|t| \geq \alpha$, it follows that $\phi_\alpha(t)$ is zero outside the interval $a - \alpha < t < b + \alpha$. Furthermore, the right-hand side of (4) is actually an integral with finite limits whose integrand is continuous with respect to (t, τ) and has a partial derivative with respect to t that is also continuous with respect to (t, τ). Hence, we may differentiate (4) under the integral sign with respect to t (Titchmarsh [1], p. 59). Since $\gamma_\alpha(t)$ is a testing function, the new integral will satisfy the same conditions. In fact, we may repeatedly differentiate (4) under the integral sign to show that $\phi_\alpha(t)$ is infinitely smooth. Hence, $\phi_\alpha(t)$ is in $\mathfrak{D}$.

Now, consider

$$|f(t) - \phi_\alpha(t)| = \left| \int_{-\infty}^{\infty} [f(t) - f(\tau)] \gamma_\alpha(t-\tau)\, d\tau \right|$$
$$\leq \int_{-\infty}^{\infty} |f(t) - f(\tau)| \gamma_\alpha(t-\tau)\, d\tau \qquad (5)$$

Since $f(t)$ is continuous in the closed interval $a - \alpha \leq t \leq b + \alpha$, it is uniformly continuous there (Widder [2], p. 172). In other words, for any $\varepsilon > 0$ there exists an $\eta > 0$ such that $|f(t) - f(\tau)| < \varepsilon$ for all pairs of t and τ in this interval that satisfy $|t - \tau| < \eta$. Moreover, $f(t)$ is zero outside the interval $a < t < b$, so that this condition is satisfied for all t and τ such that $|t - \tau| < \eta$. Hence, for all t we have from (5) and the fact that $\gamma_\alpha(t - \tau)$ is zero for $|t - \tau| \geq \alpha$ that, whenever $\alpha < \eta$,

$$|f(t) - \phi_\alpha(t)| \leq \varepsilon \int_{-\infty}^{\infty} \gamma_\alpha(t-\tau)\, d\tau = \varepsilon$$

This establishes our original statement.

It follows from our construction of the approximating function $\phi_\alpha(t)$ that, when $f(t)$ is zero outside the interval $a < t < b$, then $\phi_\alpha(t)$ will be zero outside the interval $a - \alpha < t < b + \alpha$, *where we are free to choose the positive number α as small as we wish.*

We can also show that, *if $f(t)$ is infinitely smooth for all t* (such as the function e^{st}, where s is a complex number), *then for any given finite interval $a \leq t \leq b$ there exists a testing function in $\mathfrak{D}$ that is identical to $f(t)$ over this interval.*

As a first step, a testing function $\phi_1(t)$ in $\mathfrak{D}$ that is identically equal to one over $a \leq t \leq b$ will be constructed. Let $h(t)$ be a continuous function that equals one over the interval $a - \alpha \leq t \leq b + \alpha$, where $\alpha > 0$, and let $h(t) = 0$ outside some larger finite interval. Then $\phi_1(t)$ is taken to be

$$\phi_1(t) = \int_{-\infty}^{\infty} h(\tau) \gamma_\alpha(t - \tau)\, d\tau \qquad (6)$$

where $\gamma_\alpha(t)$ is defined as before. Indeed, for $a \leq t \leq b$, (6) becomes

$$\phi_1(t) = \int_{t-\alpha}^{t+\alpha} \gamma_\alpha(t - \tau)\, d\tau = 1$$

Moreover, we have already shown [in our discussion of (4)] that a function of the form (6) is a testing function in $\mathfrak{D}$. Now, if $f(t)$ is infinitely smooth, then $f(t)\phi_1(t)$ will also be infinitely smooth. In addition, it is zero outside some finite interval. Hence, $f(t)\phi_1(t)$ is in $\mathfrak{D}$ and is identical to $f(t)$ for $a \leq t \leq b$. This is what we wished to show.

Thus we see that testing functions in $\mathfrak{D}$ may have a wide variety of forms.

We shall also need the concept of convergence in the space $\mathfrak{D}$, or simply *convergence in* $\mathfrak{D}$. A sequence of testing functions $\{\phi_\nu(t)\}_{\nu=1}^{\infty}$ is

Definition and basic properties

said *to converge in* $\mathfrak{D}$ *to zero* if all $\phi_\nu(t)$ are identically zero outside a fixed finite interval, if the sequence converges uniformly to zero, and if for each positive integer k the sequence of derivatives of order k, $\{\phi_\nu^{(k)}(t)\}_{\nu=1}^\infty$, also converges uniformly to zero. In other words, in addition to the condition that all $\phi_\nu(t)$ be zero outside some finite interval, we require that, for any given $\varepsilon > 0$ and any given nonnegative integer k, there exist a number N_k such that, for every $\nu \geq N_k$,

$$|\phi_\nu^{(k)}(t)| \leq \varepsilon \qquad (7)$$

where (7) is to hold for all values of t. The numbers N_k may vary with k. [We mean by $\phi^{(k)}(t)$ the kth derivative $d^k\phi/dt^k$ when $k = 1, 2, 3, \ldots$ and the function $\phi(t)$ itself when $k = 0$.]

As an example, the sequence $\{\zeta(t)/\nu\}$, where $\zeta(t)$ is given by (1), converges in $\mathfrak{D}$ to zero as $\nu \to \infty$. On the other hand, the sequence $\{\zeta(t/\nu)/\nu\}$ does not converge in $\mathfrak{D}$, even though it and all its derivatives converge uniformly to zero, since there does not exist a fixed finite interval outside which all the $\phi_\nu(t)$ are zero.

A sequence of testing functions $\{\phi_\nu(t)\}_{\nu=1}^\infty$ is said *to converge in* $\mathfrak{D}$ if the $\phi_\nu(t)$ are all in $\mathfrak{D}$, if they are all zero outside some fixed finite interval I, and if for every fixed nonnegative integer k the sequence $\{\phi_\nu^{(k)}(t)\}_{\nu=1}^\infty$ converges uniformly for $-\infty < t < \infty$.

Let $\phi(t)$ be the limit function of the sequence $\{\phi_\nu(t)\}_{\nu=1}^\infty$. The uniformity of the convergences ensures that, for each k, $\phi^{(k)}(t)$ is continuous and is the limit of $\{\phi_\nu^{(k)}(t)\}_{\nu=1}^\infty$ (Carslaw [1], p. 161). Since $\phi(t)$ is also zero outside the same finite interval I, $\phi(t)$ is in $\mathfrak{D}$. We can therefore conclude that the limit of every sequence that converges in $\mathfrak{D}$ is also in $\mathfrak{D}$. We shall refer to this property of $\mathfrak{D}$ by saying that $\mathfrak{D}$ is *closed under convergence*.

(Throughout this book we shall associate with each new space of functions or distributions no more than one concept of convergence. By saying that such a space is "closed under convergence," we shall mean that all its convergent sequences have limits that are also in the space. Actually, this property of being "closed under convergence" is related to the *completeness* of these spaces. In order to define completeness properly, one must introduce the concepts of "Cauchy sequences" and "neighborhoods" for these spaces. In this book, we are going to avoid such discussions, and we shall therefore refrain from using the word "complete.")

An equivalent criterion for the convergence in $\mathfrak{D}$ of $\{\phi_\nu\}_{\nu=1}^\infty$ to ϕ is that all the ϕ_ν be in $\mathfrak{D}$ and be zero outside a fixed finite interval and that the sequence of differences $\{\phi_\nu - \phi\}_{\nu=1}^\infty$ converge in $\mathfrak{D}$ to zero.

In these definitions of convergence in $\mathfrak{D}$, we are free to replace the sequence $\{\phi_\nu\}_{\nu=1}^\infty$ by a set $\{\phi_\nu\}_{\nu\to\infty}$, where the index ν tends continuously to infinity or perhaps to some finite limit. (In this regard, see Prob. 3.)

We shall usually deal with convergent sequences, but we shall also consider convergent directed sets occasionally.

Another concept that we shall employ subsequently is that of the support of a testing function $\phi(t)$. The support is the closure of the set E of all points where $\phi(t)$ is different from zero. In other words, it is E plus all the limit points of E. [For example, the support of the testing function (1) is the closed bounded interval, $-1 \leq t \leq 1$.] Thus, a testing function in $\mathfrak{D}$ is simply an infinitely smooth function whose support is a closed bounded set.

PROBLEMS

1 Show that every derivative of the function given by (1) is zero at $t = -1$ and at $t = 1$.

2 Construct an explicit testing function in $\mathfrak{D}$ that equals one for $-1 \leq t \leq 1$, equals zero for $t \leq -2$ and for $2 \leq t$, and is monotonically increasing for $-2 \leq t \leq -1$ and monotonically decreasing for $1 \leq t \leq 2$.

3 Let $\{\phi_\nu\}_{\nu \to \infty}$ be a set of testing functions in $\mathfrak{D}$, where the index ν approaches infinity through a continuous set of values. Show that $\{\phi_\nu\}_{\nu \to \infty}$ converges in $\mathfrak{D}$ to some limit ϕ if and only if:
1. All ϕ_ν have their supports contained in a fixed finite interval I.
2. Every sequence $\{\phi_{\nu_k}\}_{k=1}^\infty$ with $\nu_k \to \infty$ obtained from this set converges in $\mathfrak{D}$ to the same limit ϕ.

★1.3 DISTRIBUTIONS

A *functional* is a rule that assigns a number to every member of a certain set of functions. For our purposes, the set of functions will be taken to be the space $\mathfrak{D}$ and we shall consider functionals that assign a complex number to every member of $\mathfrak{D}$. Denoting a functional by the symbol f, we designate the number that f assigns to a particular testing function ϕ by $\langle f, \phi \rangle$.

Distributions, which we shall describe in this section, are functionals on the space $\mathfrak{D}$ that possess, in addition, two essential properties. The first of these is *linearity*. A functional f on $\mathfrak{D}$ is said to be linear if, for any two testing functions ϕ_1 and ϕ_2 in $\mathfrak{D}$ and any complex number α, the following conditions are satisfied:

$$\langle f, \phi_1 + \phi_2 \rangle = \langle f, \phi_1 \rangle + \langle f, \phi_2 \rangle$$
$$\langle f, \alpha\phi_1 \rangle = \alpha \langle f, \phi_1 \rangle \qquad (1)$$

The second property is *continuity*. A functional f on $\mathfrak{D}$ is said to be continuous if, for any sequence of testing functions $\{\phi_\nu(t)\}_{\nu=1}^\infty$ (or,

Sec. 1.3   **Definition and basic properties**   7

equivalently, for any directed set $\{\phi_\nu\}_{\nu\to\infty}$ that converges in $\mathfrak{D}$ to $\phi(t)$, the sequence of numbers $\{\langle f, \phi_\nu\rangle\}_{\nu=1}^\infty$ (or the directed set $\{\langle f, \phi_\nu\rangle\}_{\nu\to\infty}$) converges to the number $\langle f, \phi\rangle$ in the ordinary sense. If f is known to be linear, the definition of continuity may be somewhat simplified. In this case, f will be continuous if the numerical sequence $\{\langle f, \phi_\nu\rangle\}_{\nu=1}^\infty$ converges to zero whenever the sequence $\{\phi_\nu\}_{\nu=1}^\infty$ converges in $\mathfrak{D}$ to zero.

Thus, we may state the following definition of a distribution defined over the one-dimensional real euclidean space $\mathfrak{R}^1$:

A continuous linear functional on the space $\mathfrak{D}$ is a distribution.

The space of all such distributions is denoted by $\mathfrak{D}'$. $\mathfrak{D}'$ is called the *dual* (or *conjugate*) *space of* $\mathfrak{D}$.

One way to generate distributions is as follows. Let $f(t)$ be a locally integrable function (i.e., a function that is integrable in the Lebesgue sense over every finite interval). Corresponding to $f(t)$, we can define a distribution f through the convergent integral

$$\langle f, \phi\rangle = \langle f(t), \phi(t)\rangle \triangleq \int_{-\infty}^{\infty} f(t)\phi(t)\, dt \tag{2}$$

(We shall use the symbol $\triangleq$ when we wish to emphasize that a certain equation is a definition.) Actually, the limits on this integral can be altered to finite values, since $\phi(t)$ has a bounded support. Clearly, (2) is a linear functional. That it is also continuous can be shown in the following way. Let the sequence of testing functions $\{\phi_\nu(t)\}_{\nu=1}^\infty$ converge in $\mathfrak{D}$ to $\phi(t)$. Next, we observe that

$$\begin{aligned}|\langle f, \phi\rangle - \langle f, \phi_\nu\rangle| &= \left|\int_{-\infty}^{\infty} f(t)[\phi(t) - \phi_\nu(t)]\, dt\right| \\ &\leq \int_a^b |f(t)|\, |\phi(t) - \phi_\nu(t)|\, dt\end{aligned} \tag{3}$$

where the finite numbers a and b are so chosen that $\phi(t)$ and $\phi_\nu(t)$ are identically zero outside the interval $a < t < b$. Given an $\varepsilon > 0$, there exists an N such that, for all $\nu \geq N$, $|\phi(t) - \phi_\nu(t)| < \varepsilon$. Upon substituting this inequality into the right-hand side of (3) and noting that

$$\int_a^b |f(t)|\, dt$$

is a finite number, we see that

$$\lim_{\nu\to\infty} |\langle f, \phi\rangle - \langle f, \phi_\nu\rangle| = 0$$

which is what we wished to prove.

Distributions that can be generated through (2) from locally integrable functions are called *regular distributions*. It is a fact that *two continuous functions that produce the same regular distribution are identical.* In other words, if $f(t)$ and $g(t)$ are two continuous functions that differ somewhere, then the numbers $\langle f, \phi \rangle$ and $\langle g, \phi \rangle$ must also differ for at least one $\phi(t)$ in $\mathfrak{D}$. For there will be an interval $a < t < b$ in which $f(t) - g(t)$ is always positive or always negative, and by choosing

$$\phi(t) = \begin{cases} 0 & t \leq a \text{ and } t \geq b \\ \exp\left(-\dfrac{1}{t-a} - \dfrac{1}{b-t}\right) & a < t < b \end{cases} \tag{4}$$

we obtain

$$\langle f, \phi \rangle - \langle g, \phi \rangle \neq 0$$

It now follows that each testing function in $\mathfrak{D}$ uniquely determines a regular distribution in $\mathfrak{D}'$ and is, in turn, uniquely determined by this regular distribution. Upon identifying these two quantities, we may say that $\mathfrak{D}$ is a linear subspace of $\mathfrak{D}'$. In symbols, $\mathfrak{D} \subset \mathfrak{D}'$.

On the other hand, if $f(t)$ is merely a locally integrable function, the regular distribution corresponding to it certainly determines the function $f(t)$ uniquely within every open interval where it is known to be continuous. This follows directly from the argument given above. However, the regular distribution does not determine $f(t)$ at points where $f(t)$ is discontinuous. In fact, we may alter the values of $f(t)$ on a set of measure zero without altering the regular distribution. More generally, *if $f(t)$ and $g(t)$ are locally integrable and if their corresponding regular distributions agree* (that is, $\langle f, \phi \rangle = \langle g, \phi \rangle$ *for all ϕ in $\mathfrak{D}$), then $f(t)$ and $g(t)$ differ at most on a set of measure zero.* This result is a consequence of the fact that the space $\mathfrak{D}$ includes all functions of the form $[\phi(t)]^{1/n}$, where n is any positive integer and $\phi(t)$ is given by (4). (The positive nth root is understood.) Indeed, we have

$$0 = \langle f, \phi^{1/n} \rangle - \langle g, \phi^{1/n} \rangle = \int_a^b [f(t) - g(t)] \phi^{1/n}(t)\, dt \tag{5}$$

Now, for all $n > 1$, $\phi^{1/n}(t)$ is nonnegative and less than some fixed constant M, so that the magnitude of the integrand in (5) is no larger than the integrable function

$$M|f(t) - g(t)|$$

Hence, by Lebesgue's general theorem of convergence (Kestelman [1], p. 141), we may interchange the limit process $n \to \infty$ with the integration in (5). Since, for $a < t < b$, $[\phi(t)]^{1/n} \to 1$, we obtain

$$\int_a^b [f(t) - g(t)]\, dt = 0$$

Since this holds for all a and b, it follows that $f(t) = g(t)$ almost everywhere (Kestelman [1], p. 155).

Thus, we have shown that a regular distribution determines the function producing it almost everywhere. Actually, the values of a function on a set of measure zero are unimportant so far as Lebesgue integration is concerned, and this attitude is also usually adopted in physical applications. All locally integrable functions that differ at most on a set of measure zero determine an equivalence class of functions, and these functions produce the same regular distribution. Hence, without ambiguity we may consider an equivalence class of functions and its regular distribution as being the same entity.

We shall at times use the same expression to denote a regular distribution and a function that generates it. For example, t^2 will denote both the function and the corresponding distribution. An even more ambiguous situation can arise when the function is constant. For instance, the symbol c can represent the number c, the function that equals c for all values of its argument, or the corresponding regular distribution. For the sake of brevity, we shall not always specify explicitly which of these meanings such a symbol has. The context in which it is used should clarify any possible ambiguity.

Similarly, we shall on occasion refer to "a locally integrable function $f(t)$ in $\mathfrak{D}'$." In this case it will be understood that $f(t)$ is the regular distribution corresponding to the function.

In certain parts of modern analysis, such as in functional analysis, it has become essential to distinguish between the value of a function at a point t and the function itself. In these cases, it is conventional to let $f(t)$ be the numerical value corresponding to the value t and to let f denote the function (i.e., the rule that assigns numbers to numbers). This is in contrast to classical analysis, where such a distinction was not needed. Nor will our symbolism make this distinction. In general, both $f(t)$ and f will represent a regular distribution or a function. However, at times the symbols $f(0)$, $f(T)$, $f(c)$, etc., will be used to denote numerical values. Again, our intent should be discernible from the context in which these symbols are used.

The importance of the class of distributions stems from the fact that not only does it include representations of locally integrable functions (i.e., regular distributions) but, in addition, it contains many other entities that are not regular distributions. Moreover, many operations, such as integration, differentiation, and other limiting processes that were originally developed for functions, can be extended to these new entities. It should be mentioned, however, that other operations such as the multiplication of functions $f(t)g(t)$ or the formation of composite functions $f(g(t))$ cannot be extended, in general, to all distributions. This constitutes a disadvantage of distribution theory.

An example of a distribution that is not a regular distribution is the so-called "delta function" δ, which is defined by the equation

$$\langle \delta, \phi \rangle \triangleq \phi(0) \tag{6}$$

Clearly, (6) is a continuous linear functional on $\mathfrak{D}$. However, this distribution cannot be obtained from a locally integrable function through the use of (2). Indeed, if there were such a function $\delta(t)$, then we would have

$$\phi(0) = \int_{-\infty}^{\infty} \delta(t)\phi(t)\,dt \tag{7}$$

for all $\phi(t)$ in $\mathfrak{D}$. If $\phi(t)$ is chosen equal to $\zeta(t/a)$, where $a > 0$ and $\zeta(t)$ is given by Sec. 1.2, Eq. (1), Eq. (7) becomes

$$\frac{1}{e} = \int_{-a}^{a} \delta(t) \exp \frac{1}{a^2 - t^2}\,dt \tag{8}$$

If $\delta(t)$ were a locally integrable function, Lebesgue's general theorem of convergence would show that the right-hand side of (8) converges to zero as $a \to 0$. This would be a contradiction. Hence, δ is not a regular distribution. Actually, its name "delta function" is a misnomer and, for this reason, we shall refer to it as the *delta functional*.

All distributions that are not regular are called *singular distributions*.

Before leaving this section, let us say a few more words about notation. If f is any regular or singular distribution defined on the space $\mathfrak{D}$ of testing functions, where the independent variable is t, f will also be designated by $f(t)$, the symbol heretofore used for regular distributions or ordinary functions. As was indicated before, when f is a regular distribution, this alternative notation does not lead to any essential ambiguity in view of the one-to-one relation between an equivalence class of functions (composed of all functions differing only on a set of measure zero) and the corresponding regular distribution. On the other hand, if f is a singular distribution, the notation $f(t)$ is merely a convenient symbolism. It has the advantage of displaying the variable t that occurs in the testing functions. This will be useful when we define for distributions such operations as convolution, Fourier and Laplace transformation, and certain changes of variable. Henceforth, when we designate a singular distribution by $f(t)$, it will be understood that this is not a function or a regular distribution. [Some authors, for example, B. Friedman [1], refer to $f(t)$ in this case as a *symbolic function*.] Similarly, we may at times say that "the distribution $f(t)$ is defined over the t axis" without implying that $f(t)$ is a function or a regular distribution.

Let us give an illustration. For the delta functional δ, we shall also use the symbol $\delta(t)$. One often finds in the literature the expression

$$\phi(0) = \int_{-\infty}^{\infty} \delta(t)\phi(t)\,dt \tag{9}$$

It is understood that the right-hand side of (9) has no meaning other than that given to it by the left-hand side. By formally performing the manipulations that are valid for ordinary functions, we can define a new singular distribution $\delta(t - \tau)$ through the following symbolism:

$$\begin{aligned}\langle \delta(t - \tau), \phi(t) \rangle &= \int_{-\infty}^{\infty} \delta(t - \tau)\phi(t)\, dt \\ &= \int_{-\infty}^{\infty} \delta(x)\phi(x + \tau)\, dx \\ &= \phi(\tau) \end{aligned} \quad (10)$$

It is the equation

$$\langle \delta(t - \tau), \phi(t) \rangle = \phi(\tau)$$

that is taken as a definition of $\delta(t - \tau)$. The intervening formal manipulations in (10) merely demonstrate a consistency with what we would do if we were dealing with functions.

This ability of $\delta(t - \tau)$ to pick out the value of $\phi(t)$ at the point $t = \tau$ is sometimes called the *sifting property* of the delta functional (Van der Pol and Bremmer [1], p. 61).

Example 1.3-1 We end this section with a description of a somewhat crude way of thinking about the delta functional. As we have already stated, the delta functional is not a function and therefore cannot be plotted. However, a common practice in the technical literature is to treat $\delta(t)$ as if it were a function; witness (9) and (10). More exactly, one commonly thinks not in terms of $\delta(t)$ but in terms of a function that approximates $\delta(t)$ in some sense. There is value in this sort of intuitive thinking. It not only gives one a better grasp of the concept of the delta functional but also provides a method for discovering other singular distributions, as we shall see. Nevertheless, we must not ignore the fact that these considerations are not rigorous; indeed, all our conjectures will be proved subsequently in terms of distribution theory.

Consider once again the function $\gamma_\alpha(t)$, defined by (1) and (2) of the preceding section. As was shown in that section, for every continuous function f the integral

$$\int_{-\infty}^{\infty} f(t)\gamma_\alpha(t)\, dt \quad (11)$$

converges to $f(0)$ as $\alpha \to 0$. In other words, for α sufficiently small the regular distribution $\gamma_\alpha(t)$ behaves like the delta functional where now we can replace the space $\mathfrak{D}$ by the larger space of all continuous functions. Indeed, we can predict this behavior just from the shape of $\gamma_\alpha(t)$, which is indicated in Fig. 1.3-1. For small α, $\gamma_\alpha(t)$ is a very sharp pulse of unit area whose base is the interval $-\alpha < t < \alpha$. Because f is continuous, it is almost equal to $f(0)$ over $-\alpha < t < \alpha$ when α is chosen small enough.

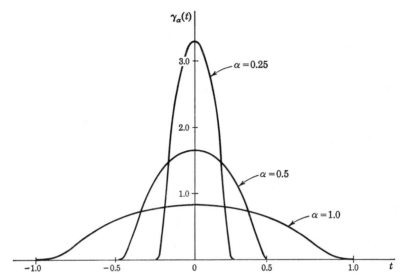

Fig. 1.3-1 Approximations to $\delta(t)$ by the function $\gamma_\alpha(t)$ for $\alpha = 0.25, 0.5, 1.0$.

Thus, without committing much error, (11) can be replaced by

$$\int_{-\infty}^{\infty} f(0)\gamma_\alpha(t)\,dt = f(0)$$

Actually, the form of the approximating pulse is not important so long as it is nonnegative and has a unit area concentrated mainly over a very small interval around the origin. In Sec. 2.2 we shall define precisely what we mean by the "convergence of a sequence of distributions," and in Sec. 2.3 we shall indicate a larger class of pulses that converge in this sense to $\delta(t)$.

Example 1.3-2 This heuristic conception of the delta functional leads to useful conjectures. Consider

$$\lambda_\alpha(t) \triangleq \int_{-\infty}^{t} \gamma_\alpha(x)\,dx$$

which is plotted in Fig. 1.3-2. As $\alpha \to 0$, $\lambda_\alpha(t)$ converges to the Heaviside unit step function:

$$1_+(t) \triangleq \begin{cases} 0 & t < 0 \\ \frac{1}{2} & t = 0 \\ 1 & t > 0 \end{cases}$$

Since our approximation $\gamma_\alpha(t)$ to the delta functional is the derivative of an approximation $\lambda_\alpha(t)$ to the step function $1_+(t)$, we can expect that $\delta(t)$

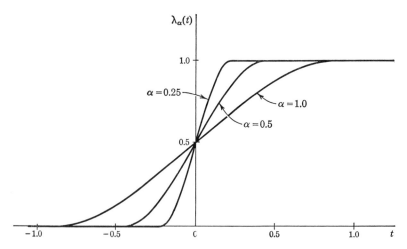

Fig. 1.3-2 Approximations to $1_+(t)$ by the function $\lambda_\alpha(t)$ for $\alpha = 0.25, 0.5, 1.0$.

is in some sense the derivative of $1_+(t)$. This is truly so, as we shall show after we define distributional differentiation. (See Example 2.4-1.)

Moreover, we can conjecture a new singular distribution, the first derivative $\delta^{(1)}(t)$ of the delta functional, and we can get an idea of how it behaves simply by differentiating our approximation $\gamma_\alpha(t)$ to the delta functional. The derivatives of $\gamma_\alpha^{(1)}(t)$ are plotted in Fig. 1.3-3. Integra-

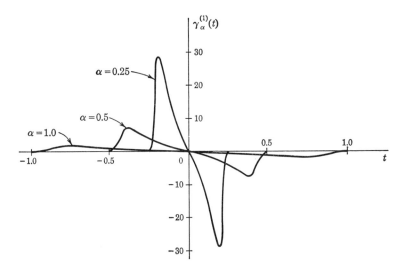

Fig. 1.3-3 Approximations to $\delta^{(1)}(t)$ by the function $\gamma_\alpha^{(1)}(t)$ for $\alpha = 0.25, 0.5, 1.0$.

tion by parts shows that for every ϕ in $\mathfrak{D}$

$$\int_{-\infty}^{\infty} \gamma_\alpha^{(1)}(t)\phi(t)\, dt = -\int_{-\infty}^{\infty} \gamma_\alpha(t)\phi^{(1)}(t)\, dt$$

As we have seen, the right-hand side converges to $-\phi^{(1)}(0)$. Therefore, the following definition suggests itself:

$$\langle \delta^{(1)}(t),\, \phi(t) \rangle \triangleq -\phi^{(1)}(0) \tag{12}$$

Actually, the rigorous definition of distributional differentiation will be constructed in just this way (see Sec. 2.4), and (12) will be a consequence of it. Also, the "double pulses" of Fig. 1.3-3 can be used as crude pictures of $\delta^{(1)}(t)$, since it can be shown that they converge in the distributional sense to $\delta^{(1)}(t)$.

PROBLEMS

1 Which of the following expressions define a distribution in $\mathfrak{D}'$? Here, ϕ is an arbitrary testing function in $\mathfrak{D}$. In those cases where a functional f on $\mathfrak{D}$ exists, establish whether it is linear and/or continuous.

a. $\langle f, \phi \rangle = \sum_{\nu=0}^{n} \phi^{(\nu)}(0)$

b. $\langle f, \phi \rangle = \sum_{\nu=0}^{\infty} \phi^{(\nu)}(0)$

c. $\langle f, \phi \rangle = \sum_{\nu=0}^{\infty} \phi^{(\nu)}(\nu)$

d. $\langle f, \phi \rangle = \sum_{\nu=0}^{\infty} \phi(\tau_\nu)$, where $\lim_{\nu \to \infty} \tau_\nu = 1$

e. $\langle f, \phi \rangle = [\phi(0)]^2$

f. $\langle f, \phi \rangle = \sup \phi(t)$

g. $\langle f, \phi \rangle = \int_{-\infty}^{\infty} |\phi(t)|\, dt$

h. $\langle f, \phi \rangle = \int_{0}^{1} \phi^{(k)}(t)\, dt \qquad k = $ any positive integer

i. $\langle f, \phi \rangle = $ supremum of the support of ϕ

2 One can define distributions (i.e., continuous linear functionals) on spaces of testing functions that are larger than $\mathfrak{D}$. But then the resulting space of distributions is usually smaller than $\mathfrak{D}'$. For example, let $\mathfrak{D}_m$ denote the space of functions $\phi(t)$ of the one-dimensional real variable t which have the following properties:

1. Each $\phi(t)$ and every one of its derivatives $\phi^{(k)}(t)$ up to and including the order m are continuous everywhere. (For $m = 0$, this merely requires that ϕ be continuous.)

2. Every ϕ has a bounded support. A sequence of such functions $\{\phi_\nu\}_{\nu=1}^{\infty}$ is said to converge in $\mathfrak{D}_m$ if the supports of the ϕ_ν are contained in a fixed finite interval and if each sequence of derivatives $\{\phi_\nu^{(k)}(t)\}_{\nu=1}^{\infty}$ ($k = 0, 1, \ldots, m$) converges

uniformly for all t. Finally, let $\mathfrak{D}'_m$ denote the space of continuous linear functionals on $\mathfrak{D}_m$. (Linearity and continuity are defined in the usual way.) Verify the following assertions:

a. The limit function ϕ of the sequence $\{\phi_\nu\}_{\nu=1}^{\infty}$ is in $\mathfrak{D}_m$ (that is, $\mathfrak{D}_m$ is closed under convergence).

b. The delta functional δ is in $\mathfrak{D}'_0$.

c. The functional f, defined by

$$\langle f, \phi \rangle \triangleq (-1)^k \phi^{(k)}(0)$$

is in $\mathfrak{D}'_k$. (As we shall see later on, f is the kth-order distributional derivative of δ.)

d. $\mathfrak{D}'_m$ is a proper subspace of $\mathfrak{D}'$ (that is, for each finite m there is a distribution in $\mathfrak{D}'$ that is not in $\mathfrak{D}'_m$).

e. For $p > m$, $\mathfrak{D}'_m$ is a proper subspace of $\mathfrak{D}'_p$.

3 There is another type of distribution, called the *bounded distribution*, that can be extended onto a space of testing functions $\mathfrak{D}_{L_1}$ which is larger than $\mathfrak{D}$. Still restricting t to being a real one-dimensional variable, we define $\mathfrak{D}_{L_1}$ as the space of infinitely smooth functions $\phi(t)$ such that, for each nonnegative integer k, $\phi^{(k)}(t)$ is absolutely integrable over $-\infty < t < \infty$. A sequence $\{\phi_\nu(t)\}_{\nu=1}^{\infty}$ is said to converge in $\mathfrak{D}_{L_1}$ if each ϕ_ν is in $\mathfrak{D}_{L_1}$ and if for each nonnegative integer k

$$\int_{-\infty}^{\infty} |\phi_\nu^{(k)}(t) - \phi_\mu^{(k)}(t)| \, dt \to 0$$

as ν and μ go to infinity independently.

a. Show that the limit function of $\{\phi_\nu\}_{\nu=1}^{\infty}$ exists and is also in $\mathfrak{D}_{L_1}$. In other words, $\mathfrak{D}_{L_1}$ is also closed under convergence. HINT: First, establish that, for each nonnegative integer k, $\{\phi_\nu^{(k)}\}_{\nu=1}^{\infty}$ converges uniformly for all t.

$\mathfrak{B}'$ is the space of continuous linear functionals on $\mathfrak{D}_{L_1}$, where linearity and continuity are defined in the customary way. The elements of $\mathfrak{B}'$ are the so-called "bounded distributions."

b. Verify that the delta functional and the functionals defined in part c of Prob. 2 are in $\mathfrak{B}'$.

c. Show that $\mathfrak{B}'$ is a proper subspace of $\mathfrak{D}'$.

4 The second derivative $\delta^{(2)}(t)$ of the delta functional is a distribution that assigns to each ϕ in $\mathfrak{D}$ the value

$$\langle \delta^{(2)}(t), \phi(t) \rangle = \phi^{(2)}(0)$$

Show that, as $\alpha \to 0$, the numbers $\langle \gamma_\alpha^{(2)}(t), \phi(t) \rangle$ converge to $\phi^{(2)}(0)$. Because of this, we might say that $\gamma_\alpha^{(2)}(t)$ is a crude approximation to $\delta^{(2)}(t)$. Sketch $\gamma_\alpha^{(2)}(t)$ for some small value of α.

★1.4 PSEUDOFUNCTIONS, HADAMARD'S FINITE PART, AND CAUCHY'S PRINCIPAL VALUE

The only singular distributions that we have discussed so far are the commonly used delta functional and its first and second derivatives. We shall now present some other simple examples of singular distributions. They are generated by Hadamard's "finite part" of a divergent integral, a concept that will be discussed in a more general fashion in Sec. 2.5.

Consider for the moment the function $1_+(t)t^{-\frac{3}{2}}$, where for definiteness we take $t^{-\frac{3}{2}}$ as the positive square root of t^{-3} (that is, $t^{-\frac{3}{2}}$ is positive for t positive). This function does not define a regular distribution, since the integral

$$\int_0^\infty t^{-\frac{3}{2}}\phi(t)\,dt \qquad \phi \in \mathfrak{D} \tag{1}$$

is in general divergent. However, Hadamard has suggested a technique for extracting a finite part from such divergent integrals (Hadamard [1], pp. 133–141), and it turns out that this finite part defines a singular distribution. The procedure is as follows:

Let

$$\phi(t) \triangleq \phi(0) + t\psi(t) \tag{2}$$

Since $\phi(t)$ is infinitely smooth, $\psi(t)$ is a continuous function for all $t \neq 0$ and can be extended so that it is continuous even at $t = 0$. This can be seen from Taylor's formula with remainder (Widder [2], p. 43). If b is a real number so large that $\phi(t) = 0$ for $t > b$, (1) becomes

$$\langle 1_+(t)t^{-\frac{3}{2}}, \phi(t)\rangle = \lim_{\varepsilon \to 0+} \int_\varepsilon^b t^{-\frac{3}{2}}\phi(t)\,dt$$

$$= \lim_{\varepsilon \to 0+} \left[\frac{2\phi(0)}{\sqrt{\varepsilon}} - \frac{2\phi(0)}{\sqrt{b}} + \int_\varepsilon^b \frac{\psi(t)}{\sqrt{t}}\,dt\right] \tag{3}$$

If $\phi(0) \neq 0$, the first term inside the brackets diverges as $\varepsilon \to 0+$, whereas the third term remains finite because $\psi(t)$ is continuous at $t = 0$. The next step is simple. We just throw away the divergent term. The result is *Hadamard's finite part* of the divergent integral (1), and we denote it as follows:

$$\text{Fp} \int_0^\infty t^{-\frac{3}{2}}\phi(t)\,dt \triangleq \int_0^b \frac{\psi(t)}{\sqrt{t}}\,dt - \frac{2\phi(0)}{\sqrt{b}} \tag{4}$$

For our purposes, the important thing is that (4) defines a continuous linear functional on $\mathfrak{D}$. Its linearity is clear. Moreover, in the interval $0 < t < b$ we have the inequality

$$|\psi(t)| = \left|\frac{\phi(t) - \phi(0)}{t}\right| = \left|\frac{1}{t}\int_0^t \phi^{(1)}(x)\,dx\right| \leq \sup_{0<t<b} |\phi^{(1)}(t)|$$

Thus,

$$\left|\int_0^b \frac{\psi(t)}{\sqrt{t}}\,dt\right| \leq 2\sqrt{b} \sup_{0<t<b} |\phi^{(1)}(t)|$$

Hence, if $\{\phi_\nu\}_{\nu=1}^\infty$ converges in $\mathfrak{D}$ to zero, we see from (4) that

$$\left\{\text{Fp}\int_0^\infty t^{-\frac{3}{2}}\phi_\nu(t)\,dt\right\}_{\nu=1}^\infty$$

converges to zero, which demonstrates the continuity of our functional.

This distribution is denoted by
$$\text{Pf } t^{-3/2}1_+(t) \tag{5}$$
and it assigns the value (4) to each ϕ in $\mathfrak{D}$. Following Schwartz, we call (5) a *pseudofunction* (Schwartz [1], vol. I, p. 41). The symbol Pf in (5) stands for *pseudofunction*. By combining (2) with (4), we obtain

$$\begin{aligned}\langle \text{Pf } t^{-3/2}1_+(t), \phi(t)\rangle &= \lim_{\varepsilon \to 0+} \left[\int_\varepsilon^\infty t^{-3/2}\phi(t)\,dt - \frac{2\phi(0)}{\sqrt{\varepsilon}} \right] \\ &= \int_0^\infty t^{-3/2}[\phi(t) - \phi(0)]\,dt \end{aligned} \tag{6}$$

By the same analysis, we can construct the pseudofunction
$$\text{Pf } t^\beta 1_+(t)$$
where $-2 < \beta < -1$ and t^β is restricted to one of its branches (i.e., for $t > 0$, $\arg t^\beta = 2n\pi\beta$ with n a fixed integer). In this case, we have

$$\begin{aligned}\langle \text{Pf } t^\beta 1_+(t), \phi(t)\rangle &\triangleq \text{Fp} \int_0^\infty t^\beta \phi(t)\,dt \\ &\triangleq \lim_{\varepsilon \to 0+} \left[\int_\varepsilon^\infty t^\beta \phi(t)\,dt + \frac{\phi(0)\varepsilon^{\beta+1}}{\beta+1} \right] \\ &= \int_0^\infty t^\beta [\phi(t) - \phi(0)]\,dt \end{aligned} \tag{7}$$

For the pseudofunction $\text{Pf } |t|^\beta 1_+(-t)$ $(-2 < \beta < -1)$ we can similarly show that

$$\begin{aligned}\langle \text{Pf } |t|^\beta 1_+(-t), \phi(t)\rangle &\triangleq \text{Fp} \int_{-\infty}^0 |t|^\beta \phi(t)\,dt \\ &= \lim_{\varepsilon \to 0+} \left[\int_{-\infty}^{-\varepsilon} |t|^\beta \phi(t)\,dt + \frac{\phi(0)\varepsilon^{\beta+1}}{\beta+1} \right] \\ &= \int_{-\infty}^0 |t|^\beta [\phi(t) - \phi(0)]\,dt \end{aligned} \tag{8}$$

Finally, we define the pseudofunction $\text{Pf } |t|^\beta$ by
$$\text{Pf } |t|^\beta \triangleq \text{Pf } |t|^\beta 1_+(-t) + \text{Pf } t^\beta 1_+(t) \tag{9}$$
where t^β ($t > 0$) and $|t|^\beta$ ($t < 0$) are restricted to the same branch.

Let us now turn to the case where $\beta = -1$. Our previous expressions have to be altered somewhat. For instance, consider the (in general) divergent integral
$$\int_0^\infty \frac{\phi(t)}{t}\,dt \qquad \phi \in \mathfrak{D} \tag{10}$$

In this case, the expression corresponding to (3) is
$$\left\langle \frac{1_+(t)}{t}, \phi(t) \right\rangle = \lim_{\varepsilon \to 0+} \left[\phi(0) \log b - \phi(0) \log \varepsilon + \int_\varepsilon^b \psi(t)\,dt \right]$$

Upon discarding the divergent term $\phi(0)\log\varepsilon$, we obtain the definition

$$\left\langle \operatorname{Pf} \frac{1_+(t)}{t}, \phi(t) \right\rangle \triangleq \int_0^b \psi(t)\,dt + \phi(0)\log b \tag{11}$$

which can be converted into

$$\left\langle \operatorname{Pf} \frac{1_+(t)}{t}, \phi(t) \right\rangle = \lim_{\varepsilon \to 0+} \left[\int_\varepsilon^\infty \frac{\phi(t)}{t}\,dt + \phi(0)\log\varepsilon \right]$$
$$= \int_0^1 \frac{\phi(t) - \phi(0)}{t}\,dt + \int_1^\infty \frac{\phi(t)}{t}\,dt \tag{12}$$

As before, it can be shown that (11) defines a continuous linear functional on $\mathfrak{D}$ and hence a pseudofunction.

Similarly, we may define the pseudofunctions $\operatorname{Pf} 1_+(-t)/|t|$ and $\operatorname{Pf} 1/|t|$ through

$$\left\langle \operatorname{Pf} \frac{1_+(-t)}{|t|}, \phi(t) \right\rangle \triangleq \lim_{\varepsilon \to 0+} \left[\int_{-\infty}^{-\varepsilon} \frac{\phi(t)}{|t|}\,dt + \phi(0)\log\varepsilon \right]$$
$$= \int_{-1}^0 \frac{\phi(t) - \phi(0)}{|t|}\,dt + \int_{-\infty}^{-1} \frac{\phi(t)}{|t|}\,dt \tag{13}$$

and

$$\operatorname{Pf} \frac{1}{|t|} \triangleq \operatorname{Pf} \frac{1_+(-t)}{|t|} + \operatorname{Pf} \frac{1_+(t)}{t} \tag{14}$$

As a final example of a singular distribution, we shall take up the *Cauchy principal value* of the (usually) divergent integral

$$\int_{-\infty}^\infty \frac{\phi(t)}{t}\,dt \qquad \phi \in \mathfrak{D}$$

By definition, this is the finite quantity

$$\operatorname{Pv} \int_{-\infty}^\infty \frac{\phi(t)}{t}\,dt \triangleq \lim_{\varepsilon \to 0+} \left(\int_{-\infty}^{-\varepsilon} + \int_\varepsilon^\infty \right) \frac{\phi(t)}{t}\,dt \tag{15}$$

It can be seen that the limit in (15) exists by noting that it is the difference between (12) and (13). In fact, (15) also defines a distribution, which we denote by $\operatorname{Pv} 1/t$. It is the pseudofunction

$$\operatorname{Pv} \frac{1}{t} = \operatorname{Pf} \frac{1_+(t)}{t} - \operatorname{Pf} \frac{1_+(-t)}{|t|} \tag{16}$$

Note that the Cauchy principal value of a divergent integral need not exist even when the integral has a finite part. For example,

$$\operatorname{Pv} \int_{-\infty}^\infty \frac{\phi(t)}{|t|}\,dt \triangleq \lim_{\varepsilon \to 0+} \left(\int_{-\infty}^{-\varepsilon} + \int_\varepsilon^\infty \right) \frac{\phi(t)}{|t|}\,dt$$

does not exist, if $\phi(0) \neq 0$, whereas

$$\text{Fp} \int_{-\infty}^{\infty} \frac{\phi(t)}{|t|} dt = \lim_{\varepsilon \to 0+} \left[\left(\int_{-\infty}^{-\varepsilon} + \int_{\varepsilon}^{\infty} \right) \frac{\phi(t)}{|t|} dt + 2\phi(0) \log \varepsilon \right]$$

does exist.

Example 1.4-1 For the sake of obtaining a better intuitive grasp of these pseudofunctions, let us construct a crude approximation to Pf $1_+(t)/t$ in a heuristic way, as we did for $\delta(t)$ and $\delta^{(1)}(t)$ at the end of the preceding section. It is a fact that this pseudofunction is the distributional derivative of the regular distribution $1_+(t) \log t$. Thus, if we construct a smooth approximation to $1_+(t) \log t$ and then plot its slope, we might expect to obtain an approximation to Pf $1_+(t)/t$.

This is indicated in Fig. 1.4-1. In part a of the figure the solid line A

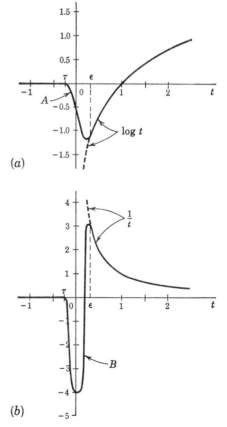

Fig. 1.4-1 (a) The dashed line represents the function $1_+(t) \log t$, and the solid line A is a smooth approximation to it. (b) The dashed line represents the function $1_+(t)/t$, and the solid line B is a plot of the slope of line A. Line B represents an approximation to the pseudofunction Pf $1_+(t)/t$. Note the sharp negative pulse in B near the origin.

represents a smooth approximation to the function $1_+(t) \log t$, which is indicated as a dashed line. Curve A replaces the infinite discontinuity in $1_+(t) \log t$ by a smooth transition which leaves the t axis at $t = \tau$ and joins the $1_+(t) \log t$ curve at $t = \varepsilon$. The corresponding slopes are indicated in Fig. 1.4-1b. Note that the approximation B to the pseudofunction Pf $1_+(t)/t$ follows the function $1_+(t)/t$ except near the origin, where B has a sharp negative pulse.

The presence of this negative pulse can be related to expression (12) in the following way. The total area under the curve B over the interval $\tau \leq t \leq \varepsilon$ is the negative quantity $\log \varepsilon$, since this is the total increment in the curve A for this interval. Let us apply B as a regular distribution to some testing function $\phi(t)$ in $\mathfrak{D}$. If ε and τ have exceedingly small magnitudes, then, over $\tau \leq t \leq \varepsilon$, $\phi(t)$ is essentially constant at the value $\phi(0)$. Therefore, the transition portion of the B curve will approximately supply the number

$$\phi(0) \log \varepsilon \tag{17}$$

to $\langle B, \phi \rangle$. This corresponds to the second term in the middle part of (12). The first term,

$$\int_\varepsilon^\infty \frac{\phi(t)}{t} \, dt \tag{18}$$

is produced by the portion of the B curve to the right of $t = \varepsilon$. Assuming that $\phi(t)$ is nonnegative for all t and that $\phi(0) > 0$, we readily see that (17) approaches $-\infty$, whereas (18) approaches $+\infty$ as $\varepsilon \to 0$. However, (17) and (18) diverge in such a way that their sum remains finite.

PROBLEMS

1 Show that (11) is a continuous linear functional on $\mathfrak{D}$.

2 Derive (8) and (13).

3 Give a direct proof that (16) is a distribution by showing that

$$\lim_{\nu \to \infty} \lim_{\varepsilon \to 0+} \left(\int_{-\infty}^{-\varepsilon} + \int_\varepsilon^\infty \right) \frac{\phi_\nu(t)}{t} \, dt = 0$$

whenever $\{\phi_\nu\}_{\nu=1}^\infty$ converges in $\mathfrak{D}$ to zero.

4 Develop the finite part of the generally divergent integral

$$\int_0^\infty t^{-\frac{3}{2}} \phi(t) \, dt \qquad \phi \in \mathfrak{D}$$

and show that it defines a pseudofunction. HINT: Set

$$\phi(t) = \phi(0) + t\phi^{(1)}(0) + t^2 \psi(t)$$

and then discard the divergent terms that arise.

5 Repeat Prob. 4 for the integral

$$\int_0^\infty t^{-2}\phi(t)\,dt$$

6 Show that Pf $1_+(t)/t$ is in $\mathfrak{D}_1'$, a space that is defined in Sec. 1.3, Prob. 2. Also, show that it is in $\mathfrak{G}'$, the space described in Sec. 1.3, Prob. 3.

7 **a.** Sketch an approximation to the pseudofunction Pf $t^{-\frac{2}{3}}1_+(t)$.
b. It is a fact that the pseudofunction obtained in Prob. 4, which we denote by Pf $t^{-\frac{5}{3}}1_+(t)$, is the distributional derivative of

$$-\tfrac{2}{3}\text{ Pf }t^{-\frac{2}{3}}1_+(t)$$

[See Sec. 2.5, Eq. (9).] On the basis of this observation, sketch an approximation to Pf $t^{-\frac{5}{3}}1_+(t)$.

★1.5 TESTING FUNCTIONS AND DISTRIBUTIONS OF SEVERAL VARIABLES

In most of the discussions in the later part of this book we shall be dealing with distributions that are defined over the real line $\mathfrak{R}^1$. On occasion, however, we shall make use of distributions over multi-dimensional spaces, as, for example, when we consider the convolution of distributions. The ideas developed heretofore can be extended with almost no alteration to this multidimensional case.

Let $t_1, t_2, \ldots, t_n$ be n real one-dimensional variables. Taking these variables as coordinates in n-dimensional real euclidean space $\mathfrak{R}^n$, we shall denote an arbitrary point in this space by t. The testing functions are now those complex-valued functions of the n real variables $t_1, t_2, \ldots, t_n$, which vanish outside a bounded domain of the euclidean space $\mathfrak{R}^n$ and for which all partial derivatives exist for all t. As before, the testing functions are not all zero outside the same bounded domain of $\mathfrak{R}^n$. We shall denote testing functions either by $\phi(t)$ or by $\phi(t_1, t_2, \ldots, t_n)$. The condition on the derivatives of the testing functions can be stated alternatively as follows. For each set of n nonnegative integers, $k \triangleq \{k_1, k_2, \ldots, k_n\}$, whose sum is denoted by $\hat{k}$, the partial derivative of order $\hat{k}$,

$$D^k\phi(t) \triangleq \frac{\partial^{k_1+k_2+\cdots+k_n}}{\partial t_1^{k_1}\,\partial t_2^{k_2}\cdots\partial t_n^{k_n}}\phi(t_1, t_2, \ldots, t_n) \tag{1}$$

exists and is continuous everywhere. As in the one-dimensional case, such functions will be said to be *infinitely smooth*. Note that, because of the continuity of all the partial derivatives of ϕ, the order of differentiation in (1) can be changed in any fashion.

We shall also denote the space of all testing functions on the n-dimensional euclidean space $\mathfrak{R}^n$ by the symbol $\mathfrak{D}$, and the value of n should be clear from the context whenever n is a specified integer. $\mathfrak{D}$ is a *linear space*. (See Appendix A.)

An example of a testing function of n independent variables is given by

$$\zeta(t) = \begin{cases} 0 & |t| \geq 1 \\ \exp \dfrac{1}{|t|^2 - 1} & |t| < 1 \end{cases} \tag{2}$$

where now

$$|t| = \sqrt{t_1{}^2 + t_2{}^2 + \cdots + t_n{}^2}$$

As before, the closure of the set of all points where a given $\phi(t)$ in $\mathfrak{D}$ is different from zero is called the support of $\phi(t)$. Thus, a testing function in $\mathfrak{D}$ is again an infinitely smooth function having a bounded support.

By applying the same arguments as those used in Sec. 1.2, we can show that *any function $f(t) = f(t_1, t_2, \ldots, t_n)$ that is continuous for all t and is zero outside a bounded domain, say $a_i \leq t_i \leq b_i$ $(i = 1, 2, \ldots, n)$, can be uniformly approximated by a testing function $\phi_\alpha(t)$ in $\mathfrak{D}$* (that is, $|f(t) - \phi_\alpha(t)| < \varepsilon$ for a preassigned $\varepsilon > 0$) *such that $\phi_\alpha(t)$ is zero outside the domain $a_i - \alpha \leq t_i \leq b_i + \alpha$ $(i = 1, 2, \ldots, n)$, where the positive number α can be chosen as small as desired. Furthermore, if*

$$f(t) = f(t_1, t_2, \ldots, t_n)$$

is an infinitely smooth function for all t, then for any bounded domain of $\mathfrak{R}^n$ there exists a testing function in $\mathfrak{D}$ that is identically equal to $f(t)$ on this domain.

The concept of *convergence in $\mathfrak{D}$* for the n-dimensional case is analogous to that for the one-dimensional case. A sequence of testing functions $\{\phi_\nu(t)\}_{\nu=1}^\infty = \{\phi_\nu(t_1, t_2, \ldots, t_n)\}_{\nu=1}^\infty$ is said to *converge in $\mathfrak{D}$ to zero* if the supports of the ϕ_ν are all contained in a fixed bounded domain in $\mathfrak{R}^n$ and if the ϕ_ν and each of their partial derivatives converge uniformly to zero as $\nu \to \infty$. The condition of the uniform convergence can also be stated as follows. Denoting a partial derivative by the symbol indicated in (1), we have that for any number $\varepsilon > 0$ and any choice of the set of nonnegative integers $k = \{k_1, k_2, \ldots, k_n\}$ there exists an integer N_k such that for every $\nu \geq N_k$ all the partial derivatives $D^k \phi_\nu(t)$ satisfy the inequality

$$|D^k \phi_\nu(t)| < \varepsilon$$

[Here, $D^k \phi_\nu(t) = \phi_\nu(t)$ if $k = \{0, 0, \ldots, 0\}$.] As before, the integers N_k may vary for different choices of k.

Furthermore, a sequence of testing functions $\{\phi_\nu(t)\}_{\nu=1}^\infty$ is said to *converge in $\mathfrak{D}$ to a function* $\phi(t)$ if the ϕ_ν are all in $\mathfrak{D}$, if their supports are contained in a fixed bounded domain in $\mathfrak{R}^n$, and if for each k the sequence $\{D^k \phi_\nu\}_{\nu=1}^\infty$ converges uniformly to $D^k \phi$ over all of $\mathfrak{R}^n$. It follows that every such limit function ϕ will also be in $\mathfrak{D}$. That is, $\mathfrak{D}$ is closed under convergence.

A rule f that assigns a complex number to each testing function $\phi(t)$ of the n-dimensional variable t is a *functional* on the space $\mathfrak{D}$. This number will be denoted as before by the symbol $\langle f, \phi \rangle$. The functional is *linear* if

$$\langle f, \alpha\phi_1 + \beta\phi_2 \rangle = \alpha\langle f, \phi_1 \rangle + \beta\langle f, \phi_2 \rangle$$

whenever ϕ_1 and ϕ_2 are functions in $\mathfrak{D}$ and α and β are numbers. It is *continuous* if, for any sequence of testing functions $\{\phi_\nu(t)\}_{\nu=1}^\infty$ that converges in $\mathfrak{D}$ to $\phi(t)$, the sequence of numbers $\{\langle f, \phi_\nu \rangle\}_{\nu=1}^\infty$ converges to the number $\langle f, \phi \rangle$. When f is linear, it is continuous if and only if the numerical sequence $\{\langle f, \phi_\nu \rangle\}_{\nu=1}^\infty$ converges to zero whenever the sequence $\{\phi_\nu\}_{\nu=1}^\infty$ converges in $\mathfrak{D}$ to zero. Now a distribution on the n-dimensional space $\mathfrak{R}^n$ is defined in the same way as it was in the one-dimensional case.

A continuous linear functional on the space $\mathfrak{D}$ of testing functions on $\mathfrak{R}^n$ is a distribution on $\mathfrak{R}^n$. The space of all such distributions is again denoted by $\mathfrak{D}'$. $\mathfrak{D}'$ is called the dual (or conjugate) space of $\mathfrak{D}$. Note that $\mathfrak{D}'$ represents a space of distributions on a particular euclidean space $\mathfrak{R}^n$; that is, n is fixed even though it may not be specified.

Analogously to the one-dimensional case, a locally integrable function $f(t)$ [that is, a function that is (Lebesgue) integrable over every bounded domain of $\mathfrak{R}^n$] generates a distribution through the multiple integral

$$\langle f, \phi \rangle = \int_{-\infty}^{\infty} \cdots \int_{-\infty}^{\infty} \int_{-\infty}^{\infty} f(t_1, \ldots, t_n)\phi(t_1, \ldots, t_n) \, dt_1 \cdots dt_n$$

$$= \int_{\mathfrak{R}^n} f(t)\phi(t) \, dt \qquad \phi \in \mathfrak{D} \qquad (3)$$

All distributions of this form will be called *regular distributions*. A regular distribution corresponds to a class of locally integrable functions, any two of which differ on at most a set of measure zero.

Any distribution that is not a regular distribution is called a *singular* one. A very common singular distribution defined over $\mathfrak{R}^n$ is the *delta functional*. It is defined by

$$\langle \delta, \phi \rangle \triangleq \phi(0)$$

where $\phi(0)$ denotes the value of $\phi(t_1, t_2, \ldots, t_n)$ at the origin of $\mathfrak{R}^n$.

In the rest of this chapter, as well as in Chaps. 2 and 4, it will be understood that we are dealing with the general case where the independent variable t has n dimensions, unless the contrary is specifically stated. In Chap. 3 and from Chap. 5 on, t will be one-dimensional.

PROBLEMS

1 Demonstrate the two assertions made in this section concerning the ability of testing functions to approximate continuous functions and to be identical to infinitely smooth functions over any bounded domain of $\mathfrak{R}^n$. HINT: This may be done by using n-dimensional convolution

$$\int_{\mathfrak{R}^n} f(\tau)g(t - \tau)\, d\tau$$

where t and τ are n-dimensional real variables and the integration is over all of $\mathfrak{R}^n$.

2 Show that (3) defines a distribution. Then verify that this distribution corresponds to an equivalence class of functions any two of which differ on no more than a set of measure zero.

3 Extend the definitions of the spaces $\mathfrak{D}_m$ and $\mathfrak{D}'_m$, which were described in Sec. 1.3, Prob. 2, to the n-dimensional case. Give an example of a distribution over $\mathfrak{R}^n$ that is in $\mathfrak{D}'_m$ but not in $\mathfrak{D}'_{m-1}$.

4 An example of a pseudofunction defined over the two-dimensional plane for the variable

$$t = (t_1, t_2) = (r \cos \theta, r \sin \theta)$$

is Pf $|t|^{-2}$. Using polar coordinates, whereby $|t| = r$ and $dt = r\, d\theta\, dr$, we define it by

$$\langle \text{Pf } |t|^{-2}, \phi(t) \rangle = \text{Fp} \int_0^\infty r^{-2} \psi(r)\, dr \qquad \phi \in \mathfrak{D}$$

where $\psi(r)$ is a testing function of the one-dimensional variable r given by

$$\psi(r) = \int_0^{2\pi} \phi(r \cos \theta, r \sin \theta) r\, d\theta$$

Verify that $\psi(r)$ is an even, infinitely smooth function of bounded support. Thus, we may write

$$\langle \text{Pf } |t|^{-2}, \phi(t) \rangle = \left\langle \text{Pf } \frac{1}{r}, \psi(r) \right\rangle$$

★1.6 EQUALITY OF DISTRIBUTIONS OVER OPEN SETS

The natural definition of *equality* between two distributions is the following. Two distributions f and g are said to be equal if

$$\langle f, \phi \rangle = \langle g, \phi \rangle \tag{1}$$

for every testing function $\phi(t)$ in $\mathfrak{D}$.

However, the concept of equality can be refined in such a way that a definite meaning becomes assigned to the statement that "two distributions are equal over an open set Ω in $\Re^n$ but not necessarily over the entire space $\Re^n$." Were we to make this statement for two functions, we would mean that the values of the functions at any point in Ω are the same. However, this interpretation cannot be made for two distributions, because distributions do not possess values at given points as functions do. On the other hand, distributions do have a specific behavior over any open set and two distributions can have the same behavior over such a set. Quite specifically, two distributions f and g are said to be equal over the open set Ω if (1) holds for every testing function $\phi(t)$ whose support is contained in Ω. Thus, two distributions can be the same over a neighborhood of a point t (i.e., over an open set that contains t), even though no meaning can be attached to "their values at the point t."

It follows that a distribution f can be equal to a regular distribution g over an open set Ω, and in such cases we shall say that "f equals the function g over the open set Ω." Here again it is understood that we are really referring to an equivalence class of functions defined on Ω by g. As an example, the delta functional $\delta(t)$ equals zero (i.e., the null function) in the open sets $-\infty < t < 0$ and $0 < t < \infty$.

We can also speak of a distribution $f(t)$ that is defined over some open set Ω for the t variable but not over all of $\Re^n$. In this case we shall simply define $f(t)$ as a continuous linear functional on those testing functions whose supports are contained in Ω. Continuity will now be defined with respect to every sequence $\{\phi_\nu\}_{\nu=1}^\infty$ that converges in $\mathfrak{D}$ to zero and all of whose terms have their supports contained in a fixed bounded closed subset of Ω.

These ideas lead to the concept of the support of a distribution, which is analogous to the support of a testing function. The union of all open sets, over each of which a distribution f equals zero, is called the *null set* of f. This null set is an open set because the union of a finite or infinite collection of open sets is still open (Kestelman [1], p. 16). The set of all other points of the space $\Re^n$ is called the *support* of f, and this support is a closed set because it is the complement of an open set, the null set (Kestelman [1], p. 20). (The support of f is the smallest closed set outside of which f equals zero. See Sec. 1.8.) Every point in the support of a distribution is called an *essential point* of that distribution. Also, if a set Θ contains the support of a distribution, that distribution is said to be *concentrated on* Θ.

Thus, the delta functional $\delta(t)$ has the origin of $\Re^n$ as its support, and is concentrated on any set that contains this point. Similarly, the distribution $|t|^2 \triangleq t_1^2 + t_2^2 + \cdots + t_n^2$ has the entire space $\Re^n$ as its support. Note that the point $t = 0$ is an essential point of $|t|^2$ even though the function equals zero there. This illustrates the fact that it is the behavior of a distribution over the vicinity of a point that determines whether the point is in the support of f.

PROBLEMS

1 Show that the support of a regular distribution on $\Re^1$ corresponding to a continuous function is the closure of the set of points where the function is different from zero.

2 Let f be a distribution defined over $\Re^n$. Let $\{\Omega_\nu\}$ be a finite collection of disjoint open sets over each of which f is equal to zero. (By disjoint we mean that none of the sets intersect.) Let ϕ and ψ be two testing functions in $\mathfrak{D}$ which differ at most over closed subsets of the Ω_ν. Show that $\langle f, \phi \rangle = \langle f, \psi \rangle$.

★1.7 SOME OPERATIONS ON DISTRIBUTIONS

The addition of distributions Let f and g be two distributions. Their sum, $f + g$, is defined as the distribution that assigns to every ϕ in $\mathfrak{D}$ the value

$$\langle f + g, \phi \rangle \triangleq \langle f, \phi \rangle + \langle g, \phi \rangle \tag{1}$$

That $f + g$ is also a distribution follows from the fact that the right-hand side of (1) defines a continuous linear functional on $\mathfrak{D}$ whenever f and g are such functionals.

The multiplication of a distribution by a constant Let α be a complex number. The multiplication of a distribution f by α is defined by

$$\langle \alpha f, \phi \rangle \triangleq \langle f, \alpha \phi \rangle \qquad \phi \in \mathfrak{D} \tag{2}$$

The right-hand side of (2) also defines a continuous linear functional on $\mathfrak{D}$.

Under the two operations just defined, the space $\mathfrak{D}'$ of distributions becomes a *linear space*. This can be verified by showing that all the axioms for a linear space, which are listed in Appendix A, are satisfied.

By choosing α to be -1, we obtain the subtraction of a distribution g from a distribution f through the addition to f of the product of -1 and g.

The shifting (or translation) of a distribution Let $f(t)$ be a distribution defined over the n-dimensional euclidean space $\Re^n$ of the variable t. Also, let τ be a given point in this space. The *shifted distribution* $f(t - \tau)$ is the distribution defined by

$$\langle f(t - \tau), \phi(t) \rangle \triangleq \langle f(t), \phi(t + \tau) \rangle \qquad \phi \in \mathfrak{D} \tag{3}$$

This is precisely the definition one might expect, since, if $f(t)$ is a locally integrable function, a simple change of variable yields

$$\int_{-\infty}^{\infty} f(t - \tau)\phi(t)\, dt = \int_{-\infty}^{\infty} f(t)\phi(t + \tau)\, dt$$

The transposition of a distribution The transpose $f(-t)$ of a distribution $f(t)$ is the distribution defined by

$$\langle f(-t), \phi(t)\rangle \triangleq \langle f(t), \phi(-t)\rangle \qquad \phi \in \mathfrak{D} \qquad (4)$$

This concept of the transpose of a distribution can be used to define even and odd distributions in the same way that even and odd functions are defined. A distribution $f(t)$ is *even* if it equals its transpose:

$$f(t) = f(-t)$$

It is *odd* if it equals the negative of its transpose:

$$f(t) = -f(-t)$$

The multiplication of the independent variable by a positive constant Given a distribution $f(t)$, we wish to define what is meant by the distribution $f(at)$, where a is a positive number and t is the n-dimensional real variable. Here, at denotes the point whose components are $at_1, at_2, \ldots, at_n$. As usual, the desired definition can be motivated by first examining what occurs when $f(t)$ is a locally integrable function. In this case, for $\phi(t)$ in $\mathfrak{D}$,

$$\int_{\mathfrak{R}^n} f(at)\phi(t)\, dt = \frac{1}{a^n} \int_{\mathfrak{R}^n} f(t)\phi\left(\frac{t}{a}\right) dt$$

Generalizing this result, $f(at)$ is defined by

$$\langle f(at), \phi(t)\rangle \triangleq \left\langle f(t), \frac{1}{a^n}\phi\left(\frac{t}{a}\right)\right\rangle \qquad a > 0 \qquad (5)$$

Note that $\phi(t/a)$ is a testing function in $\mathfrak{D}$ whenever $\phi(t)$ is one. Moreover, $f(at)$ is clearly a distribution whenever $f(t)$ is one.

The multiplication of a distribution by an infinitely smooth function An operation that would be useful in analyses involving distributions would be the multiplication of two arbitrary distributions. Unfortunately, it is not possible to define such an operation in general. It turns out that the product does not always exist within the system of distributions. As an example, for the one-dimensional variable t, let $f(t) = 1/\sqrt{|t|}$. Then, $f(t)$ represents a regular distribution as well as a locally integrable function. Now, $[f(t)]^2$ is a function of t defined for all nonzero t. But it is not integrable over any interval that includes the origin. This means that it cannot define a distribution through the expression

$$\left\langle \frac{1}{|t|}, \phi \right\rangle = \int_{-\infty}^{\infty} \frac{\phi(t)}{|t|}\, dt$$

since the integral does not converge for every ϕ in $\mathfrak{D}$. (The reader should note that here we are not dealing with the finite part of this integral.) In short, the product of $1/\sqrt{|t|}$ with itself does not exist as a distribution.

It is, however, possible to define the product of distributions in special cases. For instance, if f and g are locally integrable functions over $\mathfrak{R}^n$ and if their product fg is also locally integrable, then the product of the corresponding regular distributions exists as a regular distribution defined by

$$\langle fg, \phi \rangle = \int_{\mathfrak{R}^n} f(t)g(t)\phi(t)\, dt \qquad \phi \in \mathfrak{D}$$

A more important case arises when one of the distributions ψ is a regular distribution corresponding to an infinitely smooth function. The product of ψ with any distribution f in $\mathfrak{D}'$ exists and is defined by

$$\langle \psi f, \phi \rangle \triangleq \langle f, \psi \phi \rangle \qquad \phi \in \mathfrak{D} \tag{6}$$

To establish the consistency of this definition, we have to show that $\psi\phi$ is a testing function in $\mathfrak{D}$ and that ψf is a continuous linear functional on $\mathfrak{D}$.

For every ϕ in $\mathfrak{D}$ the function $\psi\phi$ is infinitely smooth everywhere and zero whenever ϕ is zero. Hence, $\psi\phi$ is also in $\mathfrak{D}$.

Thus, (6) defines that functional on $\mathfrak{D}$ which assigns to each ϕ in $\mathfrak{D}$ the number $\langle f, \psi\phi \rangle$. This functional is linear because, for any ϕ_1 and ϕ_2 in $\mathfrak{D}$ and any two numbers α and β, we have

$$\langle \psi f, \alpha\phi_1 + \beta\phi_2 \rangle = \langle f, \alpha\psi\phi_1 + \beta\psi\phi_2 \rangle = \langle f, \alpha\psi\phi_1 \rangle + \langle f, \beta\psi\phi_2 \rangle \\ = \alpha \langle \psi f, \phi_1 \rangle + \beta \langle \psi f, \phi_2 \rangle$$

Furthermore, it is continuous. For if $\{\phi_\nu\}_{\nu=1}^{\infty}$ is a sequence of testing functions that converges in $\mathfrak{D}$ to zero, the sequence of testing functions $\{\psi\phi_\nu\}_{\nu=1}^{\infty}$ also converges in $\mathfrak{D}$ to zero. (Prove this.) In addition,

$$\{\psi f, \phi_\nu\}_{\nu=1}^{\infty} = \{\langle f, \psi\phi_\nu \rangle\}_{\nu=1}^{\infty}$$

and, since f itself is a continuous functional, the last sequence of numbers converges to zero. This establishes everything we wanted to show.

PROBLEMS

1 Verify that the sum of two distributions, the product of a distribution by a number, the shifted distribution, and the transpose of a distribution are again distributions.

2 Verify that $\mathfrak{D}'$ is a linear space.

3 Let the real part f_R and the imaginary part f_X of an arbitrary distribution f be defined as follows. Denote the real and imaginary parts of any ϕ in $\mathfrak{D}$ by ϕ_R and ϕ_X, respectively. Then,

$$\langle f_R, \phi \rangle \triangleq \operatorname{Re} \langle f, \phi_R \rangle + i \operatorname{Re} \langle f, \phi_X \rangle$$
$$\langle f_X, \phi \rangle \triangleq \operatorname{Im} \langle f, \phi_R \rangle + i \operatorname{Im} \langle f, \phi_X \rangle$$

Show that f_R and f_X are again distributions and that f can be represented as $f = f_R + if_X$. Let the complex conjugate of f be defined by $\bar{f} \triangleq f_R - if_X$. For any complex-valued testing function in $\mathfrak{D}$, show that

$$\langle \bar{f}, \phi \rangle = \overline{\langle f, \bar{\phi} \rangle}$$

4 Let a be a negative number. Show that for t in $\mathfrak{R}^n$

$$\langle f(at), \phi(t) \rangle = \left\langle f(t), \frac{1}{|a|^n} \phi\left(\frac{t}{a}\right) \right\rangle$$

5 Let a be a positive constant and let t be one-dimensional. For

$$f(t) \triangleq \operatorname{Pf} \frac{1_+(t)}{t}$$

show that it is not correct to write

$$f(at) = \operatorname{Pf} \frac{1_+(at)}{at}$$

6 Let ψ be an infinitely smooth function and let $\{\phi_\nu\}_{\nu=1}^\infty$ converge in $\mathfrak{D}$ to zero. Show that $\{\psi\phi_\nu\}_{\nu=1}^\infty$ also converges in $\mathfrak{D}$ to zero.

7 Associativity does not always hold when infinitely smooth functions and several distributions are multiplied. Show this by computing the two products, $[\delta(t)t] \operatorname{Pv}(1/t)$ and $\delta(t)[t \operatorname{Pv}(1/t)]$, where t is one-dimensional.

8 In Sec. 1.3, Prob. 2, we described the spaces $\mathfrak{D}_m$ and $\mathfrak{D}'_m$ for a one-dimensional t. Using these spaces, define the product $f(t)g(t)$, where f is in $\mathfrak{D}'_m$ and g is in $\mathcal{E}_m$, the space of functions having continuous derivatives up to order m. Then, show that, for $\delta^{(1)}(t)$ defined by Sec. 1.3, Eq. (12), and for g in $\mathcal{E}_1$,

$$g(t)\delta^{(1)}(t) = -g^{(1)}(0)\delta(t) + g(0)\delta^{(1)}(t)$$

9 Let t and y be one-dimensional real variables with y being restricted to positive values. If $f(t)$ is a distribution over $-\infty < t < \infty$, then, for each value of y, $f(yt)$ is defined by (5). Let us define differentiation with respect to y by

$$\left\langle \frac{\partial}{\partial y} f(yt), \phi(t) \right\rangle \triangleq \frac{\partial}{\partial y} \langle f(yt), \phi(t) \rangle \qquad \phi \in \mathfrak{D}$$

Establish the following formulas:

$$\frac{\partial}{\partial y} \delta(yt) = -\frac{1}{y^2} \delta(t)$$
$$\frac{\partial}{\partial y} \delta^{(1)}(yt) = -\frac{2}{y^3} \delta^{(1)}(t)$$

For $f(t) = \operatorname{Pf} 1_+(t)/t$,

$$\frac{\partial}{\partial y} f(yt) = -\frac{1}{y^2} \operatorname{Pf} \frac{1_+(t)}{t} + \frac{1 - \log y}{y^2} \delta(t)$$

10 Under certain circumstances, one can assign a meaning to the symbol $f(g(t))$, where f is a distribution and g is an ordinary function. In particular, let h be the inverse function for g so that $x = g(t)$, $t = h(x)$. We shall again restrict x and t to being one-dimensional. Furthermore, assume that both g and h are infinitely smooth, that the first derivatives of $g(t)$ and $h(x)$ are always positive or always negative (never zero) over the respective intervals $a < t < b$ and $c < x < d$, and that g and h map these intervals onto one another. Then, if $f(x)$ is a distribution over $c < x < d$, $f(g(t))$ is defined as a distribution over $a < t < b$ as follows. For each ϕ in $\mathfrak{D}$ with support contained in $a < t < b$,

$$\langle f(g(t)), \phi(t) \rangle \triangleq \langle f(x), |h^{(1)}(x)|\phi(h(x)) \rangle \tag{7}$$

a. Show that this is precisely the expression one would have if f were a continuous function.
b. Prove that $f(g(t))$ is truly a distribution over $a < t < b$.
c. Let y and τ be real numbers with $\tau > 0$ and $y \neq 0$. Demonstrate that

$$\delta(e^{yt} - \tau) = \frac{1}{|y|\tau} \delta\left(t - \frac{\log \tau}{y}\right)$$

◆ 1.8 DISTRIBUTIONS AS LOCAL PHENOMENA

A *neighborhood* of a closed set is any open set that contains the closed set. A neighborhood of the entire space $\mathfrak{R}^n$ is $\mathfrak{R}^n$.

The principal objective of this section is to show that, for any distribution that is defined over $\mathfrak{R}^n$ and for any testing function ϕ in $\mathfrak{D}$, it is only the values that ϕ assumes over a neighborhood of the support of f which determine the number $\langle f, \phi \rangle$. Moreover, this statement holds true for every such neighborhood no matter how small. (However, it is not true in general that $\langle f, \phi \rangle$ depends only on the values that ϕ assumes over the support of f. An example of this will be given at the beginning of Sec. 3.5.)

Because of this, we can always assign a sense to the symbol $\langle f, \theta \rangle$, where f is a distribution of bounded support and θ is an infinitely smooth function over a neighborhood Ω of the support of f but is an arbitrary function or perhaps not even defined elsewhere. We do this through the definition

$$\langle f, \theta \rangle \triangleq \langle f, \phi \rangle$$

where ϕ is in $\mathfrak{D}$, is equal to θ over a (smaller) neighborhood of the support of f, and is equal to zero outside Ω. (Lemma 1 below implies that such a ϕ exists.)

In order to verify these statements, we shall have to establish

Theorem 1.8-1

If a distribution is equal to zero on every set of a collection of open sets, then it is equal to zero on the union of these sets.

Assume for the moment that this assertion is true. Let ϕ and ψ be two testing functions in $\mathfrak{D}$ that are equal over a neighborhood of the support of the distribution f but are not necessarily the same over certain parts of the null set of f. Then $\phi - \psi$ is a testing function in $\mathfrak{D}$ whose support is contained in the null set of f. By definition this null set is the union of all open sets over each of which f equals zero. Therefore, by our theorem, $\langle f, \phi - \psi \rangle = 0$. By the linearity of f, $\langle f, \phi \rangle = \langle f, \psi \rangle$. This is how the first few statements of this section follow from Theorem 1.8-1.

Since two distributions, f and g, are equal over an open set if their difference $f - g$ is zero over this set, an immediate consequence of Theorem 1.8-1 is

Corollary 1.8-1a

If two distributions are equal over each set in a collection of open sets, then they are equal over the union of these sets.

The proof of Theorem 1.8-1 is rather technical. Readers who desire a briefer introduction to distribution theory or are more interested in its applications may skip the rest of this section.

Lemma 1

If Θ is a bounded closed set in $\mathfrak{R}^n$ and Ω is a neighborhood of Θ, then there exists a testing function ϕ in $\mathfrak{D}$ such that $\phi(t) = 1$ on Θ, $\phi(t) = 0$ outside Ω, and $0 \leq \phi(t) \leq 1$ for all t.

Proof: The proof of this lemma is quite similar to some of the arguments constructed in Sec. 1.2, where we were restricting ourselves to $\mathfrak{R}^1$. We shall be more concise here.

Let t be in $\mathfrak{R}^n$, let α be a positive number, let $\zeta(t)$ be given by Sec. 1.5, Eq. (2), and let

$$\gamma_\alpha(t) = \frac{\zeta(t/\alpha)}{\int_{\mathfrak{R}^n} \zeta(t/\alpha) \, dt}$$

Ψ shall be some bounded open set which contains Θ and whose closure is contained in Ω. Let $h(t) = 1$ for each t in Ψ and let $h(t) = 0$ elsewhere. Define

$$\phi(t) \triangleq \int_{\mathfrak{R}^n} h(\tau) \gamma_\alpha(t - \tau) \, d\tau$$

By taking α sufficiently small, we can show that $\phi(t)$ possesses all the required properties. That $\phi(t)$ is infinitely smooth can be verified by differentiating under the integral sign (Hobson [1], vol. 2, p. 355). Q.E.D.

A collection of sets $\{\Lambda_\nu\}$ is said to cover another set Θ if the union of all the Λ_ν contains Θ.

Let $\{\Omega_\nu\}_{\nu=1}^\infty$ be a countable collection of bounded open sets that covers $\Re^n$ with the additional property that every bounded set in $\Re^n$ intersects at most a finite number of the Ω_ν. We call such a covering a *locally finite covering* of $\Re^n$.

As is customary, $\bar{\Lambda}$ will denote the closure of any set Λ.

Lemma 2

Let the countable collection of bounded open sets $\{\Omega_\nu\}_{\nu=1}^\infty$ be a locally finite covering of $\Re^n$. One can always choose another countable collection of open sets $\{\Lambda_\nu\}_{\nu=1}^\infty$ which covers $\Re^n$ and is such that every $\bar{\Lambda}_\nu$ is contained in the corresponding Ω_ν.

Proof: Assume that the sets $\Lambda_1, \Lambda_2, \ldots, \Lambda_{k-1}$ have been chosen such that $\bar{\Lambda}_\nu \subset \Omega_\nu$ ($\nu = 1, 2, \ldots, k-1$) and

$$\{\Lambda_1, \ldots, \Lambda_{k-1}, \Omega_k, \Omega_{k+1}, \ldots\}$$

is a locally finite covering of $\Re^n$. Let Ψ_k be the complement with respect to $\Re^n$ of

$$\Lambda_1 \cup \cdots \cup \Lambda_{k-1} \cup \Omega_{k+1} \cup \Omega_{k+2} \cup \cdots$$

Ψ_k is a closed set contained in Ω_k (Kestelman [1], pp. 16, 20). Let Λ_k be an open set such that $\Lambda_k \supset \Psi_k$ and $\bar{\Lambda}_k \subset \Omega_k$. In this way the entire covering $\{\Lambda_\nu\}_{\nu=1}^\infty$ can be chosen.

The next lemma describes a result that is known as *the partitioning of unity*.

Lemma 3

For any given locally finite countable covering $\{\Omega_\nu\}_{\nu=1}^\infty$ of $\Re^n$ where the Ω_ν are bounded open sets, there exists a collection $\{\phi_\nu\}_{\nu=1}^\infty$ of testing functions in $\mathfrak{D}$ such that:

a. $0 \leq \phi_\nu(t) \leq 1$.
b. $\phi_\nu(t)$ has its support contained in Ω_ν.
c. $\sum_{\nu=1}^\infty \phi_\nu(t) \equiv 1$.

Proof: Let $\{\Lambda_\nu\}_{\nu=1}^\infty$ be another covering of $\Re^n$ in accordance with Lemma 2. By Lemma 1 there exists, for each ν, a ψ_ν in $\mathfrak{D}$ such that $\psi_\nu(t) = 1$ on Λ_ν, $\psi_\nu(t) = 0$ outside Ω_ν, and $0 \leq \psi_\nu(t) \leq 1$ for all t. Let

$$\psi(t) = \sum_{\nu=1}^\infty \psi_\nu(t)$$

For each t there are only a finite number of nonzero terms in this sum, and at least one of these terms equals one. Therefore, $\psi(t)$ exists everywhere and is not less than one. Now, set

$$\phi_\nu(t) = \frac{\psi_\nu(t)}{\psi(t)} \qquad \nu = 1, 2, 3, \ldots$$

These functions possess the required properties. Q.E.D.

The diameter of any bounded set in $\mathfrak{R}^n$ is the least upper bound on all the distances between pairs of points of the set.

Lemma 4

Let $\mathfrak{C}_T$ be a collection of open sets whose diameters are all bounded by one and assume that $\mathfrak{C}_T$ covers $\mathfrak{R}^n$. Then, $\mathfrak{C}_T$ contains a locally finite countable covering of $\mathfrak{R}^n$.

Proof: Let $t = \{t_1, t_2, \ldots, t_n\}$ be an arbitrary point in $\mathfrak{R}^n$. $\mathfrak{R}^n$ is the union of all n-dimensional cubes of the form

$$m_k \leq t_k \leq m_k + 1 \qquad k = 0, 1, \ldots, n$$

where the m_k are integers. According to the Heine-Borel theorem (Kestelman [1], pp. 26–27), for each such cube we can choose from $\mathfrak{C}_T$ a finite subcollection that covers the cube. The union of all these chosen subcollections is a countable covering of $\mathfrak{R}^n$, which is also locally finite in view of the uniform bound on the diameters of its members. Q.E.D.

Proof of Theorem 1.8-1: We are given a collection of open sets in $\mathfrak{R}^n$ over each of which f is zero. Let Θ denote the union of this collection. (Θ is an open set.) If a distribution is defined over an open set Ω, it is certainly defined over any open subset of Ω. Therefore, we are free to replace the original collection by a collection $\mathfrak{C}$ of open sets whose diameters are all bounded by one and whose union is Θ.

Now, let ψ be a testing function in $\mathfrak{D}$ whose support Ξ is contained in Θ. We wish to show that $\langle f, \psi \rangle$ is zero.

Since Ξ is a closed set, a neighborhood Ξ_1 of Ξ can be so chosen that

$$\Xi \subset \Xi_1 \subset \bar{\Xi}_1 \subset \Theta$$

It follows that we can add to $\mathfrak{C}$ open sets, whose diameters are all bounded by one, to obtain a covering $\mathfrak{C}_T$ of $\mathfrak{R}^n$ in such a way that the added sets do not intersect Ξ_1. [To obtain the additional sets, take any covering $\{\Gamma_\nu\}$ of $\mathfrak{R}^n$, where the diameters of the open sets Γ_ν are all bounded by one. Then, replace each Γ_ν that intersects Ξ_1 by $\Gamma_\nu - (\bar{\Xi}_1 \cap \bar{\Gamma}_\nu)$.]

By Lemma 4 we can choose from $\mathfrak{C}_T$ a countable subcollection $\{\Omega_\nu\}_{\nu=1}^\infty$, which is a locally finite covering of $\mathfrak{R}^n$. In accordance with

Lemma 3, let $\{\phi_\nu\}_{\nu=1}^\infty$ be a partitioning of unity with respect to the covering $\{\Omega_\nu\}_{\nu=1}^\infty$. ψ can be decomposed into

$$\psi = \sum_{\nu=1}^\infty \psi\phi_\nu \tag{1}$$

where $\psi\phi_\nu$ is a testing function in $\mathfrak{D}$ with support in Ω_ν. Note that there are only a finite number of nonzero terms $\psi\phi_\nu$ in the summation (1) and that each such term has its support contained in one of the members of $\mathfrak{C}$. By hypothesis,

$$\langle f, \psi\phi_\nu \rangle = 0$$

and, therefore,

$$\langle f, \psi \rangle = \sum_{\nu=1}^\infty \langle f, \psi\phi_\nu \rangle = 0 \quad \text{Q.E.D.}$$

Another consequence of these results is that under certain circumstances a distribution can be constructed over all of $\mathfrak{R}^n$ from a collection of distributions that are defined only over portions of $\mathfrak{R}^n$. More precisely, we can state

Theorem 1.8-2

Let $\{\Lambda_\nu\}$ be a collection of open sets that covers $\mathfrak{R}^n$. On each Λ_ν let there be defined a distribution f_ν. Finally, whenever two sets Λ_ν and Λ_μ have a nonvoid intersection, let the corresponding distributions f_ν and f_μ be equal on this intersection. Then, there is one and only one distribution f defined over $\mathfrak{R}^n$ which is equal to f_ν over Λ_ν for every ν.

Proof: We are free to assume that all Λ_ν have diameters bounded by one because any open set not of this type can be replaced by a collection of suitably bounded open sets without upsetting the hypothesis of this theorem.

Again by Lemma 4 we can choose a countable subcollection $\{\Lambda_{\nu_k}\}_{k=1}^\infty$, which is a locally finite covering of $\mathfrak{R}^n$. Let $\{\phi_k\}_{k=1}^\infty$ be a partitioning of unity with respect to $\{\Lambda_{\nu_k}\}_{k=1}^\infty$ in accordance with Lemma 3. For any ψ in $\mathfrak{D}$ define f by

$$\langle f, \psi \rangle = \sum_{k=1}^\infty \langle f_{\nu_k}, \psi\phi_k \rangle \tag{2}$$

Note again that there are only a finite number of nonzero terms in the last summation.

f is clearly a linear functional on $\mathfrak{D}$. To show that it is continuous, let $\{\psi_\mu\}_{\mu=1}^\infty$ converge in $\mathfrak{D}$ to zero. Then, for each k, $\{\psi_\mu\phi_k\}_{\mu=1}^\infty$ also converges in $\mathfrak{D}$ to zero, and as $\mu \to \infty$

$$\langle f_{\nu_k}, \psi_\mu \phi_k \rangle \to 0$$

Therefore, as $\mu \to \infty$

$$\langle f, \psi_\mu \rangle = \sum_{k=1}^\infty \langle f_{\nu_k}, \psi_\mu \phi_k \rangle \to 0$$

Thus, f is truly a distribution over $\mathfrak{R}^n$.

We now show that $f = f_\mu$ on Λ_μ. If ψ has its support contained in Λ_μ, the support of $\psi \phi_k$ is contained in the intersection of Λ_μ and Λ_{ν_k}. By hypothesis, $f_\mu = f_{\nu_k}$ over this intersection, so that

$$\langle f, \psi \rangle = \sum_{k=1}^\infty \langle f_{\nu_k}, \psi \phi_k \rangle = \sum_{k=1}^\infty \langle f_\mu, \psi \phi_k \rangle$$
$$= \langle f_\mu, \psi \rangle$$

Finally, there can be only one distribution f over $\mathfrak{R}^n$ which is equal to f_ν over Λ_ν for every ν because of Corollary 1.8-1a. Q.E.D.

PROBLEMS

1 Assume that the set Ω in $\mathfrak{R}^n$ is the support of the distribution f and let $\theta(t)$ be an infinitely smooth function that differs from zero on every point of Ω. Show that the support of θf is precisely Ω.

2 Let f be a distribution on $\mathfrak{R}^1$. Prove that f is equal to zero on the finite *open* interval $a < t < b$ if and only if $\langle f, \phi \rangle = 0$ for every testing function ϕ in $\mathfrak{D}$ whose support is contained in the *closed* interval $a \leq t \leq b$. HINT: For the "only if" part, proceed as follows. Let c and d be two real numbers such that $a < c < d < b$. For a given ϕ in $\mathfrak{D}$ with support contained in $a \leq t \leq b$, construct two testing functions θ and ψ whose supports are contained in $a \leq t \leq d$ and $c \leq t \leq b$, respectively, such that $\phi = \theta + \psi$. Then, set

$$\phi_\nu(t) = \theta\left(t - \frac{1}{\nu}\right) + \psi\left(t + \frac{1}{\nu}\right)$$

and show that $\{\phi_\nu\}_{\nu=1}^\infty$ converges in $\mathfrak{D}$ to ϕ. Finally, invoke the continuity and linearity of f. (This procedure is taken from Dolezal [1], p. 52.)

3 Use Lemma 1 to construct a testing function of the type described in the third paragraph of this section.

2 The Calculus of Distributions

★2.1 INTRODUCTION

This chapter develops the calculus of distributions and, in particular, such concepts as the convergence of distributions, differentiation, and integration. The power of distributional analysis rests in large part on the facts that every distribution possesses derivatives of all orders and that differentiation is a continuous operation in this theory. As a consequence, distributional differentiation commutes with various limiting processes such as infinite summation and integration. This is in contrast to classical analysis wherein either such operations cannot be interchanged or the inversion of order must be justified by an additional argument.

◆2.2 CONVERGENCE OF A SEQUENCE OF DISTRIBUTIONS (CONVERGENCE IN THE SPACE $\mathfrak{D}'$)

We shall now take up the concept of convergence for a set of distributions $\{f_\nu\}$. Usually, we shall assume that the parameter ν traverses the natural numbers and increases indefinitely toward ∞. However, it may just as well traverse a continuous set of values, and it may approach some finite limit. (See Prob. 1.)

The sequence of distributions $\{f_\nu\}_{\nu=1}^\infty$ is said to *converge in* $\mathfrak{D}'$ if, for every ϕ in $\mathfrak{D}$, the sequence of numbers $\{\langle f_\nu, \phi \rangle\}_{\nu=1}^\infty$ converges in the ordinary sense of the convergence of numbers. The limit of $\{\langle f_\nu, \phi \rangle\}_{\nu=1}^\infty$, which we shall denote by $\langle f, \phi \rangle$, defines a functional f acting on the space $\mathfrak{D}$. In this case we shall also say that f is the *limit in* $\mathfrak{D}'$ of $\{f_\nu\}_{\nu=1}^\infty$ and at times write

$$\lim_{\nu \to \infty} f_\nu = f \tag{1}$$

or

$$f_\nu \to f \quad \text{as } \nu \to \infty$$

However, it is not obvious that f is a distribution, since f must be linear and continuous as a functional on $\mathfrak{D}$ for that to be true. It turns out that it is indeed true, as we shall now show. (Gelfand and Shilov [1], vol. I, p. 416, state that the following proof is due to M. S. Brodskii.)

Theorem 2.2-1

If a sequence of distributions $\{f_\nu\}_{\nu=1}^\infty$ converges in $\mathfrak{D}'$ to the functional f, then f is also a distribution. In other words, the space $\mathfrak{D}'$ is closed under convergence.

Proof: As was pointed out before, f is a functional on $\mathfrak{D}$ which assigns to each ϕ in $\mathfrak{D}$ the value

$$\lim_{\nu \to \infty} \langle f_\nu, \phi \rangle \tag{2}$$

f is linear, because for any ϕ and ψ in $\mathfrak{D}$ and for any two numbers α and β we have

$$\langle f, \alpha\phi + \beta\psi \rangle = \lim_{\nu \to \infty} \langle f_\nu, \alpha\phi + \beta\psi \rangle$$
$$= \lim_{\nu \to \infty} [\alpha \langle f_\nu, \phi \rangle + \beta \langle f_\nu, \psi \rangle]$$
$$= \alpha \langle f, \phi \rangle + \beta \langle f, \psi \rangle$$

Because f is linear, its continuity will be established if we show that the sequence $\{\langle f, \phi_\nu \rangle\}_{\nu=1}^\infty$ converges to zero for every sequence $\{\phi_\nu\}_{\nu=1}^\infty$ that converges in $\mathfrak{D}$ to zero. We shall assume the opposite and then construct a contradiction.

Assume that there exists a sequence of testing functions that converges in $\mathfrak{D}$ to zero and is such that the corresponding sequence of numbers assigned by f to the testing functions does not converge to zero. We can certainly choose a subsequence $\{\phi_\nu\}_{\nu=1}^\infty$ from the original sequence of testing functions such that

$$|\langle f, \phi_\nu \rangle| \geq c > 0 \qquad \nu = 1, 2, 3, \ldots \tag{3}$$

where c is some fixed positive number, and such that the partial derivatives of the elements of $\{\phi_\nu\}$ satisfy the inequalities

$$|D^k \phi_\nu| \leq \frac{1}{4^\nu} \qquad \hat{k} = 0, 1, \ldots, \nu \tag{4}$$

Here, D^k denotes an arbitrary partial derivative of order $\hat{k}$ and it is understood that, for each choice of ϕ_ν, (4) holds for all the partial derivatives of order no greater than ν.

Now, let $\psi_\nu = 2^\nu \phi_\nu$. By (4), the sequence $\{\psi_\nu\}_{\nu=1}^\infty$ converges in $\mathfrak{D}$ to zero, and by (3) the sequence of numbers $\{|\langle f, \psi_\nu \rangle|\}_{\nu=1}^\infty$ diverges to $+\infty$. We shall now choose a subsequence $\{\psi_\nu'\}_{\nu=1}^\infty$ from $\{\psi_\nu\}_{\nu=1}^\infty$ and a subsequence $\{f_\nu'\}_{\nu=1}^\infty$ from the original sequence $\{f_\nu\}_{\nu=1}^\infty$ of distributions, which converged in $\mathfrak{D}'$ to the functional f, in the following way.

First, choose ψ_1' such that $|\langle f, \psi_1' \rangle| > 1$. Since, for every testing function ψ, $\langle f_\nu, \psi \rangle \to \langle f, \psi \rangle$ as $\nu \to \infty$, f_1' can be so chosen that $|\langle f_1', \psi_1' \rangle| > 1$.

Now, assuming that the first $\nu - 1$ elements of these subsequences have been chosen, we select ψ_ν' as an element from $\{\psi_\nu\}_{\nu=1}^\infty$ (with index higher than those of $\psi_1', \ldots, \psi_{\nu-1}'$) such that

$$|\langle f_j', \psi_\nu' \rangle| < \frac{1}{2^{\nu-j}} \qquad j = 1, \ldots \nu - 1 \tag{5}$$

and

$$|\langle f, \psi_\nu' \rangle| > \sum_{\mu=1}^{\nu-1} |\langle f, \psi_\mu' \rangle| + \nu \tag{6}$$

The condition (5) can be satisfied, since $\{\psi_\nu\}_{\nu=1}^\infty$ converges in $\mathfrak{D}$ to zero and, for each fixed distribution f_j', $\langle f_j', \psi_\nu \rangle \to 0$ as $\nu \to \infty$. Condition (6) can also be satisfied, because $|\langle f, \psi_\nu \rangle| \to \infty$ as $\nu \to \infty$.

Since, for any testing function ϕ, $\langle f_\nu, \phi \rangle \to \langle f, \phi \rangle$ as $\nu \to \infty$, f_ν' can now be chosen as an element from $\{f_\nu\}_{\nu=1}^\infty$ (with index higher than those of $f_1', \ldots, f_{\nu-1}'$) such that

$$|\langle f_\nu', \psi_\nu' \rangle| > \sum_{\mu=1}^{\nu-1} |\langle f_\nu', \psi_\mu' \rangle| + \nu \tag{7}$$

Now that $\{\psi_\nu'\}_{\nu=1}^\infty$ and $\{f_\nu'\}_{\nu=1}^\infty$ have been specified, consider the following series:

$$\psi \triangleq \sum_{\nu=1}^\infty \psi_\nu' \tag{8}$$

We wish to show that ψ is in $\mathfrak{D}$. Consider the general remainder

$$R_n \triangleq \sum_{\nu=n}^\infty \psi_\nu'$$

Since $\{\psi'_\nu\}_{\nu=1}^\infty$ is a subsequence of $\{\psi_\nu\}_{\nu=1}^\infty$,

$$|R_n| \leq \sum_{\nu=n}^\infty |\psi'_\nu| \leq \sum_{\nu=n}^\infty |\psi_\nu| \leq \sum_{\nu=n}^\infty \frac{1}{2^\nu}$$

The last sum goes to zero as $n \to \infty$. Thus, (8) converges uniformly for all values of the independent variable t.

Similarly, consider the series of partial derivatives

$$\sum_{\nu=1}^\infty D^k \psi_\nu$$

For the general remainder and with $n \geq \hat{k}$,

$$\left| \sum_{\nu=n}^\infty D^k \psi'_\nu \right| \leq \sum_{\nu=n}^\infty |D^k \psi'_\nu| \leq \sum_{\nu=n}^\infty |D^k \psi_\nu| \leq \sum_{\nu=n}^\infty \frac{1}{2^\nu}$$

so that again we have uniform convergence. It follows that the series (8) can be differentiated term by term an arbitrary number of times (Carslaw [1], p. 161), and that ψ is, therefore, infinitely smooth. ψ also has a bounded support, since every term on the right-hand side of (8) is identically zero outside a fixed bounded domain. Hence, ψ is truly in $\mathfrak{D}$.

Finally,

$$\langle f'_\nu, \psi \rangle = \sum_{\mu=1}^{\nu-1} \langle f'_\nu, \psi'_\mu \rangle + \langle f'_\nu, \psi'_\nu \rangle + \sum_{\mu=\nu+1}^\infty \langle f'_\nu, \psi'_\mu \rangle \qquad (9)$$

By (5),

$$\left| \sum_{\mu=\nu+1}^\infty \langle f'_\nu, \psi'_\mu \rangle \right| < \sum_{\mu=\nu+1}^\infty \frac{1}{2^{\mu-\nu}} = 1 \qquad (10)$$

For any three complex numbers α, β, and γ, we have that $|\alpha + \beta + \gamma| \geq |\alpha| - |\beta| - |\gamma|$. Therefore, by (7), (9), and (10),

$$|\langle f'_\nu, \psi \rangle| \geq |\langle f'_\nu, \psi'_\nu \rangle| - \left| \sum_{\mu=1}^{\nu-1} \langle f'_\nu, \psi'_\mu \rangle \right| - \left| \sum_{\mu=\nu+1}^\infty \langle f'_\nu, \psi'_\mu \rangle \right| > \nu - 1$$

This shows that, as $\nu \to \infty$, $|\langle f'_\nu, \psi \rangle| \to \infty$, which contradicts the hypothesis that $\langle f_\nu, \psi \rangle$ converges as $\nu \to \infty$. Thus, f is a continuous linear functional on $\mathfrak{D}$. Q.E.D.

The convergence in $\mathfrak{D}'$ of an infinite series of distributions can also be defined. Let $\sum_{\nu=1}^\infty f_\nu$ be such a series. If the sequence $\{h_m\}_{m=1}^\infty$ of partial sums,

$$h_m \triangleq \sum_{\nu=1}^m f_\nu$$

converges in $\mathfrak{D}'$, then the series $\sum_{\nu=1}^{\infty} f_\nu$ is said to converge in $\mathfrak{D}'$. By Theorem 2.2-1, its sum will also be a distribution. An equivalent way of stating this definition is to say that the series converges in $\mathfrak{D}$ and has as a sum the distribution f if for every ϕ in $\mathfrak{D}$ the series of numbers $\sum_{\nu=1}^{\infty} \langle f_\nu, \phi \rangle$ converges in the ordinary sense to the value $\langle f, \phi \rangle$.

Finally, one can easily show that convergence in $\mathfrak{D}'$ is a linear operation. That is, if α and β are arbitrary numbers and if $\{f_\nu\}_{\nu=1}^{\infty}$ and $\{g_\nu\}_{\nu=1}^{\infty}$ both converge in $\mathfrak{D}'$ to f and g, respectively, then

$$\{\alpha f_\nu + \beta g_\nu\}_{\nu=1}^{\infty}$$

converges in $\mathfrak{D}'$ to $\alpha f + \beta g$.

PROBLEMS

1 Let $\{f_\nu\}_{\nu \to \infty}$ be a directed set of distributions, where ν approaches infinity through a continuous set of values. Show that $\{f_\nu\}_{\nu \to \infty}$ converges in $\mathfrak{D}'$ if and only if every sequence $\{f_{\nu_k}\}_{k=1}^{\infty}$, where $\nu_k \to \infty$, obtained from this directed set converges in $\mathfrak{D}'$.

2 Prove that convergence in $\mathfrak{D}'$ is a linear operation.

3 Let the sequence of distributions $\{f_\nu\}_{\nu=1}^{\infty}$ converge in $\mathfrak{D}'$ to f and let the numerical sequence $\{\alpha_\nu\}_{\nu=1}^{\infty}$ converge to the number α. Show that $\{\alpha_\nu f_\nu\}_{\nu=1}^{\infty}$ converges in $\mathfrak{D}'$ to αf.

4 Let the sequence of infinitely smooth functions $\{\theta_\nu\}_{\nu=1}^{\infty}$ be such that for each positive integer k the sequence $\{\theta_\nu^{(k)}(t)\}_{\nu=1}^{\infty}$ converges uniformly on every bounded t domain. Let θ be the limit function for $\{\theta_\nu\}_{\nu=1}^{\infty}$. Show that, for any distribution f, the sequence $\{\theta_\nu f\}_{\nu=1}^{\infty}$ converges in $\mathfrak{D}'$ to θf.

★2.3 SOME SPECIAL CASES OF CONVERGENCE IN $\mathfrak{D}'$

Under certain circumstances the pointwise convergence almost everywhere of a sequence of locally integrable functions is reflected in distribution theory by the convergence in $\mathfrak{D}'$ of the corresponding sequence of regular distributions. Specifically, one set of conditions under which this occurs is given by

Theorem 2.3-1

Let $\{f_\nu(t)\}_{\nu=1}^{\infty}$ be a sequence of locally integrable functions that converges pointwise almost everywhere to the function $f(t)$ and let all the functions $f_\nu(t)$ be bounded in magnitude by a locally integrable function. Then, $f(t)$ is locally integrable and the corresponding sequence of regular distributions $\{f_\nu(t)\}_{\nu=1}^{\infty}$ converges in $\mathfrak{D}'$ to the regular distribution $f(t)$.

Proof: The conclusion that $f(t)$ is locally integrable is asserted by Lebesgue's general convergence theorem (Kestelman [1], p. 141). Furthermore, for any fixed ϕ in $\mathfrak{D}$ the sequence $\{f_\nu(t)\phi(t)\}_{\nu=1}^\infty$ of locally integrable functions will converge almost everywhere to $f(t)\phi(t)$. Also, every element of this sequence is identically zero outside a fixed bounded domain and thus is integrable over all of $\mathfrak{R}^n$. Once again by Lebesgue's general convergence theorem,

$$\lim_{\nu \to \infty} \langle f_\nu(t), \phi(t) \rangle = \lim_{\nu \to \infty} \int_{\mathfrak{R}^n} f_\nu(t)\phi(t)\, dt$$
$$= \int_{\mathfrak{R}^n} f(t)\phi(t)\, dt = \langle f(t), \phi(t) \rangle \qquad \text{Q.E.D.}$$

The hypothesis of Theorem 2.3-1 will certainly be satisfied if the sequence $\{f_\nu(t)\}_{\nu=1}^\infty$ of locally integrable functions converges uniformly on every bounded domain in $\mathfrak{R}^n$ to the function $f(t)$. This constitutes a special case under Theorem 2.3-1.

It follows that, when a sequence of testing functions converges in $\mathfrak{D}$, it also converges in $\mathfrak{D}'$. (Here we are identifying the testing functions and the corresponding regular distributions, as was explained in Sec. 1.3.) Thus, not only is $\mathfrak{D}$ a subspace of $\mathfrak{D}'$ but, in addition, convergence in $\mathfrak{D}$ implies convergence in $\mathfrak{D}'$.

The type of convergence described in Theorem 2.3-1 is called *dominated convergence*, uniform convergence being a special case of it. Theorem 2.3-1 states that dominated convergence possesses an analogue in convergence in $\mathfrak{D}'$. On the other hand, it is possible for a sequence of regular distributions to converge in $\mathfrak{D}'$ to a regular distribution, whereas the corresponding sequence of functions does not converge pointwise anywhere. As an illustration, consider

Example 2.3-1 Let $f_\nu(t) = \sin \nu t$, t being one-dimensional. Then, as ν approaches infinity through a continuous set of values, the set of functions $\{f_\nu(t)\}$ does not converge pointwise anywhere except at $t = 0$. And yet the corresponding set of regular distributions converges in $\mathfrak{D}'$ to the zero distribution (i.e., the distribution that equals zero over all of $\mathfrak{R}^n$). For, when ϕ is in $\mathfrak{D}$, we may integrate by parts to write

$$\langle f_\nu, \phi \rangle = \int_{-\infty}^\infty \sin \nu t\, \phi(t)\, dt = \frac{1}{\nu} \int_{-\infty}^\infty \phi^{(1)}(t) \cos \nu t\, dt \qquad (1)$$

Since $\phi^{(1)}(t)$ is absolutely integrable over $-\infty < t < \infty$,

$$|\langle f_\nu, \phi \rangle| \leq \frac{1}{\nu} \int_{-\infty}^\infty |\phi^{(1)}(t)|\, dt \to 0$$

as $\nu \to \infty$.

The relationship between pointwise convergence and convergence in $\mathcal{D}'$ is even more complicated than this example indicates. The following two examples are of interest here.

Example 2.3-2 Let

$$f_\nu(t) = \begin{cases} 0 & |t| \geq \dfrac{1}{\nu} \\ \dfrac{\nu}{2} & |t| < \dfrac{1}{\nu} \end{cases}$$

For each ϕ in $\mathcal{D}$,

$$\langle f_\nu, \phi \rangle = \frac{\nu}{2} \int_{-1/\nu}^{1/\nu} \phi(t)\, dt \to \phi(0)$$

as $\nu \to \infty$. Thus, the regular distributions f_ν converge in $\mathcal{D}'$ to the delta functional, whereas the corresponding functions converge to zero almost everywhere (everywhere except at $t = 0$, where the limit does not exist). This pointwise limit corresponds to the zero distribution. Here, even though we have both pointwise convergence almost everywhere and convergence in $\mathcal{D}'$, the two limits do not correspond.

Example 2.3-3 Let

$$f_\nu(t) = \begin{cases} 0 & |t| \geq \dfrac{1}{\nu} \\ \nu^2 & |t| < \dfrac{1}{\nu} \end{cases}$$

For ϕ in $\mathcal{D}$, $\lim_{\nu \to \infty} \langle f_\nu, \phi \rangle$ will fail to exist whenever $\phi(0) \neq 0$. But the functions again converge to zero everywhere except at $t = 0$, at which point the limit function does not exist. Here we have an example of a directed set of absolutely integrable functions converging almost everywhere to another absolutely integrable function, while the corresponding set of regular distributions has no limit in the sense of convergence in $\mathcal{D}'$.

These examples show that pointwise convergence and convergence in $\mathcal{D}'$ are not easily related. Either one may occur without the other occurring and, even when both take place, the limits may not correspond.

The concept of convergence in $\mathcal{D}'$ allows us to assign a distributional sense to various improper integrals, which diverge in the classical sense. Let $f(\omega, t)$ be a locally integrable function of the one-dimensional real variables ω and t and consider the convergent integral

$$g_b(t) = \int_a^b f(\omega, t)\, d\omega$$

where a is fixed and finite and b is variable. As $b \to \infty$, an improper integral (which is conventionally defined as the limit of the directed set of functions $\{g_b(t)\}$) is generated. It is possible that this limit fails to exist, while at the same time the corresponding set of regular distributions converges in $\mathfrak{D}'$ to a singular distribution. This is confirmed by

Example 2.3-4 Let

$$g_b(t) = \frac{1}{\pi} \int_0^b \cos \omega t \, d\omega = \frac{\sin bt}{\pi t}$$

where ω and t are one-dimensional. For every t this fails to converge as $b \to \infty$. Yet the corresponding regular distributions converge to $\delta(t)$ because, when ϕ is in $\mathfrak{D}$,

$$\lim_{b \to \infty} \langle g_b(t), \phi(t) \rangle = \lim_{b \to \infty} \frac{1}{\pi} \int_{-\infty}^{\infty} \frac{\sin bt}{t} \phi(t) \, dt$$
$$= \phi(0)$$

(A derivation of this last well-known limit will be given in the proof of Theorem 7.2-1, to which the reader may now refer.)

Thus, as a strictly symbolic expression, we may write

$$\frac{1}{\pi} \int_0^\infty \cos \omega t \, d\omega = \delta(t)$$

Another special case of convergence in $\mathfrak{D}'$ that has practical importance is given by the next theorem, which states that a sequence of regular distributions defined over $\mathfrak{R}^n$ will converge in $\mathfrak{D}'$ to the delta functional under some rather general conditions.

Theorem 2.3-2
Let $\{f_\nu(t)\}_{\nu=1}^\infty$ be a sequence of locally integrable functions, defined over $\mathfrak{R}^n$, such that:
a. All the integrals

$$\int_{|t|<T} |f_\nu(t)| \, dt \qquad \nu = 1, 2, 3, \ldots$$

are uniformly bounded by a constant K for all ν, where T is some fixed positive number.
b. $\{f_\nu(t)\}_{\nu=1}^\infty$ converges uniformly to zero on every bounded set of the form $0 < \tau \leq |t| \leq 1/\tau < \infty$.
c. For every finite positive value of τ the sequence of numbers

$$\left\{ \int_{|t| \leq \tau} f_\nu(t) \, dt \right\}_{\nu=1}^\infty$$

converges to 1.

Then, the sequence of regular distributions $\{f_\nu(t)\}_{\nu=1}^\infty$ converges in $\mathfrak{D}'$ to $\delta(t)$.

Proof: Let ϕ be in $\mathfrak{D}$. We have

$$\langle f_\nu, \phi \rangle = \int_{\mathfrak{R}^n} f_\nu(t)\phi(t)\, dt = \int_{|t|\leq\tau} \phi(0) f_\nu(t)\, dt$$
$$+ \int_{|t|\leq\tau} [\phi(t) - \phi(0)] f_\nu(t)\, dt + \int_{|t|>\tau} \phi(t) f_\nu(t)\, dt \quad (2)$$

Since all the first partial derivatives of $\phi(t)$ are bounded for all t, the constant M can be chosen so large that

$$|\phi(t) - \phi(0)| \leq |t| M$$

Then, the second integral on the right-hand side of (2) is bounded by

$$\tau M \int_{|t|\leq\tau} |f_\nu(t)|\, dt \quad (3)$$

and, when $\tau < T$, (3) is in turn bounded by $\tau M K$. If, for an arbitrary $\varepsilon > 0$, we set τ less than $\varepsilon/3MK$ as well as less than T, an upper bound on the second integral on the right-hand side of (2) is found to be $\varepsilon/3$.

Now, the third integral on the right-hand side of (2) converges to zero as $\nu \to \infty$. This is because $\phi(t)$, being in $\mathfrak{D}$, has a bounded support and, by condition b of the hypothesis, the $f_\nu(t)$ converge uniformly to zero over that part of the support of ϕ that is contained in $|t| > \tau$. Thus, there is an N_1 such that for $\nu \geq N_1$ the magnitude of this third integral is bounded by $\varepsilon/3$.

Finally, by condition c of the hypothesis, the first integral on the right-hand side of (2) converges to $\phi(0)$. In other words, there is an N_2 such that, for $\nu \geq N_2$, the difference between the first integral and $\phi(0)$ is bounded by $\varepsilon/3$.

Since $\langle \delta, \phi \rangle = \phi(0)$, we have shown that, for $\nu \geq \sup\{N_1, N_2\}$ and τ as chosen above,

$$|\langle f_\nu, \phi \rangle - \langle \delta, \phi \rangle| \leq \varepsilon$$

so that the proof is complete.

A consequence of Theorem 2.3-2 is

Corollary 2.3-2a

If $f(t)$ is an absolutely integrable function over all of $\mathfrak{R}^n$, if

$$\int_{\mathfrak{R}^n} f(t)\, dt = 1$$

and if $|f(t)|$ decreases to zero faster than $1/|t|^n$ as $|t| \to \infty$ [that is, $|f(t)| = o(1/|t|^n)$ as $|t| \to \infty$], then the set of regular distributions $\{\nu^n f(\nu t)\}_{\nu=1}^\infty$ converges in $\mathfrak{D}'$ to $\delta(t)$ as $\nu \to \infty$.

It was with such sequences as those described in Corollary 2.3-2a that the delta functional was first introduced into the physical sciences (Van der Pol and Bremmer [1], pp. 62–64).

PROBLEMS

1 Prove Corollary 2.3-2a.

2 In Example 2.3-2 we indicated that, as $\nu \to \infty$,

$$\frac{\nu}{2} \int_{-1/\nu}^{1/\nu} \phi(t) \, dt \to \phi(0)$$

for every ϕ in $\mathfrak{D}$. Prove this.

3 Let $f_\nu(t) = \nu^a \sin \nu t$, where t is one-dimensional. Show that, as $\nu \to \infty$, the $f_\nu(t)$ converge in $\mathfrak{D}'$ to the zero distribution for each value of the real constant a.

4 Show that the following regular distributions converge in $\mathfrak{D}'$ to the delta functional as $\nu \to \infty$. Here, t is a one-dimensional variable.

a. $\dfrac{\nu}{\sqrt{\pi}} e^{-\nu^2 t^2}$

b. $\dfrac{\nu}{2} e^{-\nu|t|}$

c. $\dfrac{\nu}{\pi(1 + \nu^2 t^2)}$

Sketch some of the functions in each sequence.

5 Construct a sequence of regular distributions, defined over $\mathfrak{R}^n$, that converges in $\mathfrak{D}'$ to $\delta(t)$.

6 Establish the following distributional limit, where t and τ are one-dimensional.

$$\lim_{\tau \to 0+} \log(\tau + it) = \begin{cases} \log|t| + \dfrac{i\pi}{2} & t > 0 \\ \log|t| - \dfrac{i\pi}{2} & t < 0 \end{cases}$$

Here it is understood that in a neighborhood of $t = 0$ the limit is a regular distribution. Also, we are dealing with the principal branch of the logarithm; i.e., for $\tau > 0$

$$|\arg \log(\tau + it)| < \frac{\pi}{2}$$

7 Here is an interesting paradox that was suggested by Prof. K. Steiglitz. In electrical systems the current $j(t)$ flowing through a unit resistor equals the voltage drop $v(t)$ across it. The total heat energy E dissipated in the resistor is

$$E = \int_{-\infty}^{\infty} v(t) j(t) \, dt$$

and the total electrical charge q that flows through it is

$$q = \int_{-\infty}^{\infty} j(t) \, dt$$

Let us choose

$$j_\nu(t) = \begin{cases} 0 & t \leq 0 \\ \nu^{-\frac{3}{2}} & 0 < t < \nu \\ 0 & \nu \leq t \end{cases}$$

Then, as $\nu \to 0$, both $j_\nu(t)$ and $v_\nu(t)$ go to zero for all t; moreover,

$$q_\nu = \nu^{-\frac{1}{2}} \to 0$$

whereas

$$E_\nu = \nu^{-\frac{3}{2}} \to \infty$$

That is, not only do the current and voltage go to zero for all t, but the total charge sent through the resistor also goes to zero. Nevertheless, the resistor burns out! In terms of distribution theory, discuss this discrepancy. In particular, can products of sequences of distributions, which converge in $\mathfrak{D}'$, be defined? If so, will they also converge in $\mathfrak{D}'$ in general?

8 Again let x and t be one-dimensional variables. Show that the distribution

$$(\cos xt) \operatorname{Pv} \frac{1}{t}$$

converges in $\mathfrak{D}'$ to zero as $x \to \infty$.

9 Let $\{f_\nu\}_{\nu=1}^\infty$ be a sequence of quadratically integrable functions defined on $\mathfrak{R}^1$ which converges in the mean-square sense to the limit function f on every finite interval $a \leq t \leq b$. That is, for every choice of a and b ($a < b$),

$$\lim_{\nu \to \infty} \int_a^b |f_\nu(t) - f(t)|^2 \, dt = 0$$

Show that $\{f_\nu\}_{\nu=1}^\infty$ converges in $\mathfrak{D}'$ to f. HINT: Use Schwarz's inequality.

★2.4 THE DIFFERENTIATION OF DISTRIBUTIONS

An ordinary locally integrable function may not have a derivative at certain points or, for that matter, anywhere at all. An example of such a function of a one-dimensional t is that which equals 1 at all irrational values of t and equals zero at all rational values of t. In fact, there are functions that are continuous everywhere and yet do not have a derivative anywhere at all (Titchmarsh [1], pp. 351–353).

Distributions, on the other hand, always possess derivatives, and these derivatives are again distributions. In order to explain this statement, we must, of course, define what we mean by the "derivative of a distribution." Let us restrict ourselves for the moment to the case when the independent variable t has only one dimension. An appropriate definition can be constructed by considering a regular distribution $f(t)$ generated by a function which is differentiable everywhere and whose derivative is continuous. The derivative again generates a regular dis-

Sec. 2.4 **The calculus of distributions** 47

tribution $f^{(1)}(t)$ and, for each ϕ in $\mathfrak{D}$, an integration by parts yields

$$\langle f^{(1)}, \phi \rangle = \int_{-\infty}^{\infty} f^{(1)}(t) \phi(t)\, dt$$
$$= - \int_{-\infty}^{\infty} f(t) \phi^{(1)}(t)\, dt = \langle f, -\phi^{(1)} \rangle \qquad (1)$$

Note that $\phi^{(1)}$ is in $\mathfrak{D}$ whenever ϕ is in $\mathfrak{D}$. Thus, a knowledge of f (and, therefore, of $\langle f, -\phi^{(1)} \rangle$) determines $\langle f^{(1)}, \phi \rangle$. In other words, (1) defines $f^{(1)}$ as a functional on $\mathfrak{D}$. This result is generalized in the following definition.

The first derivative $f^{(1)}(t)$ of any distribution $f(t)$, where t is one-dimensional, is the functional on $\mathfrak{D}$ given by

$$\langle f^{(1)}(t), \phi(t) \rangle = \langle f(t), -\phi^{(1)}(t) \rangle \qquad \phi \in \mathfrak{D} \qquad (2)$$

At times, the conventional notation df/dt will also be used for the derivative of a distribution defined over $\mathfrak{R}^1$.

A simple illustration is provided by the first derivative of the delta functional $\delta^{(1)}$, which is defined by the equation

$$\langle \delta^{(1)}, \phi \rangle = \langle \delta, -\phi^{(1)} \rangle = -\phi^{(1)}(0)$$

Another illustration is given by

Example 2.4-1 The unit step function $1_+(t)$ is the function that equals zero for $t < 0$, $\tfrac{1}{2}$ for $t = 0$, and 1 for $t > 0$. Its first distributional derivative is $\delta(t)$. For, with ϕ in $\mathfrak{D}$,

$$\langle 1_+^{(1)}(t), \phi(t) \rangle = \langle 1_+(t), -\phi^{(1)}(t) \rangle$$
$$= - \int_0^{\infty} \phi^{(1)}(t)\, dt$$
$$= \phi(0) = \langle \delta(t), \phi(t) \rangle$$

On the other hand, the ordinary derivative of $1_+(t)$ is the function that is zero everywhere except at the origin, where it does not exist.

When the independent variable t has n dimensions, it is the partial derivatives that are defined in a fashion analogous to (2).

The first-order partial derivatives $\partial f/\partial t_i$ ($i = 1, 2, \ldots, n$) of any distribution f defined over $\mathfrak{R}^n$ are the functionals on $\mathfrak{D}$ given by

$$\left\langle \frac{\partial f}{\partial t_i}, \phi \right\rangle = \left\langle f, -\frac{\partial \phi}{\partial t_i} \right\rangle \qquad i = 1, 2, \ldots, n;\ \phi \in \mathfrak{D} \qquad (3)$$

The previous definition for the one-dimensional case is simply a special case of this one. The properties possessed by a partial derivative in the n-dimensional case, which we shall now discuss, will also be possessed by the derivative in the one-dimensional case.

Theorem 2.4-1

A first-order partial derivative of a distribution is again a distribution.

Proof: The partial derivative $\partial f/\partial t_i$, as defined by (3), is clearly a linear functional on $\mathfrak{D}$. To show that it is also continuous, let $\{\phi_\nu\}_{\nu=1}^\infty$ be a sequence of testing functions that converges in $\mathfrak{D}$ to zero. This means $\{\partial \phi_\nu/\partial t_i\}_{\nu=1}^\infty$ also converges in $\mathfrak{D}$ to zero. Since f is continuous,

$$\left\langle \frac{\partial f}{\partial t_i}, \phi_\nu \right\rangle = \left\langle f, -\frac{\partial \phi_\nu}{\partial t_i} \right\rangle \to 0$$

as $\nu \to \infty$. This establishes the theorem.

If an ordinary function is differentiable everywhere and its derivative is bounded, then this derivative corresponds to the distributional derivative. In general, however, an ordinary derivative and a distributional derivative of a function are not the same, as is indicated in Example 2.4-1. From now on, whenever we speak of a derivative, we shall mean the distributional derivative unless the contrary is specifically stated. (On the other hand, an infinitely smooth function is, by definition, infinitely differentiable in the ordinary sense.)

The operation of differentiation can be repeated indefinitely, and thus any distribution has derivatives of all orders, which are themselves distributions. We refer to this fact by saying that *the space $\mathfrak{D}'$ is closed under the operation of differentiation*. It is clear that this is also a property of the space $\mathfrak{D}$.

Theorem 2.4-2

The order of differentiation of the higher-order partial derivatives of a distribution defined over $\mathfrak{R}^n$ can be changed at random. For instance,

$$\frac{\partial^2 f}{\partial t_i \, \partial t_k} = \frac{\partial^2 f}{\partial t_k \, \partial t_i} \tag{4}$$

This fact is again in contrast to classical analysis, in which such a change is not always possible.

Proof: When ϕ is in $\mathfrak{D}$, it has continuous partial derivatives of all orders, and hence

$$\frac{\partial^2 \phi}{\partial t_i \, \partial t_k} = \frac{\partial^2 \phi}{\partial t_k \, \partial t_i}$$

in the ordinary sense. Thus,

$$\left\langle \frac{\partial^2 f}{\partial t_i \, \partial t_k}, \phi \right\rangle = \left\langle -\frac{\partial f}{\partial t_k}, \frac{\partial \phi}{\partial t_i} \right\rangle = \left\langle f, \frac{\partial^2 \phi}{\partial t_k \, \partial t_i} \right\rangle$$
$$= \left\langle f, \frac{\partial^2 \phi}{\partial t_i \, \partial t_k} \right\rangle = \left\langle -\frac{\partial f}{\partial t_i}, \frac{\partial \phi}{\partial t_k} \right\rangle = \left\langle \frac{\partial^2 f}{\partial t_k \, \partial t_i}, \phi \right\rangle$$

This establishes (4). The general statement follows immediately.

Since the order of differentiation is of no consequence, a given partial differential operator D^k, when acting on a distribution, is sufficiently specified by writing

$$D^k = \prod_{i=1}^{n} \left(\frac{\partial}{\partial t_i}\right)^{k_i} \tag{5}$$

where the order in which the differentiations $\partial/\partial t_i$ are taken need not be stated. Letting $\hat{k} \triangleq \sum_{i=1}^{n} k_i$, we have

$$\langle D^k f, \phi \rangle = \langle f, (-1)^{\hat{k}} D^k \phi \rangle \tag{6}$$

Once again, the delta functional, now defined over $\mathfrak{R}^n$, provides a simple illustration:

$$\langle D^k \delta, \phi \rangle = \langle \delta, (-1)^{\hat{k}} D^k \phi \rangle = (-1)^{\hat{k}} D^k \phi(t) \Big|_{t=0} \tag{7}$$

The rule for the differentiation of the product of a distribution f and a function ψ, which is infinitely smooth, is the same as that for the product of two differentiable functions:

$$\frac{\partial}{\partial t_i}(\psi f) = \psi \frac{\partial f}{\partial t_i} + f \frac{\partial \psi}{\partial t_i} \tag{8}$$

This is established as follows. For any ϕ in $\mathfrak{D}$,

$$\left\langle \frac{\partial}{\partial t_i}(\psi f), \phi \right\rangle = \left\langle \psi f, -\frac{\partial \phi}{\partial t_i} \right\rangle = \left\langle f, -\psi \frac{\partial \phi}{\partial t_i} \right\rangle$$

$$= \left\langle f, -\frac{\partial}{\partial t_i}(\psi \phi) \right\rangle + \left\langle f, \phi \frac{\partial \psi}{\partial t_i} \right\rangle$$

$$= \left\langle \frac{\partial f}{\partial t_i}, \psi \phi \right\rangle + \left\langle f, \phi \frac{\partial \psi}{\partial t_i} \right\rangle$$

$$= \left\langle \psi \frac{\partial f}{\partial t_i}, \phi \right\rangle + \left\langle f \frac{\partial \psi}{\partial t_i}, \phi \right\rangle$$

Two important properties of the differentiation of distributions are given by

Theorem 2.4-3

Differentiation is a continuous linear operation in the space $\mathfrak{D}'$ in the following sense:

Linearity For any two distributions f and g and for any number α,

$$D^k(f + g) = D^k f + D^k g$$

and

$$D^k(\alpha f) = \alpha D^k f$$

Continuity For any sequence of distributions $\{f_\nu\}_{\nu=1}^\infty$ that converges in $\mathfrak{D}'$ to a distribution f, the corresponding sequence of partial derivatives $\{D^k f_\nu\}_{\nu=1}^\infty$ also converges in $\mathfrak{D}'$ to $D^k f$.

Proof: Linearity is easily verified. For continuity, let ϕ be in $\mathfrak{D}$. Then $D^k \phi$ is also in $\mathfrak{D}$ and as $\nu \to \infty$

$$\langle D^k f_\nu, \phi \rangle = \langle f_\nu, (-1)^k D^k \phi \rangle \to \langle f, (-1)^k D^k \phi \rangle$$
$$= \langle D^k f, \phi \rangle$$

which proves the theorem.

Example 2.4-2 We have seen in Example 2.3-4 that the functions $(\sin bt)/\pi t$ (b and t being one-dimensional) converge in $\mathfrak{D}'$ to $\delta(t)$ as $b \to \infty$. It follows from Theorem 2.4-3 that as $b \to \infty$

$$\frac{d}{dt}\left(\frac{\sin bt}{\pi t}\right) = \frac{bt \cos bt - \sin bt}{\pi t^2} \to \delta^{(1)}(t)$$

Corollary 2.4-3a

Let

$$f = \sum_{\nu=1}^\infty f_\nu$$

be any series of distributions that converges in $\mathfrak{D}'$. *Such a series may be differentiated term by term to obtain*

$$D^k f = \sum_{\nu=1}^\infty D^k f_\nu$$

where the differentiated series again converges in $\mathfrak{D}'$.

This result constitutes another advantage of distribution theory over classical analysis, in which the term-by-term differentiation of an infinite series can be performed only when the series satisfies conditions in addition to its mere convergence.

Example 2.4-3 Let $f(t)$ be a bounded function on $\mathfrak{R}^1$ that is piecewise-continuous and has a piecewise-continuous first derivative in the following way. The points t_ν ($\nu = \ldots, -2, -1, 0, 1, 2, \ldots$; $t_\nu < t_{\nu+1}$), at which $f(t)$ or $f^{(1)}(t)$ is discontinuous or $f^{(1)}(t)$ fails to exist, are finite in number in every finite interval. At each such point $f(t)$ has at most a finite jump

$$\Delta f \triangleq f(t_\nu +) - f(t_\nu -)$$

and its right-hand and left-hand derivatives both exist. Then we may

define the continuous function $f_c(t)$ through
$$f(t) \triangleq f_c(t) - \sum_{\nu=-1}^{-\infty} \Delta f_\nu \, 1_+(t_\nu - t) + \sum_{\nu=0}^{\infty} \Delta f_\nu \, 1_+(t - t_\nu)$$
The infinite series certainly converges in $\mathfrak{D}'$, since in every finite interval it possesses only a finite number of nonzero terms. By differentiating term by term and invoking the result developed in Example 2.4-1, we obtain
$$f^{(1)}(t) = f_c^{(1)}(t) + \sum_{\nu=-\infty}^{\infty} \Delta f_\nu \delta(t - t_\nu)$$
where $f_c^{(1)}(t)$ is a locally integrable function. Here again, distributional differentiation generates a delta functional at each ordinary discontinuity.

Corollary 2.4-3a provides a means of extending the applicability of trigonometric series to cases where the coefficients of the series not only do not decrease in magnitude as the harmonic index increases but in fact increase. Such series may represent distributions rather than functions, even though their individual terms are ordinary exponential functions.

Corollary 2.4-3b

Let t be one-dimensional. A trigonometric series
$$\sum_{\nu=-\infty}^{\infty} b_\nu e^{i\nu t} \tag{9}$$
where $|b_\nu| < M|\nu|^k$ for $\nu \neq 0$, M and k being real constants, converges in $\mathfrak{D}'$. Its sum equals $b_0 + g^{(p)}(t)$, where p is a nonnegative integer not less than $k + 2$ and $g(t)$ is the continuous periodic function given by the trigonometric series
$$g(t) = \sum_{\substack{\nu=-\infty \\ \nu \neq 0}}^{\infty} \frac{b_\nu}{(i\nu)^p} e^{i\nu t} \tag{10}$$

Proof: Since $|b_\nu/(i\nu)^p| \leq M|\nu|^{-2}$, the series (10) converges uniformly for all t. Hence, $g(t)$ is truly a continuous periodic function.

By Theorem 2.3-1, (10) also converges in $\mathfrak{D}'$, so that, by Corollary 2.4-3a, this series may be differentiated p times term by term. Upon adding the constant b_0, we obtain the series (9), which again by Corollary 2.4-3a converges in $\mathfrak{D}'$ to $b_0 + g^{(p)}(t)$. Q.E.D.

Example 2.4-4 Let us determine what the trigonometric series
$$f(t) = \sum_{\nu=-\infty}^{\infty} e^{i\nu t} \tag{11}$$

represents (t still being one-dimensional). According to Corollary 2.4-3b, the series converges in $\mathfrak{D}'$ and the distribution $f(t)$ equals $1 + g^{(2)}(t)$, where $g(t)$ is the continuous periodic function of period 2π given by

$$g(t) = - \sum_{\substack{\nu = -\infty \\ \nu \neq 0}}^{\infty} \frac{1}{\nu^2} e^{i\nu t} \tag{12}$$

It so happens that the last series represents a periodic function whose graph is a parabola in the interval $0 < t < 2\pi$:

$$g(t) = \frac{\pi^2}{6} - \frac{1}{2}(t - \pi)^2 \qquad 0 < t < 2\pi \tag{13}$$

[Indeed, (12) is the Fourier series expansion of (13).] This function and its first and second derivatives are sketched in Fig. 2.4-1. As was shown in Example 2.4-3, $g^{(2)}(t)$ possesses delta functionals at those points where $g^{(1)}(t)$ has ordinary discontinuities. Hence,

$$g^{(2)}(t) = -1 + \sum_{\nu = -\infty}^{\infty} 2\pi \delta(t - 2\pi\nu)$$

Thus, (11) represents an infinite series of shifted delta functionals:

$$f(t) = \sum_{\nu = -\infty}^{\infty} 2\pi \delta(t - 2\pi\nu)$$

The reasoning used in this example and in the proof of Corollary 2.4-3b can be formalized in the following criterion for convergence in $\mathfrak{D}'$. This criterion is sufficient for such convergence but not necessary.

Corollary 2.4-3c

Let $\{g_\nu(t)\}_{\nu=1}^{\infty}$ be a sequence of distributions defined over $\mathfrak{R}^n$ with the following property. There exists a finite set k of nonnegative integers such that $g_\nu(t)$ equals $D^k f_\nu(t)$ ($\nu = 1, 2, 3, \ldots$), where the $f_\nu(t)$ are regular distributions and the sequence $\{f_\nu(t)\}_{\nu=1}^{\infty}$ converges to a regular distribution $f(t)$ in the manner described in Theorem 2.3-1 (that is, by dominated convergence). Then, $\{g_\nu(t)\}_{\nu=1}^{\infty}$ converges in $\mathfrak{D}'$ to $D^k f(t)$.

The classical definition of a first-order partial derivative of an ordinary function f defined over $\mathfrak{R}^n$ is

$$\frac{\partial f}{\partial t_i} = \lim_{\Delta t_i \to 0} \frac{f(t_1, \ldots, t_i + \Delta t_i, \ldots, t_n) - f(t_1, \ldots, t_i, \ldots, t_n)}{\Delta t_i} \tag{14}$$

This expression is still valid when f is a distribution over $\mathfrak{R}^n$, so long as the limit process is interpreted in the sense of convergence in $\mathfrak{D}'$. We shall

Sec. 2.4   The calculus of distributions   53

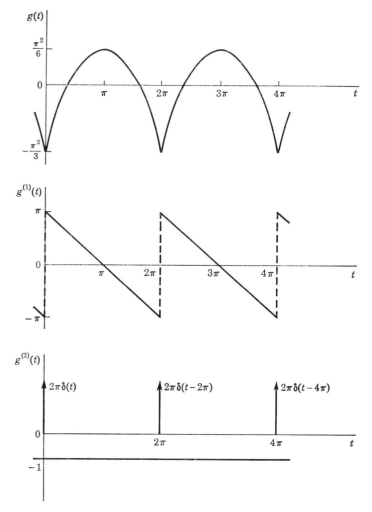

Fig. 2.4-1 Sketches of the function $g(t)$, given by (12) and (13) in Sec. 2.4, and its first and second derivatives. In the sketch of $g^{(2)}(t)$ the arrows represent delta functionals.

verify this in the case where $n = 1$ and leave the more general verification, which proceeds in the same way, as an exercise.

We wish to show that

$$\frac{f(t + \Delta t) - f(t)}{\Delta t}$$

converges in $\mathfrak{D}'$ to $f^{(1)}(t)$ as $\Delta t \to 0$. By letting ϕ be in $\mathfrak{D}$ and using the rule for the shifting of a distribution [see Sec. 1.7, Eq. (3)], we may write

$$\lim_{\Delta t \to 0} \left\langle \frac{f(t + \Delta t) - f(t)}{\Delta t}, \phi(t) \right\rangle = \lim_{\Delta t \to 0} \left\langle f(t), \frac{\phi(t - \Delta t) - \phi(t)}{\Delta t} \right\rangle \quad (15)$$

If we could show that the testing functions in the right-hand side of (15) converge in $\mathfrak{D}$ to $-\phi^{(1)}(t)$, it would follow from the continuity of $f(t)$ that (15) equals

$$\left\langle f(t), \lim_{\Delta t \to 0} \frac{\phi(t - \Delta t) - \phi(t)}{\Delta t} \right\rangle = \langle f(t), -\phi^{(1)}(t) \rangle$$
$$= \langle f^{(1)}(t), \phi(t) \rangle$$

and thus our assertion would be established.

Hence, to complete our argument, we must show that the functions

$$\psi_{\Delta t}(t) \triangleq \frac{1}{\Delta t}[\phi(t - \Delta t) - \phi(t)] + \phi^{(1)}(t)$$

converge in $\mathfrak{D}$ to zero as $\Delta t \to 0$. Observe that, for $|\Delta t| \leq 1$, all the $\psi_{\Delta t}$ have their supports contained in a fixed finite interval. Thus, it is sufficient to show that for each derivative of the $\psi_{\Delta t}$, say the kth derivative, we have uniform convergence to zero as $\Delta t \to 0$. Now,

$$\psi_{\Delta t}^{(k)}(t) \triangleq \frac{1}{\Delta t}[\phi^{(k)}(t - \Delta t) - \phi^{(k)}(t)] + \phi^{(k+1)}(t)$$
$$= \frac{1}{\Delta t} \int_{t-\Delta t}^{t} [\phi^{(k+1)}(t) - \phi^{(k+1)}(x)]\, dx$$
$$= \frac{1}{\Delta t} \int_{t-\Delta t}^{t} \int_{x}^{t} \phi^{(k+2)}(y)\, dy\, dx$$

By letting the constant M be a bound on $|\phi^{(k+2)}(t)|$, we obtain for all t that

$$|\psi_{\Delta t}^{(k)}(t)| \leq \frac{M}{|\Delta t|} \left| \int_{t-\Delta t}^{t} \int_{x}^{t} dy\, dx \right|$$
$$= \frac{M|\Delta t|}{2} \to 0$$

as $\Delta t \to 0$. This is what we wished to show.

PROBLEMS

1 Let $f(t)$ be a function defined over $\mathfrak{R}^n$ and such that all of its first-order partial derivatives exist and are continuous everywhere. Show that (3) holds when $\langle f, \phi \rangle$ is taken to be the ordinary integral of $f\phi$ over $\mathfrak{R}^n$.

2 Verify that differentiation of distributions is a linear operation.

3 Establish the following formulas wherein α is a complex number, a is a real number, ν is a positive integer, and t is one-dimensional.

$$(\sin at)\delta^{(1)}(t) = -a\delta(t)$$

$$\frac{d^2}{dt^2}[1_+(t)\sin at] = a\delta(t) - a^2 1_+(t)\sin at$$

$$e^{\alpha t}\delta^{(\nu)}(t-a) = e^{\alpha a}\sum_{k=0}^{\nu}\binom{\nu}{k}(-\alpha)^{\nu-k}\delta^{(k)}(t-a)$$

4 Let the symbol $\check{\ }$ denote the transpose operation [i.e., for any distribution f, $\check{f}(t) = f(-t)$]. Show that for the case of a one-dimensional t

$$[f^{(1)}(t)]^{\vee} = -[\check{f}(t)]^{(1)}$$

This demonstrates that the operations of differentiation and transposition do not commute.

5 Let f be a distribution defined over $\mathcal{R}^1$. Show that, for ϕ in $\mathcal{D}$ and for a a real positive number,

$$\left\langle \frac{d}{dt}f(at),\, \phi(t)\right\rangle = \left\langle f(t),\, -\frac{d}{dt}\phi\left(\frac{t}{a}\right)\right\rangle$$

What is the corresponding expression if a is a real negative number?

6 Construct sequences that converge to $\delta^{(1)}(t)$ and $\delta^{(2)}(t)$ (t being one-dimensional) by using the functions given in Example 2.3-4 and Sec. 2.3, Prob. 4. For each sequence sketch some of the functions.

7 For the case where t is n-dimensional show that (14) is a valid expression when the limit is taken in the sense of convergence in $\mathcal{D}'$.

8 Develop once again the results of Example 2.4-3 by applying integration by parts to

$$\int_I f^{(1)}(t)\phi(t)\,dt$$

over each interval I that has two successive points of discontinuity as its end points.

9 Let t be one-dimensional. Assuming that the function $f(t)$ and its first n derivatives are bounded and piecewise-continuous in the manner described in Example 2.4-3, develop the general expression for the nth derivative of the regular distribution corresponding to $f(t)$.

10 Show that, for a one-dimensional t,

$$t^n\delta^{(m)}(t) = \begin{cases} 0 & m < n \\ (-1)^n n!\,\delta(t) & m = n \\ (-1)^n \dfrac{m!}{(m-n)!}\,\delta^{(m-n)}(t) & m > n \end{cases}$$

11 Show that, for t in $\mathcal{R}^1$,

$$\delta^{(k)}(at+b) = \frac{1}{a^k|a|}\,\delta^{(k)}\left(t+\frac{b}{a}\right) \qquad k = 0, 1, 2, \ldots$$

where a and b are real constants and $a \neq 0$. What is the corresponding formula when t is in $\mathcal{R}^n$, b is a fixed point of $\mathcal{R}^n$, and a is a nonzero one-dimensional constant?

12 Show that two finite linear combinations of the delta functional and its derivatives, defined over $\mathcal{R}^1$, are equal if and only if the corresponding coefficients are equal. That is, prove that

$$\sum_{\nu=0}^{n} \alpha_\nu \delta^{(\nu)} = \sum_{\nu=0}^{n} \beta_\nu \delta^{(\nu)}$$

where the α_ν and the β_ν are constants, if and only if $\alpha_\nu = \beta_\nu$ for all ν.

13 With t again one-dimensional, a special case of Bessel's differential equation is

$$tf^{(2)}(t) + f^{(1)}(t) + tf(t) = 0 \tag{16}$$

An ordinary function that satisfies this equation for all t and has continuous derivatives of all orders is Bessel's function of first kind and zero order, $J_0(t)$. (See Example 8.2-3 for a series expansion of this function.) Show that in the distributional sense $J_0(t)1_+(t)$ is also a solution of (16) for all t.

14 Compute the first four distributional derivatives of the following quantities, wherein t is one-dimensional.

  a. $|\sin t|$
  b. $t \sin t$
  c. $|t \sin t|$
  d. $|t| \sin t$

HINT: In some cases it may be more convenient to represent the function as a product of a step function and a function that is infinitely smooth.

15 By using convergence in $\mathcal{D}'$ and assuming that t and x are one-dimensional variables, find the following limits:

  a. $\lim\limits_{x \to 0+} \dfrac{\delta(t+x) - \delta(t)}{x}$

  b. $\lim\limits_{x \to 0+} \dfrac{\delta(t+2x) - 2\delta(t+x) + \delta(t)}{x}$

  c. $\lim\limits_{x \to 0+} \dfrac{\delta(t+2x) - 2\delta(t+x) + \delta(t)}{x^2}$

  d. $\lim\limits_{x \to 0+} g_x(t)$, where

$$g_x(t) = \begin{cases} \dfrac{1}{x^2} & -x < t < 0 \\ -\dfrac{1}{x^2} & 0 < t < x \\ 0 & \text{otherwise} \end{cases}$$

16 With t in $\mathcal{R}^n$, consider the function

$$1_+(t) \triangleq 1_+(t_1)1_+(t_2) \cdots 1_+(t_n)$$

Show that

$$\frac{\partial^n}{\partial t_1 \, \partial t_2 \cdots \partial t_n} 1_+(t) = \delta(t)$$

17 Another way to define the product of two distributions is as follows (Halperin [1], p. 8): Let the distribution f on $\mathcal{R}^1$ be the mth derivative of a continuous function h and let g be a function on $\mathfrak{R}^1$ which is m times differentiable (in the ordinary sense) and has continuous derivatives up to the mth order. Then we define gf by

$$gf \triangleq (gh)^{(m)} + \sum_{k=1}^{m} (-1)^k \binom{m}{k} (g^{(k)}h)^{(m-k)}$$

Prove that the right-hand side of this expression defines gf as a distribution in $\mathfrak{D}'_m$, the space defined in Sec. 1.3, Prob. 2. Then, show that this definition is a special case of the definition

$$\langle gf, \phi \rangle \triangleq \langle f, g\phi \rangle$$

where ϕ is in $\mathfrak{D}_m$. (The latter definition is the one implied in Sec. 1.7, Prob. 8.)

2.5 HADAMARD'S FINITE PART AND SOME PSEUDOFUNCTIONS GENERATED BY IT

Distributions generated by Hadamard's finite part of a divergent integral, some simple examples of which were given in Sec. 1.4, are called *pseudofunctions*. They arise naturally when certain distributions are differentiated. As an example, consider the regular distribution $1_+(t)t^\alpha$, where $-1 < \alpha < 0$ and t is one-dimensional. (Actually, t^α is multivalued. For $t > 0$, its various values are $|t|^\alpha e^{i2n\pi\alpha}$, where n traverses all the integers. In this application it is understood that n is fixed.) The derivative of $1_+(t)t^\alpha$ is defined by

$$\left\langle \frac{d}{dt}[1_+(t)t^\alpha], \phi(t) \right\rangle = \langle 1_+(t)t^\alpha, -\phi^{(1)}(t) \rangle$$

$$= -\lim_{\varepsilon \to 0+} \int_\varepsilon^\infty t^\alpha \phi^{(1)}(t)\, dt \qquad \phi \in \mathfrak{D}$$

Through an integration by parts, the last expression becomes

$$\lim_{\varepsilon \to 0+} \left[\alpha \int_\varepsilon^\infty t^{\alpha-1}\phi(t)\, dt + \varepsilon^\alpha \phi(\varepsilon) \right]$$

Since $\phi(\varepsilon) = \phi(0) + O(\varepsilon)$ for $\varepsilon \to 0+$, this may also be written as

$$\lim_{\varepsilon \to 0+} \left[\alpha \int_\varepsilon^\infty t^{\alpha-1}\phi(t)\, dt + \varepsilon^\alpha \phi(0) \right]$$

which is simply the definition for Pf $\alpha t^{\alpha-1}1_+(t)$, a pseudofunction introduced in Sec. 1.4, Eq. (7). Thus, we have shown that distributionally

$$\frac{d}{dt}[t^\alpha 1_+(t)] = \text{Pf } \alpha t^{\alpha-1}1_+(t)$$

We shall discuss such pseudofunctions in this section in some detail for the case when t is one-dimensional. A general technique for constructing pseudofunctions is the following. Assume that the function $f(t)$ is locally integrable except in a neighborhood of the point $t = a$. With $\phi(t)$ being in $\mathfrak{D}$, consider the integral

$$\int_{a+\varepsilon}^{b} f(t)\phi(t)\,dt \qquad b > a,\, \varepsilon > 0$$

This is a function of ε which tends (at times) to ∞ as $\varepsilon \to 0+$. Assume that it is possible to subtract from it a finite linear combination of negative powers of ε and positive powers of $\log \varepsilon$ such that the remaining function of ε approaches a finite limit that is independent of the method by which it is obtained. This limit is defined as the *finite part* of the integral

$$\int_{a}^{b} f(t)\phi(t)\,dt$$

We must, of course, show that the finite part defines a distribution.

We shall now take up specific cases. Consider first of all the generally divergent integral

$$\int_{a}^{b} (t - a)^{\alpha - k}\phi(t)\,dt \qquad a < b \tag{1}$$

where $\phi(t)$ is in $\mathfrak{D}$, k is a positive integer, and $-1 < \alpha < 0$. It is also understood that $(t - a)^{\alpha - k}$ is restricted to one of its branches [that is, $\arg(t - a)^{\alpha - k} = 2n\pi\alpha$, where n is some fixed integer]. To extract the finite part of the integral, expand $\phi(t)$ by using a Taylor's formula with remainder (Widder [2], p. 43):

$$\phi(t) = \phi(a) + \frac{\phi^{(1)}(a)}{1!}(t - a) + \cdots + \frac{\phi^{(k-1)}(a)}{(k-1)!}(t-a)^{k-1} + (t - a)^{k}\psi(t) \tag{2}$$

$\psi(t)$ is continuous for all t and is given by

$$\psi(t) = \frac{\int_{a}^{t} \phi^{(k)}(\tau)(t - \tau)^{k-1}\,d\tau}{(k - 1)!(t - a)^{k}}$$

The substitution of (2) into

$$\lim_{\varepsilon \to 0+} \int_{a+\varepsilon}^{b} (t - a)^{\alpha - k}\phi(t)\,dt$$

yields

$$\lim_{\varepsilon \to 0+} [I(\varepsilon) + H(\varepsilon)]$$

where

$$I(\varepsilon) = -\sum_{\nu = 0}^{k-1} \frac{\phi^{(\nu)}(a)\varepsilon^{\alpha - k + 1 + \nu}}{\nu!(\alpha - k + 1 + \nu)} \tag{3}$$

and

$$H(\varepsilon) = \int_{a+\varepsilon}^{b} (t-a)^{\alpha}\psi(t)\,dt + \sum_{\nu=0}^{k-1} \frac{\phi^{(\nu)}(a)(b-a)^{\alpha-k+1+\nu}}{\nu!(\alpha-k+1+\nu)} \qquad (4)$$

As $\varepsilon \to 0+$, $I(\varepsilon)$ diverges in general, whereas $H(\varepsilon)$ always converges. The finite part of the generally divergent integral (1) is defined as $\lim_{\varepsilon \to 0+} H(\varepsilon)$. It is denoted by the insertion of the symbol Fp before the integral sign.

$$\text{Fp} \int_{a}^{b} (t-a)^{\alpha-k}\phi(t)\,dt \triangleq \lim_{\varepsilon \to 0+} H(\varepsilon) = H(0) \qquad a < b \qquad (5)$$

Now, (5) defines a distribution in $\mathfrak{D}'$. Indeed, its linearity is clear and its continuity will follow if we can show that, with ψ_μ related to ϕ_μ according to (2), $\{\psi_\mu(t)\}_{\mu=1}^{\infty}$ converges uniformly to zero over $a < t < b$ whenever $\{\phi_\mu(t)\}_{\mu=1}^{\infty}$ converges in $\mathfrak{D}$ to zero. But this follows immediately from Taylor's formula with remainder, which shows that

$$|\psi_\mu(t)| \le \frac{1}{k!} \sup |\phi_\mu^{(k)}(\tau)|$$

where $|\tau - a| \le |t - a|$.

The pseudofunction defined by (5) is denoted by

Pf $(t-a)^{\alpha-k}1_+(t-a)1_+(b-t)$

If we eliminate $\psi(t)$ from (4) by using (2), we obtain

$$\langle \text{Pf } (t-a)^{\alpha-k}1_+(t-a)1_+(b-t),\ \phi(t)\rangle$$
$$= \lim_{\varepsilon \to 0+}\left[\int_{a+\varepsilon}^{b}(t-a)^{\alpha-k}\phi(t)\,dt + \sum_{\nu=0}^{k-1}\frac{\phi^{(\nu)}(a)\varepsilon^{\alpha-k+1+\nu}}{\nu!(\alpha-k+1+\nu)}\right] \qquad (6)$$

By setting $b = \infty$ in either (5) or (6), we obtain the definition of the pseudofunction Pf $(t-a)^{\alpha-k}1_+(t-a)$:

$$\langle \text{Pf } (t-a)^{\alpha-k}1_+(t-a),\ \phi(t)\rangle$$
$$\triangleq \lim_{\varepsilon \to 0+}\left[\int_{a+\varepsilon}^{\infty}(t-a)^{\alpha-k}\phi(t)\,dt + \sum_{\nu=0}^{k-1}\frac{\phi^{(\nu)}(a)\varepsilon^{\alpha-k+1+\nu}}{\nu!\,\alpha-k+1+\nu)}\right]$$
$$= \int_{a}^{\infty}(t-a)^{\alpha-k}\left[\phi(t) - \sum_{\nu=0}^{k-1}\frac{\phi^{(\nu)}(a)}{\nu!}(t-a)^{\nu}\right]dt \qquad (7)$$

The differentiation formulas for these pseudofunctions are

$$\frac{d}{dt}\text{Pf }(t-a)^{\alpha-k}1_+(t-a)1_+(b-t)$$
$$= \text{Pf }(\alpha-k)(t-a)^{\alpha-k-1}1_+(t-a)1_+(b-t)$$
$$\quad - (b-a)^{\alpha-k}\delta(t-b) \qquad (8)$$

$$\frac{d}{dt}\text{Pf }(t-a)^{\alpha-k}1_+(t-a) = \text{Pf }(\alpha-k)(t-a)^{\alpha-k-1}1_+(t-a) \qquad (9)$$

Formula (8) is essentially the same as (9) except for the additional delta functional produced by the ordinary discontinuity in

$$\text{Pf } (t-a)^{\alpha-k}1_+(t-a)1_+(b-t)$$

at $t = b$. These expressions agree with the ordinary rule for the differentiation of $(t-a)^n$, except for the Pf notation. Indeed, we may allow k to be a negative integer, in which case the Pf notation is not needed. Similarly, if k is zero, we may drop the Pf notation from the left-hand sides.

To establish (9), we may proceed as follows:

$$\langle \text{Pf } (\alpha-k)(t-a)^{\alpha-k-1}1_+(\ -a), \phi(t)\rangle$$
$$= \lim_{\varepsilon \to 0+}\left[\int_{a+\varepsilon}^{\infty}(\alpha-k)('-a)^{\alpha-k-1}\phi(t)\,d \right.$$
$$\left. + (\alpha-k)\sum_{\nu=0}^{k}\frac{\phi^{(\nu)}(a)\varepsilon^{\alpha-k+\nu}}{\nu!(\alpha-\ +\nu)}\right]$$

An integration by parts converts the right-hand side into

$$\lim_{\varepsilon \to 0+}\left[-\int_{a+\varepsilon}^{\infty}(t-a)^{\alpha-}\phi^{(1)}(t)\,dt - \sum_{\nu=0}^{k-1}\frac{\phi^{(\nu+1)}(a)\varepsilon^{\alpha-k+1+\nu}}{\nu!(\alpha-k+1+\nu)} + R(\varepsilon)\right] \tag{10}$$

where

$$R(\varepsilon) = -\varepsilon^{\alpha-}\phi(a+\varepsilon) + \sum_{\nu=0}^{k}\frac{\phi^{(\nu)}(a)}{\nu!}\varepsilon^{\alpha-k+}$$

But

$$\phi(t) = \sum_{\nu=0}^{k}\frac{\phi^{(\nu)}(a)}{\nu!}(t-a)^\nu + (\ -a)^{k+1}\lambda(t)$$

where $\lambda(t) = O(1)$ as $t \to a$. So, setting $t = a + \varepsilon$, we see that

$$R(\varepsilon) = -\lambda(a+\varepsilon)\varepsilon^{\alpha+1} \to 0$$

as $\varepsilon \to 0$. Thus, (10) is equivalent to

$$\langle \text{Pf } (t-a)^{\alpha-k}1_+(t-a), -\phi^{(1)}(t)\rangle$$

By the differentiation formula for distributions, this proves (9).

A similar analysis establishes the following formula for the pseudofunction:

$$\text{Pf } |t-b|^{\alpha-k}1_+(t-a)1_+(b-t)$$

where $a < b$, $-1 < \alpha < 0$, $k = 1, 2, 3, \ldots$, and one of the branches of $|t - b|^{\alpha-k}$ is understood.

$$\langle \text{Pf } |t - b|^{\alpha-k} 1_+(t - a) 1_+(b - t),\, \phi(t) \rangle$$
$$= \lim_{\varepsilon \to 0+} \left[\int_a^{b-\varepsilon} |t - b|^{\alpha-k} \phi(t)\, dt + \sum_{\nu=0}^{k-1} \frac{(-1)^\nu \phi^{(\nu)}(b) \varepsilon^{\alpha-k+1+\nu}}{\nu!(\alpha - k + 1 + \nu)} \right] \quad (11)$$

The corresponding differentiation formula is

$$\frac{d}{dt} \text{Pf } |t - b|^{\alpha-k} 1_+(t - a) 1_+(b - t)$$
$$= \text{Pf } (k - \alpha)|t - b|^{\alpha-k-1} 1_+(t - a) 1_+(b - t)$$
$$+ (b - a)^{\alpha-k} \delta(t - a) \quad (12)$$

The formula for $\text{Pf } |t - b|^{\alpha-k} 1_+(b - t)$ is the same as (11) except that the lower limit a on the integral is set equal to $-\infty$; its differentiation formula is

$$\frac{d}{dt} \text{Pf } |t - b|^{\alpha-k} 1_+(b - t) = \text{Pf } (k - \alpha)|t - b|^{\alpha-k-1} 1_+(b - t)$$

More general forms of pseudofunctions arise from the finite part of the (generally) divergent integral

$$\int_a^b (t - a)^{\alpha-k} g(t) \phi(t)\, dt \qquad a < b;\, \phi \in \mathfrak{D} \quad (13)$$

where $g(t)$ is a locally integrable function that possesses in some interval $a \leq t < T$ continuous derivatives up to and including the order k. In this case,

$$(t - a)^{\alpha-k} g(t) = g(a+)(t - a)^{\alpha-k} + g^{(1)}(a+)(t - a)^{\alpha-k+1} + \cdots$$
$$+ \frac{g^{(k-1)}(a+)}{(k-1)!} (t - a)^{\alpha-1} + (t - a)^\alpha h(t) \quad (14)$$

Here, $h(t)$ is continuous for $a \leq t < T$ and is locally integrable elsewhere, so that $(t - a)^\alpha h(t)$ is integrable over $a < t < b$. If we apply our previous analysis to each term in (14), we shall obtain a sum of pseudofunctions plus a regular distribution, and this sum in turn defines the pseudofunction generated by (13). In fact, we may write

$$\text{Pf } (t - a)^{\alpha-k} g(t) 1_+(t - a) 1_+(b - t)$$
$$= \sum_{\nu=0}^{k-1} \text{Pf } \frac{g^{(\nu)}(a+)}{\nu!} (t - a)^{\alpha-k+\nu} 1_+(t - a) 1_+(b - t)$$
$$+ (t - a)^\alpha h(t) 1_+(t - a) 1_+(b - t) \qquad a < b \quad (15)$$

This formula is still valid when $b = \infty$.

Similar results may be obtained under a somewhat relaxed hypothesis on $g(t)$. It need only possess the expansion

$$g(t) = \sum_{\nu=0}^{k-1} c_\nu (t-a)^\nu + (t-a)^k h(t)$$

where $(t-a)^\alpha h(t)$ is merely locally integrable and the c_ν are constants. If we denote $(t-a)^{\alpha-k} g(t)$ by $f(t)$, we obtain

$$f(t) = \sum_{\nu=0}^{k-1} c_\nu (t-a)^{\alpha-k+\nu} + (t-a)^\alpha h(t)$$

and this expansion may be used to compute the pseudofunction arising from

$$\text{Fp} \int_a^b f(t) \phi(t)\, dt \tag{16}$$

Of course, the upper limit b might be the singular point for $f(t)$, and in that case we would need an expansion of the form

$$f(t) = \sum_{\nu=0}^{k-1} c_\nu |t-b|^{\alpha-k+\nu} + |t-b|^\alpha h(t) \tag{17}$$

Let us now consider what happens when the nonintegral exponent $\alpha - k$ is replaced by a negative integer. A somewhat different pseudofunction is generated by the finite part of

$$\int_a^b \frac{\phi(t)}{(t-a)^k}\, dt \qquad k = 1, 2, 3, \ldots\,; \phi \in \mathfrak{D} \tag{18}$$

If (2) is substituted into (18), we shall obtain

$$\lim_{\varepsilon \to 0+} \int_{a+\varepsilon}^b \frac{\phi(t)}{(t-a)^k}\, dt = \lim_{\varepsilon \to 0+} [I(\varepsilon) + H(\varepsilon)]$$

where now

$$I(\varepsilon) = \sum_{\nu=0}^{k-2} \frac{\phi^{(\nu)}(a)}{\nu!(k-1-\nu)\varepsilon^{k-1-\nu}} - \frac{\phi^{(k-1)}(a)}{(k-1)!} \log \varepsilon$$

and

$$H(\varepsilon) = \int_{a+\varepsilon}^b \psi(t)\, dt - \sum_{\nu=0}^{k-2} \frac{\phi^{(\nu)}(a)}{!(k-1-\nu)(b-a)^{k-1-\nu}}$$

$$+ \frac{\phi^{(k-1)}(a)}{(k-1)!} \log (b-a) \tag{19}$$

$I(\varepsilon)$ diverges in general, and $H(\varepsilon)$ always converges as $\varepsilon \to 0+$. Moreover, the argument following (5) again shows that $H(0) = \lim_{\varepsilon \to 0+} H(\varepsilon)$

defines a distribution or, more particularly, a pseudofunction, which we denote by

$$\text{Pf}\,\frac{1_+(t-a)1_+(b-t)}{(t-a)^k}$$

Upon eliminating $\psi(t)$ from (19) by using (2), we obtain

$$\left\langle \text{Pf}\,\frac{1_+(t-a)1_+(b-t)}{(t-a)^k}, \phi(t) \right\rangle = \lim_{\varepsilon \to 0+} \left[\int_{a+\varepsilon}^{b} \frac{\phi(t)}{(t-a)^k}\,dt \right. \\ \left. - \sum_{\nu=0}^{k-2} \frac{\phi^{(\nu)}(a)}{\nu!(k-1-\nu)\varepsilon^{k-1-\nu}} + \frac{\phi^{(k-1)}(a)}{(k-1)!} \log \varepsilon \right] \quad (20)$$

and this is the expression that replaces (6) when the exponent $\alpha - k$ is replaced by $-k$. (The summation $\sum_{\nu=0}^{k-2}$ is absent when $k = 1$.)

The expression corresponding to (11) can be found to be

$$\left\langle \text{Pf}\,\frac{1_+(t-a)1_+(b-t)}{|t-b|^k}, \phi(t) \right\rangle = \lim_{\varepsilon \to 0+} \left[\int_{a}^{b-\varepsilon} \frac{\phi(t)}{|t-b|^k}\,dt \right. \\ \left. - \sum_{\nu=0}^{k-2} \frac{(-1)^\nu \phi^{(\nu)}(b)}{\nu!(k-1-\nu)\varepsilon^{k-1-\nu}} + \frac{(-1)^{k-1}\phi^{(k-1)}(b)}{(k-1)!} \log \varepsilon \right] \quad (21)$$

In contrast to the previous case, the differentation formulas for the present case of negative integral exponents now contain an additional delta functional or one of its derivatives at the singularity point. For example, through an integration by parts, we may write

$$\left\langle \text{Pf}\,\frac{-k 1_+(t-a)}{(t-a)^{k+1}}, \phi(t) \right\rangle = \lim_{\varepsilon \to 0+} \left[-\int_{a+\varepsilon}^{\infty} \frac{\phi^{(1)}(t)}{(t-a)^k}\,dt \right. \\ \left. + \sum_{\nu=0}^{k-2} \frac{\phi^{(\nu+1)}(a)}{\nu!(k-1-\nu)\varepsilon^{k-1-\nu}} - \frac{\phi^{(k)}(a)}{(k-1)!} \log \varepsilon + R(\varepsilon) \right] \quad (22)$$

where

$$R(\varepsilon) = -\frac{\phi(a+\varepsilon)}{\varepsilon^k} + \sum_{\nu=0}^{k-1} \frac{\phi^{(\nu)}(a)}{\nu!\varepsilon^{k-\nu}}$$

The expansion of $\phi(t)$ into a linear combination of powers of $(t - a)$ now shows that $R(\varepsilon) \to -\phi^{(k)}(a)/k!$ as $\varepsilon \to 0+$. The right-hand side of (22) is therefore equivalent to

$$\left\langle \text{Pf}\,\frac{1_+(t-a)}{(t-a)^k}, -\phi^{(1)}(t) \right\rangle - \frac{\phi^{(k)}(a)}{k!}$$

Thus,

$$\frac{d}{dt} \text{Pf} \frac{1_+(t-a)}{(t-a)^k} = \text{Pf} \frac{-k1_+(t-a)}{(t-a)^{k+1}} + (-1)^k \frac{\delta^{(k)}(t-a)}{k!} \tag{23}$$

Similarly, it can be shown that

$$\frac{d}{dt} \text{Pf} \frac{1_+(b-t)}{(t-b)^k} = \text{Pf} \frac{-k1_+(b-t)}{(t-b)^{k+1}} - (-1)^k \frac{\delta^{(k)}(t-b)}{k!} \tag{24}$$

Under the assumptions made earlier on the integrability of the function $g(t)$ and its derivatives in the vicinity of the singular point, the expression (15) still holds true when α is replaced by zero. Similarly, (16) again defines a pseudofunction when α equals zero in the corresponding expansions for $f(t)$ so long as $h(t)$ is integrable over $a < t < b$.

In all these formulas, b may equal ∞ when a is the singular point and a may equal $-\infty$ when b is the singular point.

These results are readily extendable to pseudofunctions having several and perhaps an infinite number of singular points. Indeed, if there are only a finite number of such points in every finite interval, we can equate such a pseudofunction $j(t)$ to a sum of pseudofunctions each of which has a support concentrated on an interval between a pair of consecutive singular points. Since in any finite interval the sum will have only a finite number of nonzero terms, it will converge in $\mathfrak{D}'$ to $j(t)$. The following example illustrates the technique.

Example 2.5-1 The function $1/(\sin t)$ has the singular points $t = n\pi$ ($n = 0, \pm 1, \pm 2, \ldots$). We define $h(t)$ by

$$\frac{1}{\sin t} \triangleq \frac{1}{t} + h(t)$$

In the interval $0 \leq t \leq \pi/2$, $h(t)$ is continuous and $h(0) = 0$. This leads to the pseudofunction

$$f_1(t) = \text{Pf} \frac{1_+(t)1_+(\pi/2 - t)}{\sin t}$$
$$= \text{Pf} \frac{1_+(t)1_+(\pi/2 - t)}{t} + 1_+(t)1_+\left(\frac{\pi}{2} - t\right)h(t)$$

Similarly, for the interval $-\pi/2 \leq t \leq 0$ we obtain the pseudofunction

$$f_2(t) = \text{Pf} \frac{1_+(\pi/2 + t)1_+(-t)}{\sin t}$$
$$= \text{Pf} \frac{1_+(\pi/2 + t)1_+(-t)}{t} + 1_+\left(\frac{\pi}{2} + t\right)1_+(-t)h(t)$$

Sec. 2.5   The calculus of distributions   65

By setting $f = f_1 + f_2$ and applying (20) and (21), we obtain

$$\langle f, \phi \rangle = \langle f_1 + f_2, \phi \rangle = \mathrm{Fp} \int_{-\pi/2}^{\pi/2} \frac{\phi(t)}{\sin t} dt$$
$$= \lim_{\varepsilon \to 0+} \left(\int_{-\pi/2}^{-\varepsilon} + \int_{\varepsilon}^{\pi/2} \right) \frac{\phi(t)}{t} dt + \int_{-\pi/2}^{\pi/2} h(t)\phi(t) \, dt$$

Note that it is the Cauchy principal part of the improper integral that arises here. Finally, the periodicity of the sine function allows us to write the pseudofunction generated by

$$\mathrm{Fp} \int_{-\infty}^{\infty} \frac{\phi(t)}{\sin t} dt \qquad \phi \in \mathfrak{D}$$

as the following series:

$$\mathrm{Pf} \frac{1}{\sin t} = \sum_{\nu=-\infty}^{\infty} (-1)^\nu f(t - \nu\pi)$$

In this section we have considered only those pseudofunctions that diverge near their singularity points as some power of $1/t$. There are still other types of pseudofunctions. (See, for instance, Lavoine [1].) Here is an example of one of them.

Example 2.5-2 Let us extract the finite part of the generally divergent integral

$$\int_0^\infty \frac{\log t}{t} \phi(t) \, dt \qquad \phi \in \mathfrak{D} \tag{25}$$

Proceeding as usual, we set $\phi(t) = \phi(0) + t\psi(t)$ and replace the upper limit by b, a number so large that $\phi(t) = 0$ for $t > b$. An integration by parts shows that

$$2 \int_\varepsilon^b \frac{\log t}{t} dt = (\log b)^2 - (\log \varepsilon)^2$$

Hence, (25) equals

$$\lim_{\varepsilon \to 0+} \left\{ \int_\varepsilon^b \psi(t) \log t \, dt + \frac{\phi(0)}{2} [(\log b)^2 - (\log \varepsilon)^2] \right\}$$

Upon discarding the divergent term $(\log \varepsilon)^2$, we obtain the definition of the pseudofunction:

$$\mathrm{Pf} \frac{1_+(t) \log t}{t} \tag{26}$$

66 Distribution theory and transform analysis   Chap. 2

$$\left\langle \text{Pf}\, \frac{1_+(t)\log t}{t}, \phi(t) \right\rangle \triangleq \lim_{\varepsilon \to 0+} \left[\int_\varepsilon^\infty \frac{\log t}{t} \phi(t)\, dt + \frac{\phi(0)}{2}(\log \varepsilon)^2 \right]$$

One can show in the usual way that (26) is a distribution.

PROBLEMS

1 Develop formulas (11) and (21).

2 Develop the defining expressions for the following pseudofunctions. Specify the various branches of any multivalued expressions.

a. $\text{Pf}\, \dfrac{1_+(-t)}{t^{\frac{3}{2}}}$

b. $\text{Pf}\, \dfrac{\log t}{t^2} 1_+(t)$

c. $\text{Pf}\, \dfrac{\log t}{t^{\frac{3}{2}}} 1_+(t)$

d. $\text{Pf}\, \dfrac{1_+(t)}{e^t - 1}$

e. $\text{Pf}\, \dfrac{1}{\sinh t}$

f. $\text{Pf tan } t$

3 Establish the differentiation formula (12). [HINT: One way to do this is to show first that

$$\text{Pf } |t - b|^{\alpha-k} 1_+(t - a) 1_+(b - t)$$

is the transpose of

$$\text{Pf } (t + b)^{\alpha-k} 1_+(t + b) 1_+(-a - t)$$

and then use (8) and the formula of Sec. 2.4, Prob. 4.] Repeat this development for (24).

4 Let ϕ be an arbitrary testing function in $\mathfrak{D}$. For what values of the integer k does the following Cauchy principal value exist?

$$\text{Pv} \int_{-\infty}^\infty t^k \phi(t)\, dt \triangleq \lim_{\varepsilon \to 0+} \left(\int_{-\infty}^{-\varepsilon} + \int_\varepsilon^\infty \right) t^k \phi(t)\, dt$$

5 Show that

$$\frac{d}{dt}[1_+(t) \log t] = \text{Pf}\, \frac{1_+(t)}{t}$$

$$\frac{d}{dt} \log |\cos t| = -\text{Pf tan } t$$

$$\frac{d}{dt} \text{Pf tan } t = \text{Pf }(\sec t)^2$$

6 By using the limit of Sec. 2.3, Prob. 6, show that

$$\lim_{\tau \to 0+} \frac{1}{\tau + it} = \pi \delta(t) + \text{Pv} \frac{1}{it}$$

and, more generally, that

$$\lim_{\tau \to 0+} \frac{1}{(\tau + it)^n} = \frac{i^{n-1}\pi}{(n-1)!} \delta^{(n-1)}(t) + \text{Pf} \frac{1}{(it)^n} \qquad n = 1, 2, 3, \ldots$$

7 Let $\psi(t)$ be a function that is infinitely smooth. Show that it is valid to write

$$\psi(t) \text{ Pf} \frac{1}{|t|^\eta} = \text{Pf} \frac{\psi(t)}{|t|^\eta}$$

where η is any real number.

8 Let k be a positive integer. Find a distribution $f(t)$ that satisfies the equation $t^k f(t) = 1$ for all t.

9 Repeat Prob. 8 for the case where k is a positive number that is not an integer.

★2.6 THE PRIMITIVES OF DISTRIBUTIONS DEFINED OVER $\mathfrak{R}^1$

A *primitive* (or *indefinite integral*) of a continuous function is another function whose derivative equals the original function. The concept of the primitive may be extended to distributions. Every distribution defined on $\mathfrak{R}^1$ has a primitive. Moreover, as in the case of functions, there are an infinity of primitives belonging to each distribution and any two such primitives differ by a constant distribution. The purpose of this section is to prove these remarks. We shall restrict ourselves to the case where the independent variable has one dimension.

Since a primitive is a type of inverse of a derivative, one might expect that the definition of the derivative of a distribution, given by

$$\langle g^{(1)}, \phi \rangle \triangleq \langle g, -\phi^{(1)} \rangle \qquad \phi \in \mathfrak{D}$$

might be used to define the primitive $f^{(-1)} \triangleq g$ of the distribution $f \triangleq g^{(1)}$. However, the expression

$$\langle f^{(-1)}, \phi^{(1)} \rangle = \langle f, -\phi \rangle \qquad \phi \in \mathfrak{D} \tag{1}$$

is not adequate as a definition of the primitive $f^{(-1)}$, since (1) defines $f^{(-1)}$ as a functional only on that subspace $\mathcal{K}$ of $\mathfrak{D}$ whose elements $\phi^{(1)}$ are the derivatives of testing functions ϕ in $\mathfrak{D}$. That is, since $\mathcal{K}$ is a proper subspace of $\mathfrak{D}$ (that is, there are testing functions in $\mathfrak{D}$ that are not in $\mathcal{K}$), (1) does not define $f^{(-1)}$ over all of $\mathfrak{D}$. In order to obtain a proper definition of $f^{(-1)}$, we must develop a means of extending the definition of the primitive $f^{(-1)}$ from the space $\mathcal{K}$ onto the space $\mathcal{D}$. If turns out that there is no unique way to do this, as is reflected in the fact that there are many primitives for a given distribution.

In the subsequent development we shall need

Lemma 1

Let ϕ_0 be a fixed testing function in $\mathcal{D}$ that satisfies

$$\int_{-\infty}^{\infty} \phi_0(t)\, dt = 1 \tag{2}$$

Then every testing function ϕ in $\mathcal{D}$ may be decomposed uniquely according to

$$\phi = k\phi_0 + \chi \tag{3}$$

where χ is in $\mathcal{K}$ and the constant k is given by

$$k = \int_{-\infty}^{\infty} \phi(t)\, dt \tag{4}$$

Proof: With ϕ and ϕ_0 known, χ is uniquely specified by (3). It is obviously in $\mathcal{D}$. We shall have established that χ is in $\mathcal{K}$ when we show that its definite integral

$$\psi(t) = \int_{-\infty}^{t} \chi(x)\, dx \tag{5}$$

is in $\mathcal{D}$. Clearly, $\psi(t)$ is infinitely smooth and identically zero to the left of the supports of ϕ and ϕ_0. Also, it is equal to a constant to the right of the supports of ϕ and ϕ_0, and this constant is zero since

$$\begin{aligned}\psi(\infty) &= \int_{-\infty}^{\infty} [\phi(x) - k\phi_0(x)]\, dx \\ &= k - k = 0 \qquad \text{Q.E.D.}\end{aligned}$$

By use of the notation employed in (3) and (5), the primitive $f^{(-1)}$ is defined as a functional on $\mathcal{K}$ through

$$\langle f^{(-1)}, \chi \rangle = \langle f, -\psi \rangle \qquad \chi \in \mathcal{K} \tag{6}$$

If $f^{(-1)}$ is also specified for any single testing function ϕ_0 that satisfies (2) (such a ϕ_0 cannot be in $\mathcal{K}$), it becomes defined on all of $\mathcal{D}$. That is, given any other ϕ in $\mathcal{D}$, we may assume a value for $\langle f^{(-1)}, \phi_0 \rangle$ and we may use the decomposition of Lemma 1 to define $f^{(-1)}$ as that functional on $\mathcal{D}$ given by

$$\begin{aligned}\langle f^{(-1)}, \phi \rangle &\triangleq k\langle f^{(-1)}, \phi_0 \rangle + \langle f^{(-1)}, \chi \rangle \\ &= k\langle f^{(-1)}, \phi_0 \rangle - \langle f, \psi \rangle\end{aligned} \tag{7}$$

If ϕ is in $\mathcal{K}$, $k = 0$ and (7) becomes identical to (6).

Since $f^{(-1)}$ is a linear functional on $\mathcal{K}$ [this can be seen from (6)], it is also a linear functional on $\mathfrak{D}$. For if ϕ_1 and ϕ_2 are any two testing functions in $\mathfrak{D}$ with the decompositions

$$\phi_\nu = k_\nu \phi_0 + \chi_\nu$$
$$k_\nu = \int_{-\infty}^{\infty} \phi_\nu(t)\, dt \qquad \nu = 1, 2$$

and if α and β are any two numbers, we have that

$$\langle f^{(-1)}, \alpha\phi_1 + \beta\phi_2 \rangle = \langle f^{(-1)}, (\alpha k_1 + \beta k_2)\phi_0 + \alpha\chi_1 + \beta\chi_2 \rangle$$
$$= (\alpha k_1 + \beta k_2)\langle f^{(-1)}, \phi_0 \rangle + \alpha\langle f^{(-1)}, \chi_1 \rangle + \beta\langle f^{(-1)}, \chi_2 \rangle$$
$$= \alpha\langle f^{(-1)}, \phi_1 \rangle + \beta\langle f^{(-1)}, \phi_2 \rangle$$

Furthermore, we can show that $f^{(-1)}$ is continuous on $\mathfrak{D}$. Let the sequence

$$\{\phi_\nu\}_{\nu=1}^{\infty} = \{k_\nu\phi_0 + \chi_\nu\}_{\nu=1}^{\infty}$$

converge in $\mathfrak{D}$ to zero. Then, the numbers

$$k_\nu = \int_{-\infty}^{\infty} \phi_\nu(t)\, dt$$

converge to zero, and it follows that the χ_ν converge in $\mathfrak{D}$ to zero. Since all the χ_ν have their supports contained within a fixed finite interval, it follows that the $\psi_\nu(t)$, which are given by

$$\psi_\nu(t) = \int_{-\infty}^{t} \chi_\nu(x)\, dx$$

also converge in $\mathfrak{D}$ to zero. Since f is continuous on $\mathfrak{D}$, the numbers

$$\langle f^{(-1)}, \phi_\nu \rangle = k_\nu \langle f^{(-1)}, \phi_0 \rangle - \langle f, \psi_\nu \rangle$$

converge to zero as $\nu \to \infty$. Hence, $f^{(-1)}$ is also continuous on $\mathfrak{D}$.

The linearity and continuity of $f^{(-1)}$ show that *every primitive of a distribution is again a distribution.*

We now verify that *the derivative of any primitive of a distribution equals the distribution.* Indeed, if ϕ is an arbitrary testing function in $\mathfrak{D}$,

$$\langle (f^{(-1)})^{(1)}, \phi \rangle = \langle f^{(-1)}, -\phi^{(1)} \rangle$$

Since $\phi^{(1)}$ is in $\mathcal{K}$, we may use (6) to obtain

$$\langle (f^{(-1)})^{(1)}, \phi \rangle = \left\langle f, \int_{-\infty}^{t} \phi^{(1)}(x)\, dx \right\rangle = \langle f, \phi \rangle$$

which is what was asserted.

The particular primitive that is obtained for a given distribution f depends upon the choice of the numerical value $\langle f^{(-1)}, \phi_0 \rangle$. Any two primitives of f will differ by the constant distribution c, where the number c is the difference of the two choices of $\langle f^{(-1)}, \phi_0 \rangle$. To see this, let $\langle f_1^{(-1)}, \phi_0 \rangle$ and $\langle f_2^{(-1)}, \phi_0 \rangle$ be two different numerical values which generate the two different primitives $f_1^{(-1)}$ and $f_2^{(-1)}$ for the same distribution f. Then, for ϕ in $\mathfrak{D}$,

$$\langle f_1^{(-1)}, \phi \rangle - \langle f_2^{(-1)}, \phi \rangle = k\langle f_1^{(-1)}, \phi_0 \rangle - \langle f, \psi \rangle - k\langle f_2^{(-1)}, \phi_0 \rangle + \langle f, \psi \rangle$$
$$= ck = c \int_{-\infty}^{\infty} \phi(t)\, dt = \langle c, \phi \rangle$$

where

$$c = \langle f_1^{(-1)}, \phi_0 \rangle - \langle f_2^{(-1)}, \phi_0 \rangle \tag{8}$$

We may summarize these results as follows.

Theorem 2.6-1

Every distribution over $\mathfrak{R}^1$ has an infinity of primitives defined by (7) and the decomposition (3). Each primitive is also a distribution. The difference between any two primitives of a given distribution is the constant distribution given by (8).

If follows that we may construct a primitive of a primitive to obtain by repetition the higher-order primitives of a distribution f. For a fixed order, say p, $f(t)$ has an infinity of primitives of order p, and the difference between any two of them is a polynomial in t whose degree is not greater than $p - 1$. (See, for example, Prob. 2.)

Subsequently, we shall make use of the following generalization of the formula for integration by parts. Let λ be a function that is infinitely smooth, let f be any distribution, and let c be a constant distribution. Then, we may write

$$(\lambda f^{(1)})^{(-1)} = \lambda f - (\lambda^{(1)} f)^{(-1)} + c \tag{9}$$

as can easily be verified by differentiating both sides of (9). By proper choice of the primitives in (9), the constant distribution c can be set equal to the zero distribution.

All these ideas may be extended to distributions defined over $\mathfrak{R}^n$. (See, for instance, Schwartz [1], vol. 1, pp. 55–57.)

Finally, let us note that another means of obtaining the primitives of a distribution is provided by the convolution process, which will be discussed in a subsequent chapter. See Sec. 5.5, Prob. 2, for the appropriate formula.

PROBLEMS

1 By using (7), find that primitive of $\delta(t)$ for which $\langle \delta^{(-1)}, \phi_0 \rangle$ is the constant c when $\phi_0(t)$ is an even function.

2 By using (7), show that every primitive of the constant distribution c_1 is a distribution of the form $c_1 t + c_2$, where c_2 is also a constant distribution.

3 Develop an expression analogous to (7) that may be used to define the second-order primitives of any distribution. HINT: First show that any testing function ϕ in $\mathfrak{D}$ may be uniquely decomposed into

$$\phi = k_0 \phi_0 + k_1 \phi_1 + \chi$$

where ϕ_0 is a fixed element in $\mathfrak{D}$, ϕ_1 is a fixed element in $\mathfrak{K}$, χ is the second derivative of some element of $\mathfrak{D}$, and

$$\int_{-\infty}^{\infty} \phi_0(t)\, dt = 1$$

$$\int_{-\infty}^{\infty} dt \int_{-\infty}^{t} \phi_1(\tau)\, d\tau = 1$$

$$k_0 = \int_{-\infty}^{\infty} \phi(t)\, dt$$

$$k_1 = \int_{-\infty}^{\infty} dt \int_{-\infty}^{t} [\phi(\tau) - k_0 \phi_0(\tau)]\, d\tau$$

4 If a distribution $f(t)$ has a bounded support, we can define its definite integral over the entire t axis by

$$\int_{-\infty}^{\infty} f(t)\, dt \triangleq \langle f(t), \lambda(t) \rangle$$

where $\lambda(t)$ is in $\mathfrak{D}$ and is identically equal to one over a neighborhood of the support of f. Furthermore, even if f is not of bounded support, we can at times assign a meaning to its definite integrals in the following way. Assume that on the intervals $a - \varepsilon < t < a + 3\varepsilon$ and $b - 3\varepsilon < t < b + \varepsilon$, $f(t)$ is a regular distribution corresponding to a continuous function. Let $\mu(t)$ and $\lambda(t)$ be testing functions in $\mathfrak{D}$ such that $\mu(t) + \lambda(t) = 1$ over $a \leq t \leq b$, $\lambda(t)$ has its support contained in $a + \varepsilon < t < b - \varepsilon$, and $\lambda(t) = 1$ over $a + 2\varepsilon < t < b - 2\varepsilon$. Then our definition is

$$\int_a^b f(t)\, dt \triangleq \langle f(t), \lambda(t) \rangle + \int_a^b f(t)\mu(t)\, dt$$

Note that this yields the usual definite integral when $f(t)$ is a regular distribution for all t. Then evaluate the following:

a. $\int_{-1}^{1} \delta(t)\, dt$

b. $\int_{-1}^{1} \delta^{(1)}(t)\, dt$

c. $\int_{-1}^{1} \operatorname{Pf} \frac{1_+(t)}{t}\, dt$

d. $\int_{-1}^{1} \operatorname{Pf} \frac{1_+(t)}{t^2}\, dt$

e. $\int_a^b \operatorname{Pv} \frac{1}{t}\, dt \qquad a < 0 < b$

5 Show that the distributional primitives of a locally integrable function f are none other than the ordinary primitives of the function. HINT: Show that, in the distributional sense of equality and differentiation,

$$f(t) = \frac{d}{dt} g(t)$$

where $g(t)$ is the regular distribution corresponding to the function

$$\int_a^t f(x)\,dx + c$$

Here, a is a real number and c is a complex number.

◆2.7 CONTINUITY AND DIFFERENTIABILITY WITH RESPECT TO A PARAMETER UPON WHICH THE TESTING FUNCTIONS DEPEND

Let $h(t, \tau)$ be a locally integrable function of the n-dimensional variable t and of the one-dimensional parameter τ. Consider the ordinary integral

$$g(\tau) = \int_{\mathcal{R}^n} h(t, \tau)\,dt \tag{1}$$

In the classical theory of integration $g(\tau)$ will be continuous or differentiable when certain conditions are imposed on $h(t, \tau)$. (A lucid discussion of these conditions for the Riemann integral is given by Carslaw [1], chap. 6.)

A similar situation arises in distribution theory when the testing functions depend upon an additional parameter. When $f(t)$ is a distribution and $\phi(t, \tau)$ is a testing function of t that depends in addition upon τ,

$$g(\tau) \triangleq \langle f(t),\, \phi(t, \tau) \rangle \tag{2}$$

is a sort of generalization of (1). In this section we shall present conditions on $\phi(t, \tau)$ sufficient to ensure that (2) is continuous or differentiable with respect to τ.

Let $\phi(t, \tau)$ be a function of the n-dimensional variable t and the one-dimensional variable τ. Furthermore, assume that, for each fixed value of τ in some set of values for τ, $\phi(t, \tau)$ is a testing function of t [that is, $\phi(t, \tau) \in \mathcal{D}_t$, where $\mathcal{D}_t$ denotes the space $\mathcal{D}$ of testing functions whose independent variables are t]. Now, if $f(t)$ is a distribution defined over $\mathcal{R}^n$, then (2) is an ordinary function of τ defined over the given set of values for τ. The first theorem states conditions on $\phi(t, \tau)$ under which $g(\tau)$ is continuous.

Theorem 2.7-1

Let $\phi(t, \tau)$ possess the following two properties:

a. As τ traverses a neighborhood I of the point τ_0, the supports of all the $\phi(t, \tau)$, considered as functions of t alone, are contained in a fixed bounded domain in $\mathcal{R}^n$.

b. Let D_t^k denote an arbitrary partial derivative with respect to the components of t but not with respect to τ. Every partial derivative

$D_t^k \phi(t, \tau)$ is continuous as a function of (t, τ) for t in $\Re^n$ and τ in I. Then, if $f(t)$ is any distribution, the function $g(\tau) \triangleq \langle f(t), \phi(t, \tau) \rangle$ is continuous at $\tau = \tau_0$.

Proof: When τ is restricted to I, every partial derivative $D_t^k \phi(t, \tau)$ is zero for all values of t outside a fixed closed bounded domain Ω in $\Re^n$. Hence, $D_t^k \phi(t, \tau)$ is a uniformly continuous function of (t, τ) for t in $\Re^n$ and τ in some neighborhood of τ_0 (Widder [2], p. 214). This means that, as $\tau \to \tau_0$, $D_t^k \phi(t, \tau)$ converges to $D_t^k \phi(t, \tau_0)$ uniformly for t in $\Re^n$. This is true for every k, and, therefore, $\phi(t, \tau)$ converges in $\mathfrak{D}_t$ to $\phi(t, \tau_0)$. Since $f(t)$ is a continuous functional on $\mathfrak{D}_t$, it follows that

$$\langle f(t), \phi(t, \tau) \rangle \to \langle f(t), \phi(t, \tau_0) \rangle$$

as $\tau \to \tau_0$. Thus, the function $g(\tau)$ is continuous at $\tau = \tau_0$, as was asserted.

The second theorem specifies conditions on $\phi(t, \tau)$ under which $g(\tau)$ is differentiable.

Theorem 2.7-2

Let $\phi(t, \tau)$ possess the following two properties:

a. As τ traverses a neighborhood I of the point τ_0, the supports of all the $\phi(t, \tau)$, considered as functions of t alone, are contained in a fixed bounded domain in $\Re^n$.

b. Let D_t^k again denote an arbitrary partial derivative with respect to $t_1, t_2, \ldots, t_n$ but not with respect to τ. For τ in I and for all t, each ordinary partial derivative $D_t^k \phi(t, \tau)$ exists, is continuous with respect to (t, τ), and possesses an ordinary partial derivative with respect to τ, which is also continuous with respect to (t, τ).

Then, if $f(t)$ is any distribution, the function $g(\tau) \triangleq \langle f(t), \phi(t, \tau) \rangle$ has an ordinary derivative with respect to τ at $\tau = \tau_0$ and

$$\left. \frac{dg}{d\tau} \right|_{\tau=\tau_0} = \left\langle f(t), \frac{\partial}{\partial \tau} \phi(t, \tau) \Big|_{\tau=\tau_0} \right\rangle \tag{3}$$

Proof: Consider

$$\frac{g(\tau_0 + \Delta\tau) - g(\tau_0)}{\Delta\tau} - \left\langle f(t), \frac{\partial}{\partial \tau} \phi(t, \tau) \Big|_{\tau=\tau_0} \right\rangle$$
$$= \left\langle f(t), \frac{\phi(t, \tau_0 + \Delta\tau) - \phi(t, \tau_0)}{\Delta\tau} - \frac{\partial}{\partial \tau} \phi(t, \tau) \Big|_{\tau=\tau_0} \right\rangle$$
$$\triangleq \langle f(t), \psi(t, \Delta\tau) \rangle \tag{4}$$

Here $\psi(t, \Delta\tau)$ is taken to be zero for $\Delta\tau = 0$. For each value of $\Delta\tau$ in some bounded neighborhood of zero, $\psi(t, \Delta\tau)$ is a testing function in $\mathfrak{D}_t$. [Our hypothesis b ensures that $\partial \phi(t, \tau)/\partial \tau \big|_{\tau=\tau_0}$ is infinitely smooth with respect to t. See Hobson [1], vol. 1, p. 425.] By condition b, $\psi(t, \Delta\tau)$ is a continuous function of $(t, \Delta\tau)$ for t in $\mathcal{R}^n$ and $\Delta\tau$ in some neighborhood of zero and converges to zero as $\Delta\tau \to 0$. By condition a, $\psi(t, \Delta\tau)$ is zero for t outside a fixed bounded domain in $\mathcal{R}^n$ and for $\Delta\tau$ in the afore mentioned neighborhood. It follows that $\psi(t, \Delta\tau)$ converges to zero uniformly for t in $\mathcal{R}^n$ when $\Delta\tau \to 0$ (Widder [2], p. 214).

For the same reasons, as $\Delta\tau \to 0$, $D_t^k \psi(t, \Delta\tau)$ converges to zero uniformly for t in $\mathcal{R}^n$, so that the $\psi(t, \Delta\tau)$ converge in $\mathfrak{D}_t$ to zero. Since $f(t)$ is a continuous functional on $\mathfrak{D}_t$, (4) converges to zero as $\Delta\tau \to 0$. Q.E.D.

Corollary 2.7-2a

Let x be an n-dimensional real variable and y an m-dimensional real variable. Also, let $\phi(x, y)$ be a testing function in $\mathfrak{D}$ defined over $\mathcal{R}^{n+m}$. If $f(x)$ is a distribution defined over $\mathcal{R}^n$, then

$$\theta(y) \triangleq \langle f(x), \phi(x, y) \rangle$$

is a testing function of y in $\mathfrak{D}$, and an arbitrary partial derivative $D_y^k \theta(y)$ with respect to the components of y is given by

$$D_y^k \theta(y) = D_y^k \langle f(x), \phi(x, y) \rangle = \langle f(x), D_y^k \phi(x, y) \rangle \tag{5}$$

Proof: We may identify a particular component of y, say y_k, with τ of Theorem 2.7-2 and all other components of y and those of x with the components of t in that theorem. Then, for each value of y_k, $\phi(x, y)$ satisfies the hypothesis of Theorem 2.7-2 so that

$$\frac{\partial \theta(y)}{\partial y_k} = \left\langle f(x), \frac{\partial \phi(x, y)}{\partial y_k} \right\rangle$$

By repetition of this argument, singling out for every differentiation the appropriate component of y, we see that $\theta(y)$ is infinitely smooth and that its partial derivatives are given by (5). Finally, $\phi(x, y)$ has a bounded support, which means that $\phi(x, y)$ and, therefore, $\theta(y)$ are zero whenever y lies outside some bounded domain in $\mathcal{R}^m$.

PROBLEM

1 Let t and x denote one-dimensional real variables. A distribution $f(t)$ is said to be nonnegative if $\langle f, \phi \rangle \geq 0$ for every nonnegative ϕ in $\mathfrak{D}$. A distribution $g(t)$ is said to be nondecreasing if

$$\langle g(t), \phi(t - x) \rangle$$

is a nondecreasing function of x for every nonnegative ϕ in $\mathfrak{D}$. Show that f is a nonnegative distribution if and only if a primitive $f^{(-1)}$ of f is a nondecreasing distribution.

2.8 DISTRIBUTIONS THAT DEPEND UPON A PARAMETER AND INTEGRATION WITH RESPECT TO THAT PARAMETER

A somewhat different situation arises when it is the distribution, rather than the testing function, that depends upon a parameter. In particular, let τ be a one-dimensional parameter. To each value of τ let there correspond a distribution $f_\tau(t)$ defined over some open domain Ω of the space $\mathfrak{R}^n$ of the n-dimensional variable t. In this case we shall display τ as a subscript on f and thereby specify which variable is the parameter. For each testing function $\phi(t)$ in $\mathfrak{D}_t$, whose support is contained in Ω,

$$g(\tau) \triangleq \langle f_\tau(t), \phi(t) \rangle$$

defines an ordinary function of τ.

By choosing different testing functions $\phi(t)$, we obtain different functions $g(\tau)$. If, for each $\phi(t)$ in $\mathfrak{D}_t$, the corresponding $g(\tau)$ is integrable over the interval I, the distribution is said to be integrable with respect to τ over I. This integration defines a new functional $h(t)$ on $\mathfrak{D}_t$ through

$$\langle h(t), \phi(t) \rangle = \int_I \langle f_\tau(t), \phi(t) \rangle \, d\tau \qquad \phi(t) \in \mathfrak{D}_t \tag{1}$$

Symbolically, we shall write

$$h = \int_I f_\tau \, d\tau \tag{2}$$

The convergence of the integral (2) is quite analogous to the definition of convergence in $\mathfrak{D}'$ of a set of distributions, which was discussed in Sec. 2.2. In fact, we shall also say that the integral (2) converges in $\mathfrak{D}'$ to the functional h. Moreover, h turns out to be a distribution. We shall prove this fact only in the special case where $f_\tau(t)$ depends continuously on the parameter τ.

The distribution $f_\tau(t)$ is said to be *continuous with respect to the parameter* τ at some point $\tau = \tau_1$ if, for each $\phi(t)$ in $\mathfrak{D}_t$, the function $\langle f_\tau(t), \phi(t) \rangle$ is continuous at $\tau = \tau_1$. The distribution is continuous over a set of points if the function $\langle f_\tau(t), \phi(t) \rangle$ is continuous at each point of that set.

If $f_\tau(t)$ is an ordinary function of t and τ that is continuous for t in $\mathfrak{R}^n$ and τ in some neighborhood of $\tau = \tau_1$, the regular distribution in $\mathfrak{D}'_t$ (the dual space of $\mathfrak{D}_t$) corresponding to $f_\tau(t)$ is also continuous at $\tau = \tau_1$. To see this, note that for a given $\varepsilon > 0$ there exists an η such that for $|\Delta\tau| < \eta$

$$|f_{\tau_1}(t) - f_{\tau_1+\Delta\tau}(t)| < \varepsilon$$

Moreover, by the uniform continuity of $f_\tau(t)$, we can choose η independently of t so long as t is restricted to a bounded domain in $\mathfrak{R}^n$. Since the support of $\phi(t)$ is bounded, it follows that for $|\Delta\tau| < \eta$

$$|\langle f_{\tau_1}(t), \phi(t)\rangle - \langle f_{\tau_1+\Delta\tau}(t), \phi(t)\rangle| = \left|\int_{-\infty}^{\infty}[f_{\tau_1}(t) - f_{\tau_1+\Delta\tau}(t)]\phi(t)\,dt\right|$$
$$\leq \varepsilon\int_{-\infty}^{\infty}|\phi(t)|\,dt$$

By choosing ε appropriately, we can make the last expression as small as we wish. Thus, the distribution f_τ is continuous at $\tau = \tau_1$, as was asserted.

Theorem 2.8-1

If the distribution $f_\tau(t)$ defined over $\mathfrak{R}^n$ is continuous with respect to the one-dimensional parameter τ in the closed interval $a \leq \tau \leq b$, then its integral over this interval, which is defined by

$$\langle h(t), \phi(t)\rangle \triangleq \int_a^b \langle f_\tau(t), \phi(t)\rangle\,d\tau \qquad \phi(t) \in \mathfrak{D}_t \tag{3}$$

and denoted symbolically by

$$h = \int_a^b f_\tau\,d\tau \tag{4}$$

is again a distribution.

Proof: We shall construct a sequence of distributions that converges in $\mathfrak{D}'_t$ to $h(t)$. It will then follow from the fact that $\mathfrak{D}'_t$ is closed under convergence (see Theorem 2.2-1) that $h(t)$ is also a distribution.

Divide the interval $a \leq \tau \leq b$ into m subintervals with the end points $\tau_0 = a, \tau_1, \ldots, \tau_{m-1}, \tau_m = b$ and choose an arbitrary point ξ_i from each subinterval (that is, $\tau_{i-1} \leq \xi_i \leq \tau_i$). Then, construct the Riemann integral of the function

$$g(\tau) \triangleq \langle f_\tau(t), \phi(t)\rangle$$

in the standard way by first setting up the sum

$$S_m = \sum_{i=1}^m \langle f_{\xi_i}(t), \phi(t)\rangle\,\Delta\tau_i \qquad \Delta\tau_i = \tau_i - \tau_{i-1} \tag{5}$$

and then letting $m \to \infty$ in such a fashion that

$$\sup_{1\leq i\leq m}\Delta\tau_i \to 0$$

Since $\langle f_\tau(t), \phi(t)\rangle$ is continuous for $a \leq \tau \leq b$, $\lim_{m\to\infty} S_m$ exists and is independent of the choices of the subintervals and the choices of the points ξ_i. The limit depends, of course, on the choice of ϕ, but it will exist for every ϕ in $\mathfrak{D}_t$. Hence, corresponding to (5) we have the sequence of distributions

$$\left\{\sum_{i=1}^m f_{\xi_i}(t)\,\Delta\tau_i\right\}_{m=1}^{\infty}$$

which converges in $\mathfrak{D}'_t$. Its limit in $\mathfrak{D}'_t$ does not depend upon the aforementioned choices. This sequence thus defines a distribution h, which is denoted symbolically by (4). Q.E.D.

This theorem, in addition to being extendible to the case where $\langle f_\tau(t), \phi(t) \rangle$ is merely Riemann-integrable with respect to τ, may be extended in other ways. For instance, the parameter τ may be m-dimensional ($m > 1$) and the integration may be over some domain of τ having m or fewer dimensions. We may account for the two-dimensional case for τ by assuming that τ is a complex parameter. Then, an integral over any arc C of finite length in the τ plane can be constructed according to (5) whenever $f_\tau(t)$ is continuous with respect to τ on C. (By an "arc" we mean a special type of rectifiable curve, as defined by Titchmarsh [1], p. 71.)

Moreover, the length of C may be infinite so long as

$$\langle h(t), \phi(t) \rangle \triangleq \int_C \langle f_\tau(t), \phi(t) \rangle \, d\tau \tag{6}$$

is convergent for every $\phi(t)$ in $\mathfrak{D}_t$. This is because the integration on the arc of infinite length can be considered as the limit of a sequence of integrations on finite portions of the arc. Since the limit exists for every $\phi(t)$ in $\mathfrak{D}_t$ (the limit being independent of the choices of the finite portions of the curve), we again have a sequence of distributions that converges in $\mathfrak{D}'_t$ to

$$h = \int_C f_\tau \, d\tau \tag{7}$$

Thus, we have

Corollary 2.8-1a

Let τ be a complex variable and C an arc of finite length in the τ plane. If the distribution $f_\tau(t)$ is continuous on C with respect to τ, then (6) defines a distribution h in $\mathfrak{D}'_t$. In addition, if C is infinite in length, (6) defines a distribution in $\mathfrak{D}'_t$ whenever (6) converges for every $\phi(t)$ in $\mathfrak{D}_t$.

Under the conditions of Theorem 2.8-1 or Corollary 2.8-1a, we can integrate not only $f_\tau(t)$ but also every one of its partial derivatives with respect to the components of t. More precisely, we may state

Corollary 2.8-1b

Assume the hypothesis of Corollary 2.8-1a and let h be given by (7). Then, every partial derivative $D_t^k f_\tau(t)$ of $f_\tau(t)$ with respect to the components of t can also be integrated on C and the order of integration and differentiation may be interchanged. In short,

$$D_t^k h(t) = \int_C D_t^k f_\tau(t) \, d\tau$$

Proof: For every $\phi(t)$ in $\mathfrak{D}_t$,

$$\int_C \langle D_t^k f_\tau(t), \phi(t) \rangle \, d\tau = (-1)^k \int_C \langle f_\tau(t), D_t^k \phi(t) \rangle \, d\tau$$
$$= (-1)^k \langle h(t), D_t^k \phi(t) \rangle = \langle D_t^k h(t), \phi(t) \rangle$$

and this proves the corollary.

These results provide a means of extending the Fourier integral to functions that increase not too rapidly as their arguments go to infinity.

Theorem 2.8-2

Let t and τ be one-dimensional real variables. If $k(\tau)$ is a continuous function for all τ, which is bounded by some power of $|\tau|$ as $|\tau| \to \infty$ [that is,

$$|k(\tau)| \leq M(|\tau|^p + 1)$$

where M and p are real constants], then the integral

$$\int_{-\infty}^{\infty} k(\tau) e^{-i\tau t} \, d\tau \tag{8}$$

converges in $\mathfrak{D}_t'$ to a distribution.

Proof: Consider

$$h_1(t) \triangleq \int_{|\tau|>1} \frac{k(\tau) e^{-i\tau t}}{(-i\tau)^{q+2}} \, d\tau \tag{9}$$

where q is an integer not less than p. This integral converges in the ordinary sense to yield the continuous function $h_1(t)$. Moreover, the integrand of (9) satisfies the hypothesis of Corollary 2.8-1a. Therefore, (9) also converges in $\mathfrak{D}_t'$ and its limit is the regular distribution $h_1(t)$, as can be seen from the fact that the integrations on t and τ indicated in (6) can be interchanged. According to Corollary 2.8-1b, we may differentiate (9) under the integral sign $q + 2$ times to obtain the distribution given by

$$h_1^{(q+2)}(t) = \int_{|\tau|>1} k(\tau) e^{-i\tau t} \, d\tau \tag{10}$$

Also,

$$h_2(t) \triangleq \int_{|\tau| \leq 1} k(\tau) e^{-i\tau t} \, d\tau \tag{11}$$

is an ordinary continuous function of t and hence a regular distribution. Upon adding (10) and (11), we see that (8) is a distribution, as was asserted.

This proof is longer than it need be, and we indicate a more concise argument in Prob. 3. However, it indicates an explicit means of computing the distribution (8) as the sum $h_1^{(q+2)}(t) + h_2(t)$.

In Chap. 7 we shall discuss a more general and informative way of generalizing the Fourier integral. There the Fourier transform of any distribution over $\Re^1$ will be defined and Theorem 2.8-2 will be encompassed as a special case.

PROBLEMS

1 Using the technique employed in the proof of Theorem 2.8-2, show that

$$\frac{1}{2\pi}\int_{-\infty}^{\infty} e^{-i\tau t}\, d\tau = \delta(t)$$

2 Let $k(\tau)$ be a continuous function over $a \leq \tau \leq b$ and let $g(t)$ be an arbitrary distribution over $\Re^1$. Show that

$$\int_a^b k(\tau)g(t)\, d\tau = g(t)\int_a^b k(\tau)\, d\tau$$

3 Show that, for any $\phi(t)$ in $\mathfrak{D}_t$, $|\langle e^{-i\tau t}, \phi(t)\rangle|$ goes to zero faster than any power of $1/|\tau|$ as $|\tau| \to \infty$. Then, use this fact to simplify the proof of Theorem 2.8-2 by eliminating the use of Corollary 2.8-1b.

3 Further Properties of Distributions

★3.1 INTRODUCTION

Distributions possess two quite useful properties when they are restricted to testing functions whose supports are contained in a fixed bounded interval. The first of these is a certain boundedness property, which is discussed in Sec. 3.3. The second is that on finite intervals distributions are equal to finite-order derivatives of continuous functions; this result is proved in Sec. 3.4. Sections 3.2 and 3.5 are devoted to some other characteristics that are peculiar to the delta functional and its derivatives.

We shall restrict ourselves in this chapter (as we shall again from Chap. 5 to the end of the book) to distributions that are defined over the one-dimensional euclidean space $\mathcal{R}^1$. However, the results of this chapter possess analogues for distributions having n-dimensional independent variables.

3.2 A CHARACTERIZATION OF THE DELTA FUNCTIONAL AND ITS DERIVATIVES

The process of multiplying a distribution by an infinitely smooth function leads to another means of characterizing the delta functional and its derivatives. For example, if a distribution $f(t)$ satisfies the equation

$$tf(t) = 0$$

then $f(t)$ must be the delta functional multiplied by some constant. A more general result is also true.

Theorem 3.2-1

A necessary and sufficient condition for a distribution $f(t)$ defined over $\mathcal{R}^1$ to satisfy the equation

$$t^m f(t) = 0 \tag{1}$$

where m is a positive integer, is that $f(t)$ be a linear combination of the delta functional and its derivatives of order no greater than $m - 1$. That is,

$$f = \sum_{\nu=0}^{m-1} c_\nu \delta^{(\nu)} \tag{2}$$

where the c_ν are arbitrary constants.

That (1) is necessary for (2) to hold is obvious. To prove the sufficiency of (1), let us proceed by means of two lemmas.

Lemma 1

For a testing function $\chi(t)$ in $\mathfrak{D}$ to have the form

$$\chi(t) = t^m \phi(t) \qquad \phi(t) \in \mathfrak{D} \tag{3}$$

it is necessary and sufficient that

$$\chi^{(\nu)}(0) = 0 \qquad \nu = 0, 1, \ldots, m - 1 \tag{4}$$

Proof: If (3) holds, then (4) will certainly be satisfied. Conversely, assume that (4) holds. Clearly the function $\phi(t) \triangleq \chi(t)/t^m$ ($t \neq 0$) is infinitely smooth for $t \neq 0$. Moreover, $\phi(t)$ has the same support as $\chi(t)$. Hence, we need merely verify that $\phi(t)$ can be so defined at $t = 0$ that it is infinitely smooth at that point. Now,

$$\frac{d^k \phi}{dt^k} = \sum_{\nu=0}^{k} \binom{k}{\nu} \frac{d^\nu \chi}{dt^\nu} \frac{d^{k-\nu}}{dt^{k-\nu}} \left(\frac{1}{t^m}\right) \qquad t \neq 0$$

and we may expand each derivative $d^\nu\chi/dt^\nu$ around $t = 0$ according to Taylor's formula with remainder. This will show that the asymptotic behavior of $d^k\phi/dt^k$ as $t \to 0$ is determined by the leading terms of these expansions, since the remainder terms become comparatively negligible as $t \to 0$. Thus, for every sufficiently large integer q we have that as $t \to 0$ with $t \neq 0$

$$\frac{d^k\phi}{dt^k} \sim \sum_{\nu=0}^{k} \binom{k}{\nu} \frac{d^\nu}{dt^\nu} \left[\frac{\chi^{(m)}(0)}{m!} t^m + \frac{\chi^{(m+1)}(0)}{(m+1)!} t^{m+1} + \cdots \right.$$
$$\left. + \frac{\chi^{(m+q)}(0)}{(m+q)!} t^{m+q} \right] \frac{d^{k-\nu}}{dt^{k-\nu}} \left(\frac{1}{t^m} \right)$$
$$= \frac{d^k}{dt^k} \left[\frac{\chi^{(m)}(0)}{m!} + \frac{\chi^{(m+1)}(0)}{(m+1)!} t + \cdots + \frac{\chi^{(m+q)}(0)}{(m+q)!} t^q \right]$$

Setting

$$\phi(0) \triangleq \frac{\chi^{(m)}(0)}{m!}$$

we see that $\phi(t)$ has ordinary derivatives of all orders at $t = 0$, which completes the proof.

Lemma 2

Let $\lambda(t)$ be a fixed testing function in $\mathfrak{D}$ such that $\lambda(0) = 1$ and $\lambda^{(\nu)}(0) = 0$ ($\nu = 1, 2, \ldots, m - 1$). Then any testing function $\psi(t)$ in $\mathfrak{D}$ can be uniquely decomposed according to

$$\psi(t) = \lambda(t) \sum_{\nu=0}^{m-1} \frac{1}{\nu!} \psi^{(\nu)}(0) t^\nu + \chi(t) \tag{5}$$

where $\chi(t)$ is in $\mathfrak{D}$ and satisfies (4).

Proof: Given ψ and λ, (5) uniquely determines χ as a testing function in $\mathfrak{D}$. That χ satisfies (4) can be seen by differentiating (5).

Proof of Theorem 3.2-1: As was pointed out before, we need merely show that (2) follows from (1). Let $t^m f(t) = 0$. Then, for any $\chi(t)$ of the form given in Lemma 1,

$$\langle f, \chi \rangle = \langle f(t), t^m \phi(t) \rangle = \langle t^m f(t), \phi(t) \rangle = 0 \tag{6}$$

Thus, for every ψ in $\mathfrak{D}$, we have by Lemma 2 and by (6) that

$$\langle f, \psi \rangle = \left\langle f(t), \lambda(t) \sum_{\nu=0}^{m-1} \frac{1}{\nu!} \psi^{(\nu)}(0) t^\nu + \chi(t) \right\rangle$$
$$= \sum_{\nu=0}^{m-1} \frac{1}{\nu!} \psi^{(\nu)}(0) \langle f(t), t^\nu \lambda(t) \rangle$$

Define the constants c_ν by

$$c_\nu = \frac{1}{\nu!}(-1)^\nu \langle f(t), t^\nu \lambda(t)\rangle \tag{7}$$

It follows that every solution of (1) has the form (2). Any choice of the constants c_ν will provide a solution to (1). Q.E.D.

PROBLEMS

1 Show that Theorem 3.2-1 still holds if in (1) t^m is replaced by a function $g(t)$, which is infinitely smooth and is zero only at $t = 0$, this zero being of multiplicity m [that is, $\lim_{t\to 0} g(t)/t^m$ is finite and not zero].

2 Let $f(t)$ be a distribution, and let

$(e^t - 1)(t - 1)^3(\sin 2t)f(t) = 0$

Specify as much about $f(t)$ as is possible.

3 Find all the distributions that for all t satisfy the equation $t^k f(t) = 1$. Here k is a positive integer. HINT: Use the solution of Sec. 2.5, Prob. 8.

★3.3 A LOCAL-BOUNDEDNESS PROPERTY OF DISTRIBUTIONS

Every distribution f that is defined over some neighborhood of a fixed finite closed interval I in $\Re^1$ possesses the following boundedness property. There exist a nonnegative integer r and a constant C such that, for each ϕ in $\mathfrak{D}$ whose support is in I,

$$|\langle f, \phi\rangle| \leq C \sup |\phi^{(r)}(t)|$$

C and r depend only on f and I. This fact, which we shall establish in this section, will be used in the subsequent section to show that locally every distribution is obtainable by a finite number of differentiations of a continuous function. It will also imply that a distribution has a single point as its support if and only if it is a linear combination of the delta functional and a finite number of its derivatives.

Let I be the closed finite interval $a \leq t \leq b$. Let $\mathfrak{D}_I$ be the space of all testing functions in $\mathfrak{D}$ whose supports are in I. A sequence $\{\phi_\nu\}_{\nu=1}^\infty$ of testing functions is said to converge in $\mathfrak{D}_I$ to a limit function ϕ if the ϕ_ν are in $\mathfrak{D}_I$ and the sequence converges in $\mathfrak{D}$ to ϕ. That is, the ϕ_ν have their supports in I and $\{\phi_\nu^{(k)}\}_{\nu=1}^\infty$ converges uniformly to $\phi^{(k)}$ for each nonnegative integer k. As usual, the uniformity of the convergence need hold only over $a \leq t \leq b$ and not over all the k. It follows that ϕ will also be in $\mathfrak{D}_I$.

An alternative means of describing convergence in $\mathfrak{D}_I$ will now be developed. For every ϕ in $\mathfrak{D}_I$

$$\phi^{(k)}(t) = \int_a^t \phi^{(k+1)}(x)\,dx$$

and therefore

$$|\phi^{(k)}(t)| \leq (b - a) \sup_t |\phi^{(k+1)}(t)| \tag{1}$$

Let γ_k ($k = 0, 1, 2, \ldots$) denote the real numbers depending upon ϕ that are defined by

$$\gamma_k \triangleq \gamma_k(\phi) \triangleq (b - a)^k \sup_t |\phi^{(k)}(t)| \tag{2}$$

These numbers are positive so long as ϕ is not identically zero. They are called *seminorms* for the space $\mathfrak{D}_I$. Obviously, *a sequence* $\{\phi_\nu\}_{\nu=1}^\infty$ *converges in* $\mathfrak{D}_I$ *to zero if and only if all the* ϕ_ν *are in* $\mathfrak{D}_I$ *and if for each k the sequence* $\{\gamma_k(\phi_\nu)\}_{\nu=1}^\infty$ *converges to zero.*

Furthermore, we have from (1) that

$$0 \leq \gamma_0 \leq \gamma_1 \leq \gamma_2 \leq \cdots \tag{3}$$

Consequently, *the convergence of* $\{\phi_\nu^{(r)}(t)\}_{\nu=1}^\infty$ *to zero uniformly for all t implies the same sort of convergence for* $\{\phi_\nu^{(k)}(t)\}_{\nu=1}^\infty$ ($k = 0, 1, 2, \ldots, r$).

We are now ready to establish the local-boundedness property possessed by every distribution.

Theorem 3.3-1

Let f be a distribution that is defined over some neighborhood of a fixed finite closed interval I in $\mathfrak{R}^1$. *There exist a nonnegative integer r and a finite positive constant C such that for every ϕ in* $\mathfrak{D}_I$

$$|\langle f, \phi \rangle| \leq C \sup_t |\phi^{(r)}(t)| \tag{4}$$

Both C and r depend in general upon f and I.

Proof: Let us assume that for the given f no relation such as (4) can hold. Therefore, for each positive integer ν there exists a testing function ϕ_ν in $\mathfrak{D}_I$ such that

$$|\langle f, \phi_\nu \rangle| > \nu(b-a)^\nu \sup_t |\phi_\nu^{(\nu)}(t)| = \nu\gamma_\nu(\phi_\nu) \tag{5}$$

Note that $\gamma_\nu(\phi_\nu) \geq \gamma_0(\phi_\nu) > 0$ if $\phi_\nu \not\equiv 0$. Set

$$\theta_\nu \triangleq \frac{\phi_\nu}{\nu\gamma_\nu(\phi_\nu)}$$

The θ_ν are also in $\mathfrak{D}_I$. Let k be some nonnegative integer. For $\nu \geq k$,

$$\gamma_k(\theta_\nu) \leq \gamma_\nu(\theta_\nu) = \frac{\gamma_\nu(\phi_\nu)}{\nu\gamma_\nu(\phi_\nu)} = \frac{1}{\nu}$$

Thus, for each k, the sequence $\{\gamma_k(\theta_\nu)\}_{\nu=1}^{\infty}$ converges to zero and, consequently, $\{\theta_\nu\}_{\nu=1}^{\infty}$ converges in $\mathfrak{D}_I$ to zero.

Since f is a continuous functional on $\mathfrak{D}_I$, the sequence $\{\langle f, \theta_\nu \rangle\}_{\nu=1}^{\infty}$ converges to zero. But (5) implies that

$$|\langle f, \theta_\nu \rangle| = \frac{|\langle f, \phi_\nu \rangle|}{\nu\gamma_\nu(\phi_\nu)} > 1$$

This contradiction establishes (4).

We can show that C and r depend in general upon f and I by constructing some examples.

For a given I, the distributions

$$f(t) = n\delta^{(n)}(t - \tau) \qquad \tau \in I; n = 0, 1, 2, \ldots$$

show that there exist no fixed C and r that will allow (4) to hold for all distributions.

Similarly, if I is allowed to increase indefinitely for a given f, the values of C and r may have to increase indefinitely, as is shown by the following example of a distribution:

$$f(t) = \sum_{n=1}^{\infty} n\delta^{(n)}(t - n) \tag{6}$$

That this is truly a distribution follows from the fact that for every ϕ in $\mathfrak{D}$

$$\langle f, \phi \rangle = \sum_n n(-1)^n \phi^{(n)}(n)$$

where now the summation possesses only a finite number of nonzero terms since ϕ has a bounded support. This completes the proof of Theorem 3.3-1.

The inequalities (3) and Theorem 3.3-1 show that locally (i.e., for I a fixed finite closed interval and for ϕ in $\mathfrak{D}_I$) the value of $\langle f, \phi \rangle$ depends only upon ϕ and its derivatives up to the rth order in the following sense. We can choose the testing function ϕ such that

$$\sup_t |\phi^{(k)}(t)| < \varepsilon \qquad k = 0, 1, \ldots, r$$

whereas

$$\sup_t |\phi^{(r+1)}(t)| > \eta$$

where ε can be chosen as small as we wish and η can be chosen as large as we wish. Then, by Theorem 3.3-1,

$$|\langle f, \phi \rangle| \leq C \sup_t |\phi^{(r)}(t)| < C\varepsilon$$

A more precise statement of this property will be established in Sec. 3.5.

PROBLEMS

1 Show that, if f is a fixed distribution on $\mathfrak{R}^1$ with a bounded support, there exist a nonnegative integer r and a constant C such that (4) holds for every ϕ in $\mathfrak{D}$, where now r and C do not depend on the support of ϕ, unlike the situation in Theorem 3.3-1. Then, show that f is in $\mathfrak{E}'$, the space defined in Sec. 1.3, Prob. 3.

2 Let t be n-dimensional, let $\mathfrak{D}_I$ be the space of testing functions whose supports are contained in the bounded domain I of $\mathfrak{R}^n$, and let the seminorms for $\mathfrak{D}_I$ be defined by

$$\gamma_m(\phi) = \sup_{t \in \mathfrak{R}^n, k \leq m} |D^k \phi(t)| \qquad m = 0, 1, 2, \ldots$$

where the notation for the partial derivative D^k is defined in Sec. 1.5. Generalize (4) by showing that for a given f in $\mathfrak{D}'$ and any ϕ in $\mathfrak{D}_I$,

$$|\langle f, \phi \rangle| \leq C \gamma_r(\phi)$$

where the constant C and the integer r are independent of the choice of ϕ.

♦3.4 LOCALLY EVERY DISTRIBUTION IS A FINITE-ORDER DERIVATIVE OF A CONTINUOUS FUNCTION

By using the results of the preceding section, we shall now show that locally every distribution can be obtained by differentiating (in the distributional sense) a continuous function a finite number of times. More precisely, if I is the fixed finite closed interval $a \leq t \leq b$ and if $\mathfrak{D}_I$ is the space of testing functions in $\mathfrak{D}$ whose supports are in I, then for each ϕ in $\mathfrak{D}_I$ and for a given distribution f defined over a neighborhood of I, we have

$$\langle f, \phi \rangle = (-1)^r \int_a^b h(t) \phi^{(r+2)}(t) \, dt \\ = \langle h, (-1)^r \phi^{(r+2)} \rangle = \langle h^{(r+2)}, \phi \rangle \tag{1}$$

where $h(t)$ is a continuous function and r is the integer specified in Theorem 3.3-1. We shall first prove this result in the case where r is zero by using an argument due to Pietsch (see Pietsch [1].) Then, (1) will be extended to the case where $r > 0$ by exploiting the results of Sec. 2.6, where we discussed the primitive of a distribution.

This analysis will employ some properties, given by (2) to (4) below, of the function

$$u(x, t) = \begin{cases} \dfrac{(t - b)(x - a)}{b - a} & a \leq x \leq t \leq b \\ \dfrac{(x - b)(t - a)}{b - a} & a \leq t \leq x \leq b \\ 0 & \text{elsewhere} \end{cases}$$

where x and t are one-dimensional real variables. The function $u(x, t)$ is continuous for all x and t and is zero when either x or t lies outside I or on the boundary of I. Since

$$\frac{\partial u}{\partial t} = \frac{x - a}{b - a} \qquad a \leq x < t \leq b$$

and

$$\frac{\partial u}{\partial t} = \frac{x - b}{b - a} \qquad a \leq t < x \leq b$$

it follows that

$$\left|\frac{\partial u}{\partial t}\right| \leq 1 \qquad t \neq x$$

and that

$$|u(x, t) - u(x, \tau)| \leq |t - \tau|$$
$$a \leq t \leq b, a \leq \tau \leq b, -\infty < x < \infty \quad (2)$$

By substituting $\tau = a$ into (2), we get

$$|u(x, t)| \leq t - a \leq b - a \tag{3}$$

Finally, every ϕ in $\mathfrak{D}_I$ satisfies the integral equation

$$\phi(x) = \int_a^b u(x, t) \, \phi^{(2)}(t) \, dt \tag{4}$$

This can be shown by breaking the interval of integration into $a \leq t \leq x$ and $x \leq t \leq b$ and then integrating by parts twice in both of the resulting integrals.

Now, for a fixed I, let $\mathfrak{D}_{0I}$ be the linear space of all continuous functions $j(t)$ on $\mathfrak{R}^1$ that are identically zero outside I. A sequence of functions $\{j_\nu\}_{\nu=1}^\infty$ is said to converge in $\mathfrak{D}_{0I}$ to a limit function j if each j_ν is in $\mathfrak{D}_{0I}$ and if $\sup_t |j(t) - j_\nu(t)| \to 0$ as $\nu \to \infty$. In other words, the functions $j_\nu(t)$ in $\mathfrak{D}_{0I}$ converge uniformly to $j(t)$. It follows that $j(t)$ will also be in $\mathfrak{D}_{0I}$.

To each function j in $\mathfrak{D}_{0I}$ we may assign a number $\|j\|$, which is called its *norm* and is defined by

$$\|j\| \triangleq \sup_t |j(t)|$$

Clearly, the sequence $\{j_\nu(t)\}_{\nu=1}^\infty$ will converge in $\mathfrak{D}_{0I}$ to $j(t)$ if and only if each j_ν is in $\mathfrak{D}_{0I}$ and the sequence of norms $\{\|j_\nu - j\|\}_{\nu=1}^\infty$ converges to zero.

Lemma 1

$\mathfrak{D}_I$ *is dense in* $\mathfrak{D}_{0I}$. *That is, for any given j in $\mathfrak{D}_{0I}$ and an arbitrary $\varepsilon > 0$, there exists a function ϕ in $\mathfrak{D}_I$ such that*

$$\|\phi - j\| = \sup_t |\phi(t) - j(t)| < \varepsilon$$

Proof: This proof is quite similar to the argument, given in Sec. 1.2, which showed that j could be approximated uniformly for all t by a testing function in $\mathfrak{D}$. Here, in addition, we must show that the approximating testing function has its support contained in I.

Let I be the interval $a \leq t \leq b$, let I_α be the interval $a + \alpha \leq t \leq b - \alpha$, let $I_{2\alpha}$ be the interval $a + 2\alpha \leq t \leq b - 2\alpha$, and let $I_{3\alpha}$ be the interval $a + 3\alpha \leq t \leq b - 3\alpha$, where $0 < \alpha < (b - a)/6$. Also, let $\gamma_\alpha(t)$ be the nonnegative testing function given by Sec. 1.2, Eq. (2), which we repeat here:

$$\gamma_\alpha(t) = \frac{\zeta(t/\alpha)}{\int_{-\infty}^\infty \zeta(t/\alpha)\, dt}$$

where

$$\zeta(t) = \begin{cases} 0 & |t| \geq 1 \\ \exp \dfrac{1}{t^2 - 1} & |t| < 1 \end{cases}$$

The integral of $\gamma_\alpha(t)$ over any interval that contains its support is equal to 1.

Finally, let

$$\phi_\alpha(t) = \int_{a+\alpha}^{b-\alpha} \gamma_\alpha(t - \tau) j(\tau)\, d\tau \tag{5}$$

This integral may be differentiated in the ordinary sense under the integral sign any number of times (Titchmarsh [1], p. 59), which shows that $\phi_\alpha(t)$ is infinitely smooth. Moreover, $\phi_\alpha(t)$ is zero outside I since, for $t \leq a$ or $t \geq b$, $\gamma_\alpha(t - \tau)$ will be zero for $a + \alpha \leq \tau \leq b - \alpha$. Hence, $\phi_\alpha(t)$ is in $\mathfrak{D}_I$.

Furthermore, for t in $I_{2\alpha}$

$$|j(t) - \phi_\alpha(t)| = \left| \int_{a+\alpha}^{b-\alpha} \gamma_\alpha(t-\tau)[j(t) - j(\tau)]\, d\tau \right|$$
$$\leq \sup_{\substack{t \in I_{2\alpha} \\ |t-\tau|<\alpha}} |j(t) - j(\tau)| \qquad (6)$$

The last "sup" denotes the maximum value of $|j(t) - j(\tau)|$, where t varies over $I_{2\alpha}$ and, for each fixed t, τ varies over those values for which $|t - \tau| < \alpha$.

For t in $I - I_{2\alpha}$ (that is, for $a \leq t < a + 2\alpha$ or $b - 2\alpha < t \leq b$),

$$|\phi_\alpha(t)| \leq \sup_{\tau \in I - I_{3\alpha}} |j(\tau)|$$

and

$$|j(t)| \leq \sup_{\tau \in I - I_{3\alpha}} |j(\tau)|$$

from which we get

$$|j(t) - \phi_\alpha(t)| \leq 2 \sup_{\tau \in I - I_{3\alpha}} |j(\tau)| \qquad (7)$$

By the uniform continuity of $j(\tau)$ and the fact that $j(a) = j(b) = 0$, we can conclude from (6) and (7) the following. Given an $\varepsilon > 0$, there exists a δ such that, for $0 < \alpha < \delta$,

$$|j(t) - \phi_\alpha(t)| < \varepsilon$$

This completes the proof of the lemma.

It also follows that, as α converges to zero, the $\phi_\alpha(t)$ converge in $\mathfrak{D}_{0I}$ to $j(t)$.

Now let f be a distribution over $\mathfrak{R}^1$ for which the integer r, specified in Theorem 3.3-1, is zero. That is, for every ϕ in $\mathfrak{D}_I$,

$$|\langle f, \phi \rangle| \leq C \sup_t |\phi(t)| \qquad (8)$$

Since $\mathfrak{D}_I$ is dense in $\mathfrak{D}_{0I}$, f can be extended continuously and uniquely onto the space $\mathfrak{D}_{0I}$ in the following way. Let ϕ be an arbitrary element in $\mathfrak{D}_{0I}$ and let $\{\phi_\nu\}_{\nu=1}^\infty$ be a sequence of elements in $\mathfrak{D}_I$ that converges in $\mathfrak{D}_{0I}$ to ϕ. Then, we define $\langle f, \phi \rangle$ by

$$\langle f, \phi \rangle \triangleq \lim_{\nu \to \infty} \langle f, \phi_\nu \rangle$$

That the right-hand side exists follows from (8) and the fact that

$$|\langle f, \phi_\nu \rangle - \langle f, \phi_\mu \rangle| = |\langle f, \phi_\nu - \phi_\mu \rangle| \leq C \sup_t |\phi_\nu(t) - \phi_\mu(t)| \to 0$$

as ν and μ go to infinity separately.

This extension of f will still satisfy the inequality (8) for every ϕ in $\mathfrak{D}_{0I}$. Moreover, since f is a continuous linear functional on $\mathfrak{D}_I$, it will also be continuous and linear over $\mathfrak{D}_{0I}$. Here, the continuity of f over $\mathfrak{D}_{0I}$ means the following. For each sequence $\{\phi_\nu\}_{\nu=1}^\infty$ that converges in $\mathfrak{D}_{0I}$ to ϕ, the sequence of numbers $\{\langle f, \phi_\nu\rangle\}_{\nu=1}^\infty$ converges to $\langle f, \phi\rangle$. (The proof of the linearity and continuity of f on $\mathfrak{D}_{0I}$ is left as an exercise.)

Let us now observe that, as a function of x, $u(x, t)$ is in $\mathfrak{D}_{0I}$ for each fixed value of t. Hence, we can define a function of t through

$$h(t) = \langle f(x), u(x, t)\rangle$$

This function is zero for $t \leq a$ and for $t \geq b$, since $u(x, t)$ is zero for these values of t. Moreover, by the use of (2) and (8), we have

$$|h(t) - h(\tau)| = |\langle f(x), u(x, t) - u(x, \tau)\rangle|$$
$$\leq C \sup_x |u(x, t) - u(x, \tau)|$$
$$\leq C|t - \tau| \qquad a \leq t \leq b, a \leq \tau \leq b$$

Thus, $h(t)$ is a continuous function for all t and $h(a) = h(b) = 0$.

We are finally ready to show that for each ϕ in $\mathfrak{D}_I$

$$\langle f, \phi\rangle = \int_a^b h(t)\phi^{(2)}(t)\, dt \tag{9}$$

Decompose I into a set of subintervals with end points $a \triangleq t_0, t_1, \ldots, t_n \triangleq b$. Since $u(x, t)$, $\phi^{(2)}(t)$, and $h(t)$ are uniformly continuous with respect to t, the decomposition of I can be made so fine that, on every subinterval $t_{i-1} \leq t \leq t_i$ ($i = 1, 2, \ldots, n$), the difference between any two values of $h(t)\phi^{(2)}(t)$ or of $u(x, t)\phi^{(2)}(t)$ becomes less than some given $\varepsilon > 0$. So, choosing a point τ_i from each such subinterval (that is, $t_{i-1} \leq \tau_i \leq t_i$), we see from (4) that

$$\left| \phi(x) - \sum_{i=1}^n u(x, \tau_i)\phi^{(2)}(\tau_i)(t_i - t_{i-1}) \right| \leq \varepsilon(b - a)$$

As a function of x, the quantity inside the magnitude signs on the left-hand side is an element in $\mathfrak{D}_{0I}$. By applying the continuous linear functional f to this element and using (8) (which, as stated above, holds over $\mathfrak{D}_{0I}$), we get

$$\left| \langle f, \phi\rangle - \sum_{i=1}^n h(\tau_i)\phi^{(2)}(\tau_i)(t_i - t_{i-1}) \right|$$
$$\leq C \sup_x \left| \phi(x) - \sum_{i=1}^n u(x, \tau_i)\phi^{(2)}(\tau_i)(t_i - t_{i-1}) \right|$$
$$\leq C\varepsilon(b - a) \tag{10}$$

Moreover, for the afore-mentioned decomposition of I,

$$\left| \int_a^b h(t)\phi^{(2)}(t)\, dt - \sum_{i=1}^n h(\tau_i)\phi^{(2)}(\tau_i)(t_i - t_{i-1}) \right| \leq \varepsilon(b-a) \tag{11}$$

By using (10) and (11), we finally obtain the inequality

$$\left| \langle f, \phi \rangle - \int_a^b h(t)\phi^{(2)}(t)\, dt \right| \leq \varepsilon(C+1)(b-a)$$

Thus, since ε may be chosen arbitrarily small, we have obtained the desired representation (9).

Since ϕ is identically zero for t outside I, we may use our customary notation to represent (9) by

$$\langle f, \phi \rangle = \langle h, \phi^{(2)} \rangle = \langle h^{(2)}, \phi \rangle \qquad \phi \in \mathfrak{D}_I$$

[Because $h(t)$ is continuous for all t, $h^{(2)}(t)$ can contain delta functionals but no derivatives of delta functionals. Prove this.]

As a summary of these arguments, we have

Theorem 3.4-1

Let I be a fixed finite closed interval in $\mathfrak{R}^1$ and let f be a distribution defined over a neighborhood of I. Also, assume that, for every ϕ in $\mathfrak{D}_I$,

$$|\langle f, \phi \rangle| \leq C \sup_t |\phi(t)|$$

where C is a constant independent of ϕ. Then,

$$\langle f, \phi \rangle = \langle h^{(2)}, \phi \rangle \qquad \phi \in \mathfrak{D}_I \tag{12}$$

where $h(t)$ is a continuous function for all t.

Alternatively, we may say that, over the open interval $a < t < b$, f equals $h^{(2)}$.

Note that in our construction

$$h(t) = 0 \qquad t \leq a,\, t \geq b \tag{13}$$

However, in (12) we can replace $h(t)$ by $h(t) + k_0 + k_1 t$, where k_0 and k_1 are arbitrary constants. Thus, the restriction (13) is not necessary for the representation (12).

This result can be extended to the case where the lowest possible value for the integer r of Theorem 3.3-1 is greater than zero. The general result is

Theorem 3.4-2

Locally, every distribution is a finite-order derivative of a continuous function. More precisely, let I be a fixed finite closed interval in $\mathfrak{R}^1$ and let f be a distribution defined over a neighborhood of I. There exist a nonnegative integer r (as specified in Theorem 3.3-1) and a continuous function $h(t)$ such that

$$\langle f, \phi \rangle = \langle h^{(r+2)}, \phi \rangle \tag{14}$$

for every ϕ in $\mathfrak{D}_I$.

Here again the continuous function $h(t)$ may be so chosen that it is identically zero outside I, but this is not necessary.

Proof: We shall construct an inductive proof, but first we shall establish a certain property of the primitives of f.

According to Theorem 3.3-1, there exists an r such that

$$|\langle f, \phi \rangle| \leq C \sup_t |\phi^{(r)}(t)| \qquad \phi \in \mathfrak{D}_I \tag{15}$$

where C is a constant independent of ϕ. Let $\mathfrak{K}_I$ be the space of all derivatives of testing functions in $\mathfrak{D}_I$. $\mathfrak{K}_I$ is a proper subspace of $\mathfrak{D}_I$. Also, let $f^{(-1)}$ denote a particular first-order primitive of f. Since

$$\langle f, \phi \rangle = \langle f^{(-1)}, -\phi^{(1)} \rangle$$

we may set $\chi = -\phi^{(1)}$ to obtain from (15)

$$|\langle f^{(-1)}, \chi \rangle| \leq C \sup_t |\chi^{(r-1)}(t)| \tag{16}$$

where χ is an arbitrary testing function in $\mathfrak{K}_I$. We shall now show that (16) can be extended to

$$|\langle f^{(-1)}, \phi \rangle| \leq C_1 \sup_t |\phi^{(r-1)}(t)| \tag{17}$$

where ϕ is an arbitrary testing function in $\mathfrak{D}_I$ and C_1 is a constant independent of ϕ.

As was discussed in Sec. 2.6, the primitive $f^{(-1)}$ is defined over $\mathfrak{D}_I$ as follows. Choose a fixed ϕ_0 in $\mathfrak{D}_I$ that satisfies

$$\int_{-\infty}^{\infty} \phi_0(t)\, dt = 1$$

and choose a fixed value K for the arbitrary parameter $\langle f^{(-1)}, \phi_0 \rangle$. Next, decompose any ϕ in $\mathfrak{D}_I$ into

$$\phi = k\phi_0 + \chi$$

where

$$k = \int_{-\infty}^{\infty} \phi(t)\, dt$$

Here, χ will be in $\mathcal{K}_I$. Then, the definition of $f^{(-1)}$ over $\mathfrak{D}_I$ is

$$\langle f^{(-1)}, \phi \rangle \triangleq \langle f^{(-1)}, k\phi_0 + \chi \rangle$$
$$= kK + \langle f^{(-1)}, \chi \rangle \qquad \phi \in \mathfrak{D}_I$$

We may employ (16) and the fact that

$$|k| \leq (b-a) \sup_t |\phi(t)| = (b-a) \sup_t \left| \int_a^t \phi^{(1)}(x)\, dx \right|$$
$$\leq (b-a)^2 \sup_t |\phi^{(1)}(t)| \leq \cdots \leq (b-a)^r \sup_t |\phi^{(r-1)}(t)|$$

to establish (17), as follows:

$$|\langle f^{(-1)}, \phi \rangle| \leq |Kk| + C \sup_t |\chi^{(r-1)}(t)|$$
$$= |Kk| + C \sup_t |\phi^{(r-1)}(t) - k\phi_0^{(r-1)}(t)|$$
$$\leq |K|(b-a)^r \sup_t |\phi^{(r-1)}(t)| + C \sup_t |\phi^{(r-1)}(t)|$$
$$\quad + C(b-a)^r \sup_t |\phi^{(r-1)}(t)| \sup_t |\phi_0^{(r-1)}(t)|$$
$$= C_1 \sup_t |\phi^{(r-1)}(t)|$$

where C_1 is a constant independent of ϕ. This establishes (17).

We now proceed with our inductive proof. If (15) holds for $r = 1$, then (17) also holds for $r = 1$ and we may invoke Theorem 3.4-1 to write

$$\langle f, \phi \rangle = \langle f^{(-1)}, -\phi^{(1)} \rangle = \langle h^{(2)}, -\phi^{(1)} \rangle$$
$$= \langle h^{(3)}, \phi \rangle \tag{18}$$

where h is the usual continuous function. This shows that the assertion "(15) implies (14)" is true for $r = 1$.

More generally, assume that the assertion "(15) implies (14)" has been established for r replaced by $r - 1$. If (15) holds for the given distribution f, then (17), which has the form of (15) with r replaced by $r - 1$, also is valid and our preceding assertion allows us to write

$$\langle f, \phi \rangle = \langle f^{(-1)}, -\phi^{(1)} \rangle = \langle h^{(r+1)}, -\phi^{(1)} \rangle$$
$$= \langle h^{(r+2)}, \phi \rangle$$

This completes the proof of Theorem 3.4-2.

We shall now show that, if f has a bounded support, it is a finite-order derivative of a continuous function not only locally but also for $-\infty < t < \infty$.

Corollary 3.4-2a

Let f be a distribution defined over $\mathfrak{R}^1$ with a bounded support. There exist a nonnegative integer r and a continuous function $h(t)$ such that $f(t) = h^{(r+2)}(t)$ for all t.

Proof: Let Ω denote the support of f, which by hypothesis is a bounded set. Let I_1 be an open interval $c < t < d$ that contains Ω and let I be the closed interval $c - \alpha \leq t \leq d + \alpha$, where α is a fixed positive number.

By the procedure described in Sec. 1.2, a testing function λ in $\mathfrak{D}$ that equals one over I_1 and is identically zero outside I can be constructed. Then, for every ϕ in $\mathfrak{D}$, $\lambda\phi$ is also in $\mathfrak{D}$ and has a support contained in I. Since $\langle f, \phi \rangle$ depends only on the values of ϕ in a neighborhood of the support Ω of f (as was pointed out in Sec. 1.8), $\langle f, \phi \rangle = \langle f, \lambda\phi \rangle$. Hence, by Theorem 3.4-2,

$$\langle f, \phi \rangle = \langle p^{(r+2)}, \lambda\phi \rangle = \langle \lambda p^{(r+2)}, \phi \rangle$$

where $p(t)$ is a continuous function for all t.

Now we shall set $h(t)$ equal to one of the $(r + 2)$nd-order primitives of $\lambda p^{(r+2)}$. To show that $h(t)$ is also a continuous function, first note that formula (9), Sec. 2.6, for integration by parts (with $c = 0$) yields

$$\begin{aligned}(\lambda p^{(r+2)})^{(-1)} &= \lambda p^{(r+1)} - (\lambda^{(1)} p^{(r+1)})^{(-1)} \\ &= \lambda p^{(r+1)} - \lambda^{(1)} p^{(r)} + \lambda^{(2)} p^{(r-1)} \\ &\quad - \cdots + (-1)^{r+1} \lambda^{(r+1)} p + (-1)^{r+2} (\lambda^{(r+2)} p)^{(-1)}\end{aligned} \quad (19)$$

We may take another primitive of (19) and apply the same sort of expansion to each term on the right-hand side of (19) that contains a derivative of p. On continuing this process, we see that $h \triangleq (\lambda p^{(r+2)})^{(-r-2)}$ is a finite linear combination of the following distributions:

$$(\lambda^{(k)} p)^{(-k)} \qquad k = 0, 1, \ldots, r + 2$$

Since $\lambda^{(k)} p$ is a continuous function, so are all its primitives. Hence, h is continuous. Q.E.D.

These results lead to the concept of the *order of a distribution*. For a given I and a given f, there will be a smallest nonnegative integer r_m such that, for all ϕ in $\mathfrak{D}_I$,

$$\langle f, \phi \rangle = \langle h^{(r_m+2)}, \phi \rangle \tag{20}$$

where h is some continuous function. This r_m is the order of the distribution on the interval I. (This definition is not consistent in the literature. For other definitions, see Schwartz [1], vol. I, p. 26, and Halperin [1], p. 10.) Thus, on any interval the delta functional is of order zero, whereas on any interval that contains the origin in its interior the mth derivative of the delta functional is of order m. [The reader should not infer that the order r_m is also the smallest nonnegative integer r that can be used in Sec. 3.3, Eq. (4). There exist distributions for which r_m is smaller than the smallest permissible r for Sec. 3.3, Eq. (4).]

Sec. 3.4   Further properties of distributions   95

We can also speak of the order of a distribution on all of $\mathcal{R}^1$. Every distribution has either a finite order or an infinite order on $\mathcal{R}^1$. When the order is finite, it is defined as the smallest nonnegative integer r_m for which (20) holds for all ϕ in $\mathcal{D}$. If no such r_m exists, the order of f is infinite.

According to Corollary 3.4-2a, every distribution with a bounded support is of finite order on $\mathcal{R}^1$. On the other hand, there are distributions without bounded support which are still of finite order on $\mathcal{R}^1$. Any regular distribution is an example of such a zero-order distribution. Finally, an example of an infinite-order distribution is given by Sec. 3.3, Eq. (6).

PROBLEMS

1 Let f be a given distribution over $\mathcal{R}^1$ that satisfies (8) for every ϕ in $\mathcal{D}_I$, I being a fixed finite closed interval. Let f be extended continuously onto $\mathcal{D}_{0I}$. Show that this extension is a continuous linear functional on $\mathcal{D}_{0I}$.

2 What is the order on $\mathcal{R}^1$ of the following distributions? In (d) and (e), q denotes a positive integer.

a. $\operatorname{Pv} \dfrac{1}{t}$

b. $\operatorname{Pf} \dfrac{1_+(t)}{t^{\frac{3}{2}}}$

c. $\operatorname{Pf} \tan t$

d. $\displaystyle\sum_{\nu=0}^{\infty} \nu^q e^{i\nu t}$

e. $\displaystyle\int_{-\infty}^{\infty} \tau^q e^{-i\tau t} \, d\tau$

3 a. We asserted (just before the statement of Theorem 3.4-1) that, if $h(t)$ is continuous for all t, then $h^{(2)}(t)$ can contain delta functionals but no derivatives of delta functionals. Prove this.

b. Verify that the continuous function $h(t)$ described in Theorem 3.4-2 can always be so chosen that it is zero outside the interval I.

4 Find a continuous function $h(t)$ such that for $-1 < t < 1$

$$\operatorname{Pf} \frac{1_+(t)}{t} = h^{(2)}(t)$$

in two ways. First, integrate twice by parts the definition of Pf $1_+(t)/t$. Second, evaluate

$$h(t) = \left\langle \operatorname{Pf} \frac{1_+(t)}{t}, u(x,t) \right\rangle$$

[$u(x,t)$ was defined at the beginning of this section.] Show that the two results you obtain are the same except for a linear additive term. Then, find two other continuous functions whose second and third derivatives are Pv $1/t$ and Pf $1_+(t)/t^2$, respectively.

5 By constructing an example, show that the function h and the integer r of Theorem 3.4-2 depend, in general, upon the choice of the interval I.

6 Let I denote the interval $a \leq t \leq b$. Assume that, over neighborhoods of the points $t = a$ and $t = b$, the (in general) singular distribution $f(t)$ is regular and corresponds to a continuous function. We may define the definite integral of $f(t)$ over $a \leq t \leq b$ by

$$\int_a^b f(t)\,dt \triangleq h^{(r+1)}(b) - h^{(r+1)}(a) \tag{21}$$

where $h(t)$ is the continuous function specified in Theorem 3.4-2. Show that this definition is equivalent to that given in Sec. 2.6, Prob. 4. Then, compute the definite integrals of that problem by using (21).

7 Show that, if f is of order m for all t, then f is in $\mathfrak{D}'_{m+2}$, where $\mathfrak{D}'_{m+2}$ is defined in Sec. 1.3, Prob. 2.

8 As an application of the Hahn-Banach theorem (Taylor [1], p. 186), one can say that, if a linear functional f is defined over some linear subspace $\mathfrak{IC}$ of $\mathfrak{D}_I$ and satisfies (8) for all ϕ in $\mathfrak{IC}$, then it can be extended onto all of $\mathfrak{D}_I$ in such a way that (8) is still satisfied for all ϕ in $\mathfrak{D}_I$. Use this result in conjunction with Theorem 3.4-1 to prove Theorem 3.4-2 in a brief way.

3.5 ONLY FINITE LINEAR COMBINATIONS OF THE DELTA FUNCTIONAL AND ITS DERIVATIVES ARE CONCENTRATED ON A POINT

We showed in Sec. 1.8 that for a given distribution f the value $\langle f, \phi \rangle$ ($\phi \in \mathfrak{D}$) depends only on the values that ϕ assumes over a neighborhood of the support of f. We cannot say, however, that $\langle f, \phi \rangle$ depends only on the values that ϕ assumes just on the support of f. As an example, $\langle \delta^{(1)}, \phi \rangle = -\phi^{(1)}(0)$ and $\phi^{(1)}(0)$ is not determined by $\phi(0)$. It is instead determined by the values of $\phi(t)$ over some neighborhood of $t = 0$. Nevertheless, on any fixed finite closed interval I and for ϕ in $\mathfrak{D}_I$, $\langle f, \phi \rangle$ is specified by the values that $\phi, \phi^{(1)}, \ldots, \phi^{(r_m+2)}$ assume on only the support of f in I. Here, r_m is the order of f on I. (We have already indicated a looser form of this assertion at the end of Sec. 3.3.)

In order to show this, let us first note that, on any open interval $c < t < d$ that is contained in I and in the null set N of f (assuming that some of the null set does occur in I), f equals the zero distribution. In the representation (20) of Sec. 3.4 the function $h(t)$ is one of the $(r_m + 2)$nd-order primitives of $f(t)$ and is therefore, on $c < t < d$, equal to a polynomial in t of degree no larger than $r_m + 1$. Since the null set N of any distribution is an open set, it can be expressed as the union of a finite or denumerably infinite set of nonoverlapping open intervals, which we denote by $\{c_i < t < d_i\}$ (Titchmarsh [1], pp. 321–322). Furthermore, all the end points c_i and d_i must be in the support of f, since N is open. Hence, for all ϕ in $\mathfrak{D}_I$

$$\langle f, \phi \rangle = \sum_i (-1)^{r_m} \int_{c_i}^{d_i} h(t) \phi^{(r_m+2)}(t) \, dt$$

$$+ (-1)^{r_m} \int_{I-N} h(t) \phi^{(r_m+2)}(t) \, dt \quad (1)$$

where the second integration is over the set of points $I - N$ consisting of the support points of f in I. By repeatedly integrating by parts and using the fact that, over each interval of N, $h(t)$ is a polynomial in t of degree no larger than $r_m + 1$, we see that the integrals over the intervals $c_i < t < d_i$ depend only on the values $\phi^{(\nu)}(c_i)$ and $\phi^{(\nu)}(d_i)$ ($\nu = 0, 1, \ldots, r_m + 1$), so far as their dependence on ϕ is concerned. Moreover, the integral over $I - N$ depends upon the values of $\phi^{(r_m+2)}$ on $I - N$. Thus, we have demonstrated

Theorem 3.5-1

For any given distribution f over $\mathfrak{R}^1$ and for any finite closed interval I in $\mathfrak{R}^1$, the values of $\langle f, \phi \rangle$ for all ϕ in $\mathfrak{D}_I$ depend only on the values that $\phi, \phi^{(1)}, \ldots, \phi^{(r_m+2)}$ assume on $I - N$, the support of f in I. Here, r_m is the order of f on I.

In certain cases, it is possible to strengthen this theorem by terminating the sequence $\{\phi^{(\nu)}\}$ after the r_mth-order derivative rather than after the $(r_m + 2)$nd-order derivative, as we shall see.

If f has a bounded support, we can choose I large enough to contain a neighborhood of the support of f and then conclude (through an argument similar to the proof of Corollary 3.4-2a) that Theorem 3.5-1 holds true for all ϕ in $\mathfrak{D}$. Thus, if a distribution f is concentrated on a single point, say τ, $\langle f, \phi \rangle$ is a linear combination of the values $\phi(\tau), \phi^{(1)}(\tau), \ldots, \phi^{(r_m+2)}(\tau)$ for each ϕ in $\mathfrak{D}$, where r_m now denotes the order of f on $\mathfrak{R}^1$.

We can improve this conclusion by eliminating $\phi^{(r_m+1)}(\tau)$ and $\phi^{(r_m+2)}(\tau)$, as follows. When f is concentrated on τ, our previous arguments show that $\langle f, \phi \rangle$ is represented by

$$\langle f, \phi \rangle = \int_{-\infty}^{\tau} p_L(t) \phi^{(r_m+2)}(t) \, dt + \int_{\tau}^{\infty} p_R(t) \phi^{(r_m+2)}(t) \, dt \quad (2)$$

Here, $p_L(t)$ and $p_R(t)$ are polynomials in t of degree no larger than $r_m + 1$. By the continuity of $h(t)$, $p_L(\tau) = p_R(\tau)$. By employment of this condition and repeated integration by parts, (2) becomes

$$\langle f, \phi \rangle = [p_R^{(1)}(\tau) - p_L^{(1)}(\tau)] \phi^{(r_m)}(\tau) - [p_R^{(2)}(\tau) - p_L^{(2)}(\tau)] \phi^{(r_m-1)}(\tau)$$
$$+ \cdots + (-1)^{r_m} [p_R^{(r_m+1)}(\tau) - p_L^{(r_m+1)}(\tau)] \phi(\tau)$$

This means that f is a finite linear combination of the shifted delta functional and its derivatives up to and including the order r_m. This is a consequence of the simple assumption that f is concentrated on a point. The converse of this is obviously true, and thus we have

Theorem 3.5-2

A necessary and sufficient condition for a distribution $f(t)$ on $\mathfrak{R}^1$ to have a support consisting of a single point τ is that it be a finite sum

$$f(t) = \sum_{\nu=0}^{r_m} a_\nu \delta^{(\nu)}(t - \tau)$$

where the a_ν are constants and $a_{r_m} \neq 0$. Here, r_m is the order of the distribution.

PROBLEMS

1 Demonstrate the following. A necessary and sufficient condition for a distribution on $\mathfrak{R}^1$ to have a support consisting of isolated points (i.e., around each support point there is a neighborhood in which no other support points exist) is that the distribution be a finite or infinite linear combination of the delta functional and its derivatives shifted onto these points such that in each open interval only a finite number of nonzero terms occur.

2 For an unknown distribution f, let r of Theorem 3.3-1 be zero. By using Theorem 3.2-1 show that f has just the origin as its support only if it equals $C\delta(t)$, where C is some constant. HINT: By the argument given in Sec. 3.4, f is a continuous linear functional on the space $\mathfrak{D}_{0I}$. Show that, for any fixed ϕ in $\mathfrak{D}$ and for I a closed interval that includes the origin and the support of ϕ, a sequence of continuous functions in $\mathfrak{D}_{0I}$ can be so chosen that each function has its support contained in the null set of f and that the sequence converges uniformly to $t\phi(t)$. Then show that (1) of Sec. 3.2 is satisfied.

4 Distributions of Slow Growth

★4.1 INTRODUCTION

An important type of distribution, namely, the distribution of slow growth, arises naturally in the development of the distributional Fourier and Laplace transformations. The distributions of slow growth comprise a proper subspace of $\mathcal{D}'$, but on the other hand, they can be defined as continuous linear functionals on a class of testing functions that is wider than $\mathcal{D}$. This extended class of testing functions is discussed in the next section, and the distributions of slow growth are defined in Sec. 4.3. The bulk of the results that are important to us are obtained in these next two sections. Sections 4.4 and 4.5 are devoted to two special theorems that we shall occasionally utilize.

★4.2 THE SPACE $\mathcal{S}$ OF TESTING FUNCTIONS OF RAPID DESCENT

As usual, let $t \triangleq \{t_1, t_2, \ldots, t_n\}$ be the n-dimensional real variable and let $|t|$ denote

$$\sqrt{t_1{}^2 + t_2{}^2 + \cdots + t_n{}^2}$$

$\mathfrak{S}$ is the space of all complex-valued functions $\phi(t)$ that are infinitely smooth and are such that, as $|t| \to \infty$, they and all their partial derivatives decrease to zero faster than every power of $1/|t|$.

This required behavior as $|t| \to \infty$ can also be stated in the following alternative way. When t is one-dimensional, every function $\phi(t)$ in $\mathfrak{S}$ satisfies the infinite set of inequalities

$$|t^m \phi^{(k)}(t)| \leq C_{mk} \qquad -\infty < t < \infty \tag{1}$$

where m and k run through all nonnegative integers. Here the C_{mk} are constants (with respect to t) which depend upon m and k. When t is n-dimensional, the requirement is that, for every set of nonnegative integers $m, k_1, k_2, \ldots, k_n$,

$$|t|^m \left| \frac{\partial^{k_1+k_2+\cdots+k_n}}{\partial t_1{}^{k_1} \partial t_2{}^{k_2} \cdots \partial t_n{}^{k_n}} \phi(t_1, t_2, \ldots, t_n) \right| \leq C_{mk_1k_2\cdots k_n} \tag{2}$$

over all of $\mathfrak{R}^n$, where the quantity on the right-hand side of (2) is a constant with respect to t but depends upon the choices of the $m, k_1, \ldots, k_n$. Because of the continuity of all the partial derivatives of $\phi(t)$, the order of differentiation in (2) may be changed in any fashion.

For the sake of simplicity, we shall use the symbolism

$$k \triangleq \{k_1, k_2, \ldots, k_n\}$$

and

$$D^k \triangleq \frac{\partial^{k_1+k_2+\cdots+k_n}}{\partial t_1 \, \partial t_2 \, \cdots \, \partial t_n}$$

and we shall replace (2) by the shorthand notation

$$|t|^m |D^k \phi(t)| \leq C_{mk} \tag{3}$$

The elements of $\mathfrak{S}$ are called *testing functions of rapid descent*. $\mathfrak{S}$ is a linear space. (See Appendix A.) If ϕ is in $\mathfrak{S}$, every one of its partial derivatives is again in $\mathfrak{S}$. Furthermore, all the testing functions in $\mathfrak{D}$ are also in $\mathfrak{S}$. However, there are testing functions in $\mathfrak{S}$ that are not in $\mathfrak{D}$ as, for example,

$$\exp(-t_1{}^2 - t_2{}^2 - \cdots - t_n{}^2)$$

Thus, $\mathfrak{D}$ is a proper subspace of $\mathfrak{S}$.

A sequence of functions $\{\phi_\nu(t)\}_{\nu=1}^\infty$ is said to *converge in* $\mathfrak{S}$ if every function $\phi_\nu(t)$ is in $\mathfrak{S}$ and if, for each set of nonnegative integers m and $k \triangleq \{k_1, k_2, \ldots, k_n\}$, the sequence $\{|t|^m D^k \phi_\nu(t)\}_{\nu=1}^\infty$ converges uniformly over all of $\mathfrak{R}^n$. The convergence must be uniform for any given set of integers m and k, but it need not be uniform over all such sets of integers.

A slightly different but entirely equivalent way to define convergence in $\mathcal{S}$ of a sequence $\{\phi_\nu(t)\}_{\nu=1}^\infty$ is to require:

1. That every $\phi_\nu(t)$ be in $\mathcal{S}$.
2. That for any given set of nonnegative integers m and k

$$|t|^m |D^k \phi_\nu(t)| \leq C_{mk} \tag{4}$$

for all ν and all t in $\mathfrak{R}^n$, where the constants C_{mk} are independent of t and ν.

3. That for each k the sequence $\{D^k \phi_\nu(t)\}_{\nu=1}^\infty$ converge uniformly over every bounded domain of $\mathfrak{R}^n$.

By either of these definitions it follows that the limit function $\phi(t)$ of $\{\phi_\nu(t)\}_{\nu=1}^\infty$ will also be in $\mathcal{S}$, so that $\mathcal{S}$ is closed under convergence. This is because the uniformity of the convergences ensures that every partial derivative $D^k \phi(t)$ exists, is continuous everywhere, and is the limit of $\{D^k \phi_\nu(t)\}_{\nu=1}^\infty$ and because (4) will still be satisfied by $\phi(t)$.

Also, $\{\phi_\nu(t)\}_{\nu=1}^\infty$ will converge in $\mathcal{S}$ to $\phi(t)$ if and only if $\{\phi_\nu(t) - \phi(t)\}_{\nu=1}^\infty$ converges in $\mathcal{S}$ to the zero function and every $\phi_\nu(t)$ is in $\mathcal{S}$.

Another useful fact is that convergence in $\mathfrak{D}$ implies convergence in $\mathcal{S}$. For if $\{\phi_\nu(t)\}_{\nu=1}^\infty$ converges in $\mathfrak{D}$, all the ϕ_ν are identically zero outside some bounded domain in $\mathfrak{R}^n$, say, outside the circle $|t| = T$. Hence, for each k and m and with ϕ as the limit function

$$|t|^m |D^k \phi_\nu(t) - D^k \phi(t)| \tag{5}$$

is identically zero for $|t| \geq T$ and is therefore bounded by

$$T^m \sup_t |D^k \phi_\nu(t) - D^k \phi(t)|$$

over all of $\mathfrak{R}^n$. The convergence in $\mathfrak{D}$ implies that the last set of numbers converges to zero as $\nu \to \infty$. Thus, (5) converges to zero uniformly over $\mathfrak{R}^n$. The conditions for convergence in $\mathcal{S}$ are thereby fulfilled.

Theorem 4.2-1

The space $\mathfrak{D}$ is dense in the space $\mathcal{S}$ in the sense that for each ϕ in $\mathcal{S}$ there exists a sequence $\{\phi_\nu(t)\}_{\nu=1}^\infty$ with every $\phi_\nu(t)$ in $\mathfrak{D}$ which converges in $\mathcal{S}$ to $\phi(t)$.

Proof: Let

$$\xi(t) = \begin{cases} 0 & |t| \geq 1 \\ \exp \dfrac{|t|^2}{|t|^2 - 1} & |t| < 1 \end{cases}$$

and consider $\xi(t/\nu)$. Upon setting $\lambda_k(t) = D^k\xi(t)$, we have that

$$D^k\xi\left(\frac{t}{\nu}\right) = \nu^{-\hat{k}}\lambda_k\left(\frac{t}{\nu}\right)$$

where $\hat{k} \triangleq k_1 + k_2 + \cdots + k_n$. Now, $\{\xi(t/\nu)\}_{\nu=1}^{\infty}$ converges to one and, for $\hat{k} \geq 1$, $\{D^k\xi(t/\nu)\}_{\nu=1}^{\infty}$ converges to zero, these convergences being uniform over every bounded domain of $\mathfrak{R}^n$. Furthermore, $\xi(t/\nu)$ and $D^k\xi(t/\nu)$ are uniformly bounded with respect to ν over all of $\mathfrak{R}^n$.

Now, let $\phi(t)$ be an arbitrary testing function in $\mathcal{S}$ and set

$$\phi_\nu(t) = \phi(t)\xi\left(\frac{t}{\nu}\right)$$

For every ν, $\phi_\nu(t)$ is in $\mathfrak{D}$. Also, for m an arbitrary nonnegative integer, $|t|^m\phi(t)$ is bounded over all of $\mathfrak{R}^n$. Thus $|t|^m\phi(t)\xi(t/\nu)$ is uniformly bounded with respect to ν over all of $\mathfrak{R}^n$ and, as $\nu \to \infty$, these functions converge uniformly over every bounded domain of $\mathfrak{R}^n$ to $|t|^m\phi(t)$.

Moreover, for $\hat{k} \geq 1$, $D^k[\phi(t)\xi(t/\nu)]$ is a sum of terms one of which is $\xi(t/\nu)D^k\phi(t)$ and all the others of which are of the form $D^p\xi(t/\nu)D^q\phi(t)$, where $\hat{p} \triangleq p_1 + p_2 + \cdots + p_n \geq 1$ and $p + q = k$. Since $|t|^m D^q\phi(t)$ is bounded over all of $\mathfrak{R}^n$, $|t|^m D^k[\phi(t)\xi(t/\nu)]$ is uniformly bounded with respect to ν over all of $\mathfrak{R}^n$ and, as $\nu \to \infty$, the corresponding sequence converges uniformly over every bounded domain of $\mathfrak{R}^n$ to $|t|^m D^k\phi(t)$. Thus, $\{\phi_\nu(t)\}_{\nu=1}^{\infty}$ truly converges in $\mathcal{S}$ to $\phi(t)$, as was asserted.

PROBLEMS

1 By constructing examples, show that the C_{mk} in (1) cannot be replaced by a single constant for which (1) will hold for all pairs of nonnegative integers m and k.

2 Show that the two definitions of convergence in $\mathcal{S}$, given in this section, are equivalent.

3 Convince yourself that $\mathcal{S}$ is closed under the following operations. [That is, for $\phi(t)$ in $\mathcal{S}$, show that the resulting function is also in $\mathcal{S}$.]
 a. Shifting by the real amount τ: $\phi(t + \tau)$
 b. Transposition: $\phi(-t)$
 c. Multiplication of the independent variable by a positive constant a: $\phi(at)$
 d. Multiplication by a polynomial $p(t)$: $p(t)\phi(t)$

★4.3 THE SPACE $\mathcal{S}'$ OF DISTRIBUTIONS OF SLOW GROWTH

A distribution f is said to be *of slow growth* if it is a continuous linear functional on the space $\mathcal{S}$ of testing functions of rapid descent. (Such distributions are also called *temperate* or *tempered* distributions.) That is, a distribution f of slow growth is a rule that assigns a number $\langle f, \phi \rangle$ to each ϕ in $\mathcal{S}$ in such a way that the following conditions are fulfilled.

Linearity If ϕ_1 and ϕ_2 are in $\mathcal{S}$ and if α is a number, then

$$\langle f, \phi_1 + \phi_2 \rangle = \langle f, \phi_1 \rangle + \langle f, \phi_2 \rangle$$

and

$$\langle f, \alpha\phi_1 \rangle = \alpha \langle f, \phi_1 \rangle$$

Continuity If $\{\phi_\nu\}_{\nu=1}^\infty$ is any sequence that converges in $\mathcal{S}$ to zero, then

$$\lim_{\nu \to \infty} \langle f, \phi_\nu \rangle = 0$$

(As usual, in this continuity requirement we may replace the sequences $\{\phi_\nu\}_{\nu=1}^\infty$ by nondenumerable directed sets that converge in $\mathcal{S}$ to zero.)

The space of all distributions of slow growth is denoted by $\mathcal{S}'$. $\mathcal{S}'$ is also called the *dual* (or *conjugate*) *space of* $\mathcal{S}$.

Assume still that f denotes a distribution of slow growth. Since $\mathcal{D}$ is a subspace of $\mathcal{S}$, $\langle f, \phi \rangle$ is defined whenever ϕ is in $\mathcal{D}$. f is clearly linear as a functional on $\mathcal{D}$. Also, convergence in $\mathcal{D}$ implies convergence in $\mathcal{S}$ and $\{\langle f, \phi_\nu \rangle\}_{\nu=1}^\infty$ therefore converges to zero whenever $\{\phi_\nu\}_{\nu=1}^\infty$ converges in $\mathcal{D}$ to zero. Thus, f is also a distribution in $\mathcal{D}'$. Furthermore, a knowledge of $\langle f, \phi \rangle$, as ϕ traverses only the space $\mathcal{D}$, uniquely determines $\langle f, \phi \rangle$ for all ϕ in $\mathcal{S}$. This is because $\mathcal{D}$ is dense in $\mathcal{S}$ and each $\langle f, \phi \rangle$ ($\phi \in \mathcal{S}$) is the limit of every sequence $\{\langle f, \phi_\nu \rangle\}_{\nu=1}^\infty$, where the ϕ_ν are all in $\mathcal{D}$ and converge in $\mathcal{S}$ to ϕ. In summary, $\mathcal{S}'$ *is a subspace of* $\mathcal{D}'$; *moreover, if* f_1 *and* f_2 *are both in* $\mathcal{S}'$ *and if* $\langle f_1, \phi \rangle = \langle f_2, \phi \rangle$ *for every* ϕ *in* $\mathcal{D}$, *then* $\langle f_1, \phi \rangle = \langle f_2, \phi \rangle$ *for every* ϕ *in* $\mathcal{S}$.

$\mathcal{S}'$ is a *proper* subspace of $\mathcal{D}'$; that is, there are distributions in $\mathcal{D}'$ that are not in $\mathcal{S}'$. For instance, the series

$$g(t) = \sum_{\mu=1}^\infty e^{\mu^2} \delta(t - \mu) \qquad t \in R' \tag{1}$$

defines a distribution in $\mathcal{D}'$. Indeed, given any ϕ in $\mathcal{D}$,

$$\langle g, \phi \rangle = \sum_{\mu=1}^\infty e^{\mu^2} \phi(\mu) \tag{2}$$

The last series possesses only a finite number of nonzero terms and therefore converges. On the other hand, there are testing functions in $\mathcal{S}$, such as $\phi(t) = \exp(-t^2)$, for which the series (2) does not converge. Hence, (1) is not a distribution of slow growth.

We shall show in Sec. 5.6 that $\mathcal{S}'$ is dense in $\mathcal{D}'$.

In order for a locally integrable function $f(t)$ to assign a finite number $\langle f, \phi \rangle$ to every testing function ϕ in $\mathcal{S}$ through the expression

$$\langle f, \phi \rangle \triangleq \int_{-\infty}^{\infty} f(t)\phi(t)\, dt \tag{3}$$

the behavior of $f(t)$ as $|t| \to \infty$ must be so restricted that the integral converges for all ϕ in $\mathcal{S}$. This is certainly assured if $f(t)$ satisfies the condition

$$\lim_{t \to \infty} |t|^{-N} f(t) = 0 \tag{4}$$

for some integer N. Functions that satisfy (4) are said to be *functions of slow growth*. Every locally integrable function of slow growth defines a regular distribution of slow growth through (3). The proof of this is left as an exercise.

Since each testing function in $\mathcal{S}$ certainly satisfies (4), it generates a regular distribution of slow growth.

Another fact, which can be readily proved, is that *every distribution in $\mathcal{D}'$ with a bounded support is of slow growth*. Thus, the delta functional and its derivatives are distributions of slow growth.

A mode of convergence is assigned to the space $\mathcal{S}'$ as follows. A sequence of distributions $\{f_\nu\}_{\nu=1}^{\infty}$ is said to *converge in* $\mathcal{S}'$ if every f_ν is in $\mathcal{S}'$ and if, for every ϕ in $\mathcal{S}$, the sequence of numbers $\{\langle f_\nu, \phi \rangle\}_{\nu=1}^{\infty}$ converges. The limits of all such sequences of numbers, as ϕ traverses $\mathcal{S}$, define a functional on $\mathcal{S}$. Clearly, if $\{f_\nu\}_{\nu=1}^{\infty}$ converges in $\mathcal{S}'$, it also converges in $\mathcal{D}'$, since $\mathcal{D} \subset \mathcal{S}$ and $\mathcal{S}' \subset \mathcal{D}'$. By the fact that $\mathcal{D}'$ is closed under convergence (see Theorem 2.2-1), the limit f of $\{f_\nu\}_{\nu=1}^{\infty}$ is also a distribution in $\mathcal{D}'$. The stronger conclusion, that f is in $\mathcal{S}'$, is also true. In other words, *the space $\mathcal{S}'$ is also closed under convergence*. The proof of this fact is almost identical to that of Theorem 2.2-1. [The only change is that sequences of testing functions must now converge in $\mathcal{S}$. For instance, if a sequence converges in $\mathcal{S}$ to zero, a subsequence $\{\phi_\nu\}_{\nu=1}^{\infty}$ can be so chosen that it satisfies the inequalities

$$|t|^m |D^k \phi_\nu| \leq \frac{1}{4^\nu} \qquad k, m = 0, 1, \ldots, \nu$$

which replace those of (4) in Sec. 2.2.]

As usual, an infinite series is said to converge in $\mathcal{S}'$ if its partial sums converge in $\mathcal{S}'$. Also, the above definition and all the subsequent discussions that apply to sequences $\{f_\nu\}_{\nu=1}^{\infty}$ convergent in $\mathcal{S}'$ can be extended to nondenumerable directed sets $\{f_\nu\}_{\nu \to \infty}$.

There is a one-to-one correspondence between the testing functions in $\mathcal{S}$ and the regular distributions in $\mathcal{S}'$ that are generated by them. We can therefore identify such testing functions and distributions and say that $\mathcal{S}$ is a subspace of $\mathcal{S}'$. Moreover, if a sequence of testing functions converges in $\mathcal{S}$, the corresponding sequence of regular distributions converges in $\mathcal{S}'$. For if θ is any element of $\mathcal{S}$ and if $\{\phi_\nu\}_{\nu=1}^\infty$ converges in $\mathcal{S}$ to ϕ, then

$$|\langle \phi_\nu, \theta \rangle - \langle \phi, \theta \rangle| \leq \int_{-\infty}^{\infty} |\phi_\nu(t) - \phi(t)| \, |\theta(t)| \, dt \to 0$$

as $\nu \to \infty$ because the $|\phi_\nu(t) - \phi(t)|$ converge uniformly to zero for $-\infty < t < \infty$.

As a summary of the results of this nature that we have so far established for the spaces $\mathcal{D}$, $\mathcal{D}'$, $\mathcal{S}$, and $\mathcal{S}'$, let us note that

$$\mathcal{D} \subset \mathcal{S} \subset \mathcal{S}' \subset \mathcal{D}' \tag{5}$$

and that, when a sequence converges in one of these spaces, it also converges in every space occurring in (5) to the right of the original space.

Since $\mathcal{S}'$ is a subspace of $\mathcal{D}'$, it follows that all the operations that were defined for distributions in $\mathcal{D}'$ also apply to distributions in $\mathcal{S}'$. However, the application of some operation to a distribution in $\mathcal{S}'$ need not result in a distribution that is also in $\mathcal{S}'$. When a given operation does produce distributions of slow growth from distributions of slow growth, the space $\mathcal{S}'$ is said to be *closed under that operation*. The following is a list of such operations. The section in which the operation was first defined is listed in parentheses. These definitions should now be adjusted to allow the testing functions to traverse $\mathcal{S}$ rather than just $\mathcal{D}$.

1. Addition of distributions (Sec. 1.7)
2. Multiplication of a distribution by a constant (Sec. 1.7)
3. Shifting of a distribution (Sec. 1.7)
4. Transposition of a distribution (Sec. 1.7)
5. Multiplication of the independent variable by a positive constant (Sec. 1.7)
6. Differentiation of a distribution (Sec. 2.4)
7. Construction of a primitive (Sec. 2.6)

Note that $\mathcal{S}'$ is a linear space because it is closed under the first two operations listed here and because the axioms listed in Appendix A are satisfied.

That these operations carry any distribution of slow growth into another one of slow growth is readily shown. For instance, consider differentiation and let f be in $\mathcal{S}'$. If ϕ is in $\mathcal{S}$, then $\partial \phi / \partial t_i$ is also in $\mathcal{S}$.

For ϕ_1 and ϕ_2 in $\mathcal{S}$ and for any two numbers α and β,

$$\left\langle \frac{\partial f}{\partial t_i}, \alpha\phi_1 + \beta\phi_2 \right\rangle = \left\langle f, -\alpha\frac{\partial \phi_1}{\partial t_i} - \beta\frac{\partial \phi_2}{\partial t_i} \right\rangle$$
$$= \alpha\left\langle f, -\frac{\partial \phi_1}{\partial t_i} \right\rangle + \beta\left\langle f, -\frac{\partial \phi_2}{\partial t_i} \right\rangle$$
$$= \alpha\left\langle \frac{\partial f}{\partial t_i}, \phi_1 \right\rangle + \beta\left\langle \frac{\partial f}{\partial t_i}, \phi_2 \right\rangle$$

Thus, $\partial f/\partial t_i$ is a linear functional on $\mathcal{S}$.

$\partial f/\partial t_i$ is also a continuous functional on $\mathcal{S}$. If $\{\phi_\nu\}_{\nu=1}^\infty$ converges in $\mathcal{S}$ to zero, then $\{\partial \phi_\nu/\partial t_i\}_{\nu=1}^\infty$ also converges in $\mathcal{S}$ to zero. Thus, as $\nu \to \infty$

$$\left\langle \frac{\partial f}{\partial t_i}, \phi_\nu \right\rangle = \left\langle f, -\frac{\partial \phi_\nu}{\partial t_i} \right\rangle \to 0$$

since f is continuous on $\mathcal{S}$. We have thereby shown that $\partial f/\partial t_i$ is also in $\mathcal{S}'$, which is what we wished to do.

Moreover, differentiation is a continuous linear operation on the space $\mathcal{S}'$. By linearity we mean as usual that, if f and g are in $\mathcal{S}'$ and α and β are any two numbers,

$$\frac{\partial}{\partial t_i}(\alpha f + \beta g) = \alpha\frac{\partial f}{\partial t_i} + \beta\frac{\partial g}{\partial t_i}$$

By continuity we mean that, if $\{f_\nu\}_{\nu=1}^\infty$ converges in $\mathcal{S}'$ to f, then $\{\partial f_\nu/\partial t_i\}_{\nu=1}^\infty$ also converges in $\mathcal{S}'$ to $\partial f/\partial t_i$. Indeed, the last assertion follows from the relation

$$\left\langle \frac{\partial f_\nu}{\partial t_i}, \phi \right\rangle = \left\langle f_\nu, -\frac{\partial \phi}{\partial t_i} \right\rangle \to \left\langle f, -\frac{\partial \phi}{\partial t_i} \right\rangle = \left\langle \frac{\partial f}{\partial t_i}, \phi \right\rangle \qquad \phi \in \mathcal{S}$$

An operation under which $\mathcal{S}'$ is not closed is the multiplication of a distribution by a function that is infinitely smooth. For example, let t be one-dimensional and consider

$$f(t) = \sum_{\mu=1}^\infty \delta(t - \mu) \tag{6}$$

This distribution is in $\mathcal{S}'$. (Show this.) Multiplying it by $\exp t^2$, we get (1), which, as we noted before, is not a distribution of slow growth.

However, if we restrict somewhat the class of infinitely smooth functions, the space $\mathcal{S}'$ will be closed under multiplication by these functions. In particular, let $\mathcal{O}_M$ be the space of all infinitely smooth functions such that they and all their derivatives are of slow growth. That is, $\theta(t)$ is in $\mathcal{O}_M$ if it is infinitely smooth and if for each n-tuple $k = \{k_1, \ldots, k_n\}$ of

Sec. 4.3   ***Distributions of slow growth***

nonnegative integers there exists an integer N_k for which

$$\frac{D^k \theta(t)}{(1 + |t|^2)^{N_k}}$$

is bounded for all t.

Now, let θ be a fixed function in $\mathcal{O}_M$. If ϕ is a testing function in $\mathcal{S}$, then $\theta\phi$ is clearly in $\mathcal{S}$. Furthermore, if $\{\phi_\nu\}_{\nu=1}^{\infty}$ converges in $\mathcal{S}$ to zero, then $\{\theta\phi_\nu\}_{\nu=1}^{\infty}$ also converges in $\mathcal{S}$ to zero.

We can now conclude that, *if f is in $\mathcal{S}'$, then θf is also in $\mathcal{S}'$*. Indeed, for ϕ in $\mathcal{S}$, $\langle \theta f, \phi \rangle = \langle f, \theta \phi \rangle$, so that θf is defined as a functional on $\mathcal{S}$. By the comments of the preceding paragraph, it is readily shown that f is both a linear and a continuous functional on $\mathcal{S}$. Thus, we see that *the space $\mathcal{S}'$ is closed under the operation of multiplication by functions in* $\mathcal{O}_M$.

A useful criterion for convergence in $\mathcal{S}'$, which corresponds somewhat to that of Corollary 2.4-3c, is stated by

Theorem 4.3-1

Let $\{g_\nu(t)\}_{\nu=1}^{\infty}$ be a sequence of distributions with the following properties:

a. There exists an n-tuple $k = \{k_1, \ldots, k_n\}$ of nonnegative integers such that $g_\nu(t)$ equals $D^k f_\nu(t)$ ($\nu = 1, 2, 3, \ldots$), where the $f_\nu(t)$ are ordinary continuous functions (more precisely, the regular distributions corresponding to such functions) which are all bounded in magnitude by a fixed polynomial $p(t)$.

b. The sequence of functions $\{f_\nu(t)\}_{\nu=1}^{\infty}$ converges to a limit function $f(t)$ uniformly over every bounded t domain.

Then, $\{g_\nu(t)\}_{\nu=1}^{\infty}$ converges in $\mathcal{S}'$ to $D^k f(t)$.

Proof: By the uniform convergence of the $f_\nu(t)$ on every bounded t domain, the limit function $f(t)$ is continuous everywhere. Moreover, the fact that $|f_\nu(t)| \leq p(t)$ ($\nu = 1, 2, \ldots$) implies that $|f(t)| \leq p(t)$. Certainly, then, f is in $\mathcal{S}'$.

Now, for any fixed ϕ in $\mathcal{S}$,

$$|\langle f - f_\nu, \phi \rangle| = \left| \int_{\mathcal{R}^n} [f(t) - f_\nu(t)] \phi(t)\, dt \right|$$
$$\leq \int_{\mathcal{R}^n} |f(t) - f_\nu(t)| |\phi(t)|\, dt \leq 2 \int_{\mathcal{R}^n} p(t)|\phi(t)|\, dt < \infty$$

Hence, given an $\varepsilon > 0$, there exists a $T > 0$ such that

$$\left| \int_{|t|>T} [f(t) - f_\nu(t)] \phi(t)\, dt \right| < \varepsilon$$

for all ν. There also exists an integer N such that, for $\nu \geq N$,

$$\left| \int_{|t|<T} [f(t) - f_\nu(t)] \phi(t)\, dt \right| \leq \int_{|t|<T} |\phi(t)|\, dt \sup_{|t|<T} |f(t) - f_\nu(t)| < \varepsilon$$

We have shown that

$$\lim_{\nu \to \infty} \langle f_\nu, \phi \rangle = \langle f, \phi \rangle$$

In other words, $\{f_\nu\}_{\nu=1}^\infty$ converges in $\mathcal{S}'$ to f.

Our conclusion now follows by the continuity of differentiation on the space $\mathcal{S}'$.   Q.E.D.

PROBLEMS

1 Show that (6) is a distribution of slow growth.

2 Is

$$\sum_{\nu=0}^\infty \delta^{(\nu)}(t - \nu) \qquad t \in R'$$

a distribution of slow growth?

3 Show that (3) defines a regular distribution of slow growth whenever $f(t)$ is a locally integrable function of slow growth.

4 $e^t \cos e^t$, where $t \in R'$, is not a function of slow growth. Show that it is a distribution of slow growth. This example shows that a regular distribution of slow growth need not correspond to a function of slow growth. However, it is a fact that every distribution of slow growth (singular as well as regular) can be obtained by differentiating in the distributional sense some continuous function of slow growth a finite number of times. (See Schwartz [1], vol. II, p. 95.)

5 Is the pseudofunction Pv $1/t$ a distribution of slow growth? Prove your answer.   Here, $t \in R'$.

6 Show that the space $\mathcal{S}'$ is closed under the first five operations listed in this section. Are the second through the fifth operations linear and continuous on the space $\mathcal{S}'$? (The linearity and continuity of these operations are defined in the same way as they are for differentiation.)

7 In the case where the independent variable has but one dimension, show that $\mathcal{S}'$ is closed under the construction of a primitive. HINT: Show that any ϕ in $\mathcal{S}$ can be decomposed uniquely according to Lemma 1 of Sec. 2.6, where now χ and ϕ_0 are in general in $\mathcal{S}$. Then, $f^{(-1)}$ is defined by (7) of Sec. 2.6, where ψ is given by (5) of Sec. 2.6. Show that ψ is in $\mathcal{S}$. Finally, show that $f^{(-1)}$ is a continuous linear functional on $\mathcal{S}$.

8 Prove that the space $\mathcal{S}'$ is closed under convergence.

9 The hypothesis of Theorem 4.3-1 can be weakened considerably. For example, we need merely assume that all the $f_\nu(t)$ are locally integrable, are bounded in magnitude by $p(t)$, and converge to $f(t)$ almost everywhere. Prove this by invoking some standard theorems.

10 Let t be one-dimensional, let b be a real number, and let $f(t)$ be a distribution of bounded support. Show that

$$\sum_{\nu=-\infty}^{\infty} f(t - \nu b)$$

converges in $\mathcal{S}'$.

11 Show that $\mathcal{B}'$ is a proper subspace of $\mathcal{S}'$. (The space $\mathcal{B}'$ is defined in Sec. 1.3, Prob. 3.)

◆4.4 A BOUNDEDNESS PROPERTY FOR DISTRIBUTIONS OF SLOW GROWTH

We saw in Sec. 3.3 that every distribution f defined over $\mathcal{R}^1$ possesses the following property. For any given finite closed interval I there exist a nonnegative integer r and a constant C such that for all testing functions ϕ in $\mathcal{D}_I$

$$|\langle f, \phi \rangle| \leq C \sup_t |\phi^{(r)}(t)|$$

Here, C and r depend only on f and I and not on ϕ. This result is no longer true for all f in $\mathcal{D}'$ if we allow I to be an infinite interval. However, all distributions of slow growth do possess a boundedness property of this sort that holds over the infinite interval, and, moreover, in this case ϕ may be a testing function in $\mathcal{S}$ and not merely in $\mathcal{D}$. Namely, for a given f in $\mathcal{S}'$ and for all ϕ in $\mathcal{S}$

$$|\langle f, \phi \rangle| \leq C \sup_t |(1 + t^2)^r \phi^{(r)}(t)| \tag{1}$$

where the constant C and the nonnegative integer r depend only on f. The proof of (1) is quite similar to the analysis presented in Sec. 3.3. Here, however, we must construct a set of seminorms that is suitable for convergence in $\mathcal{S}$ rather than for convergence in $\mathcal{D}_I$.

We shall again restrict ourselves to the one-dimensional case for the independent variable. For each pair of nonnegative integers m and k let

$$\rho_{m,k} \triangleq \sup_t |(1 + t^2)^m \phi^{(k)}(t)| \qquad \phi \in \mathcal{S} \tag{2}$$

These finite nonnegative numbers satisfy the inequalities

$$\rho_{m,k} \leq \rho_{p,k} \qquad p = m, m+1, m+2, \ldots \tag{3}$$

$$\rho_{m,k} \leq \frac{\pi}{2} \rho_{m+1,k+1} \tag{4}$$

$$\rho_{m,k} \leq \rho_{m+2,k+3} \tag{5}$$

as we shall now show. The first one is obvious.

To establish (4), first assume that $t \leq 0$. Then,

$$\begin{aligned}
|(1+t^2)^m \phi^{(k)}(t)| &= \left|(1+t^2)^m \int_{-\infty}^{t} \phi^{(k+1)}(\tau)\, d\tau\right| \\
&\leq (1+t^2)^m \int_{-\infty}^{t} \frac{\rho_{m+1,k+1}}{(1+\tau^2)^{m+1}}\, d\tau \\
&\leq \rho_{m+1,k+1} \int_{-\infty}^{t} \frac{d\tau}{1+\tau^2} \leq \frac{\pi}{2}\rho_{m+1,k+1}
\end{aligned}$$

When $t \geq 0$, we may similarly write

$$\begin{aligned}
|(1+t^2)^m \phi^{(k)}(t)| &= \left|(1+t^2)^m \int_{t}^{\infty} \phi^{(k+1)}(\tau)\, d\tau\right| \\
&\leq \frac{\pi}{2}\rho_{m+1,k+1}
\end{aligned}$$

These inequalities imply (4).

The proof of (5) is similar. Upon integrating by parts twice, we have

$$\phi^{(k)}(t) = \tfrac{1}{2}\int_{-\infty}^{t} (t-\tau)^2 \phi^{(k+3)}(\tau)\, d\tau$$

and for $t \leq 0$ we may write

$$\begin{aligned}
|(1+t^2)^m \phi^{(k)}(t)| &= \left|\tfrac{1}{2}(1+t^2)^m \int_{-\infty}^{t} (t-\tau)^2 \phi^{(k+3)}(\tau)\, d\tau\right| \\
&\leq \tfrac{1}{2}(1+t^2)^m \int_{-\infty}^{t} (t-\tau)^2 \frac{\rho_{m+2,k+3}}{(1+\tau^2)^{m+2}}\, d\tau \\
&\leq \frac{\rho_{m+2,k+3}}{2}\int_{-\infty}^{t}\frac{\tau^2}{(1+\tau^2)^2}\, d\tau \\
&\leq \frac{\rho_{m+2,k+3}}{2}\int_{-\infty}^{t}\frac{d\tau}{1+\tau^2} \leq \rho_{m+2,k+3}
\end{aligned}$$

For $t \geq 0$ we can get the same result by working with

$$\phi^{(k)}(t) = \tfrac{1}{2}\int_{\infty}^{t} (t-\tau)^2 \phi^{(k+3)}(\tau)\, d\tau$$

This establishes (5).

Now, let n be an integer no less than m or k. For $m \leq k$, (3) and (4) imply that

$$\rho_{m,k} \leq \rho_{k,k} \leq \left(\frac{\pi}{2}\right)^n \rho_{n,n} \tag{6}$$

On the other hand, for $m > k$, (5) shows that

$$\rho_{m,k} \leq \rho_{m+2(m-k),k+3(m-k)} = \rho_{3m-2k,3m-2k} \tag{7}$$

Finally, (6) and (7) together imply that for any pair of nonnegative integers m and k,

$$\rho_{m,k} \leq \left(\frac{\pi}{2}\right)^n \rho_{n,n} \triangleq \gamma_n \qquad (8)$$

where now $n = \sup(k, 3m - 2k)$. In addition, according to (4),

$$0 \leq \gamma_0 \leq \gamma_1 \leq \gamma_2 \leq \cdots$$

The numbers $\gamma_n = \gamma_n(\phi)$ depend upon the choice of ϕ. They are called *seminorms* for the space $\mathcal{S}$. Moreover, in view of (8), convergence in $\mathcal{S}$ can be stated quite simply in terms of these seminorms. A sequence $\{\phi_\nu\}_{\nu=1}^\infty$ *converges in* $\mathcal{S}$ *to zero if and only if each* ϕ_ν *is in* $\mathcal{S}$ *and the sequence* $\{\gamma_n(\phi_\nu)\}_{\nu=1}^\infty$ *converges to zero for each nonnegative integer* n.

By using this result, we can establish the boundedness property for distributions in $\mathcal{S}'$, which we again state by the following theorem. Its proof is precisely the same as that of Theorem 3.3-1 except that now the seminorms

$$\gamma_n \triangleq \left(\frac{\pi}{2}\right)^n \sup_t |(1 + t^2)^n \phi^{(n)}(t)| \qquad \phi \in \mathcal{S}$$

should be employed.

Theorem 4.4-1

For each distribution f of slow growth there exist a constant C and a nonnegative integer r such that the inequality (1) is fulfilled for every ϕ in $\mathcal{S}$. C and r depend only on f and not on ϕ.

◆4.5 A DIFFERENTIABILITY PROPERTY FOR THE APPLICATION OF A DISTRIBUTION IN $\mathcal{S}'_\tau$ TO A TESTING FUNCTION IN $\mathcal{S}_{t,\tau}$

A result that we shall require subsequently is an analogue to Corollary 2.7-2a. We shall continue to assume that the real variables t and τ are one-dimensional. The symbols $\mathcal{S}_t$, $\mathcal{S}_\tau$, and $\mathcal{S}_{t,\tau}$ will denote the spaces of testing functions of rapid descent defined over the euclidean spaces of the variables t, τ, and (t, τ), respectively. Similarly, $\mathcal{S}'_t$, $\mathcal{S}'_\tau$, and $\mathcal{S}'_{t,\tau}$ will denote the spaces of distributions of slow growth defined over these same euclidean spaces, respectively.

Theorem 4.5-1

If $g(\tau)$ is in $\mathcal{S}'_\tau$ and $\phi(t, \tau)$ is in $\mathcal{S}_{t,\tau}$, then

$$\psi(t) \triangleq \langle g(\tau), \phi(t, \tau) \rangle$$

is in $\mathcal{S}_t$ *and*

$$\frac{d^k}{dt^k}\psi(t) = \left\langle g(\tau), \frac{\partial^k}{\partial t^k}\phi(t,\tau)\right\rangle$$

Proof: For some fixed t, consider

$$\frac{\psi(t+\Delta t)-\psi(t)}{\Delta t} - \left\langle g(\tau), \frac{\partial}{\partial t}\phi(t,\tau)\right\rangle = \langle g(\tau), \theta_{\Delta t}(\tau)\rangle \tag{1}$$

where

$$\theta_{\Delta t}(\tau) \triangleq \begin{cases} \dfrac{\phi(t+\Delta t, \tau) - \phi(t,\tau)}{\Delta t} - \dfrac{\partial}{\partial t}\phi(t,\tau) & \Delta t \neq 0 \\ 0 & \Delta t = 0 \end{cases} \tag{2}$$

We shall first show that, as $\Delta t \to 0$, $\theta_{\Delta t}(\tau)$ converges in $\mathcal{S}_\tau$ to zero. Indeed, for each fixed Δt, $\theta_{\Delta t}(\tau)$ is infinitely smooth for all τ. Also, $\theta_{\Delta t}(\tau)$ is continuous with respect to τ and Δt over every closed bounded domain $a \leq \tau \leq b$ and $|\Delta t| \leq c$, because $\phi(t,\tau)$ is infinitely smooth. Thus, $\theta_{\Delta t}(\tau)$ is *uniformly* continuous over the given domain. Furthermore, $\lim_{\Delta t \to 0} \theta_{\Delta t}(\tau) = 0$, and therefore $\theta_{\Delta t}(\tau)$ converges to zero uniformly over every finite τ interval as $\Delta t \to 0$. Similar reasoning shows that the derivatives $\theta_{\Delta t}^{(k)}(\tau)$ also converge to zero uniformly over every finite τ interval as $\Delta t \to 0$.

It is also a fact that for each pair of nonnegative integers m and k

$$\tau^m \frac{\partial^k}{\partial \tau^k} \theta_{\Delta t}(\tau) \tag{3}$$

is bounded for $-\infty < \tau < \infty$ and for Δt in any finite interval. This follows directly from (2) and the facts that $\phi(t,\tau)$ is in $\mathcal{S}_{t,\tau}$ and that

$$\left|\frac{\partial^k}{\partial \tau^k}[\phi(t+\Delta t, \tau) - \phi(t,\tau)]\right| \leq |\Delta t| \sup_{t-|\Delta t| \leq x \leq t+|\Delta t|} \left|\frac{\partial^{k+1}}{\partial \tau^k \partial x}\phi(x,\tau)\right|$$

Thus, $\theta_{\Delta t}(\tau)$ does truly converge in $\mathcal{S}_\tau$ to zero.

Since $g(\tau)$ is a continuous functional, (1) converges to zero. Hence,

$$\frac{d\psi}{dt} = \left\langle g(\tau), \frac{\partial \phi(t,\tau)}{\partial t}\right\rangle$$

Upon repeated application of this argument, it is seen that $\psi(t)$ is infinitely smooth and that

$$\frac{d^k \psi(t)}{dt^k} = \left\langle g(\tau), \frac{\partial^k \phi(t,\tau)}{\partial t^k}\right\rangle \tag{4}$$

We have yet to show that $\psi(t)$ is bounded according to (1) of Sec. 4.2.

Distributions of slow growth

For a given choice of the nonnegative integers p and q and for a fixed t, set

$$\lambda_t(\tau) \triangleq t^p \frac{\partial^q}{\partial t^q} \phi(t, \tau)$$

As t varies through all real values, an infinite set $\{\lambda_t(\tau)\}$ of testing functions in $\mathcal{S}_\tau$ is generated. Since $\phi(t,\tau)$ is in $\mathcal{S}_{t,\tau}$, $\lambda_t(\tau)$ satisfies for each nonnegative integer r the inequality

$$|(1 + \tau^2)^r \lambda_t^{(r)}(\tau)| \leq K_{rpq}$$

where the constant K_{rpq} is independent of t. It follows now from Theorem 4.4-1 that there exist a constant C and a fixed r such that

$$|t^p \psi^{(q)}(t)| = \left| \left\langle g(\tau), t^p \frac{\partial^q}{\partial t^q} \phi(t, \tau) \right\rangle \right| = |\langle g(\tau), \lambda_t(\tau) \rangle|$$
$$\leq C \sup_\tau |(1 + \tau^2)^r \lambda_t^{(r)}(\tau)| \leq C K_{rpq} \triangleq B_{pq}$$

where B_{pq} is a constant depending only on p and q. Thus, $\psi(t)$ does satisfy (1) of Sec. 4.2 and our proof is complete.

With some minor modifications the argument leading up to (4) remains valid when $\phi(t, \tau)$ is replaced by $\phi(t - \tau)$. [We need only revise the argument for the boundedness of (3) for all τ and for Δt in any finite interval.] Thus, the first part of this proof also establishes

Corollary 4.5-1a

If $g(\tau)$ is in $\mathcal{S}'_\tau$ and $\phi(x)$ is in $\mathcal{S}_x$, then $\psi(t) \triangleq \langle g(\tau), \phi(t - \tau) \rangle$ is a function that is infinitely smooth and

$$\frac{d^k \psi}{dt^k} = \left\langle g(\tau), \frac{\partial^k}{\partial t^k} \phi(t - \tau) \right\rangle$$

PROBLEM

1 Prove Corollary 4.5-1a by showing that (3) is bounded for all τ and for Δt in any finite interval.

5 Convolution

★5.1 INTRODUCTION

When considered from a distributional point of view, convolution turns out to be a very general process. Various types of differential equations, difference equations, and integral equations are all special cases of convolution equations. We shall investigate the convolution process in this chapter and take up the problem of solving convolution equations in the next one. The first two sections of this chapter are devoted to a discussion of the direct product of distributions, a concept that is used in the definition of distributional convolution.

From this point on through the rest of the book, we shall always restrict the various real independent variables, such as t, τ, x, y, and ω, to being one-dimensional. The symbols $\phi(t)$, $f(\tau)$, $h(x)$, etc., will denote functions or distributions that are defined over $\mathcal{R}^1$ or a portion of $\mathcal{R}^1$. Functions or distributions defined over part or all of $\mathcal{R}^2$ will be denoted by $f(t, \tau)$, $g(x, y)$, etc. If we do not specifically indicate the contrary, it will be understood that the functions and distributions under consideration are defined only over $\mathcal{R}^1$.

In order to specify the particular variables that comprise a euclidean space, we shall attach these variables as subscripts to the symbol $\Re$. For example, $\Re_t$ is the one-dimensional euclidean space consisting of all real values for t; $\Re_{x,y}$ is the two-dimensional euclidean space composed of all real pairs (x, y). Similarly, when such subscripts appear on the symbols for spaces of functions or distributions, they will denote the independent variables on which the elements of these spaces are defined. Thus, $\mathfrak{D}_\tau$ is the space $\mathfrak{D}$ of testing functions that are defined over $\Re_\tau$, and $\mathcal{S}'_{t,\tau}$ is the space $\mathcal{S}'$ of distributions of slow growth defined over $\Re_{t,\tau}$. (This notation has already been employed in the preceding sections.)

★5.2 THE DIRECT PRODUCT OF DISTRIBUTIONS

As was mentioned in the preceding section, the *direct product* (or *tensor product*) of distributions is an operation that arises in the development of convolution. In fact, the definition of convolution is based squarely on that of the direct product, and some properties of the direct product carry over to convolution. For this reason we shall discuss the direct product in some detail.

This product is an operation that combines a distribution $f(t)$ in $\mathfrak{D}'_t$ with a distribution $g(\tau)$ in $\mathfrak{D}'_\tau$ to obtain a distribution in $\mathfrak{D}'_{t,\tau}$, which is denoted by $f(t) \times g(\tau)$, in the following way. If $\phi(t, \tau)$ is an element of $\mathfrak{D}_{t,\tau}$, then $\langle g(\tau), \phi(t, \tau) \rangle$ is clearly a function of t. In fact, it is a testing function in $\mathfrak{D}_t$ according to Corollary 2.7-2a. Upon applying $f(t)$ to this testing function, we obtain the definition of the direct product:

$$\langle f(t) \times g(\tau), \phi(t, \tau) \rangle \triangleq \langle f(t), \langle g(\tau), \phi(t, \tau) \rangle \rangle \tag{1}$$

So far, (1) merely defines the direct product $f(t) \times g(\tau)$ as a functional on the space $\mathfrak{D}_{t,\tau}$. We shall now prove

Theorem 5.2-1

The direct product $f(t) \times g(\tau)$ of two distributions $f(t)$ and $g(\tau)$ is a distribution in $\mathfrak{D}'_{t,\tau}$.

Proof: Let us point out again that $\langle g(\tau), \phi(t, \tau) \rangle$ is in $\mathfrak{D}_t$ whenever $\phi(t, \tau)$ is in $\mathfrak{D}_{t,\tau}$, according to Corollary 2.7-2a. Clearly, $f(t) \times g(\tau)$ is a linear functional. In order to establish its continuity, we shall show that the sequence of numbers

$$\{\langle f(t), \langle g(\tau), \phi_\nu(t, \tau) \rangle \rangle\}_{\nu=1}^\infty$$

converges to zero whenever $\{\phi_\nu(t, \tau)\}_{\nu=1}^\infty$ is a sequence of testing functions that converges in $\mathfrak{D}_{t,\tau}$ to zero. Since $f(t)$ is a distribution (i.e., a *continuous* linear functional on $\mathfrak{D}_t$), this will certainly be the case if we can show that the testing functions

$$\psi_\nu(t) \triangleq \langle g(\tau), \phi_\nu(t, \tau) \rangle$$

converge in $\mathfrak{D}_t$ to zero.

Note that all the $\psi_\nu(t)$ have their supports contained within a fixed bounded interval in $\mathfrak{R}_t$, since the $\phi_\nu(t, \tau)$ have this property in $\mathfrak{R}_{t,\tau}$. It remains to show that the $\psi_\nu^{(k)}(t)$ uniformly converge to zero over all of $\mathfrak{R}_t$ for each fixed nonnegative integer k. Assume the opposite; then there is at least one k for which an $\varepsilon > 0$ and a sequence of points $\{t_\nu\}_{\nu=1}^\infty$ exist such that

$$|\psi_\nu^{(k)}(t_\nu)| \geq \varepsilon$$

By Corollary 2.7-2a, this can be written as

$$|\psi_\nu^{(k)}(t_\nu)| = \left| \left\langle g(\tau), \frac{\partial^k}{\partial t^k} \phi_\nu(t, \tau) \right|_{t=t_\nu} \right\rangle \right| \geq \varepsilon \tag{2}$$

On the other hand, under the assumption that $\{\phi_\nu(t, \tau)\}_{\nu=1}^\infty$ converges in $\mathfrak{D}_{t,\tau}$ to zero, the functions

$$\lambda_\nu(\tau) \triangleq \frac{\partial^k}{\partial t^k} \phi_\nu(t, \tau) \bigg|_{t=t_\nu}$$

which are testing functions in $\mathfrak{D}_\tau$, converge in $\mathfrak{D}_\tau$ to zero. Since $g(\tau)$ is a distribution,

$$\psi_\nu^{(k)}(t_\nu) = \langle g(\tau), \lambda_\nu(\tau) \rangle \to 0 \tag{3}$$

as $\nu \to \infty$. Equations (2) and (3) comprise a contradiction, and therefore this theorem is proved.

There is a corresponding result for distributions of slow growth. The direct product of two distributions of slow growth is still defined by (i), where now $\phi(t, \tau)$ traverses the space $\mathcal{S}_{t,\tau}$.

Theorem 5.2-2

The direct product of two distributions of slow growth is another distribution of slow growth. That is, if $f(t)$ is in $\mathcal{S}'_t$ and $g(\tau)$ is in $\mathcal{S}'_\tau$, then $f(t) \times g(\tau)$ is in $\mathcal{S}'_{t,\tau}$.

The proof of this theorem is almost identical to that of Theorem 5.2-1, except that now we rely on Theorem 4.5-1 rather than on Corollary 2.7-2a. Some simple examples follow.

Example 5.2-1 The direct product of the delta functional over $\mathfrak{R}^1$ with itself yields the delta functional over $\mathfrak{R}^2$. That is,

$$\delta(t) \times \delta(\tau) = \delta(t, \tau)$$

because with ϕ in $\mathfrak{D}_{t,\tau}$

$$\langle \delta(t) \times \delta(\tau), \phi(t, \tau)\rangle = \langle \delta(t), \phi(t, 0)\rangle = \phi(0, 0)$$
$$= \langle \delta(t, \tau), \phi(t, \tau)\rangle$$

Example 5.2-2 If $f(t)$ and $g(\tau)$ are two locally integrable functions, then their ordinary product $f(t)g(\tau)$ is locally integrable over $\mathfrak{R}_{t,\tau}$. Letting $f(t)$ and $g(\tau)$ also denote the regular distributions corresponding to these functions, we find that the direct product $f(t) \times g(\tau)$ is the regular distribution in $\mathfrak{D}'_{t,\tau}$ corresponding to the function $f(t)g(\tau)$. For, with ϕ in $\mathfrak{D}_{t,\tau}$

$$\langle f(t) \times g(\tau), \phi(t, \tau)\rangle = \langle f(t), \langle g(\tau), \phi(t, \tau)\rangle\rangle$$
$$= \int_{-\infty}^{\infty} f(t) \left[\int_{-\infty}^{\infty} g(\tau)\phi(t, \tau) \, d\tau \right] dt$$

Since $f(t)g(\tau)\phi(t, \tau)$ is locally integrable and of bounded support on $\mathfrak{R}_{t,\tau}$, we may write the last iterated integral as the following double integral (Kestelman [1], p. 206):

$$\int_{-\infty}^{\infty} \int_{-\infty}^{\infty} f(t)g(\tau)\phi(t, \tau) \, d\tau \, dt = \langle f(t)g(\tau), \phi(t, \tau)\rangle$$

This confirms our assertion.

This example also shows that, in this case at least, the direct product is commutative [that is, $f(t) \times g(\tau) = g(\tau) \times f(t)$], since by the same argument $g(\tau) \times f(t)$ is also the regular distribution corresponding to $f(t)g(\tau)$.

The next example shows another case where the direct product commutes.

Example 5.2-3

$$\delta(t) \times 1_+(\tau) = 1_+(\tau) \times \delta(t)$$

because, with ϕ in $\mathfrak{D}_{t,\tau}$,

$$\langle \delta(t) \times 1_+(\tau), \phi(t, \tau)\rangle = \left\langle \delta(t), \int_0^{\infty} \phi(t, \tau) \, d\tau \right\rangle = \int_0^{\infty} \phi(0, \tau) \, d\tau$$
$$= \langle 1_+(\tau), \phi(0, \tau)\rangle = \langle 1_+(\tau) \times \delta(t), \phi(t, \tau)\rangle$$

Indeed, it will be proved in the next section that the direct product is always commutative. In order to show this, we shall make use of the special form that the defining expression (1) takes when $\phi(t, \tau)$ is of the form $\psi(t)\theta(\tau)$, where ψ and θ are testing functions in $\mathfrak{D}_t$ and $\mathfrak{D}_\tau$, respectively. In this case, we have

$$\langle f(t) \times g(\tau), \psi(t)\theta(\tau)\rangle = \langle f(t), \langle g(\tau), \psi(t)\theta(\tau)\rangle\rangle$$
$$= \langle f(t), \psi(t)\langle g(\tau), \theta(\tau)\rangle\rangle = \langle f(t), \psi(t)\rangle\langle g(\tau), \theta(\tau)\rangle \quad (4)$$

By a similar manipulation, it follows that

$$\langle g(\tau) \times f(t), \psi(t)\theta(\tau)\rangle = \langle g(\tau), \theta(\tau)\rangle\langle f(t), \psi(t)\rangle$$
$$= \langle f(t) \times g(\tau), \psi(t)\theta(\tau)\rangle \tag{5}$$

PROBLEMS

1 By using Theorem 3.3-1 develop an alternative proof of the latter part of Theorem 5.2-1; i.e., show that $\{\psi_\nu(t)\}_{\nu=1}^\infty$ converges in $\mathfrak{D}_t$ to zero whenever $\{\phi_\nu(t,\tau)\}_{\nu=1}^\infty$ converges in $\mathfrak{D}_{t,\tau}$ to zero.

2 Prove Theorem 5.2-2.

3 Show that the direct product is linear in the following sense. If α and β are arbitrary numbers and if f_1, f_2, g_1, and g_2 are arbitrary distributions defined over $\mathfrak{R}^1$, then

$$f_1 \times (\alpha g_1 + \beta g_2) = \alpha f_1 \times g_1 + \beta f_1 \times g_2$$
$$(\alpha f_1 + \beta f_2) \times g_1 = \alpha f_1 \times g_1 + \beta f_2 \times g_1$$

4 Find an expression for

$$\left\langle \operatorname{Pf} \frac{1_+(t)}{t} \times \operatorname{Pf} \frac{1_+(\tau)}{\tau}, \phi(t,\tau) \right\rangle$$

in terms of certain integrations and limits on $\phi(t,\tau)$.

★5.3 THE SUPPORT, COMMUTATIVITY, AND ASSOCIATIVITY OF THE DIRECT PRODUCT

The support of the direct product If Ω_a is a set of points in $\mathfrak{R}_t$ and Ω_b is a set of points in $\mathfrak{R}_\tau$, their cartesian product $\Omega_a \times \Omega_b$ is defined to be the set of those points in $\mathfrak{R}_{t,\tau}$ for which t is in Ω_a and τ is in Ω_b.

Theorem 5.3-1

The support of the direct product of two distributions is the cartesian product of their supports. That is, if Ω_f is the support of the distribution $f(t)$ and Ω_g is the support of the distribution $g(\tau)$, the support of $f(t) \times g(\tau)$ is the set $\Omega_f \times \Omega_g$.

Proof: First let us assume that the point t_0 lies outside the support of $f(t)$. We wish to show that every point (t_0, τ_0) also lies outside the support of $f(t) \times g(\tau)$, no matter what value τ_0 takes. By the definition of the support of a distribution (Sec. 1.6), there will be some neighborhood Ω_{t_0} of the point t_0 such that $\langle f(t), \theta(t)\rangle$ will be zero for every testing function $\theta(t)$ whose support is contained within Ω_{t_0}. If $\phi(t, \tau)$ is any testing function whose support lies in the strip $\Omega_{t_0} \times \mathfrak{R}_\tau$ [that is, $\phi(t, \tau)$ is zero whenever t is not in Ω_{t_0}], then

$$\theta(t) \triangleq \langle g(\tau), \phi(t, \tau) \rangle$$

is a testing function of t, according to Corollary 2.7-2a, and its support is contained in Ω_{t_0}. Therefore,

$$\langle f(t) \times g(\tau), \phi(t, \tau) \rangle = \langle f(t), \theta(t) \rangle = 0$$

Hence, (t_0, τ_0) is not in the support of $f(t) \times g(\tau)$.

Similarly, if τ_0 is outside the support of $g(\tau)$, there is some neighborhood Ω_{τ_0} of τ_0 such that $\langle g(\tau), \psi(\tau) \rangle = 0$ for every testing function $\psi(\tau)$ whose support is contained in Ω_{τ_0}. Therefore, when $\phi(t, \tau)$ has its support contained in $\mathfrak{R}_t \times \Omega_{\tau_0}$,

$$\langle f(t) \times g(\tau), \phi(t, \tau) \rangle = \langle f(t), \langle g(\tau), \phi(t, \tau) \rangle \rangle$$
$$= \langle f(t), 0 \rangle = 0$$

Thus, in this case too, (t_0, τ_0) is not in the support of $f(t) \times g(\tau)$.

On the other hand, let both t_0 and τ_0 be in the supports Ω_f and Ω_g of $f(t)$ and $g(\tau)$, respectively. Then, for every two-dimensional neighborhood in $\mathfrak{R}_{t,\tau}$ of the point (t_0, τ_0), a testing function $\psi(t)\theta(\tau)$ can be so chosen that its support is contained in this neighborhood and $\langle f(t) \times g(\tau), \psi(t)\theta(\tau) \rangle$ is not zero. This is because the right-hand side of (4) of the preceding section can be made different from zero by choosing $\psi(t)$ and $\theta(\tau)$ appropriately. Thus, the support of $f(t) \times g(\tau)$ is precisely $\Omega_f \times \Omega_g$, as was asserted.

The commutativity of the direct product As was shown in some examples in the preceding section, the direct product is commutative at least in certain cases. Actually, this is true in general and, in order to show this, we shall need

Lemma 1

The space of all testing functions of the form

$$\phi(t, \tau) = \sum_\nu \psi_\nu(t)\theta_\nu(\tau) \tag{1}$$

where the $\psi_\nu(t)$ are in $\mathfrak{D}_t$, the $\theta_\nu(\tau)$ are in $\mathfrak{D}_\tau$, and the summation has a finite number of terms, is dense in $\mathfrak{D}_{t,\tau}$.

Proof: Let the support of a given testing function $\phi(t, \tau)$ be contained in the square $|t| \leq a$ and $|\tau| \leq a$. Then, according to an extension by Borel (Borel [1], pp. 73–78) of a theorem of Weierstrass, a sequence

$\{p_m(t, \tau)\}_{m=1}^{\infty}$ of polynomials of the variables t and τ can be found which converges uniformly on the square $|t| \leq 2a$ and $|\tau| \leq 2a$ to $\phi(t, \tau)$ and each of whose partial derivatives converges uniformly on the same square to the same partial derivative of $\phi(t, \tau)$.

Now, let $\rho(t)$ be a testing function that equals one for $|t| \leq a$ and vanishes for $|t| \geq 2a$. Then, $p_m(t, \tau)\rho(t)\rho(\tau)$ is a testing function in $\mathfrak{D}_{t,\tau}$ which has the form (1). Furthermore, the sequence $\{p_m(t, \tau)\rho(t)\rho(\tau)\}_{m=1}^{\infty}$ converges in $\mathfrak{D}_{t,\tau}$ to $\phi(t, \tau)\rho(t)\rho(\tau) = \phi(t, \tau)$. Q.E.D.

Theorem 5.3-2

The direct product of two distributions is commutative:

$$f(t) \times g(\tau) = g(\tau) \times f(t)$$

That is, for every testing function $\phi(t,\tau)$ in $\mathfrak{D}_{t,\tau}$ we have

$$\langle f(t), \langle g(\tau), \phi(t, \tau)\rangle\rangle = \langle g(\tau), \langle f(t), \phi(t, \tau)\rangle\rangle \tag{2}$$

Proof: As was stated by Theorem 5.2-1, both $f(t) \times g(\tau)$ and $g(\tau) \times f(t)$ are distributions and hence continuous functionals on $\mathfrak{D}_{t,\tau}$. Thus, to prove this theorem, we need merely establish (2) for the subspace of all testing functions of the form (1), since this subspace is dense in $\mathfrak{D}_{t,\tau}$. But now, by employing (4) and (5) of Sec. 5.2, we get

$$\left\langle f(t) \times g(\tau), \sum_{\nu} \psi_\nu(t)\theta_\nu(\tau) \right\rangle = \sum_{\nu} \langle f(t) \times g(\tau), \psi_\nu(t)\theta_\nu(\tau)\rangle$$
$$= \sum_{\nu} \langle f(t), \psi_\nu(t)\rangle\langle g(\tau), \theta_\nu(\tau)\rangle$$

and similarly

$$\left\langle g(\tau) \times f(t), \sum_{\nu} \psi_\nu(t)\theta_\nu(\tau) \right\rangle = \sum_{\nu} \langle f(t), \psi_\nu(t)\rangle\langle g(\tau), \theta_\nu(\tau)\rangle$$

This establishes the theorem.

Since $\mathfrak{D}_{t,\tau}$ is dense in $\mathcal{S}_{t,\tau}$, we can readily extend this result to distributions of slow growth and testing functions of rapid descent.

Corollary 5.3-2a

Equation (2) still holds when $f(t)$ is in $\mathcal{S}'_t$, $g(\tau)$ is in $\mathcal{S}'_\tau$, and $\phi(t, \tau)$ is in $\mathcal{S}_{t,\tau}$.

A useful consequence of Theorem 5.3-2 provides an analogue to Fubini's theorem; in particular, an integration with respect to one component variable in a two-dimensional euclidean space and an application of a distribution to a testing function with respect to the other component variable can be interchanged under certain conditions.

Corollary 5.3-2b
Let $\phi(t, \tau)$ be a testing function in $\mathfrak{D}_{t,\tau}$ (or in $\mathcal{S}_{t,\tau}$) and let $g(\tau)$ be a distribution in $\mathfrak{D}'_\tau$ (or in $\mathcal{S}'_\tau$). Then

$$\int_a^b \langle g(\tau), \phi(t, \tau) \rangle \, dt = \left\langle g(\tau), \int_a^b \phi(t, \tau) \, dt \right\rangle \tag{3}$$

Here the limits $a = -\infty$ and $b = +\infty$ are also permissible.

NOTE: This result is also analogous to Corollary 2.7-2a, where differentiation is now replaced by integration.

Proof: Let $1_{a,b}(t)$ denote both the function

$$1_{a,b}(t) = \begin{cases} 0 & t \leq a \\ 1 & a < t < b \\ 0 & b \leq t \end{cases}$$

and the corresponding regular distribution. Then

$$\langle 1_{a,b}(t) \times g(\tau), \phi(t, \tau) \rangle = \langle 1_{a,b}(t), \langle g(\tau), \phi(t, \tau) \rangle \rangle$$
$$= \int_a^b \langle g(\tau), \phi(t, \tau) \rangle \, dt$$

and

$$\langle g(\tau) \times 1_{a,b}(t), \phi(t, \tau) \rangle = \left\langle g(\tau), \int_a^b \phi(t, \tau) \, dt \right\rangle$$

The commutativity of the direct product completes the proof.

The associativity of the direct product One can construct a distribution over the three-dimensional euclidean space $\mathfrak{R}_{t,\tau,x}$ from the three distributions $f(t)$, $g(\tau)$, $h(x)$ by iterating the direct product. That is, for ϕ in $\mathfrak{D}_{t,\tau,x}$, we set up the definitions

$$\langle [f(t) \times g(\tau)] \times h(x), \phi(t, \tau, x) \rangle \triangleq \langle f(t) \times g(\tau), \langle h(x), \phi(t, \tau, x) \rangle \rangle \tag{4}$$

and

$$\langle f(t) \times [g(\tau) \times h(x)], \phi(t, \tau, x) \rangle \triangleq \langle f(t), \langle g(\tau) \times h(x), \phi(t, \tau, x) \rangle \rangle \tag{5}$$

(This procedure can be continued to obtain definitions for the n-dimensional euclidean space.) The right-hand sides of these expressions have a sense because, by Corollary 2.7-2a,

$$\langle h(x), \phi(t, \tau, x) \rangle$$

is in $\mathfrak{D}_{t,\tau}$ and

$$\langle g(\tau) \times h(x), \phi(t, \tau, x) \rangle$$

is in $\mathfrak{D}_t$. Moreover, by an easy extension of Theorem 5.2-1, we can show

that both
$$[f(t) \times g(\tau)] \times h(x) \tag{6}$$
and
$$f(t) \times [g(\tau) \times h(x)] \tag{7}$$
are distributions in $\mathcal{D}'_{t,\tau,x}$. Finally, by expanding the direct products inside the right-hand sides of (4) and (5) according to the definition of the direct product, we find that (4) and (5) become the same quantity, namely,

$$\langle f(t), \langle g(\tau), \langle h(x), \phi(t, \tau, x)\rangle\rangle\rangle$$

Thus, we have

Theorem 5.3-3
The direct product is associative. That is, (6) and (7) are the same distribution.

PROBLEMS

1 If $f(t)$ and $g(\tau)$ are distributions in $\mathcal{D}'_t$ and $\mathcal{D}'_\tau$, respectively, show that
$$\frac{\partial^{k+p}}{\partial t^k \, \partial \tau^p} [f(t) \times g(\tau)] = \frac{\partial^k f(t)}{\partial t^k} \times \frac{\partial^p g(\tau)}{\partial \tau^p}$$

2 Show that (6) and (7) are distributions in $\mathcal{D}'_{t,\tau,x}$.

★5.4 THE CONVOLUTION OF DISTRIBUTIONS

Let $f(t)$ and $g(t)$ be two continuous functions with bounded support. Their convolution produces a third function $h(t)$, which is denoted by $f * g$ and defined by

$$h(t) \triangleq f(t) * g(t) \triangleq \int_{-\infty}^{\infty} f(\tau) g(t - \tau) \, d\tau$$

Thus, when dealing with ordinary functions, convolution is customarily interpreted as follows. To obtain the value of $h(t)$ at some fixed point t, one of the functions (it does not matter which, as one can easily show) is transposed, then shifted by the amount t, and then multiplied by the other function. Finally, the resulting product is integrated over the entire range of the independent variable.

Such a procedure cannot be used when f and g are arbitrary distributions because, for one reason, two distributions cannot be multiplied in general. In order to extend the convolution process to distributions, we have to interpret it in another way. We can achieve our objective by viewing the resulting function $h(t)$ as a regular distribution.

Still assuming that $f(t)$ and $g(t)$ are continuous functions with bounded supports and letting ϕ be in $\mathfrak{D}$, we may write

$$\langle h, \phi \rangle = \langle f * g, \phi \rangle \\ = \int_{-\infty}^{\infty} dt \int_{-\infty}^{\infty} f(\tau) g(t - \tau) \phi(t) \, d\tau$$

Since the integrand of the last integral is continuous and has a bounded support on the (t, τ) plane, it may be written as a double integral. Thereupon, by applying the change of variable $\tau = x$ and $t = x + y$ and noting that the corresponding jacobian determinant equals one, we obtain

$$\langle f * g, \phi \rangle = \int_{-\infty}^{\infty} \int_{-\infty}^{\infty} f(x) g(y) \phi(x + y) \, dx \, dy \tag{1}$$

The last expression has a form that is similar to that of the direct product of two regular distributions. Thus the rule that defines the convolution $f * g$ of two distributions $f(t)$ and $g(t)$ is suggested by this expression to be

$$\langle f * g, \phi \rangle \triangleq \langle f(t) \times g(\tau), \phi(t + \tau) \rangle \\ \triangleq \langle f(t), \langle g(\tau), \phi(t + \tau) \rangle \rangle \tag{2}$$

However, a problem arises in this case. Even though the function $\phi(t + \tau)$ is infinitely smooth, it is not a testing function, since its support is not bounded in the (t, τ) plane. Actually, its support is contained in an infinite strip of finite width that runs parallel to the line $t + \tau = 0$. A meaning can still be assigned to the right-hand side of (2) if the supports of f and g are suitably restricted. In particular, if the support of $f(t) \times g(\tau)$ intersects the support of $\phi(t + \tau)$ in a bounded set, say Ω, we can replace the right-hand side of (2) by

$$\langle f(t) \times g(\tau), \lambda(t, \tau) \phi(t + \tau) \rangle \tag{3}$$

where $\lambda(t, \tau)$ is some testing function in $\mathfrak{D}_{t,\tau}$ that is equal to one over some neighborhood of Ω. (In general, the choice of λ has to be altered as ϕ traverses $\mathfrak{D}$.) Since $\lambda(t, \tau) \phi(t + \tau)$ will also be a testing function in $\mathfrak{D}_{t,\tau}$, (3) and, therefore, (2) serve to define $f * g$ in this case as a functional over all ϕ in $\mathfrak{D}$. This replacement is legitimate because the values of a testing function outside some neighborhood of the support of $f(t) \times g(\tau)$ can be altered at will without affecting the value assigned by $f(t) \times g(\tau)$ to that testing function. (See Sec. 1.8.)

We have yet to determine under what conditions the intersection of the supports of $f(t) \times g(\tau)$ and $\phi(t + \tau)$ is always bounded for all ϕ in $\mathfrak{D}$ and whether $f * g$ is a distribution. This is resolved by

Theorem 5.4-1

Let f and g be two distributions over $\mathbb{R}^1$ and let their convolution $f * g$ be defined by (2), where the right-hand side of (2) is understood to be (3). Then, $f * g$ will exist as a distribution over $\mathbb{R}^1$ under any one of the following conditions:

a. Either f or g has a bounded support.
b. Both f and g have supports bounded on the left [i.e., there exists some constant T_1 such that $f(t) = g(t) = 0$ for $t < T_1$].
c. Both f and g have supports bounded on the right [i.e., there exists some constant T_2 such that $f(t) = g(t) = 0$ for $t > T_2$].

Proof: Let Ω_f and Ω_g be the supports of $f(t)$ and $g(\tau)$, respectively. Under condition 1, $\Omega_f \times \Omega_g$ is contained in either a horizontal or a vertical strip of finite width in the (t, τ) plane. Under condition 2, $\Omega_f \times \Omega_g$ is contained in a quarter-plane lying above some horizontal line and to the right of some vertical line in the (t, τ) plane. Finally, under the third condition, $\Omega_f \times \Omega_g$ is contained in a quarter-plane lying below some horizontal line and to the left of some vertical line in the (t, τ) plane. Under every one of these conditions, the intersection of $\Omega_f \times \Omega_g$ with the support of $\phi(t + \tau)$, where ϕ is in $\mathfrak{D}$, will be a bounded set. Hence, the definition of convolution is applicable, and it specifies $f * g$ as a functional on $\mathfrak{D}$.

This functional is a linear one on $\mathfrak{D}$, since the direct product is a distribution. Second, if $\{\phi_\nu(t)\}_{\nu=1}^\infty$ is a sequence of testing functions that converges in $\mathfrak{D}_t$ to zero, a fixed $\lambda(t, \tau)$ can be so chosen that it equals one over a neighborhood of each intersection of $\Omega_f \times \Omega_g$ with the support of every $\phi_\nu(t + \tau)$. Furthermore,

$$\{\lambda(t, \tau)\phi_\nu(t + \tau)\}_{\nu=1}^\infty$$

converges in $\mathfrak{D}_{t,\tau}$ to zero. The fact that the direct product is a distribution again indicates that the numbers

$$\langle f * g, \phi_\nu \rangle \triangleq \langle f(t) \times g(\tau), \lambda(t, \tau)\phi_\nu(t + \tau)\rangle$$

converge to zero as $\nu \to \infty$. Thus, $f * g$ is truly a distribution over $\mathbb{R}^1$. Q.E.D.

Note that, if none of the three conditions stated in Theorem 5.4-1 are fulfilled, the intersection of the support of $f(t) \times g(\tau)$ with that of any testing function other than the identically zero one will not be bounded and, therefore, our method of assigning a sense to the right-hand side of (2) by replacing it by (3) will not be valid. Hence, in this and the next three sections, it will be understood that at least one of these three conditions is satisfied whenever we speak of the convolution of two distributions. The reader should not infer, however, that the convolution of distributions cannot be defined under any other conditions. If the supports of the distributions are not restricted but, instead, sufficiently strong restrictions are placed on the behavior of the distributions as their arguments approach infinity, then the convolution of distributions can still be defined. (See, for instance, Schwartz [1], vol. II, pp. 102–104.) In Sec. 5.7 we shall present another circumstance under which convolution will exist.

Since the direct product is commutative, it follows from (2) that the same property holds for convolution. Indeed,

$$\langle f * g, \phi \rangle = \langle f(t) \times g(\tau), \phi(t + \tau) \rangle$$
$$= \langle g(\tau) \times f(t), \phi(t + \tau) \rangle = \langle g * f, \phi \rangle$$

Thus, we have

Corollary 5.4-1a

The convolution of two distributions is commutative:

$$f * g = g * f$$

That is, for every ϕ in $\mathfrak{D}$,

$$\langle f(t), \langle g(\tau), \phi(t + \tau) \rangle \rangle = \langle g(\tau), \langle f(t), \phi(t + \tau) \rangle \rangle$$

We mention in passing that the linearity of the direct product (see Sec. 5.2, Prob. 3) implies the linearity of the convolution process. That is, if α and β are arbitrary constants and if f, g, and h are distributions such that f can be convolved with both g and h separately, then

$$f * (\alpha g + \beta h) = \alpha f * g + \beta f * h$$

Theorem 5.4-2

Let Ω_f and Ω_g be the respective supports of the distributions f and g, which are defined over $\mathfrak{R}^1$, and let $\Omega_f + \Omega_g$ be the set in $\mathfrak{R}^1$ each of whose points can be written as the sum of a point in Ω_f and a point in Ω_g. Then, the support of $f * g$ is contained in $\Omega_f + \Omega_g$.

Proof: Let Ω be the complement of the closed set $\Omega_f + \Omega_g$. Ω is therefore an open set. We shall show that, if a testing function ϕ has its support contained in Ω, then

$$\langle f * g, \phi \rangle = \langle f(t) \times g(\tau), \phi(t + \tau) \rangle = 0 \tag{4}$$

According to Theorem 5.3-1, the support of $f(t) \times g(\tau)$ in the (t, τ) plane is the closed set $\Omega_f \times \Omega_g$, which by definition is the set of points (t, τ) such that t is in Ω_f and τ is in Ω_g. Thus, if (t, τ) is in $\Omega_f \times \Omega_g$, $t + \tau$ is in $\Omega_f + \Omega_g$. On the other hand, the support of ϕ is assumed to be contained in the open set Ω, which does not intersect $\Omega_f + \Omega_g$. Thus, the support of $\phi(t + \tau)$ in the (t, τ) plane is contained in the set defined by $t + \tau \in \Omega$. Hence, the support of $\phi(t + \tau)$ is in the null set of $f(t) \times g(\tau)$. This proves (4) and the theorem.

Corollary 5.4-2a

If both of the supports of the distributions f and g are either (a) bounded or (b) bounded on the left or (c) bounded on the right, then the support of $f * g$ is respectively either (a) bounded or (b) bounded on the left or (c) bounded on the right.

Another fact, which we mention here even though we shall need it only in a subsequent chapter, is that, *if f and g are distributions of slow growth with supports bounded on the left, then $f * g$ is also a distribution of slow growth with support bounded on the left.* This is an immediate consequence of Theorem 5.2-2 and Corollary 5.4-2a. It should also be noted here that, when the testing function $\phi(t)$ is in $\mathcal{S}_t$, our convention for assigning a sense to (2) should be altered somewhat. In that case, $\lambda(t, \tau)$ is taken as an infinitely smooth function that equals one over a neighborhood of the support of $f(t) \times g(\tau)$ and is identically zero outside some larger region having the form $T \leq t$ and $T \leq \tau$, where T is finite. The function $\lambda(t, \tau)\phi(t + \tau)$ will now be in $\mathcal{S}_{t,\tau}$, so that (3) will have a sense.

A number of examples of the convolution of two distributions follow.

Example 5.4-1 If f and g are locally integrable functions whose supports satisfy one of the conditions stated in Theorem 5.4-1, then their distributional convolution $h \triangleq f * g$ is a regular distribution corresponding to a locally integrable function that is given almost everywhere by

$$h(t) = \int_{-\infty}^{\infty} f(\tau) g(t - \tau) \, d\tau \tag{5}$$

This can be established by reversing the argument given near the beginning of this section. More specifically, for ϕ in $\mathcal{D}$,

$$\langle h, \phi \rangle = \langle f * g, \phi \rangle = \int_{-\infty}^{\infty} \int_{-\infty}^{\infty} f(x) g(y) \phi(x + y) \, dx \, dy$$

and the last double integral converges absolutely, since $f(x)g(y)\phi(x + y)$ is locally integrable and has a bounded support in the (x, y) plane. By again using the change of variable $x = \tau$ and $y = t - \tau$ and applying Fubini's theorem (Kestelman [1], p. 206) to alter the resulting double integral into an iterated one, we get

$$\langle h, \phi \rangle = \int_{-\infty}^{\infty} \phi(t) \left[\int_{-\infty}^{\infty} f(\tau) g(t - \tau) \, d\tau \right] dt$$

Thus, $h(t)$ is given by the inner integral of the last expression. It exists almost everywhere and is locally integrable (Titchmarsh [1], p. 391).

Example 5.4-2 The convolution of the delta functional with any distribution yields that distribution again; the convolution of the mth derivative of the delta functional with any distribution yields the mth derivative of that distribution. In symbols,

$$\delta * f = f \tag{6}$$
$$\delta^{(m)} * f = f^{(m)} \tag{7}$$

Note that these convolutions are valid for every distribution f in $\mathfrak{D}'$ because $\delta^{(m)}$ has a bounded support. The more general expression (7) may be justified as follows. For every ϕ in $\mathfrak{D}$,

$$\langle \delta^{(m)} * f, \phi \rangle = \langle f * \delta^{(m)}, \phi \rangle = \langle f(t), \langle \delta^{(m)}(\tau), \phi(t + \tau) \rangle \rangle$$
$$= \langle f(t), (-1)^m \phi^{(m)}(t) \rangle = \langle f^{(m)}(t), \phi(t) \rangle$$

An important consequence of (7) is that every linear differential operator with constant coefficients can be represented as a convolution. That is, with the a_ν being constants, we have

$$a_n f^{(n)} + a_{n-1} f^{(n-1)} + \cdots + a_0 f$$
$$= (a_n \delta^{(n)} + a_{n-1} \delta^{(n-1)} + \cdots + a_0 \delta) * f$$

Note that this statement could not be made if we restricted ourselves to the ordinary convolution of functions.

Example 5.4-3 The convolution of the shifted delta functional with any distribution in $\mathfrak{D}'$ yields the distribution shifted by the same amount:

$$\delta(t - a) * f(t) = f(t - a)$$

More generally,

$$\delta^{(m)}(t - a) * f(t) = f^{(m)}(t - a) \tag{8}$$

Indeed, for every ϕ in $\mathfrak{D}$,

$$\langle \delta^{(m)}(t - a) * f(t), \phi(t) \rangle = \langle f(t), \langle \delta^{(m)}(\tau - a), \phi(t + \tau) \rangle \rangle$$
$$= \langle f(t), \langle \delta^{(m)}(\tau), \phi(t + \tau + a) \rangle \rangle = \langle f(t), (-1)^m \phi^{(m)}(t + a) \rangle$$
$$= \langle f^{(m)}(t), \phi(t + a) \rangle = \langle f^{(m)}(t - a), \phi(t) \rangle$$

Here again, the extension of convolution to distributions allows us to consider any linear difference expression having constant coefficients as a convolution, something that we could not do with ordinary convolution. In particular, if a_ν and t_ν are complex and real constants, respectively, then

$$a_n f(t - t_n) + a_{n-1} f(t - t_{n-1}) + \cdots + a_0 f(t - t_0)$$
$$= [a_n \delta(t - t_n) + a_{n-1} \delta(t - t_{n-1}) + \cdots + a_0 \delta(t - t_0)] * f(t)$$

More generally, a linear differential-difference expression with constant coefficients is a distributional convolution:

$$\sum_{\mu=0}^{m} \sum_{\nu=0}^{n} a_{\mu\nu} f^{(\mu)}(t - t_\nu) = \Big[\sum_{\mu=0}^{m} \sum_{\nu=0}^{n} a_{\mu\nu} \delta^{(\mu)}(t - t_\nu) \Big] * f(t)$$

The convolution of three (or more) distributions can be defined by iterating the definition of the convolution of two distributions. Thus, if f, g, and h are distributions over $\mathcal{R}^1$, $f * (g * h)$ can be assigned a meaning by first convolving g with h to obtain the distribution $g * h$ and then by convolving $g * h$ with f. Of course, the first or the second of these two convolutions may not exist, so that $f * (g * h)$ may itself not exist. On the other hand, $f * (g * h)$ will certainly exist as a distribution if at each convolution at least one of the conditions cited in Theorem 5.4-1 is fulfilled.

Moreover, even if $f * (g * h)$ does exist as a distribution, it need not be the same as $(f * g) * h$, which may be another distribution or perhaps nonexistent. In other words, *convolution is not in general an associative process*.

Example 5.4-4 As usual, 1_+ denotes the conventional unit step function. Let 1 and 0 be the functions that everywhere equal one and zero, respectively. Then, using the same symbols to represent the corresponding regular distributions, we have

$$1 * (\delta^{(1)} * 1_+) = 1 * \delta = 1$$

and

$$(1 * \delta^{(1)}) * 1_+ = 0 * 1_+ = 0$$

Moreover, $(1 * 1_+) * \delta^{(1)}$ is nonexistent, since $1 * 1_+$ does not exist.

As with the existence of the convolution of two distributions, certain restrictions imposed on the supports of three or more distributions will ensure that they may be convolved in any order without affecting the final result.

Theorem 5.4-3

For the convolution of three or more distributions, let at least one of the following conditions be fulfilled:

Sec. 5.4   ***Convolution***   **129**

a. *The supports of all the distributions, except for at most one of them, are bounded.*
b. *All the supports are bounded on the left.*
c. *All the supports are bounded on the right.*
Then this convolution is associative and its unique result exists as a distribution.

Proof: We shall restrict ourselves to the case of three distributions, f, g, and h. The extension to still more distributions proceeds in exactly the same way.

The following argument relies once again on the concept of the direct product. Under any one of the above conditions, the support of the distribution $g * h$ will be such that it can be convolved with f to obtain still another distribution. For any ϕ in $\mathfrak{D}$, we have, therefore,

$$\langle f * (g * h), \phi \rangle = \langle f(t) \times [g * h(\tau)], \phi(t + \tau) \rangle$$
$$= \langle f(t), \langle g * h(\tau), \phi(t + \tau) \rangle \rangle \tag{9}$$

Here the symbol $g * h(\tau)$ signifies that the distribution $g * h$ has τ as its independent variable. For each fixed value of t, the following expansion can be made:

$$\langle g * h(\tau), \phi(t + \tau) \rangle = \langle g(\tau) \times h(x), \phi(t + \tau + x) \rangle$$
$$= \langle g(\tau), \langle h(x), \phi(t + \tau + x) \rangle \rangle$$

Upon inserting this into (9), we finally obtain

$$\langle f * (g * h), \phi \rangle = \langle f(t), \langle g(\tau), \langle h(x), \phi(t + \tau + x) \rangle \rangle \rangle$$
$$= \langle f(t) \times [g(\tau) \times h(x)], \phi(t + \tau + x) \rangle \tag{10}$$

Note that the support of $f(t) \times [g(\tau) \times h(x)]$ in the (t, τ, x) plane has a bounded intersection with the support of any $\phi(t + \tau + x)$ if at least one of the three conditions of this theorem is fulfilled. Thus, the right-hand side of (10) is understood to be the same as the expression obtained by replacing $\phi(t + \tau + x)$ by a testing function in $\mathfrak{D}_{t,\tau,x}$ that is identical to $\phi(t + \tau + x)$ over a neighborhood of this bounded intersection.

Since the direct product is associative, we may replace $f(t) \times [g(\tau) \times h(x)]$ by $[f(t) \times g(\tau)] \times h(x)$ and reverse the argument of the preceding paragraph (where now we also employ the commutativity of the direct product and of convolution) to obtain

$$\langle f * (g * h), \phi \rangle = \langle [f(t) \times g(\tau)] \times h(x), \phi(t + \tau + x) \rangle$$
$$= \langle h(x) \times [f(t) \times g(\tau)], \phi(t + \tau + x) \rangle$$
$$= \langle h * (f * g), \phi \rangle$$
$$= \langle (f * g) * h, \phi \rangle$$

Thus, we have shown that the distribution $f * (g * h)$ is identical to $(f * g) * h$.   Q.E.D.

PROBLEMS

1 Let Ω_f and Ω_g be the supports of the distributions f and g, respectively. Show that at least one of the three conditions of Theorem 5.4-1 is satisfied if and only if the intersection of the line $t + \tau = 0$ with $\Omega_f \times \Omega_g$ is a bounded set.

2 Let f and g be distributions of slow growth with supports bounded on the right. What can be said about $f * g$? Discuss any alterations you may need in the definition of $\langle f * g, \phi \rangle$, where ϕ is in $\mathcal{S}$.

3 Repeat the preceding problem in the case where the distribution f has a bounded support and g is an arbitrary distribution in $\mathcal{S}'$.

4 Establish the following distributional convolution formulas:

a. $[t 1_+(t)] * [e^t 1_+(t)] = (e^t - t - 1) 1_+(t)$

b. $[1_+(t) \sin t] * [1_+(t) \cos t] = \frac{1}{2} 1_+(t) t \sin t$

c. $[f(t) 1_+(t)] * [1_+(t)]^{*n} = \dfrac{1_+(t)}{(n-1)!} \displaystyle\int_0^t (t - \tau)^{n-1} f(\tau)\, d\tau$

{Here $f(t)$ is assumed to be a locally integrable function and $[1_+(t)]^{*n}$ denotes the convolution of $1_+(t)$ with itself $n - 1$ times.}

d. $1_+(t) * \operatorname{Pf} \dfrac{1_+(t)}{t} = 1_+(t) \log t$

e. $\delta^{(1)}(t) * \operatorname{Pv} \dfrac{1}{t} = \operatorname{Pf} \dfrac{1_+(-t)}{t^2} - \operatorname{Pf} \dfrac{1_+(t)}{t^2}$

f. $\left[\displaystyle\sum_{\nu=0}^{\infty} \delta^{(\nu)}(t-\nu)\right] * \left[\displaystyle\sum_{\nu=0}^{\infty} \delta(t-\nu)\right] = \displaystyle\sum_{\nu=0}^{\infty} \sum_{\mu=0}^{\nu} \delta^{(\mu)}(t-\nu)$

g. $\left[\displaystyle\sum_{\nu=0}^{\infty} \delta^{(\nu)}(t-\nu)\right]^{*2} = \displaystyle\sum_{\nu=0}^{\infty} (\nu + 1) \delta^{(\nu)}(t-\nu)$

h. $\operatorname{Pf} \dfrac{1_+(t)}{t} * \operatorname{Pf} \dfrac{1_+(t)}{t} = \dfrac{d^2}{dt^2} \displaystyle\int_0^t \log \tau \log (t - \tau)\, d\tau \, 1_+(t)$

i. $\operatorname{Pf} \dfrac{1_+(t)}{t^{\frac{1}{2}}} * [1_+(t) e^t] = -2\, 1_+(t) \left[\dfrac{1}{\sqrt{t}} + e^t \sqrt{\pi}\, \operatorname{erf} \sqrt{t} \right]$

Here, $\operatorname{erf} s$ is the error function:

$$\operatorname{erf} s \triangleq \int_0^{s^2} \dfrac{e^{-z}}{\sqrt{\pi z}}\, dz$$

(See Jahnke, Emde, and Losch [1], p. 26. Their notation differs from ours by a factor of $2/\sqrt{\pi}$.)

5 Show that there does not exist any distribution g such that for all distributions f of bounded support

$$g(t) * f(t) = t^m f^{(k)}(t) \tag{11}$$

where m is a positive integer and k is a nonnegative integer. This result shows that linear differential expressions with nonconstant polynomial coefficients cannot be written as convolutions. HINT: Set $f = \delta$ and determine g from (11). Then, set

$$f(t) = 1_+(t-1)1_+(2-t)$$

and construct a contradiction.

6 Similarly, show that a linear difference expression with nonconstant polynomial coefficients cannot be written as a convolution by proving that there does not exist any distribution g such that, for all distributions f with bounded support,

$$g(t) * f(t) = t^m f(t-a)$$

where m is a positive integer and a is a constant.

7 Let f and g be distributions of bounded support. Demonstrate the following. If f and g are both even or both odd, then $f * g$ is even. If f is even and g is odd, then $f * g$ is odd.

8 For k a positive integer, show that

$$t^k(f*g) = \sum_{\nu=0}^{k} \binom{k}{\nu} (t^\nu f) * (t^{k-\nu} g)$$

9 Let a and b be fixed real numbers. Show that

$$\int_a^b \delta^{(1)}(t-\tau)\, d\tau = \delta^{(1)}(t) * [1_+(t-a) - 1_+(t-b)]$$

where the left-hand side is interpreted in accordance with Sec. 2.8.

★5.5 SOME OPERATIONS ON THE CONVOLUTION PROCESS

The shifting (or translation) of a convolution The shifting of the convolution of two distributions can be accomplished simply by shifting either one of the distributions. That is, if $h(t) = f(t) * g(t)$, then $h(t-a)$ is given by

$$h(t-a) = f(t-a) * g(t) = f(t) * g(t-a) \tag{1}$$

To show this, let ϕ be in $\mathfrak{D}$; then

$$\langle h(t-a), \phi(t) \rangle = \langle h(t), \phi(t+a) \rangle = \langle f(t), \langle g(\tau), \phi(t+\tau+a) \rangle \rangle$$
$$= \langle f(t), \langle g(\tau-a), \phi(t+\tau) \rangle \rangle = \langle f(t) * g(t-a), \phi(t) \rangle$$

By using the commutativity of convolution, the other part of (1) can be established in a similar way. Example 5.4-3 provides an illustration of this result.

The differentiation of a convolution Similarly, a convolution may be differentiated by differentiating either one of the distributions in it. In symbols,

$$\frac{d}{dt}(f*g) = f*\frac{dg}{dt} = \frac{df}{dt}*g \tag{2}$$

Indeed, let ϕ be in $\mathfrak{D}$. By using the identity

$$\frac{d\phi(t+\tau)}{d(t+\tau)} = \frac{\partial\phi(t+\tau)}{\partial\tau}$$

we can write

$$\left\langle \frac{d}{dt}(f*g), \phi \right\rangle = \left\langle f*g, -\frac{d\phi}{dt} \right\rangle = \left\langle f(t), \left\langle g(\tau), -\frac{d\phi(t+\tau)}{d(t+\tau)} \right\rangle \right\rangle$$

$$= \left\langle f(t), \left\langle \frac{dg(\tau)}{d\tau}, \phi(t+\tau) \right\rangle \right\rangle = \left\langle f*\frac{dg}{dt}, \phi \right\rangle$$

The combination of this result with the commutativity of convolution yields the other part of (2):

$$\frac{d}{dt}(f*g) = \frac{d}{dt}(g*f) = g*\frac{df}{dt} = \frac{df}{dt}*g$$

By applying (2) repeatedly, we get the more general result

$$\frac{d^m}{dt^m}(f*g) = \frac{d^p f}{dt^p} * \frac{d^q g}{dt^q} \qquad p+q = m;\, p,q = 0,1,2,\ldots \tag{3}$$

The regularization of a distribution An operation that one encounters at times in the analysis of distributions is the convolution of a distribution with some testing function in $\mathfrak{D}$. This operation, which is called the *regularization of a distribution*, converts the distribution into a function that is infinitely smooth.

Theorem 5.5-1

Let f be in $\mathfrak{D}'$ and let ϕ be in $\mathfrak{D}$. Then $h \triangleq f * \phi$ is an ordinary function which is given by

$$h(t) = \langle f(\tau), \phi(t-\tau) \rangle \tag{4}$$

Moreover, $h(t)$ is infinitely smooth and

$$\frac{d^k}{dt^k} h(t) = \left\langle f(\tau), \frac{\partial^k}{\partial t^k} \phi(t-\tau) \right\rangle \tag{5}$$

Proof: First, consider the quantity $\langle f(\tau), \phi(t-\tau) \rangle$. It follows directly from Theorem 2.7-2 that this is a function which has ordinary derivatives of all orders for all t and that

$$\frac{d^k}{dt^k} \langle f(\tau), \phi(t-\tau) \rangle = \left\langle f(\tau), \frac{\partial^k}{\partial t^k} \phi(t-\tau) \right\rangle$$

Hence, if (4) is true, (5) is certainly true.

Second, let θ be an arbitrary testing function in $\mathfrak{D}$; then

$$\langle h, \theta \rangle = \langle f * \phi, \theta \rangle = \langle f(\tau), \langle \phi(t), \theta(t+\tau) \rangle \rangle$$
$$= \langle f(\tau), \langle \theta(t), \phi(t-\tau) \rangle \rangle = \langle f(\tau) \times \theta(t), \phi(t-\tau) \rangle$$

We can assign a sense to the last expression in the usual way because the intersection of the support of $f(\tau) \times \theta(t)$ with that of $\phi(t-\tau)$ is a bounded set. By the commutativity of the direct product, we get

$$\langle h, \theta \rangle = \langle \theta(t) \times f(\tau), \phi(t-\tau) \rangle$$
$$= \langle \theta(t), \langle f(\tau), \phi(t-\tau) \rangle \rangle = \langle \langle f(\tau), \phi(t-\tau) \rangle, \theta(t) \rangle$$

which establishes (4) and the theorem.

Example 5.5-1 For an illustration of the regularization process, let

$$f(t) = \text{Pf} \frac{1_+(t)}{t}$$

and let

$$\phi(t) = \gamma_\alpha(t) \qquad \alpha > 0$$

where $\gamma_\alpha(t)$ is defined by (1) and (2) of Sec. 1.2. Then the regularization $h = f * \phi$ is found to be the following: For $t \leq -\alpha$, $h(t) = 0$. For $-\alpha < t < \alpha$,

$$h(t) = \frac{1}{A} \text{Fp} \int_0^{t+\alpha} \frac{1}{x} \exp \frac{\alpha^2}{(t-x)^2 - \alpha^2} \, dx$$
$$= \frac{1}{A} \int_0^{t+\alpha} \psi(x) \, dx + \frac{1}{A} \left(\exp \frac{\alpha^2}{t^2 - \alpha^2} \right) \log (t+\alpha)$$

where

$$\psi(x) = \frac{1}{x} \left[\exp \frac{\alpha^2}{(t-x)^2 - \alpha^2} - \exp \frac{\alpha^2}{t^2 - \alpha^2} \right]$$

and

$$A = \alpha \int_{-1}^{1} \exp \frac{1}{x^2 - 1} \, dx$$

Finally, for $\alpha \leq t$,

$$h(t) = \frac{1}{A} \int_{t-\alpha}^{t+\alpha} \frac{1}{x} \exp \frac{\alpha^2}{(t-x)^2 - \alpha^2} \, dx$$

With $\alpha = \frac{1}{4}$, the function $h(t)$ has been plotted in Fig. 5.5-1 as the solid curve. The dashed curve in that figure is a plot of $1/t$. Note that the

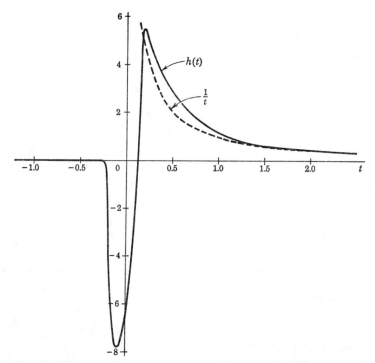

Fig. 5.5-1 The regularization $h(t)$ of Pf $1_+(t)/t$ by the testing function $\gamma_\alpha(t)$, where $\alpha = \frac{1}{4}$.

regularization $h(t)$ approximates the function $1_+(t)/t$ except in the vicinity of the origin, where it has a large negative pulse. In fact, as we saw in Example 1.4-1, the form of $h(t)$ is just what one would expect of an ordinary function that approximates (in a crude sense) the pseudofunction Pf $1_+(t)/t$. We have here an interesting correspondence, namely, ϕ approximates the delta functional (see Example 1.3-1) and at the same time the regularization $h = f * \phi$ approximates f. There is a rigorous justification for this. It is simply a reflection of the fact that convolution is a continuous operation, a subject that we take up in the next section.

PROBLEMS

1 Establish the formulas for the shifting and for the differentiation of a convolution by using the results of Examples 5.4-2 and 5.4-3.

2 Let $\lambda(t)$ be an infinitely smooth function that equals one for $t \geq T > 0$

and equals zero for $t \leq -T < 0$. Show that

$$1_+ * (\lambda f) + (-\check{1}_+) * (f - \lambda f) \tag{6}$$

is a primitive of f. [Here, $\check{1}_+$ denotes the transpose $1_+(-t)$ of $1_+(t)$.] Thus, this formula provides another means of computing the primitives of f. Every other primitive of f differs from (6) by some constant distribution.

3 Let $f(t)$ be a distribution of bounded support and let $p(t)$ be a function that is infinitely smooth but not necessarily of bounded support. Show that

$$f(t) * p(t) = \langle f(\tau), p(t - \tau) \rangle \triangleq \langle f(\tau), \lambda(\tau) p(t - \tau) \rangle$$

where λ is a testing function in $\mathfrak{D}$ that equals one over a neighborhood of the support of f. Then show that, if p is a polynomial of degree n, $f * p$ is also a polynomial of degree n given by

$$f(t) * p(t) = \sum_{\nu=0}^{n} \frac{t^\nu}{\nu!} \langle f(\tau), p^{(\nu)}(-\tau) \rangle$$

4 In the following, J_0 and J_1 represent the Bessel functions of first kind and zero and first order, respectively (Jahnke, Emde, and Losch [1]). J_0 is an even function, whereas J_1 is an odd function. Use the identities

$$[J_0(t)1_+(t)] * [J_0(t)1_+(t)] = 1_+(t) \sin t$$
$$J_0^{(1)}(t) = -J_1(t)$$
$$J_1^{(1)}(t) = J_0(t) - \frac{J_1(t)}{t}$$

to derive the following equations:

$$J_0(t) - \int_0^t J_1(\tau) J_0(t - \tau) \, d\tau = \cos t$$
$$\int_0^t J_1(\tau) J_1(t - \tau) \, d\tau = 2J_1(t) - \sin t$$
$$\int_0^t \frac{J_1(\tau) J_0(t - \tau)}{\tau} \, d\tau = J_1(t)$$

★5.6 THE CONTINUITY OF THE CONVOLUTION PROCESS

It is convenient at this point to introduce some specific notations for certain subspaces of $\mathfrak{D}'$. $\mathcal{E}'$ will denote the (linear) space of all distributions having bounded supports. A sequence $\{f_\nu\}_{\nu=1}^\infty$ will be said to converge in $\mathcal{E}'$ to a limit f if it converges in $\mathfrak{D}'$ to f and if all the f_ν have their supports contained in one fixed bounded interval I. Clearly, the limit distribution f will also be in $\mathcal{E}'$ and will have its support contained in I.

Similarly, $\mathfrak{D}'_R$ (or $\mathfrak{D}'_L$) will denote the (linear) space of all distributions whose supports are bounded on the left (or, respectively, on the right). We shall refer to the elements of $\mathfrak{D}'_R$ as *right-sided distributions* and to the elements of $\mathfrak{D}'_L$ as *left-sided distributions*. A sequence $\{f_\nu\}_{\nu=1}^\infty$ will be said to converge in $\mathfrak{D}'_R$ (or in $\mathfrak{D}'_L$) to a limit f if it converges in $\mathfrak{D}'$ to f and if all the f_ν have their supports bounded on the left (or, respectively, on the right) by a fixed bound. Here again, the limit f will be in $\mathfrak{D}'_R$ (or, respectively, in $\mathfrak{D}'_L$) with the same bound on its support.

Theorem 5.6-1

Let the sequence of distributions $\{f_\nu\}_{\nu=1}^\infty$ converge in $\mathfrak{D}'$ to f and let g be a fixed distribution in $\mathfrak{D}'$. Then $\{f_\nu * g\}_{\nu=1}^\infty$ converges in $\mathfrak{D}'$ to $f * g$ if at least one of the following conditions is fulfilled:
a. *g is in $\mathcal{E}'$.*
b. *$\{f_\nu\}_{\nu=1}^\infty$ converges in $\mathcal{E}'$.*
c. *$\{f_\nu\}_{\nu=1}^\infty$ converges in $\mathfrak{D}'_R$ and g is also in $\mathfrak{D}'_R$.*
d. *$\{f_\nu\}_{\nu=1}^\infty$ converges in $\mathfrak{D}'_L$ and g is also in $\mathfrak{D}'_L$.*

Proof: With ϕ in $\mathfrak{D}$, let us consider

$$\langle f_\nu * g, \phi \rangle = \langle f_\nu(t), \langle g(\tau), \phi(t+\tau) \rangle \rangle \triangleq \langle f_\nu(t), \psi(t) \rangle$$

Under condition a,

$$\psi(t) \triangleq \langle g(\tau), \phi(t+\tau) \rangle$$

is infinitely smooth according to Theorem 2.7-2. Furthermore, $\psi(t)$ has a bounded support, since for all sufficiently large values of $|t|$ the supports of $g(\tau)$ and $\phi(t+\tau)$ do not intersect. Thus, $\psi(t)$ is in $\mathfrak{D}$. By the definition of convergence in $\mathfrak{D}'$,

$$\langle f_\nu * g, \phi \rangle = \langle f_\nu, \psi \rangle \to \langle f, \psi \rangle = \langle f * g, \phi \rangle \tag{1}$$

as $\nu \to \infty$, which is what we wished to show.

Under condition b, $\psi(t)$ is infinitely smooth but not necessarily of bounded support. However, it can be replaced by a testing function $\theta(t)$ in $\mathfrak{D}$ that is identical to $\psi(t)$ over a neighborhood of the finite interval in which the f_ν have their supports. Thus, (1) again holds and we arrive once more at the desired conclusion.

Under condition c, let the supports of $g(t)$ and all the $f_\nu(t)$ be contained in $T_0 \leq t < \infty$. The function $\psi(t)$ is again infinitely smooth. Moreover, it has a support bounded on the right (say, by $t = T_1$), since $g(\tau)$ is identically zero for $\tau < T_0$. Here we may again replace $\psi(t)$ by a testing function $\theta(t)$ in $\mathfrak{D}$ that is identical to $\psi(t)$ over $T_0 < \varepsilon \leq t < \infty$, where $\varepsilon > 0$. [$\theta(t)$ may be zero for all t if $T_1 < T_0$.] The argument now proceeds as before.

The proof for condition d is similar to that for condition c.

The continuity of convolution is a useful property. For instance, we can employ it to prove that the space $\mathfrak{D}$ is dense in the space $\mathfrak{D}'$.

Theorem 5.6-2

$\mathfrak{D}$ is dense in $\mathfrak{D}'$. That is, for each distribution f in $\mathfrak{D}'$ there exists a sequence of testing functions in $\mathfrak{D}$ whose corresponding regular distributions converge in $\mathfrak{D}'$ to f.

Proof: Let

$$\zeta(t) = \begin{cases} 0 & |t| \geq 1 \\ \exp\dfrac{1}{t^2 - 1} & |t| < 1 \end{cases}$$

and let

$$\eta_\nu(t) = \frac{\nu\zeta(\nu t)}{\int_{-\infty}^{\infty} \zeta(t)\,dt}$$

Then $\{\eta_\nu(t)\}_{\nu=1}^\infty$ is a sequence of testing functions in $\mathfrak{D}$ which, according to Corollary 2.3-2a, converges in $\mathfrak{D}'$ to the delta functional $\delta(t)$. Moreover, this convergence is in $\mathcal{E}'$, since the supports of the $\eta_\nu(t)$ contract to the origin.

Let f be an arbitrary distribution in $\mathfrak{D}'$. The regularizations $f(t) * \eta_\nu(t)$ are infinitely smooth, and by the continuity of convolution, they converge in $\mathfrak{D}'$ to $f * \delta = f$. However, the $f(t) * \eta_\nu(t)$ are not necessarily of bounded support. In view of this, let $\lambda(t)$ be a testing function in $\mathfrak{D}$ that equals one for $|t| \leq 1$ and is identically zero for $|t| \geq 2$. Then

$$\left\{ \lambda\left(\frac{t}{\nu}\right)[f(t) * \eta_\nu(t)] \right\}_{\nu=1}^\infty$$

is a sequence of testing functions in $\mathfrak{D}$ that converges in $\mathfrak{D}'$ to f.   Q.E.D.

Since

$$\mathfrak{D} \subset \mathcal{S} \subset \mathcal{S}' \subset \mathfrak{D}'$$

it now follows that $\mathcal{S}$ and $\mathcal{S}'$, as well as $\mathfrak{D}$, are dense in $\mathfrak{D}'$.

PROBLEM

1 Prove that every distribution f in $\mathcal{E}'$ is the limit in $\mathcal{E}'$ of a sequence $\{f_\nu\}_{\nu=1}^\infty$ with the following property. Each f_ν is a finite sum of the form

$$f_\nu(t) = \sum_{\mu=1}^{m_\nu} a_{\mu\nu}\delta(t - \tau_{\mu\nu}) \tag{2}$$

where the $a_{\mu\nu}$ are complex numbers and the $\tau_{\mu\nu}$ are real numbers. [In other words, the (linear) space of all distributions of the form (2) is dense in $\mathcal{E}'$.] HINT: Let $\theta_\alpha = \gamma_\alpha * f$, where γ_α is defined by Sec. 1.2, Eq. (2), and α is a real positive number.

Show that θ_α converges in $\mathcal{E}'$ to f as $\alpha \to 0$. Then let $a \le t \le b$ be a finite interval that contains the support of $\theta_\alpha(t)$. Set $T = b - a$. Show that

$$\frac{T}{n} \sum_{\nu=1}^{n} \theta_\alpha\left(a + \frac{\nu T}{n}\right) \delta\left(t - a - \frac{\nu T}{n}\right)$$

converges in $\mathcal{E}'$ to θ_α as $n \to \infty$. Finally, combine these two results to prove the above assertion.

5.7 THE CONVOLUTION OF A DISTRIBUTION IN $\mathcal{S}'$ WITH A TESTING FUNCTION IN $\mathcal{S}$

When the supports of the distributions $f(t)$ and $g(t)$ are entirely unbounded, our previously used definition of their convolution $f * g$ is no longer applicable. In fact, $f * g$ does not even exist, in general. However, if their behavior as $|t| \to \infty$ is sufficiently restricted, then their convolution may be defined in a somewhat different way. We shall assume that one of the distributions, say f, is of slow growth ($f \in \mathcal{S}'$) and that the other one, g, is a testing function of rapid descent ($g \in \mathcal{S}$). These restrictions are stronger than they need be; the convolution $f * g$ can be defined under still weaker hypotheses. (See Schwartz [1], vol. II, pp. 102–104.) However, the conditions given here are all that we shall need in our subsequent applications.

As in Sec. 5.4, it is natural to try to define $f * g$ by

$$\begin{aligned}\langle f * g, \phi \rangle &= \langle f(t) \times g(\tau), \phi(t + \tau)\rangle \\ &= \langle f(t), \langle g(\tau), \phi(t + \tau)\rangle\rangle \end{aligned} \quad (1)$$

where ϕ traverses $\mathcal{D}$. The direct product $f(t) \times g(\tau)$ is a distribution in $\mathcal{S}'_{t,\tau}$, according to Theorem 5.2-2, but $\phi(t + \tau)$ is not a testing function in $\mathcal{S}_{t,\tau}$. Moreover, we cannot use the device employed in Sec. 5.4 to assign a meaning to the right-hand side of (1), since the intersection of the supports of $f(t) \times g(\tau)$ and of $\phi(t + \tau)$ is not in general bounded. Instead, we may proceed as follows.

Let $\lambda(\tau)$ be a testing function in $\mathcal{D}$ that equals one over some neighborhood of the support of $\phi(\tau)$. Then, proceeding formally, we may rewrite (1) in the following way:

$$\begin{aligned}\langle f * g, \phi\rangle &= \langle f(t) \times g(\tau), \lambda(t + \tau)\phi(t + \tau)\rangle \\ &= \langle f(t), \langle \lambda(t + \tau)g(\tau), \phi(t + \tau)\rangle\rangle \\ &= \langle f(t), \langle \phi(\tau), \lambda(\tau)g(\tau - t)\rangle\rangle \\ &= \langle f(t) \times \phi(\tau), \lambda(\tau)g(\tau - t)\rangle \end{aligned} \quad (2)$$

Since $f(t) \times \phi(\tau)$ is in $\mathcal{S}'_{t,\tau}$, the last expression will have a sense if we can show that $\lambda(\tau)g(\tau - t)$ is a testing function in $\mathcal{S}_{t,\tau}$. Over the (t, τ) plane,

$\lambda(\tau)g(\tau - t)$ is clearly infinitely smooth. Moreover, the support of $\lambda(\tau)g(\tau - t)$ is contained in some strip of finite width running parallel to the t axis. Let us specify this strip by $a \leq \tau \leq b$. Since g is in $\mathcal{S}$, it follows that, as $|t| \to \infty$, $g(\tau - t)$ goes to zero faster than any power of $1/|t|$ uniformly for $a \leq \tau \leq b$. The same thing may be said of any derivative of $g(\tau - t)$, and consequently $\lambda(\tau)g(\tau - t)$ is truly a testing function in $\mathcal{S}_{t,\tau}$.

Thus, when f is in $\mathcal{S}'$ and g is in $\mathcal{S}$, their convolution $f * g$ is defined as a functional on $\mathcal{D}$ by

$$\langle f * g, \phi \rangle \triangleq \langle f(t) \times \phi(\tau), \lambda(\tau)g(\tau - t) \rangle \qquad \phi \in \mathcal{D} \tag{3}$$

where $\lambda(\tau)$ is a testing function in $\mathcal{D}$ that equals one over a neighborhood of the support of $\phi(\tau)$. Note that, in general, we must choose a new $\lambda(\tau)$ for each new choice of $\phi(\tau)$.

We have yet to show that $f * g$ is a distribution. It is certainly linear as a functional on $\mathcal{D}$. (See Sec. 5.2, Prob. 3.) Furthermore, let us define $\psi_\nu(t)$ by

$$\psi_\nu(t) \triangleq \int_{-\infty}^{\infty} \phi_\nu(t + \tau)g(\tau)\, d\tau$$

where $\phi_\nu(t)$ is in $\mathcal{D}$ and is identically zero for, say, $|t| \geq a > 0$. Then, for any two nonnegative integers m and k we may differentiate under the integral sign to write

$$|t^m \psi_\nu^{(k)}(t)| = \left| t^m \int_{-t-a}^{-t+a} \phi_\nu^{(k)}(t + \tau)g(\tau)\, d\tau \right|$$
$$\leq 2a[\sup_t |\phi_\nu^{(k)}(t)|]\, |t|^m \sup_{|\tau| \geq |t| - a} |g(\tau)| \tag{4}$$

Since $g(\tau)$ is of rapid descent

$$|t|^m \sup_{|\tau| \geq |t| - a} |g(\tau)|$$

remains bounded for all t. Therefore, if $\{\phi_\nu\}_{\nu=1}^{\infty}$ is a sequence that converges in $\mathcal{D}$ to zero, the corresponding sequence $\{\psi_\nu\}_{\nu=1}^{\infty}$ converges in $\mathcal{S}$ to zero. Finally, (3) can be converted into

$$\langle f * g, \phi_\nu \rangle = \langle f(t), \langle \phi_\nu(t + \tau), g(\tau) \rangle \rangle$$
$$= \langle f(t), \psi_\nu(t) \rangle$$

and, since f is in $\mathcal{S}'$, the last quantity converges to zero as $\nu \to \infty$. Thus, $f * g$ is truly a distribution.

We can say even more: $f * g$ is a function that is infinitely smooth. For, by the commutativity of the direct product (see Corollary 5.3-2a), (3) may be replaced by

$$\langle f * g, \phi \rangle = \langle \phi(\tau), \langle f(t), \lambda(\tau)g(\tau - t) \rangle \rangle$$
$$= \langle \phi(\tau), \langle f(t), g(\tau - t) \rangle \rangle$$

which shows that
$$f * g(\tau) = \langle f(t), g(\tau - t) \rangle \tag{5}$$
(The symbolism on the left-hand side denotes that $f * g$ has τ as its independent variable.) Our assertion now follows from Corollary 4.5-1a. In fact,
$$\frac{d^k}{d\tau^k}(f * g) = \left\langle f(t), \frac{\partial^k}{\partial \tau^k} g(\tau - t) \right\rangle \tag{6}$$
(The argument of this paragraph does not depend upon the fact that $f * g$ is a distribution, and thus it provides another proof of this fact.)

We shall also show that $f * g$ is a function of slow growth. To do so, we shall need the inequality
$$1 + t^2 < 2(1 + \tau^2)[1 + (\tau - t)^2] \tag{7}$$
which holds over the entire (t, τ) plane. Indeed, if t is held constant, the right-hand side has precisely one minimum, which occurs at $\tau = t/2$. This minimum value is
$$2 + t^2 + \frac{t^4}{8}$$
a quantity obviously larger than the left-hand side of (7). Thus, (7) holds on any vertical line (where t is a constant) and, since we may choose t arbitrarily, (7) holds for all t and τ.

Now, according to Theorem 4.4-1, for each f in $\mathcal{S}'$ there are a constant C and a nonnegative integer r such that
$$|\langle f(t), \phi(t) \rangle| \leq C \sup_t |(1 + t^2)^r \phi^{(r)}(t)|$$
for all ϕ in $\mathcal{S}$. Upon setting $v = f * g$, we get
$$|v(\tau)| = |\langle f(t), g(\tau - t) \rangle|$$
$$\leq C \sup_t \left| (1 + t^2)^r \frac{\partial^r}{\partial t^r} g(\tau - t) \right|$$
By letting p be an integer and using (7), we may write
$$\frac{|v(\tau)|}{(1 + \tau^2)^p} \leq \frac{2^r C}{(1 + \tau^2)^p} \sup_t \left| (1 + \tau^2)^r [1 + (\tau - t)^2]^r \frac{\partial^r}{\partial t^r} g(\tau - t) \right| \tag{8}$$
Now
$$\left| [1 + (\tau - t)^2]^r \frac{\partial^r}{\partial t^r} g(\tau - t) \right| = |[1 + (\tau - t)^2]^r g^{(r)}(\tau - t)|$$
and, since g is in $\mathcal{S}$, this quantity is bounded for all t and τ. Consequently, for $p \geq r$ the right-hand side of (8) is bounded for all t and τ, which shows

that $v(\tau)$ is a function of slow growth. Since $(f*g)^{(k)} = f*g^{(k)}$, every derivative of $f*g$ is also a function of slow growth.

Similarly, if $\{\phi_\nu\}_{\nu=1}^{\infty}$ converges in $\mathcal{S}$ to g, then with $v_\nu \triangleq f*\phi_\nu$ we may write, as before,

$$\frac{|v_\nu(\tau) - v(\tau)|}{(1+\tau^2)^p}$$
$$\leq \frac{2^r C}{(1+\tau^2)^p} \sup_t \left| (1+\tau^2)^r [1+(\tau-t)^2]^r \frac{\partial^r}{\partial t^r} [\phi_\nu(\tau-t) - g(\tau-t)] \right|$$

Here, as $\nu \to \infty$, the functions

$$[1+(\tau-t)^2]^r \frac{\partial^r}{\partial t^r} [\phi_\nu(\tau-t) - g(\tau-t)]$$

converge to zero uniformly over the entire (t, τ) plane. Thus, we see that, for $p \geq r$, the

$$\frac{v_\nu(\tau)}{(1+\tau^2)^p}$$

converge to the bounded function

$$\frac{v(\tau)}{(1+\tau^2)^p}$$

uniformly for all τ.

We summarize these conclusions by

Theorem 5.7-1

If f is in $\mathcal{S}'$, g is in $\mathcal{S}$, and $f*g$ is defined by (3), then $f*g$ and all its derivatives are functions of slow growth. Furthermore, $f*g$ is given by (5) and its derivatives by (6). Finally, let $\{\phi_\nu\}_{\nu=1}^{\infty}$ be a sequence of testing functions that converges in $\mathcal{S}$ to g and let $v \triangleq f*g$ and $v_\nu \triangleq f*\phi_\nu$; then there exists a nonnegative integer r such that, for each integer $p \geq r$, the sequence

$$\left\{ \frac{v_\nu(\tau)}{(1+\tau^2)^p} \right\}_{\nu=1}^{\infty}$$

converges uniformly for $-\infty < \tau < \infty$ to the bounded function

$$\frac{v(\tau)}{(1+\tau^2)^p}$$

Let us finally note that the convolution defined in this section is continuous in the following sense. If $\{g_\nu\}_{\nu=1}^{\infty}$ converges in $\mathcal{S}$ to g, then with f in $\mathcal{S}'$ we have that $\{f*g_\nu\}_{\nu=1}^{\infty}$ converges in $\mathcal{S}'$ to $f*g$. This follows directly from the last statement of the theorem.

Throughout the remainder of this book, whenever we say that a convolution $f * g$ exists, we shall mean that f and g satisfy at least one of the following conditions:

1. Either f or g is in $\mathcal{E}'$.
2. Both f and g are in $\mathcal{D}'_R$.
3. Both f and g are in $\mathcal{D}'_L$.
4. f is in $\mathcal{S}'$ and g is in $\mathcal{S}$.

As we mentioned before, we can define convolution under still other conditions, but we shall not do so.

PROBLEMS

1 If f is in $\mathcal{S}'$ and g is in $\mathcal{S}$, their convolution is clearly commutative in the sense that

$$\langle f(t), g(\tau - t) \rangle = \langle f(\tau - t), g(t) \rangle$$

Show that, if h is also in $\mathcal{S}$, the following associativity also holds:

$$(f * g) * h = f * (g * h)$$

2 Show that, if $\{g_\nu(t)\}_{\nu=1}^{\infty}$ converges in $\mathcal{S}_t$ to $g(t)$, then $\{\lambda(\tau)g_\nu(\tau - t)\}_{\nu=1}^{\infty}$, where $\lambda(\tau)$ is in $\mathcal{D}$, converges in $\mathcal{S}_{t,\tau}$ to $\lambda(\tau)g(\tau - t)$.

3 Verify that convolution is linear in the following sense. If f_1 and f_2 are in $\mathcal{S}'$, if g_1 and g_2 are in $\mathcal{S}$, and if α and β are complex numbers, then

$$f_1 * (\alpha g_1 + \beta g_2) = \alpha f_1 * g_1 + \beta f_1 * g_2$$

and

$$(\alpha f_1 + \beta f_2) * g_1 = \alpha f_1 * g_1 + \beta f_2 * g_1$$

4 In Sec. 1.3, Prob. 3, we discussed the testing-function space $\mathcal{D}_{L_1}$ and its dual space $\mathcal{B}'$ of bounded distributions. We can define the convolution $f * g$ of a distribution f in $\mathcal{B}'$ with a given element g in $\mathcal{D}_{L_1}$ through

$$\langle f * g, \phi \rangle \triangleq \langle f(t), \langle g(\tau), \phi(t + \tau) \rangle \rangle$$

where ϕ traverses $\mathcal{D}_{L_1}$. Verify that this definition makes sense by proving that

$$\langle g(\tau), \phi(t + \tau) \rangle$$

is in $\mathcal{D}_{L_1}$. Then, demonstrate that $f * g$ is also in $\mathcal{B}'$.

5.8 CONVOLUTION OPERATORS

Another interesting result is the identity between a certain type of operator and the convolution process. An *operator in* $\mathcal{D}'$ is a rule $\mathfrak{R}$ that assigns one or more elements v in the distribution space $\mathfrak{D}'$ to each element f in some subset of $\mathfrak{D}'$. This subset is called the *domain* of the operator and is denoted by $D(\mathfrak{N})$. The set of all distributions v corresponding to those in $D(\mathfrak{N})$ is called the *range* of the operator and is denoted by $R(\mathfrak{N})$. The notation for this correspondence is

$$v = \mathfrak{N}f \qquad f \in D(\mathfrak{N}) \qquad v \in R(\mathfrak{N}) \tag{1}$$

Now if w is a fixed distribution in $\mathfrak{D}'$, the relationship $v = w * f$ defines a special type of operator in $\mathfrak{D}'$, which we may denote by $\mathfrak{N} \triangleq w *$. We shall take the domain of this operator as the set of all distributions f in $\mathfrak{D}'$ for which the convolution $w * f$ exists in the sense of Theorem 5.4-1 or 5.7-1. Such an operator will be called a *convolution operator*. Note that $w * \delta = w$; thus, w is simply the distribution that $w *$ assigns to the delta functional. From Theorem 5.4-1, we see that $D(w *)$ will always contain $\mathcal{E}'$. If w is in $\mathfrak{D}'_R$ (or in $\mathfrak{D}'_L$), then $D(w *)$ will contain $\mathfrak{D}'_R$ (or, respectively, $\mathfrak{D}'_L$). Also, if w is in $\mathcal{E}'$, then $D(w *)$ will be identical with $\mathfrak{D}'$. Finally, as was shown in the preceding section, if w is in $\mathcal{S}'$, then $D(w *)$ will contain $\mathcal{S}$.

We shall also say that $\mathfrak{N}$ is a *convolution operator over* $\mathcal{K}$, where $\mathcal{K}$ is some space of distributions, if $\mathfrak{N}f = w * f$ for all f in $\mathcal{K}$ but not necessarily for other distributions of $D(\mathfrak{N})$ that are not in $\mathcal{K}$.

We now turn our attention to four fundamental properties that may or may not be possessed by an operator in $\mathfrak{D}'$. As will be indicated in each case, *every convolution operator $w *$ possesses all of them.*

Single-valuedness *An operator $\mathfrak{N}$ in $\mathfrak{D}'$ is single-valued if $\mathfrak{N}$ assigns exactly one distribution v to each distribution f in $D(\mathfrak{N})$.*

That $w *$ is single-valued follows directly from the definition of convolution,

$$\langle w * f, \phi \rangle \triangleq \langle w(t) \times f(\tau), \phi(t + \tau) \rangle \qquad \phi \in \mathfrak{D}$$

and the fact that for the given w and f the direct product $w(t) \times f(\tau)$ is unique.

Linearity *A single-valued operator $\mathfrak{N}$ in $\mathfrak{D}'$ is linear if for every pair f_1 and f_2 in $D(\mathfrak{N})$ and for every pair of complex numbers α and β we always have that $\alpha f_1 + \beta f_2$ is also in $D(\mathfrak{N})$ and*

$$\mathfrak{N}(\alpha f_1 + \beta f_2) = \alpha \mathfrak{N} f_1 + \beta \mathfrak{N} f_2 \tag{2}$$

The linearity of the convolution process shows that $w *$ is a linear operator.

Commutativity with the shifting operator *A single-valued operator $\mathfrak{N}$ in $\mathfrak{D}'$ is said to commute with the shifting operator if the facts that $f(t)$ is in $D(\mathfrak{N})$ and $v(t) = \mathfrak{N}f(t)$ imply that $f(t + x)$ is also in $D(\mathfrak{N})$ and $v(t + x) = \mathfrak{N}f(t + x)$, where x is an arbitrary real number.*

This property was established for the convolution process in the first paragraph of Sec. 5.5.

Continuity as a mapping of $\mathcal{E}'$ into $\mathcal{D}'$ *A single-valued operator $\mathfrak{R}$ in $\mathcal{D}'$ is said to be a continuous mapping of $\mathcal{E}'$ into $\mathcal{D}'$ if its domain contains $\mathcal{E}'$ and if $\{\mathfrak{R}f_\nu\}_{\nu=1}^\infty$ converges in $\mathcal{D}'$ to $\mathfrak{R}f$ for every sequence $\{f_\nu\}_{\nu=1}^\infty$ that converges in $\mathcal{E}'$ to f.*

As was noted before, $D(w*)$ contains $\mathcal{E}'$ for every convolution operator $w*$. Also, the convergence of $\{w*f_\nu\}_{\nu=1}^\infty$ to $w*f$ was established in Theorem 5.6-1. Thus, every convolution operator is truly a continuous mapping of $\mathcal{E}'$ into $\mathcal{D}'$.

Surprisingly, the following converse proposition is also true. This remarkable fact was established by L. Schwartz. (See Schwartz [1], vol. II, pp. 17–20, 53–54, and [2], pp. 148–151.)

Theorem 5.8-1

Any operator $\mathfrak{R}$ in $\mathcal{D}'$ that possesses the above four properties must be a convolution operator over $\mathcal{E}'$; that is, there exists a unique distribution w in $\mathcal{D}'$ such that $\mathfrak{R}f = w*f$ for at least all f in $\mathcal{E}'$.

We can demonstrate the uniqueness of w right away. Assume the existence of two distributions w_1 and w_2 such that

$$\mathfrak{R}f = w_1 * f = w_2 * f$$

for every f in $\mathcal{E}'$. Then, for each ϕ in $\mathcal{D}$, we may write

$$\langle w_1(\tau), \phi(t-\tau) \rangle = \langle w_2(\tau), \phi(t-\tau) \rangle$$

since $\mathcal{D} \subset \mathcal{E}'$. As $\phi(\tau)$ traverses $\mathcal{D}$, $\phi(t-\tau)$ also traverses $\mathcal{D}$ for any given value of t. Hence, $w_1 = w_2$.

In proving the rest of Theorem 5.8-1, we shall proceed by means of two lemmas.

Lemma 1

Let $\mathfrak{R}$ be an operator in $\mathcal{D}'$ possessing the afore-mentioned four properties. If f has a bounded support and if $\mathfrak{R}f = v$, then $\mathfrak{R}f^{(r)} = v^{(r)}$ ($r = 1, 2, 3, \ldots$). That is, $\mathfrak{R}$ commutes with differentiation.

Proof: Since $\mathfrak{R}$ is linear and commutes with the shifting operator, we may write for each nonzero value of τ

$$\frac{v(t+\tau) - v(t)}{\tau} = \mathfrak{R} \frac{f(t+\tau) - f(t)}{\tau} \tag{3}$$

As was shown in Sec. 2.4, the left-hand side of (3) converges in $\mathcal{D}'$ to $v^{(1)}$

as $\tau \to 0$. Moreover, for all τ in some neighborhood of zero,

$$\frac{f(t+\tau) - f(t)}{\tau}$$

has its support in one fixed bounded interval, and hence it converges in $\mathcal{E}'$ to $f^{(1)}$ as $\tau \to 0$. By the continuity of $\mathfrak{N}$ the right-hand side of (3) converges in $\mathfrak{D}'$ to $\mathfrak{N} f^{(1)}$, and, in addition, $\mathfrak{N} f^{(1)} = v^{(1)}$. Our conclusion now follows by iteration.

Lemma 2

Every distribution f of bounded support is the limit in $\mathcal{E}'$ of a sequence $\{f_\nu\}_{\nu=1}^\infty$ with the following property. Each f_ν is a finite sum of the form

$$f_\nu(t) = \sum_{\mu=1}^{m_\nu} a_{\mu\nu} \delta^{(r)}(t - \tau_{\mu\nu}) \tag{4}$$

where the $a_{\mu\nu}$ are complex constants, the $\tau_{\mu\nu}$ are real constants, and r is the order of f.

Proof: Let I be a finite closed interval that contains the support of f. According to Corollary 3.4-2a, there exist a continuous function h and a nonnegative integer r for which $f = h^{(r+2)}$. Now, we can choose a sequence of continuous functions $\{h_\nu\}_{\nu=1}^\infty$ such that the following conditions are fulfilled:

1. Outside I, each h_ν is identical to h.
2. On I, each h_ν is composed of a finite set of confluent straight-line segments, the abscissas of their end points being $\tau_{\mu\nu}$ ($\mu = 1, 2, \ldots, m_\nu$).
3. The sequence $\{h_\nu(t)\}_{\nu=1}^\infty$ converges to $h(t)$ uniformly for $-\infty < t < \infty$. [This can be done because $h(t)$ is uniformly continuous over I.]

The third condition implies that $\{h_\nu\}_{\nu=1}^\infty$ converges in $\mathfrak{D}'$ to h. By the continuity of differentiation, $\{h_\nu^{(r+2)}\}_{\nu=1}^\infty$ converges in $\mathfrak{D}'$ to f. Finally, each $h_\nu^{(r+2)}$ has the form (4). Since all the $\tau_{\mu\nu}$ are contained in I, it follows that $\{h_\nu^{(r+2)}\}_{\nu=1}^\infty$ converges in $\mathcal{E}'$. Q.E.D.

Proof of Theorem 5.8-1: We can now show that $\mathfrak{N}$ is a convolution operator as follows. First of all, define w as the distribution that $\mathfrak{N}$ assigns to δ. Thus,

$$\mathfrak{N}\delta \triangleq w = w * \delta$$

Since $\mathfrak{N}$ commutes with the shifting operator,

$$\mathfrak{N}\delta(t - \tau) = w(t - \tau) = w(t) * \delta(t - \tau)$$

By Lemma 1,

$$\mathfrak{N}\delta^{(r)}(t - \tau) = w^{(r)}(t - \tau) = w(t) * \delta^{(r)}(t - \tau)$$

Also, by the linearity of $\Re$,

$$\Re \sum_{\mu=1}^{m_\nu} a_{\mu\nu} \delta^{(r)}(t - \tau_{\mu\nu}) = w(t) * \sum_{\mu=1}^{m_\nu} a_{\mu\nu} \delta^{(r)}(t - \tau_{\mu\nu}) \qquad (5)$$

Now choose the sequence $\{f_\nu\}_{\nu=1}^{\infty}$ as in Lemma 2. By the continuity of both $\Re$ and the convolution process, the left-hand side of (5) converges to $\Re f$, the right-hand side converges to $w * f$, and both limits are equal, which is what we wished to prove.

By combining Theorem 5.8-1 with the discussion preceding it, we immediately get

Corollary 5.8-1a

A necessary and sufficient condition for an operator $\Re$ in $\mathfrak{D}'$ to be a convolution operator over $\mathcal{E}'$ is that $\Re$ possess the properties of single-valuedness, linearity, commutativity with the shifting operator, and continuity as a mapping of $\mathcal{E}'$ into $\mathfrak{D}'$.

Theorem 5.8-1 establishes $\Re$ as a convolution operator merely over the space $\mathcal{E}'$. This result can be extended to wider spaces of distributions if we expand the continuity assumption on $\Re$. Let us first define what we shall mean by the continuity of an operator in $\mathfrak{D}'$ (in distinction to its continuity as a mapping of $\mathcal{E}'$ into $\mathfrak{D}'$).

Continuity *A single-valued operator $\Re$ in $\mathfrak{D}'$ is said to possess continuity (or to be continuous) if δ is in $D(\Re)$ and if $\Re$ satisfies those of the following conditions which apply to $w \triangleq \Re \delta$:*

1. *If w has an unbounded support, then $D(\Re)$ contains $\mathcal{E}'$ and $\{\Re f_\nu\}_{\nu=1}^{\infty}$ converges in $\mathfrak{D}'$ to $\Re f$ for every sequence $\{f_\nu\}_{\nu=1}^{\infty}$ that converges in $\mathcal{E}'$ to f.*
2. *If w has its support bounded only on the left (or only on the right), then $D(\Re)$ contains $\mathfrak{D}'_R$ (or, respectively, $\mathfrak{D}'_L$) and $\{\Re f_\nu\}_{\nu=1}^{\infty}$ converges in $\mathfrak{D}'$ to $\Re f$ for every sequence $\{f_\nu\}_{\nu=1}^{\infty}$ that converges in $\mathfrak{D}'_R$ (or, respectively, in $\mathfrak{D}'_L$) to f.*
3. *If w has a bounded support, then $D(\Re)$ is $\mathfrak{D}'$ and $\{\Re f_\nu\}_{\nu=1}^{\infty}$ converges in $\mathfrak{D}'$ to $\Re f$ for every sequence $\{f_\nu\}_{\nu=1}^{\infty}$ that converges in $\mathfrak{D}'$ to f.*
4. *If w is a distribution of slow growth, then $D(\Re)$ contains $\mathcal{S}$ and $\{\Re f_\nu\}_{\nu=1}^{\infty}$ converges in $\mathfrak{D}'$ to $\Re f$ for every sequence $\{f_\nu\}_{\nu=1}^{\infty}$ that converges in $\mathcal{S}$ to f.*

Note that the conditions that $\Re$ must satisfy depend upon the character of w. Moreover, *every* condition that applies to the distribution w must be satisfied by $\Re$ in order for $\Re$ to be called continuous. Thus, if w is in $\mathcal{S}'$ and has an unbounded support, then both condition 1 and condition 4 must be fulfilled. Also, observe that, if w is of bounded support and $\mathfrak{N}$ satisfies condition 3, then w and $\mathfrak{N}$ automatically satisfy conditions 1, 2, and 4. Finally, if w is of unbounded support and is not in $\mathcal{S}'$, then "continuity" and "continuity as a mapping of $\mathcal{E}'$ into $\mathcal{D}'$" are the same thing.

Theorem 5.8-2

A necessary and sufficient condition for an operator $\mathfrak{N}$ in $\mathcal{D}'$ to be a convolution operator $w *$ on every f for which $w * f$ exists in the sense of Theorems 5.4-1 and 5.7-1 is that $\mathfrak{N}$ possess the properties of single-valuedness, linearity, commutativity with the shifting operator, and continuity.

Proof Necessity: We have already seen that every convolution operator $w *$ is single-valued and linear and commutes with the shifting operator. Moreover, $D(w *)$ contains δ, since it contains $\mathcal{E}'$. It now follows directly from Theorem 5.6-1 and the paragraph after Theorem 5.7-1 that $w *$ possesses continuity.

Sufficiency: From the definition of continuity we have that $D(\mathfrak{N})$ always contains the space $\mathcal{E}'$. Hence, by Theorem 5.8-1, $\mathfrak{N}$ is a convolution operator $w *$ over $\mathcal{E}'$. We have to show that $\mathfrak{N}f = w * f$ for every f for which $w * f$ exists in the sense of Theorems 5.4-1 and 5.7-1.

Every distribution f is the limit in $\mathcal{D}'$ of a sequence $\{f_\nu\}_{\nu=1}^\infty$ of distributions having bounded supports. Also, if w has a bounded support, then $D(\mathfrak{N})$ is in $\mathcal{D}'$ and $w * f$ exists. Consequently, by the continuity of $\mathfrak{N}$ and by Theorem 5.6-1, we may write $\mathfrak{N}f_\nu = w * f_\nu$, $\mathfrak{N}f_\nu \to \mathfrak{N}f$ in $\mathcal{D}'$, $w * f_\nu \to w * f$ in $\mathcal{D}'$, and, thus, $\mathfrak{N}f = w * f$.

Similarly, if w is in $\mathcal{D}'_R$ (or in $\mathcal{D}'_L$), we can choose a sequence of elements from $\mathcal{E}'$ that converges in $\mathcal{D}'_R$ (or in $\mathcal{D}'_L$) to f and thereby show that $\mathfrak{N}f = w * f$.

Finally, every f in $\mathcal{S}$ is the limit in $\mathcal{S}$ of a sequence $\{f_\nu\}_{\nu=1}^\infty$ whose elements are in $\mathcal{D}$. In view of the paragraph after Theorem 5.7-1, it again follows that, if w is in $\mathcal{S}'$, then $\mathfrak{N}f = w * f$ for every f in $\mathcal{S}$.

PROBLEMS

1 Let $\mathfrak{N} \triangleq w *$ be a convolution operator, where w has its support bounded on the left. The application of $\mathfrak{N}$ to $t1_+(t)$ yields

$\mathfrak{N}[t1_+(t)] = 1_+(t) \sin t$

Find a distribution that can be w.

2 Use the assertion of Sec. 5.6, Prob. 1, to simplify the proof of Theorem 5.8-1.

6 *Convolution Equations*

★6.1 INTRODUCTION

As was mentioned at the beginning of the preceding chapter, the distributional convolution process is one that occurs quite generally in physics. A very wide variety of fundamental equations that arise in nature can be represented by the single simple form

$$f * u = g \tag{1}$$

Here f and g are given distributions and u is an unknown distribution for which this equation is to be solved. Actually, ordinary linear differential equations with constant coefficients, similar difference equations, and integral equations, as well as combination differential, difference, and integral equations, are but a few examples of the types of equation that are encompassed by (1). Various partial differential equations are also representable by (1). Moreover, sets of simultaneous equations of the afore-mentioned types can be taken into account if f, u, and g are interpreted as matrices. Thus, the convolution process, when considered from the distributional point of view, possesses great generality, and a study of the techniques for solving (1) provides powerful methods for attacking many types of practical problems. This is still another example of how distribution theory unifies and simplifies mathematical analysis.

In this short chapter we shall discuss convolution equations in general and then give an application to ordinary linear differential equations with constant coefficients. Other applications will be presented after we develop the distributional Laplace transformation, because this transformation is a powerful tool for solving such equations. (The reader can investigate this subject still further by referring to Schwartz [1], vol. II, and to Vasilach [1].)

As usual, we shall have to restrict the types of distributions that are considered in order to make sure that the various convolutions that arise possess a sense. One situation, which is encountered quite often, occurs when f, u, and g have supports that are bounded on the left. The imposition of this restriction leads to a *convolution algebra* and to a convenient technique for solving (1). As we shall see, (1) may or may not have a solution in this algebra but, if it does have one, the solution will be unique.

Another common situation arises when f has a bounded support; then the left-hand side of (1) will have a sense for all distributions u. Moreover, the support of g need not be bounded in order for such a solution to exist. If, however, we assume that g and u do have bounded supports, we shall have another convolution algebra. The technique for solving convolution equations of the previous form may again be applied to this new algebra.

On the other hand, if we do not restrict ourselves to some convolution algebra, the situation becomes more complicated. For instance, assuming f and g to be fixed distributions with bounded supports, we may seek all possible solutions u in $\mathfrak{D}'$ to (1). Now, instead of either a unique solution or no solution at all, there may be several (indeed, an infinite number of) solutions.

In the last section of this chapter we very briefly describe Mikusiński's operational calculus, in which every equation always has a unique solution. We then compare it with distribution theory.

★6.2 CONVOLUTION ALGEBRAS

A space $\mathfrak{A}'$ of distributions is said to be a *convolution algebra* if it possesses the following properties:

1. $\mathfrak{A}'$ is a linear space. (See Appendix A for a definition of a linear space.)
2. $\mathfrak{A}'$ is closed under convolution (i.e., the convolution of a finite

number of distributions in $\mathcal{C}'$ exists and yields a distribution that is again in $\mathcal{C}'$).

3. Convolution is associative for any three distributions in $\mathcal{C}'$.

For an abstract space to be an algebra, its multiplication operation, which in our case is convolution, must also possess some other properties such as distributivity with respect to addition. These additional requirements are automatically satisfied by convolution. (For an axiomatic treatment of algebras, the reader may refer to Taylor [1], p. 162.)

The particular subspace of $\mathcal{D}'$, with which we shall mainly be concerned, is the space $\mathcal{D}'_R$ of all right-sided distributions. $\mathcal{D}'_R$ is clearly a convolution algebra. Moreover, it is a *commutative* algebra, since the operation of convolution in $\mathcal{D}'_R$ is commutative. Finally, the algebra has a unit element, the delta functional, since $\delta * f = f$. (As will be shown later on, there is no other distribution in $\mathcal{D}'_R$ with this property.) In summary, $\mathcal{D}'_R$ *is a commutative convolution algebra having a unit element.* [It is also a fact that the convolution of two distributions in $\mathcal{D}'_R$ can be the zero distribution only if at least one of them is the zero distribution (Schwartz [1], vol. II, p. 29). One refers to this fact by saying that the convolution algebra $\mathcal{D}'_R$ does not contain divisors of zero.]

Because of these algebraic properties of $\mathcal{D}'_R$, one can manipulate right-sided distributions in much the same way that one manipulates numbers, except that here the multiplication of numbers is replaced by the convolution of distributions. This leads naturally to a simple means for solving a convolution equation. In particular, if a, b, and c are numbers such that $ab = c$ with a and c known ($a \neq 0$) and b unknown, one could solve for b by multiplying c by the inverse a^{-1} of a. Analogously, given the convolution equation

$$f * u = g \tag{1}$$

where f and g are known distributions in $\mathcal{D}'_R$ and u is unknown but required to be in $\mathcal{D}'_R$, it is natural to solve (1) for u by first finding a distribution that is an inverse of f in the convolution algebra $\mathcal{D}'_R$. Such an inverse, which we denote by f^{*-1}, is any element of $\mathcal{D}'_R$ such that

$$f^{*-1} * f = \delta \tag{2}$$

Then we may solve (1) by convolving both of its sides by f^{*-1} to obtain

$$u = \delta * u = f^{*-1} * f * u = f^{*-1} * g \tag{3}$$

This solution u will be a right-sided distribution, since both f^{*-1} and g are right-sided distributions.

However, in contrast to the algebra of numbers, not all nonzero right-sided distributions possess inverses. For example, if f is any testing function in $\mathcal{D}$, its convolution with any distribution will yield a function that is infinitely smooth (see Theorem 5.5-1). Thus, such a testing function is in $\mathcal{D}'_R$ but cannot possess an inverse. In this case, (1) will not have a solution for *every* g in $\mathcal{D}'_R$. However, for *certain* choices of the right-sided distribution g, (1) will possess solutions. (These solutions can be shown to be unique in $\mathcal{D}'_R$ because of the lack of divisors of zero in $\mathcal{D}'_R$.)

No element of $\mathcal{D}'_R$ can possess more than one inverse in $\mathcal{D}'_R$ and, for each right-sided distribution g, (1) can have no more than one solution in $\mathcal{D}'_R$. More precisely, we shall establish

Theorem 6.2-1

Let f be a given distribution in $\mathcal{D}'_R$. A necessary and sufficient condition for (1) to have at least one solution in $\mathcal{D}'_R$ for every g in $\mathcal{D}'_R$ is that f possess an inverse f^{*-1} in $\mathcal{D}'_R$. When f does possess an inverse in $\mathcal{D}'_R$, this inverse is unique and (1) possesses a unique solution in $\mathcal{D}'_R$, given by (3).

Proof: If (1) has at least one solution in $\mathcal{D}'_R$ for every g in $\mathcal{D}'_R$, then one of the solutions of (1) with $g = \delta$ will be one of the inverses of f. Conversely, if f has an inverse in $\mathcal{D}'_R$, (3) will be one of the solutions of (1) since the convolution of both sides of (3) by f will produce (1) again.

Finally, if f has an inverse in $\mathcal{D}'_R$, (1) can possess no more than one solution in $\mathcal{D}'_R$ for each g in $\mathcal{D}'_R$. For if u and v are both in $\mathcal{D}'_R$ and are solutions of (1), then

$$f * u = g \qquad f * v = g$$

On convolving all terms by this particular inverse f^{*-1}, we get

$$u = f^{*-1} * g \qquad v = f^{*-1} * g$$

Since convolution is a single-valued operation, $u = v$. This also implies that f^{*-1} is unique in $\mathcal{D}'_R$, since we may choose g to be δ. Q.E.D.

Thus, if f is a right-sided distribution, the problem of seeking solutions of (1) in the space $\mathcal{D}'_R$ for arbitrary g in $\mathcal{D}'_R$ is equivalent to finding an inverse of f in $\mathcal{D}'_R$.

The following theorem presents a tool for determining the inverse of f; it is useful when f can be written as the convolution of two or more right-sided distributions, each of which possesses a known inverse in $\mathcal{D}'_R$.

Theorem 6.2-2

If h and j are in $\mathcal{D}'_R$ and possess inverses in $\mathcal{D}'_R$, then

$$(h * j)^{*-1} = h^{*-1} * j^{*-1} \tag{4}$$

Proof: Since convolution in $\mathcal{D}'_R$ is associative and commutative,

$$(h * j) * (h^{*-1} * j^{*-1}) = (h * h^{*-1}) * (j * j^{*-1}) = \delta * \delta = \delta$$

This establishes (4).

If we do not require that u be in $\mathfrak{D}'_R$, then (1) may have more than one solution for the given right-sided distributions f and g. More generally, if f and g are arbitrary distributions not necessarily in $\mathfrak{D}'_R$, then (1) may have either no solution, exactly one solution, or many solutions in $\mathfrak{D}'$. When $g = \delta$, such solutions are called *elementary solutions*. Given a particular elementary solution, one can generate another one by adding to the first solution any nonzero solution (if it exists) of the homogeneous equation, $f * u = 0$. In the convolution algebra $\mathfrak{D}'_R$, the inverse f^{*-1} is that elementary solution whose support is bounded on the left. It follows from the uniqueness property of Theorem 6.2-1 that no other elementary solutions can have their supports bounded on the left.

There are other convolution algebras besides $\mathfrak{D}'_R$. For instance, the space $\mathcal{E}'$ of all distributions with bounded support is also an algebra under convolution. Indeed, $\mathcal{E}'$ is a commutative convolution algebra with unit element, the unit element again being δ. Our discussion of $\mathfrak{D}'_R$ and, in particular, Theorems 6.2-1 and 6.2-2 also apply to $\mathcal{E}'$. Moreover, since $\mathcal{E}'$ is a subspace of $\mathfrak{D}'_R$, $\mathcal{E}'$ is said to be a *subalgebra under convolution in* $\mathfrak{D}'_R$. Thus, with f and g in $\mathcal{E}'$ (f fixed, g arbitrary), the equation $f * u = g$ will have one (and the same) solution in $\mathfrak{D}'_R$ if it has one in $\mathcal{E}'$. However, if it does have a solution in $\mathfrak{D}'_R$, this solution need not be in $\mathcal{E}'$. In other words, if a convolution equation is solvable in $\mathfrak{D}'_R$, it need not be solvable in $\mathcal{E}'$ even though both f and g are in $\mathcal{E}'$.

Example 6.2-1 Consider

$$(\delta^{(1)} - \delta) * u = \delta$$

Both $\delta^{(1)} - \delta$ and δ are in $\mathcal{E}'$, but a solution is $1_+(t)e^t$, which is in $\mathfrak{D}'_R$ but not in $\mathcal{E}'$. Thus, $\delta^{(1)} - \delta$ has an inverse in $\mathfrak{D}'_R$ but does not have one in $\mathcal{E}'$.

An example of a convolution algebra that is not a subspace of $\mathfrak{D}'_R$ is the space $\mathfrak{D}'_L$ of all left-sided distributions. The analysis for $\mathfrak{D}'_R$ can be directly carried over to $\mathfrak{D}'_L$. However, a particular convolution equation may have one solution in $\mathfrak{D}'_R$ and an entirely different one in $\mathfrak{D}'_L$. Similarly, a given distribution in $\mathcal{E}'$ may have one inverse in $\mathfrak{D}'_R$ and a different one in $\mathfrak{D}'_L$.

Example 6.2-2 Let us consider the convolution equation of Example 6.2-1 again. Its solution in $\mathfrak{D}'_L$ is $-1_+(-t)e^t$. As before, it can be shown that this is the only solution in $\mathfrak{D}'_L$. Thus, the inverse of $\delta^{(1)} - \delta$ in $\mathfrak{D}'_L$ is not the same as its inverse in $\mathfrak{D}'_R$.

Another example of a convolution algebra is presented in Sec. 11.5, where the elements under consideration are the periodic distributions

Sec. 6.2 *Convolution equations* 153

having a common period T. That algebra does possess divisors of zero, so that a convolution equation may possess more than one solution in it, as we shall see.

Rather than having a single convolution equation of the form (1), one often encounters sets of simultaneous equations of the following form:

$$\begin{aligned} f_{11} * u_1 + f_{12} * u_2 + \cdots + f_{1n} * u_n &= g_1 \\ f_{21} * u_1 + f_{22} * u_2 + \cdots + f_{2n} * u_n &= g_2 \\ &\cdots \\ f_{n1} * u_1 + f_{n2} * u_2 + \cdots + f_{nn} * u_n &= g_n \end{aligned} \quad (5)$$

These simultaneous equations can be represented in the matrix form

$$\mathbf{f} * \mathbf{u} = \mathbf{g} \quad (6)$$

where $\mathbf{f}$ is the $n \times n$ matrix

$$\mathbf{f} = \begin{bmatrix} f_{11} & f_{12} & \cdots & f_{1n} \\ f_{21} & f_{22} & \cdots & f_{2n} \\ \cdots & \cdots & \cdots & \cdots \\ f_{n1} & f_{n2} & \cdots & f_{nn} \end{bmatrix}$$

and $\mathbf{u}$ and $\mathbf{g}$ are the $n \times 1$ column matrices

$$\mathbf{u} = \begin{bmatrix} u_1 \\ u_2 \\ \cdot \\ \cdot \\ \cdot \\ u_n \end{bmatrix} \quad \mathbf{g} = \begin{bmatrix} g_1 \\ g_2 \\ \cdot \\ \cdot \\ \cdot \\ g_n \end{bmatrix}$$

Here, as well as elsewhere, boldface notation is used to denote matrices.

Let us again restrict ourselves to the algebra $\mathfrak{D}'_R$; that is, the f_{ik} ($i, k = 1, 2, \ldots, n$) and the g_i ($i = 1, 2, \ldots, n$) are given right-sided distributions, and the solutions for the unknown u_i ($i = 1, 2, \ldots, n$) are also required to be in $\mathfrak{D}'_R$. Since $\mathfrak{D}'_R$ is a convolution algebra, the rules for manipulating these matrices are precisely the same as those which customarily apply, except that now all multiplications are replaced by convolutions. The reader should note that, even though the convolution of distributions in $\mathfrak{D}'_R$ is commutative, the convolution of matrices of such distributions is not, in general. Indeed, it is not even possible to convolve all such matrices. In order for $\mathbf{f} * \mathbf{h}$ to exist, where $\mathbf{f}$ and $\mathbf{h}$ are matrices with elements in $\mathfrak{D}'_R$, it is necessary that the number of columns in $\mathbf{f}$ be equal to the number of rows in $\mathbf{h}$. On the other hand, if q is a single distribution in $\mathfrak{D}'_R$, $q * \mathbf{f} \triangleq \mathbf{f} * q$ is, by definition, obtained by convolving each element in $\mathbf{f}$ with q.

To each square (that is, $n \times n$) matrix $\mathbf{f}$, we may assign a determinant, which we denote by det $\mathbf{f}$. It too is computed by incorporating the convolution process with the usual rules for expanding a determinant. It follows that det $\mathbf{f}$ will be a single distribution in $\mathcal{D}'_R$. For example,

$$\det \begin{bmatrix} f_{11} & f_{12} \\ f_{21} & f_{22} \end{bmatrix} = f_{11} * f_{22} - f_{21} * f_{12}$$

Since the rules for multiplying (in our case, convolving) two matrices are the same as those for multiplying two determinants,

$$\det (\mathbf{f} * \mathbf{h}) = (\det \mathbf{f}) * (\det \mathbf{h}) \tag{7}$$

where now $\mathbf{f}$ and $\mathbf{h}$ are two $n \times n$ matrices.

The *cofactor* c_{ik} of the element f_{ik} in the $n \times n$ matrix $\mathbf{f}$ is the product of $(-1)^{i+k}$ with the determinant of the $(n-1) \times (n-1)$ matrix obtained by deleting the ith row and the kth column of $\mathbf{f}$. Let $\mathbf{c}$ be the $n \times n$ matrix consisting of the cofactors c_{ik} arranged according to their subscripts. Then, the *adjoint* $\mathbf{a}$ of $\mathbf{f}$ is defined as the matrix transpose $\mathbf{c}^T$ of $\mathbf{c}$. That is,

$$\mathbf{a} \triangleq \mathbf{c}^T \triangleq \begin{bmatrix} c_{11} & c_{21} & \cdots & c_{n1} \\ c_{12} & c_{22} & \cdots & c_{n2} \\ \cdots & \cdots & \cdots & \cdots \\ c_{1n} & c_{2n} & \cdots & c_{nn} \end{bmatrix}$$

Finally, in our convolution algebra $\mathcal{D}'_R$, the *inverse* $\mathbf{f}^{*-1}$ of an $n \times n$ matrix $\mathbf{f}$ is defined as any matrix (necessarily $n \times n$) with elements in $\mathcal{D}'_R$ such that

$$\mathbf{f} * \mathbf{f}^{*-1} = \boldsymbol{\delta}_{n \times n} \tag{8}$$

where $\boldsymbol{\delta}_{n \times n}$ is the $n \times n$ matrix whose main-diagonal elements equal δ and whose elements off the main diagonal equal the zero distribution. Here, $\boldsymbol{\delta}_{n \times n}$ takes the place of the $n \times n$ unit matrix, and it is called the $n \times n$ *unit matrix in the convolution algebra* $\mathcal{D}'_R$. These definitions are consistent with those customarily used in matrix algebra.

It should be noted that $\mathbf{f}^{*-1}$ need not exist. This is the case, for instance, when all the elements of $\mathbf{f}$ are testing functions in $\mathcal{D}$. However, as we shall show in a moment, if $\mathbf{f}$ does have an inverse, it is unique and

$$\mathbf{f} * \mathbf{f}^{*-1} = \mathbf{f}^{*-1} * \mathbf{f}$$

In this case the solution to (6) is unique and is given by

$$\mathbf{u} = \mathbf{f}^{*-1} * \mathbf{g} \tag{9}$$

More specifically, we have the following extension to Theorem 6.2-1. (Henceforth, $\mathcal{D}'_{R, n \times m}$ will denote the space of all $n \times m$ matrices whose elements are all right-sided distributions.)

Theorem 6.2-3

Let $\mathbf{f}$ be a given matrix in $\mathcal{D}'_{R,n\times n}$. A necessary and sufficient condition for (6) to have at least one solution in $\mathcal{D}'_{R,n\times 1}$ for each $\mathbf{g}$ in $\mathcal{D}'_{R,n\times 1}$ is that the determinant of $\mathbf{f}$ possess an inverse in $\mathcal{D}'_R$. If this is the case, the inverse of $\mathbf{f}$ is unique in $\mathcal{D}'_{R,n\times n}$ and there is precisely one solution in $\mathcal{D}'_{R,n\times 1}$ for $\mathbf{u}$. This solution is given by (9), where

$$\mathbf{f}^{*-1} = (\det \mathbf{f})^{*-1} * \mathbf{a}$$

$\mathbf{a}$ *is the adjoint of* $\mathbf{f}$.

Proof: First, assume that (6) has at least one solution in $\mathcal{D}'_{R,n\times 1}$ for every $\mathbf{g}$ in $\mathcal{D}'_{R,n\times 1}$. Choose $\mathbf{g}$ such that the jth element of $\mathbf{g}$ equals δ and all other elements of $\mathbf{g}$ equal the zero distribution. If with this $\mathbf{g}$ there is more than one solution for $\mathbf{u}$ in $\mathcal{D}'_{R,n\times 1}$, choose one of them. (As we shall show later, there cannot be more than one solution.) Denote the element in the ith row of $\mathbf{u}$ by h_{ij}. Repeat this procedure as j traverses the integers from 1 to j. Finally, let $\mathbf{h}$ be the $n \times n$ matrix whose elements are h_{ij}, where, as usual, i denotes the row position and j denotes the column position of h_{ij}.

Now, by combining the ith row of $\mathbf{f}$ with the jth column of $\mathbf{h}$ according to the rules of matrix convolution, we obtain

$$\sum_{k=1}^{n} f_{ik} * h_{kj} = \begin{cases} \delta & i = j \\ 0 & i \neq j \end{cases} \tag{10}$$

because of the way $\mathbf{h}$ was constructed. In matrix notation, (10) is simply written as

$$\mathbf{f} * \mathbf{h} = \boldsymbol{\delta}_{n\times n}$$

In short, $\mathbf{f}$ does possess at least one inverse in $\mathcal{D}'_{R,n\times n}$, namely, $\mathbf{h}$. Moreover, by (7)

$$(\det \mathbf{f}) * (\det \mathbf{h}) = \det \boldsymbol{\delta}_{n\times n} = \delta$$

Hence, det $\mathbf{f}$ does have an inverse in $\mathcal{D}'_R$.

Conversely, assume that det $\mathbf{f}$ possesses an inverse in $\mathcal{D}'_R$; $(\det \mathbf{f})^{*-1}$ must be unique according to Theorem 6.2-1. An $n \times n$ matrix $\mathbf{q}$ can be constructed by convolving $(\det \mathbf{f})^{*-1}$ with every element of the adjoint $\mathbf{a}$ of $\mathbf{f}$; that is,

$$\mathbf{q} \triangleq (\det \mathbf{f})^{*-1} * \mathbf{a} \tag{11}$$

Now, by following the standard derivation of the fact that the product of a matrix with its adjoint divided by its determinant is the unit matrix (Mirsky [1], p. 88), it can be shown that

$$\mathbf{f} * \mathbf{q} = \mathbf{q} * \mathbf{f} = \boldsymbol{\delta}_{n \times n} \tag{12}$$

Thus, (11) is an inverse of $\mathbf{f}$ and the convolution of (11) with $\mathbf{f}$ is commutative. Furthermore, a solution of (6) is given by

$$\mathbf{u} = \mathbf{q} * \mathbf{g} \tag{13}$$

because

$$\mathbf{f} * \mathbf{u} = \mathbf{f} * \mathbf{q} * \mathbf{g} = \boldsymbol{\delta}_{n \times n} * \mathbf{g} = \mathbf{g}$$

Finally, if det $\mathbf{f}$ has an inverse in $\mathfrak{D}'_R$, each solution in $\mathfrak{D}'_{R, n \times 1}$ of (6) must be unique. Indeed, let $\mathbf{u}$ and $\mathbf{v}$ be two such solutions for the same $\mathbf{g}$:

$$\mathbf{f} * \mathbf{u} = \mathbf{g} \qquad \mathbf{f} * \mathbf{v} = \mathbf{g} \tag{14}$$

Now $\mathbf{q}$ may be constructed according to (11). By convolving all terms in (14) with $\mathbf{q}$ and using (12), we obtain

$$\mathbf{u} = \mathbf{q} * \mathbf{g} \qquad \mathbf{v} = \mathbf{q} * \mathbf{g}$$

Therefore, $\mathbf{u} = \mathbf{v}$. This also implies that $\mathbf{f}$ has a unique inverse, since the jth column of $\mathbf{f}^{*-1}$ is the solution $\mathbf{u}$ of $\mathbf{f} * \mathbf{u} = \mathbf{g}$, where the element in the jth row of $\mathbf{g}$ is δ and all other elements of $\mathbf{g}$ are the zero distribution. In other words, $\mathbf{f}^{*-1} = \mathbf{q} = \mathbf{h}$ and (13) is the same as (9). This completes the proof.

Here again we see that the problem of finding a solution to the set of simultaneous convolution equations (5) reverts to the problem of finding an inverse to the matrix $\mathbf{f}$, which in turn requires the determination of an inverse for det $\mathbf{f}$.

Needless to say, we can have simultaneous equations with their consequent matrix manipulations for other convolution algebras as well.

PROBLEMS

1 Give an example of a commutative convolution algebra that does not possess any inverses whatsoever. Does your example possess divisors of zero?

2 Establish (12).

3 Let

$$1_+(t) \int_0^t \cos(t - \tau) u(\tau) \, d\tau = 1_+(t) g(t)$$

where g is a differentiable (in the ordinary sense) function and where u is unknown. Find $u(t)$ for $t > 0$. HINT: Use the fact that

$$[1_+(t) \cos t]^{*-1} = \delta^{(1)}(t) + 1_+(t)$$

4 By using the fact that $\mathcal{D}'_R$ does not possess divisors of zero, show that the equation $f * u = g$, where f and g are distributions in $\mathcal{D}'_R$ and $f \neq 0$, will have either no solution or precisely one solution in $\mathcal{D}'_R$.

5 Verify that the space $\mathcal{C}_R$ of all continuous functions having their supports bounded on the left is a commutative convolution algebra without a unit element. It is known that $\mathcal{C}_R$ does not have divisors of zero. (See Mikusiński [2], chap. II.) By using this fact, show that $\mathcal{D}'_R$ also does not possess divisors of zero. HINT: Let f and g be in $\mathcal{D}'_R$, let $f * g = 0$, and let θ and ψ be nonzero testing functions in $\mathcal{D}$. Show that one of the regularizations, $f * \theta$ or $g * \psi$, is identically zero. Then show that either $f * \phi$ or $g * \phi$ is identically zero for every ϕ in $\mathcal{D}$.

6 A special form of Volterra's integral equation of the second kind is the convolution equation

$$u(t) + \int_0^t k(t - \tau)u(\tau)\, d\tau = g(t)$$

where k and g are given locally integrable functions that are zero for $t < 0$. Assuming that k is continuous except possibly for an ordinary discontinuity at $t = 0$, show that the solution is

$$u(t) = g(t) + \int_0^t h(t - \tau)g(\tau)\, d\tau$$

where $h(t)$ is also a continuous function with possibly an ordinary discontinuity at $t = 0$ and is given by

$$h = \sum_{\nu=1}^{\infty} (-1)^\nu k^{*\nu}$$

Here $k^{*\nu}$ denotes the convolution of k with itself $\nu - 1$ times (that is, $k^{*1} \triangleq k$, $k^{*2} \triangleq k * k$, $k^{*3} \triangleq k * k * k$, . . .). HINT: Show that the infinite series for h converges uniformly on the finite interval $0 \leq t \leq T$, where T is arbitrary. Then, show that $(\delta + k) * (\delta + h) = \delta$.

★6.3 AN APPLICATION TO ORDINARY LINEAR DIFFERENTIAL EQUATIONS WITH CONSTANT COEFFICIENTS

The problem of determining inverses for distributions in $\mathcal{D}'_R$ is very readily attacked by means of the distributional Laplace transformation; hence we shall postpone our general discussion of this question and the corresponding applications to various differential and difference equations until this transformation has been developed. However, mainly for the purpose of illustrating the techniques developed in the preceding section, we present in this section a preliminary discussion of the important problem of solving an ordinary linear differential equation with constant coefficients.

Let L denote the general differential operator of the form

$$L \triangleq a_n \frac{d^n}{dt^n} + a_{n-1} \frac{d^{n-1}}{dt^{n-1}} + \cdots + a_1 \frac{d}{dt} + a_0 \qquad (1)$$

where the a_ν ($\nu = 1, 2, \ldots, n$) are constants, $a_n \neq 0$, and $n \geq 1$. We wish to resolve the equation

$$Lu = g \qquad (2)$$

where g is a known distribution in $\mathfrak{D}'_R$ and u is unknown but also required to be in $\mathfrak{D}'_R$. (2) may be written as a convolution equation:

$$(L\delta) * u = g \qquad (3)$$
$$L\delta = a_n \delta^{(n)} + a_{n-1} \delta^{(n-1)} + \cdots + a_1 \delta^{(1)} + a_0 \delta \qquad (4)$$

Thus, the technique developed for the convolution algebra $\mathfrak{D}'_R$ may be applied here and, as we have shown, the problem becomes simply that of finding in $\mathfrak{D}'_R$ an inverse for $L\delta$.

Before doing this, however, let us review some classical properties of the differential equation (2). The *homogeneous equation* related to (2) is obtained by setting $g = 0$:

$$Lu = 0 \qquad (5)$$

It has an infinite number of solutions, which can be found as follows. Let D^k denote the differential operator d^k/dt^k ($k = 0, 1, 2, \ldots$). The differential operator L can be treated as a polynomial $P(D)$, which may be factored:

$$P(D) = a_n D^n + a_{n-1} D^{n-1} + \cdots + a_1 D + a_0$$
$$= a_n \prod_{\mu=1}^{q} (D - \gamma_\mu)^{k_\mu}$$

$$\sum_{\mu=1}^{q} k_\mu = n$$

The complex numbers γ_μ are the *distinct* roots of the *characteristic equation* $P(D) = 0$, and the positive integers k_μ are the multiplicities of these distinct roots. Every solution of (5) can be written in the form

$$u(t) = \sum_{\mu=1}^{q} \sum_{\nu=1}^{k_\mu} c_{\mu\nu} t^{\nu-1} e^{\gamma_\mu t} \qquad (6)$$

where the $c_{\mu\nu}$ are complex numbers. Each choice of the set of n constants $c_{\mu\nu}$ yields a solution to (5). However, if at some point $t = t_0$ the values of the function (6) and all its derivatives up to the $(n-1)$st order are specified as

$$u(t_0) = u_0 \qquad u^{(1)}(t_0) = u_1 \qquad \cdots \qquad u^{(n-1)}(t_0) = u_{n-1} \qquad (7)$$

then the solution to (5) will be uniquely determined. In other words, the *initial conditions* (7) uniquely determine the coefficients $c_{\mu\nu}$.

More generally, if g is a continuous function over some interval $a \leq t \leq b$, then any solution u that satisfies (2) will be the sum of a *particular solution* u_p and a *complementary solution* u_c. The complementary solution u_c is simply the general solution (6) of the homogeneous equation (5). The particular solution u_p is any solution of the *nonhomogeneous equation* (2). Then the sum $u_p + u_c$ is a solution of (2) and each different choice of the n constants $c_{\mu\nu}$ will yield a different solution for (2). Thus, two particular solutions differ by no more than some complementary solution. Finally, there will be one and only one solution $u_p + u_c$ that will satisfy a specified set of initial conditions (7), where now $a \leq t_0 \leq b$. That is, the set (7) will again uniquely determine the coefficients $c_{\mu\nu}$. (For a more thorough discussion of these classical results, the reader may refer to Kaplan [1], chap. 4.)

We return now to the problem of solving (2) in the case where u and g are required to be distributions in $\mathfrak{D}'_R$.

Theorem 6.3-1

The distribution $L\delta$, given by (4) with the a_ν ($\nu = 0, 1, \ldots, n; n \geq 1$) being constants and $a_n \neq 0$, has an inverse in $\mathfrak{D}'_R$. This inverse is $1_+(t)h(t)$, where $h(t)$ is that classical solution of the homogeneous equation $Lu = 0$ which satisfies the initial conditions

$$h(0) = h^{(1)}(0) = \cdots = h^{(n-2)}(0) = 0$$
$$h^{(n-1)}(0) = \frac{1}{a_n} \tag{8}$$

NOTE: $1_+(t)h(t)$ is also called the *Green's function*, or the *weighting function*, for L. We can compute $h(t)$ by substituting (8) into the form (6) (with u replaced by h) and solving for the $c_{\mu\nu}$.

Proof: The function h is infinitely smooth, and so the first n distributional derivatives of $1_+(t)h(t)$ are

$$\begin{aligned}(1_+h)^{(1)} &= 1_+h^{(1)} + \delta h(0) \\ (1_+h)^{(2)} &= 1_+h^{(2)} + \delta h^{(1)}(0) + \delta^{(1)}h(0) \\ &\cdots\cdots\cdots\cdots\cdots\cdots\cdots\cdots\cdots \\ (1_+h)^{(n-1)} &= 1_+h^{(n-1)} + \delta h^{(n-2)}(0) + \cdots + \delta^{(n-2)}h(0) \\ (1_+h)^{(n)} &= 1_+h^{(n)} + \delta h^{(n-1)}(0) + \cdots + \delta^{(n-1)}h(0)\end{aligned} \tag{9}$$

where the $h^{(\nu)}$ are continuous functions. The conditions (8) therefore yield

$$\begin{aligned}(1_+h)^{(\nu)} &= 1_+h^{(\nu)} \qquad \nu = 1, 2, \ldots, n-1 \\ (1_+h)^{(n)} &= 1_+h^{(n)} + \frac{1}{a_n}\delta\end{aligned} \tag{10}$$

Consequently,

$$(L\delta) * (1_+h) = L(1_+h) = 1_+Lh + \delta$$

But $Lh = 0$. Hence, $(L\delta) * (1_+h) = \delta$, so that 1_+h is truly the inverse in $\mathfrak{D}'_R$ of $L\delta$, which is what we wished to show.

It now follows that, if g is in $\mathfrak{D}'_R$, (2) has a unique solution in $\mathfrak{D}'_R$ given by

$$u = (1_+h) * g \tag{11}$$

Consider, once again, the classical problem of solving $Lu = g$ under the conditions that g is a continuous function and that u is to be a function that satisfies the initial conditions (7). Let us solve for u over the interval $t_0 < t < \infty$ by using our convolution algebra $\mathfrak{D}'_R$. To do this, we shall replace u, which need not be in $\mathfrak{D}'_R$, by $1_+(t - t_0)u(t)$, which is in $\mathfrak{D}'_R$. A computation similar to that indicated in equations (9) and the use of the equation $Lu(t) = g(t)$ show that

$$L[1_+(t - t_0)u(t)] = 1_+(t - t_0)g(t) + \sum_{\nu=0}^{n-1} b_\nu \delta^{(\nu)}(t - t_0) \tag{12}$$

where

$$b_\nu = a_{\nu+1}u_0 + a_{\nu+2}u_1 + \cdots + a_n u_{n-\nu-1} \tag{13}$$

The convolution of both sides of (12) by $1_+(t)h(t)$, the inverse of $L\delta$ in $\mathfrak{D}'_R$, yields

$$1_+(t - t_0)u(t) = 1_+(t - t_0)\int_{t_0}^{t} h(t - \tau)g(\tau)\,d\tau \\ + \sum_{\nu=0}^{n-1} b_\nu [1_+(t - t_0)h(t - t_0)]^{(\nu)}$$

By (10), this can be converted into

$$1_+(t - t_0)u(t) = 1_+(t - t_0)\int_{t_0}^{t} h(t - \tau)g(\tau)\,d\tau \\ + \sum_{\nu=0}^{n-1} b_\nu h^{(\nu)}(t - t_0)1_+(t - t_0) \tag{14}$$

For $t > t_0$, this expression is the solution we seek, as we shall show presently.

Before doing so, however, let us compare the classical method with the above distributional method for solving our differential equation. Classically, one would perform two distinct steps. First, a general complementary solution u_c of the form (6) and some particular solution u_p would be constructed. Second, the unknown coefficients $c_{\mu\nu}$ would then be determined by substituting the initial conditions into the sum $u_p + u_c$.

On the other hand, the method given here alters this two-step procedure somewhat. In this case, the original equation, $Lu = g$, is replaced by the distributional differential equation (12). To solve (12), we must first find the Green's function $1_+(t)h(t)$, which involves the determination of a solution $h(t)$ to $Lu = 0$ subject to the initial conditions (8). Then (12) is solved by convolving both sides of it with $1_+(t)h(t)$. In solving it, the initial conditions on $u(t)$ are taken into account automatically. As will be shown below, it is the sum of the delta functional and its derivatives on the right-hand side of (12) that introduces these initial conditions into the solution.

Let us now establish that (14) is truly the proper solution. First of all, for $t > t_0$ it satisfies the differential equation $Lu = g$. Indeed, a differentiation under the integral sign, which is permissible in this case, yields

$$\frac{d}{dt}\int_{t_0}^{t} h(t-\tau)g(\tau)\,d\tau = \int_{t_0}^{t} h^{(1)}(t-\tau)g(\tau)\,d\tau + h(0)g(t)$$

Upon repeating this differentiation and using conditions (8), we get

$$\frac{d^{\nu}}{dt^{\nu}}\int_{t_0}^{t} h(t-\tau)g(\tau)\,d\tau = \int_{t_0}^{t} h^{(\nu)}(t-\tau)g(\tau)\,d\tau$$
$$\nu = 1, 2, \ldots, n-1$$

and

$$\frac{d^n}{dt^n}\int_{t_0}^{t} h(t-\tau)g(\tau)\,d\tau = \int_{t_0}^{t} h^{(n)}(t-\tau)g(\tau)\,d\tau + \frac{g(t)}{a_n}$$

Thus,

$$L\int_{t_0}^{t} h(t-\tau)g(\tau)\,d\tau = \int_{t_0}^{t} g(\tau)Lh(t-\tau)\,d\tau + g(t)$$
$$= g(t)$$

since $Lh = 0$. Also, $Lh^{(\nu)} = 0$ for all positive integers ν. Consequently, the application of L to the right-hand side of (14) yields for $t > t_0$ simply $g(t)$, as was asserted.

Second, the right-hand side of (14) can be shown to satisfy the initial conditions (7); for the substitution of the conditions (8) into the equation $Lh^{(\nu)} = 0$ yields

$$a_n h^{(n+\nu)}(0) + a_{n-1}h^{(n+\nu-1)}(0) + \cdots + a_{n-\nu-1}h^{(n-1)}(0) = 0$$
$$\nu = 0, 1, \ldots, n-1 \quad (15)$$

Again let $t > t_0$ and differentiate (14) k times ($k = 0, 1, \ldots, n-1$) to get

$$u^{(k)} = \int_{t_0}^{t} h^{(k)}(t-\tau)g(\tau)\,d\tau + \sum_{\nu=0}^{n-1} b_{\nu}h^{(\nu+k)}(t-t_0)$$

As $t \to t_0$, this becomes

$$u^{(k)}(t_0) = 0 + \sum_{\nu=0}^{n-1} b_\nu h^{(\nu+k)}(0) \tag{16}$$

If the expressions (13) are substituted into (16) and the resulting equation is then rearranged according to the common factors u_ν and simplified through the substitution of (8), it will be found that the coefficient of each u_ν will be of the form (15), and hence zero, except for the single term $a_n h^{(n-1)}(0) u_k$. Since $h^{(n-1)}(0) = 1/a_n$, we thereby obtain the initial condition, $u^{(k)}(t_0) = u_k$.

Thus, through the use of distribution theory we have established the following classical theorem.

Theorem 6.3-2

Let $Lu = g$ be a linear differential equation with constant coefficients, where L is given by (1) and g is a continuous function for $t \geq t_0$. Also, let (7) be a prescribed set of initial conditions. For $t > t_0$ the solution to $Lu = g$ that satisfies (7) is given by

$$u(t) = \int_{t_0}^{t} h(t - \tau) g(\tau) \, d\tau + \sum_{\nu=0}^{n-1} b_\nu h^{(\nu)}(t - t_0) \tag{17}$$

where $h(t)$ is the classical solution to the homogeneous equation $Lu = 0$ that satisfies the initial conditions (8) and where the b_ν are given by (13).

The question remains whether (17) is the only solution to $Lu = g$, g still being a continuous function in $t_0 \leq t < \infty$. To be sure, Theorem 6.2-1 ensures that (14) is the only solution in $\mathfrak{D}'_R$ of the distributional differential equation (12), and the classical uniqueness theorem states that there is no other *function* that satisfies $Lu = g$ in $t_0 < t < \infty$ and the given initial conditions. But these theorems do not eliminate the possibility that there is some *distribution* (not in $\mathfrak{D}'_R$) that will satisfy $Lu = g$ over the interval $t_0 < t < \infty$. Actually, there is no such distribution. So long as g is a continuous function in $t_0 \leq t < \infty$, all solutions u must be continuous functions over $t_0 < t < \infty$. Therefore, the initial conditions will specify u uniquely.

This can be seen by considering the orders of the distributions involved in the equation $Lu = g$. The reader may recall that the order of a distribution on some closed finite interval I is the smallest nonnegative integer r for which $\langle f, \phi \rangle = \langle v^{(r+2)}, \phi \rangle$, where v is a continuous function on I and all testing functions ϕ are in $\mathfrak{D}_I$. In the following argument it will be convenient to extend this order notation. We shall say that f is of order -1 on I if it is the first derivative of a continuous function but is not itself a continuous function on I. (More precisely, we should speak of the regular distribution corresponding to a continuous function.) Also, we shall say that f is of order r ($r = -2, -3, -4, \ldots$) on I if it and its derivatives up to and including the order $-r - 2$ are continuous functions and its $(-r - 1)$st-order derivative is not a continuous function on I. Thus, under this notation f is a singular distribution on I if $r \geq 0$. (It may be either singular or regular if $r = -1$.)

It has been shown in Sec. 3.4 that on a finite closed interval I no distribution f can have an order r of $+\infty$. Hence, in our extended order notation, r can only be either a finite integer or $-\infty$, the latter occurring when f is an infinitely smooth function. Furthermore, when r is finite, each differentiation of f increases r by one and each integration decreases it by one. The order of a finite sum of distributions equals the maximum of their various orders, except when the distributions of largest order cancel each other by subtraction, yielding thereby a lower order for the sum.

From $Lu = g$ we have that

$$a_n u^{(n)} = g - a_{n-1} u^{(n-1)} - \cdots - a_0 u \tag{18}$$

Assume that I is fixed and finite and that g is of finite order q on I. Let r be the order of $u^{(n)}$ on I. If r were greater than q, the orders of $u^{(n-1)}$, $u^{(n-2)}$, ..., u and, therefore, of the right-hand side of (18) would be no greater than $r - 1$. But this would imply a contradiction, since both sides of (18) must be of the same order. Similarly, if r were less than q, then the right-hand side of (18) would be of higher order than the left-hand side, again a contradiction. Hence, $r = q$. The same argument shows that $r = q$ even when $q = -\infty$. Thus, $u^{(n)}$ and g must be of the same order on I.

This implies that, if g is a continuous function on $t_0 < t < \infty$, $u^{(n)}$ and therefore u must also be continuous functions on $t_0 < t < \infty$, as was asserted. This also shows that all the solutions of the homogeneous equation $Lu = 0$, even in the distributional sense, are functions and, by the classical uniqueness theorem, are none other than the classical solutions given by (6).

We can now see what kind of solutions we should expect when g is a given distribution of order q on I. When q is finite, u is of order $q - n$ on I. Hence, if $q \geq n$, u is a singular distribution. If $q = n - 1$, u is of order -1 on I. Finally, if $q \leq n - 2$, u and its derivatives up to and including the order $n - 2 - q$ are continuous functions on I.

By referring to (17), we can also see how initial conditions can be introduced into the problem even when $g(t)$ is a singular distribution. Assume that $g(t) = 0$ for $-\infty < t < t_0$. If follows from (11) that the solution u_g in $\mathfrak{D}'_R$ of $Lu = g$ will also be zero for $-\infty < t < t_0$. Any other solution of $Lu = g$ will fail to be in $\mathcal{D}'_R$ and will differ from the solution u_g by some solution u_δ of the homogeneous equation $Lu = 0$. This complementary solution u_δ is a function that is infinitely smooth and can be specified by initial conditions somewhere in the interval $-\infty < t < t_0$. Let

$$u_0 = u(t_0 - \varepsilon) \qquad u_1 = u^{(1)}(t_0 - \varepsilon)$$
$$\cdots \qquad u_{n-1} = u^{(n-1)}(t_0 - \varepsilon) \quad (19)$$

Here ε is a positive number that can be taken arbitrarily small. As is indicated in (17), the solution corresponding to these intial conditions is

$$u_\delta = \sum_{\nu=0}^{n-1} b_\nu h^{(\nu)}(t - t_0 + \varepsilon) \tag{20}$$

where now it is the values (19) that are substituted into (13) when computing the b_ν. Hence, the total solution, which now holds for all t, is

$$u = u_g + u_\delta = (1_+ h) * g + \sum_{\nu=0}^{n-1} b_\nu h^{(\nu)}(t - t_0 + \varepsilon) \tag{21}$$

Although u_δ is infinitely smooth, u_g may be either a function or a singular distribution, depending upon the order of g. The reader should carefully note that, if we let $\varepsilon = 0$, we may not be able to identify the initial conditions with the values of u and its derivatives at $t = t_0$ because u_g may have a discontinuity at $t = t_0$ and, indeed, may even be a singular distribution in every neighborhood of $t = t_0$.

Example 6.3-1 Consider an electrical circuit consisting of an electromotive force

$$v(t) = \sum_{\nu=0}^{\infty} \delta^{(\nu)}(t - \nu)$$

an inductance of 2 henrys, and a resistance of 1 ohm all connected in series as shown in Fig. 6.3-1. Just before $t = 0$ (that is, at $t = -\varepsilon < 0$, where ε is infinitesimally small) the current j in the circuit is 3 amperes. We wish to find the current j for all time t.

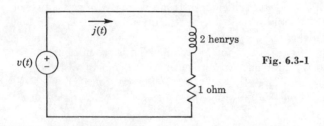

Fig. 6.3-1

Sec. 6.3   Convolution equations

The differential equation relating v to j is

$$2j^{(1)} + j = (2\delta^{(1)} + \delta) * j = v$$

According to Theorem 6.3-1, the inverse in $\mathfrak{D}'_R$ for $2\delta^{(1)} + \delta$ is $1_+ h$, where h is obtained by solving

$$2h^{(1)} + h = 0$$

under the initial condition, $h(0) = \tfrac{1}{2}$. This yields

$$h(t) = \tfrac{1}{2} e^{-t/2}$$

From (13) we get $b_0 = a_1 u_0 = 6$. So, by (21),

$$j(t) = [1_+(t) h(t)] * v(t) + b_0 h(t + \varepsilon)$$

When $\varepsilon \to 0$, $h(t + \varepsilon) \to h(t)$ and our solution becomes

$$j(t) = [\tfrac{1}{2} 1_+(t) e^{-t/2}] * \sum_{\nu=0}^{\infty} \delta^{(\nu)}(t - \nu) + 3 e^{-t/2}$$

Some computation will show that

$$j(t) = \tfrac{1}{2} \sum_{\nu=0}^{\infty} (-\tfrac{1}{2})^{\nu} 1_+(t - \nu) e^{-(t-\nu)/2}$$
$$+ \tfrac{1}{2} \sum_{\nu=1}^{\infty} \sum_{\mu=0}^{\nu-1} (-\tfrac{1}{2})^{\mu} \delta^{(\nu-1-\mu)}(t - \nu) + 3 e^{-t/2}$$

Example 6.3-2 In certain practical problems one customarily writes an integrodifferential equation to represent the system. For example, consider an electrical circuit, Fig. 6.3-2, consisting of an electromotive force v, an inductance L_1, a resistance R, and a capacitance C, all connected in series. Assume for the moment that the electromotive force $v(t)$, the current $j(t)$, and the capacitor's charge $q(t)$ are continuous functions for all t. Now, two initial conditions, namely, the current $j(t_0)$ and the charge $q(t_0)$, may be specified at some given time $t = t_0$. The integrodifferential equation that describes this circuit is

$$v(t) = L_1 \frac{dj}{dt} + Rj(t) + \frac{1}{C} \int_{t_0}^{t} j(x)\, dx + \frac{q(t_0)}{C} \tag{22}$$

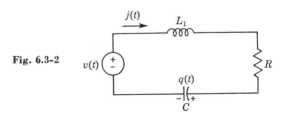

Fig. 6.3-2

Furthermore, the charge q on the capacitor is related to the current j according to

$$q(t) = \int_{t_0}^{t} j(x)\,dx + q(t_0) \tag{23}$$

The techniques for solving a differential equation may be applied here if we first substitute (23) into (22) to get

$$v = L_1 \frac{d^2q}{dt^2} + R\frac{dq}{dt} + \frac{q}{C} \tag{24}$$

Now let the variables v, q, and j be distributions. They will be related according to (24) and the equation $j = q^{(1)}$. The inverse 1_+h in $\mathfrak{D}'_R$ of

$$L_1 \delta^{(2)} + R\delta^{(1)} + \frac{1}{C}\delta$$

is specified by Theorem 6.3-1. If v has its support contained in $t_0 \le t < \infty$ and if the system is initially at rest [i.e., if $q(t) = j(t) = 0$ for $-\infty < t < t_0$], then the resulting charge $q(t)$ is given for all t by

$$q = (1_+h) * v$$

On the other hand, if the system is not initially at rest, its natural response (which is a continuous function) is specified by the initial conditions $q(t_0 - \varepsilon)$ and $j(t_0 - \varepsilon)$, where $\varepsilon > 0$; and according to (13) and (20) this natural response is

$$u_\delta(t) = [L_1 j(t_0 - \varepsilon) + Rq(t_0 - \varepsilon)]h(t - t_0 + \varepsilon) \\ + L_1 q(t_0 - \varepsilon) h^{(1)}(t - t_0 + \varepsilon)$$

Then, $q(t)$ is given for all time by

$$q = (1_+h) * v + u_\delta$$

When v is a regular distribution, we may set $\varepsilon = 0$ to get an expression corresponding to (17) for q. Finally, $j = q^{(1)}$.

PROBLEMS

1 Let a_ν and b_ν be complex constants. Find the inverses in $\mathfrak{D}'_R$ of
a. $a_1 \delta^{(1)} + a_0 \delta$
b. $a_2 \delta^{(2)} + a_0 \delta$
c. $(a_1 \delta^{(1)} + a_0 \delta) * (b_1 \delta^{(1)} + b_0 \delta)$
d. $(a_1 \delta^{(1)} + a_0 \delta)^{*m}$. Here, m is a positive integer.

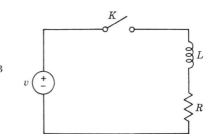

Fig. 6.3-3

2 Show that the inverse in $\mathcal{D}'_R$ of

$$a_n \delta^{(n)} + a_{n-1}\delta^{(n-1)} + \cdots + a_0 \delta$$

where the a_ν are constants, is

$$\frac{1}{a_n}[1_+(t)e^{\gamma_1 t}] * [1_+(t)e^{\gamma_2 t}] * \cdots * [1_+(t)e^{\gamma_n t}]$$

where the γ_ν are the roots of the equation

$$a_n x^n + a_{n-1}x^{n-1} + \cdots + a_0 x = 0$$

3 Consider the circuit of Fig. 6.3-3, where v is an electromotive force, K is a switch, L is an inductance, and R is a resistance. When v is a given continuous function, one can speak of the closing of the switch at some given time t_0 and the application of $v(t)1_+(t - t_0)$ to the rest of the circuit. Does this situation possess a sense when v is an arbitrary distribution in $\mathcal{D}'$? Justify your answer.

4 For $t \geq 0$ let

$$u^{(2)}(t) + 2u^{(1)}(t) + u(t) = 1_+(t) \sin t$$

and let $u(0) = 0$, $u^{(1)}(0) = 1$. Find $u(t)$ for $t > 0$.

5 In the circuit of Fig. 6.3-2 let $L_1 = 1$ henry, $R = 3$ ohms, $C = \frac{1}{2}$ farad, $q(0-) = 1$ coulomb, $j(0-) = 2$ amperes, and

$$v(t) = \delta^{(1)}(t) + 1_+(t) \sin t \qquad \text{volts}$$

Find the current $j(t)$ flowing in the circuit for all time t.

6 Determine the inverse in $\mathcal{D}'_R$ of

$$f(t) = a_n \delta^{(n)}(t) + a_{n-1}\delta^{(n-1)}(t) + \cdots + a_0 \delta(t) + a_{-1}1_+(t)$$
$$+ a_{-2}t1_+(t) + \cdots + a_{-m}\frac{t^{m-1}1_+(t)}{(m-1)!}$$

If $f * u = g$ and if g is of order r on some closed interval I, what is the order of u on this interval?

7 Let f be the distribution specified in the preceding problem and consider the equation $f * u = g$, where g is in $\mathcal{D}'_R$. This is an integrodifferential equation. How may initial conditions be introduced and how many such conditions can one specify?

8 Use the results of Probs. 6 and 7 to solve Prob. 5 without converting the integrodifferential equation into a differential equation.

6.4 MIKUSIŃSKI'S OPERATIONAL CALCULUS

Mikusiński's operational calculus, which comprises an alternative theory of generalized functions, is intimately related to the convolution process. It is, therefore, fitting to discuss it briefly at this point and to compare it with distribution theory.

For the sake of an analogy that we shall point out in a moment, consider the class of all integers. We may multiply any pair of integers, and we shall always obtain an integer as a result. This is not so with division. The quotient m/n of the arbitrary integers m and n ($n \neq 0$) does not always exist as an integer. However, if we increase our class of elements from that of the integers to that of the rational numbers, we can introduce "division by a nonzero element" as an always possible operation. Then, any equation $ab = c$, where a and c are given rational numbers and $a \neq 0$, can always be solved for the rational number b by dividing both sides by a.

Now consider the space $\mathcal{D}'_+$ of all distributions whose supports are contained in $0 \leq t < \infty$. We have seen that convolution, which replaces multiplication, can always be performed in this space. Moreover, if f and g are in $\mathcal{D}'_+$, then $f * g$ is also in $\mathcal{D}'_+$. But convolution by an inverse in $\mathcal{D}'_+$, which replaces the division process, cannot always be performed. That is, the equation

$$f * u = g \tag{1}$$

may not be solvable in $\mathcal{D}'_+$ for u. However, in analogy with the integers and the rational numbers, one might expect that new entities can be defined in such a way that (1) always has a solution in terms of these entities and that this solution corresponds to the distributional solution of (1) when such a solution exists. This is just what Mikusiński accomplishes. In the next few paragraphs we shall very briefly describe how he does it.

Let $\mathcal{C}_+$ be the space of all complex-valued functions whose supports are contained in the one-dimensional real half-line $0 \leq t < \infty$ and which are continuous for $0 < t < \infty$ and continuous from the right at $t = 0$ [that is, every a in $\mathcal{C}_+$ satisfies $a(0) = a(0+)$]. Thus, an ordinary discontinuity at the origin is allowed.

We shall now let convolution take the place of multiplication. In conformity with Mikusiński's notation, we shall denote the convolution of two arbitrary elements e and u in $\mathcal{C}_+$ by eu rather than by $e * u$. It is easily shown that eu will also be in $\mathcal{C}_+$. Now consider the convolution equation

$$eu = h \tag{2}$$

where e and h are given elements of $\mathcal{C}_+$. We shall always assume that e is different from the zero function (that is, $e \not\equiv 0$). Equation (2) does not always have a solution u in $\mathcal{C}_+$. For example, by setting both e and h equal to one for $0 \leq t < \infty$, we get

$$\int_0^t u(x)\,dx = 1 \qquad t \geq 0$$

and this is not possible for any function u.

However, $\mathcal{C}_+$ does not contain divisors of zero. That is, if eu is the zero function, then either e or u (or both) is equal to the zero function. (The proof of this is rather long. It can be found in Mikusiński [2], chap. II, and in Erdélyi [1], pp. 16–20.) Because of this, (2) has precisely one solution u in $\mathcal{C}_+$ if it has any at all. Thus, we may without ambiguity speak of the solution u (if it exists) as the division of h by e and denote it by $u = h/e$.

It is at this point that Mikusiński suggested that, by adapting a standard procedure in abstract algebra, a unique solution u can always be assigned to (2) if we define some new entities. These entities, called *convolution quotients* or *Mikusiński operators*, are defined as follows. First, consider the *ordered pair* (a, b), where a and b are required to be elements of $\mathcal{C}_+$ with $b \not\equiv 0$. Two ordered pairs (a, b) and (c, d) are said to be equivalent if $ad = bc$. The class of all ordered pairs that are equivalent to (a, b) is by definition a convolution quotient. It is denoted by a/b.

The space of all convolution quotients will be denoted by $\mathcal{Q}$. In conformity with the above definition, the convolution quotients a/b and c/d are equal if and only if $ad = cb$. The elements of $\mathcal{Q}$ are added according to

$$\frac{a}{b} + \frac{c}{d} \triangleq \frac{ad + cb}{bd}$$

and are multiplied (i.e., convolved) according to

$$\frac{a}{b}\frac{c}{d} \triangleq \frac{ac}{bd}$$

Any element a of $\mathcal{C}_+$ can be associated with the convolution quotient $(ab)/b$, and in this way $\mathcal{C}_+$ is embedded in $\mathcal{Q}$. Moreover, $\mathcal{D}'_+$ can also be embedded in $\mathcal{Q}$ as follows. Let f be in $\mathcal{D}'_+$ and let ϕ be some element of $\mathcal{D}$ whose support is contained in $0 \leq t < \infty$. Then, $f * \phi \triangleq \theta$ is an infinitely smooth function and also has its support contained in $0 \leq t < \infty$. Hence, θ/ϕ is the convolution-quotient representation of f. One can easily show that this representation is independent of the choice of ϕ.

The unit element δ in $\mathcal{Q}$ is equal to the convolution quotient c/c because

$$\frac{c}{c}\frac{a}{b} = \frac{ca}{cb} = \frac{a}{b}$$

The zero element is $0/c$, where 0 now denotes the (identically) zero function.

The multiplication of a convolution quotient a/b by a complex number α is defined by

$$\alpha \frac{a}{b} \triangleq \frac{\alpha a}{b}$$

Just as linear differential, integrodifferential, and difference equations with constant coefficients can be represented in terms of distributional convolutions (see Examples 5.4-2 and 5.4-3 and the beginning of Sec. 9.4), it is also a fact that they can be represented as equations in Q.

In view of the afore-mentioned operations in Q, one can show that Q is a linear space. (See Appendix A.) Moreover, the multiplication can be shown to be associative, commutative, and distributive with respect to addition, and it follows that Q is a commutative algebra having a unit element. (See Taylor [1], p. 162.) Because of this, the techniques described in Sec. 6.2 for solving convolution equations can be applied to the solution of equations in Q. But now *an equation in Q always has a unique solution in Q*, in contrast to the situation for the convolution algebra $\mathcal{D}'_R$. In particular, if

$$\frac{a}{b} u = \frac{c}{d}, \quad a \neq 0$$

where u is an unknown convolution quotient, then u always exists, is unique, and is given by

$$u = \frac{bc}{ad}$$

It follows that every nonzero convolution quotient has an inverse. This is an advantage of Mikusiński's operational calculus. But it is not a crucial advantage, since the equations that arise in practical problems, such as differential and difference equations, are solvable just as easily in terms of distributional convolution and Laplace transformation as they are in Mikusiński's operational calculus.

Actually, one cannot say that either distribution theory or Mikusiński's method provides the better technique. For example, there are convolution quotients corresponding to which no distributions exist. On the other hand, Mikusiński's operational calculus cannot take into account distributions whose supports are the entire real axis, $-\infty < t < \infty$ (see Mikusiński [2], pp. 356–357), nor is it suitable for dealing with generalized functions of an n-dimensional real variable that ranges over an arbitrary region in $\mathcal{R}^n$. Thus, distribution theory can handle certain types of analyses that would be very awkward, if not altogether impossible, in Mikusiński's operational calculus. This is the case, for example, with the theory of passive systems, which is developed in Chap. 10.

7 The Fourier Transformation

★7.1 INTRODUCTION

This chapter can be divided into three parts. The first part, which consists of Secs. 7.2 and 7.3, develops the ordinary Fourier transformation for functions, assuming some rather severe restrictions on these functions such as differentiability in the ordinary sense. Our purpose here is to develop only those classical results that will be needed in our subsequent discussion of the distributional Fourier transformation. For this reason we do not extend the ordinary transformation to more general types of functions such as the piecewise-differentiable ones, even though we could do so with very little additional effort. To be sure, in Theorem 7.2-2 we do develop the Fourier transformation for locally integrable functions under the sole assumption that they are absolutely integrable over $-\infty < t < \infty$, but we do this in a distributional sense and not a classical one. Moreover, when in later sections the Fourier transformation is generalized to distributions, all locally integrable functions (or, more exactly, all regular distributions) will automatically be taken into account.

The second part, which extends through Secs. 7.4 and 7.5, takes up the Fourier transformation of distributions of slow growth. This is the generalization of the Fourier transformation that Schwartz discusses (Schwartz [1], vol. II, chap. VII). Moreover, it provides the foundation for the Laplace transformation of distributions, as will be discussed in the next chapter. However, only distributions of slow growth possess Fourier transforms in the Schwartz sense. Similarly, our Laplace transformation will be defined for some but not for all distributions in $\mathfrak{D}'$.

In order to extend the Fourier transformation to all of $\mathfrak{D}'$, we must go beyond the framework of distribution theory. We do this in the third part of the chapter (Secs. 7.6 to 7.10), where a new class Z of testing functions and a new dual space Z' of continuous linear functionals on Z are introduced. The elements of Z' will be called *ultradistributions*, and they turn out to be the Fourier transforms of all the distributions in $\mathfrak{D}'$.

By using different spaces of testing functions, one can construct different types of continuous linear functionals. Some authors refer to these various types of continuous linear functionals as *generalized functions* (Gelfand and Shilov [1], A. Friedman [1]). The distributions that we have discussed and the ultradistributions that we shall discuss are examples of generalized functions.

♦7.2 THE ORDINARY FOURIER TRANSFORMATION

Assume that $f(t)$ is a locally integrable function that is absolutely integrable over $-\infty < t < \infty$:

$$\int_{-\infty}^{\infty} |f(t)|\, dt < \infty \tag{1}$$

The *Fourier transform* $\tilde{f}(\omega)$ of $f(t)$ is a function of the real one-dimensional variable ω and is defined by

$$\tilde{f}(\omega) \triangleq \int_{-\infty}^{\infty} f(t)e^{-i\omega t}\, dt \tag{2}$$

We shall refer to the operation that carries $f(t)$ into $\tilde{f}(\omega)$ as the *Fourier transformation* (and, at times, as the *direct Fourier transformation*), and we shall denote it by $\mathfrak{F}$. The notations $\mathfrak{F}f$ and $\tilde{f}$ will be used interchangeably.

Because of (1), $\tilde{f}(\omega)$ is bounded for all ω. In addition, $\tilde{f}(\omega)$ is uniformly continuous over $-\infty < \omega < \infty$. Indeed, if η is an increment in ω, we may write

$$|\tilde{f}(\omega + \eta) - \tilde{f}(\omega)| = \left| \int_{-\infty}^{\infty} f(t) e^{-i\omega t} (e^{-i\eta t} - 1) \, dt \right|$$

$$\leq \int_{-\infty}^{\infty} |f(t)| \, |e^{-i\eta t} - 1| \, dt$$

$$= \int_{-\infty}^{\infty} |f(t)| \left| 2 \sin \frac{\eta t}{2} \right| dt$$

$$\leq 2 \left(\int_{-\infty}^{-T} + \int_{T}^{\infty} \right) |f(t)| \, dt + \int_{-T}^{T} |f(t)| \left| \eta t \frac{\sin(\eta t/2)}{\eta t/2} \right| dt$$

$$\leq 2 \left(\int_{-\infty}^{-T} + \int_{T}^{\infty} \right) |f(t)| \, dt + |\eta| T \int_{-T}^{T} |f(t)| \, dt$$

Given an $\varepsilon > 0$, we can choose T so large and subsequently $|\eta|$ so small that the right-hand side of the last inequality becomes less than ε. Hence our assertion.

The main objective of this section is to establish the following result:

Theorem 7.2-1

If the locally integrable function $f(t)$ is absolutely integrable for $-\infty < t < \infty$ and differentiable in the ordinary sense throughout some open interval $a < t < b$, then

$$f(t) = \lim_{y \to \infty} \frac{1}{2\pi} \int_{-y}^{y} \tilde{f}(\omega) e^{i\omega t} \, d\omega \tag{3}$$

for $a < t < b$.

The right-hand side of (3) is called the *inverse Fourier transform* of $\tilde{f}(\omega)$ and is denoted by $\mathfrak{F}^{-1} f$, the corresponding operation now being symbolized by $\mathfrak{F}^{-1}$. Moreover, it is the Cauchy principal value of the improper integral

$$\frac{1}{2\pi} \int_{-\infty}^{\infty} \tilde{f}(\omega) e^{i\omega t} \, dt \tag{4}$$

However, in almost all of our subsequent applications of the ordinary Fourier transformation, $\tilde{f}(\omega)$ will turn out to be locally integrable and absolutely integrable over $-\infty < \omega < \infty$. In this case, (4) (as an ordinary integral) will be the same as the right-hand side of (3) and our inverse Fourier transform can be simply written as

$$f(t) = \mathfrak{F}^{-1} \tilde{f}(\omega) = \frac{1}{2\pi} \int_{-\infty}^{\infty} \tilde{f}(\omega) e^{i\omega t} \, d\omega \tag{5}$$

We shall present here a standard proof of Theorem 7.2-1. Then we shall show that (3) possesses a distributional sense under the sole assumption that $f(t)$ is locally integrable and absolutely integrable over $-\infty < t < \infty$ (see Theorem 7.2-2).

We proceed by means of three lemmas. The first two are called the Riemann-Lebesgue lemmas.

Lemma 1

If $f(t)$ is a (Lebesgue) integrable function over the finite interval $a < t < b$ and y is a real variable, then

$$\lim_{y \to \infty} \int_a^b f(t) \sin yt \, dt = 0$$

Proof: First note that the lemma is true if $f(t)$ equals a constant c; indeed,

$$\int_a^b c \sin yt \, dt = \frac{c}{y} (\cos ya - \cos yb)$$

and the right-hand side of this equation approaches zero as $y \to \infty$.

Next let $h(t)$ be a step function over $a < t < b$; that is, let $h(t)$ be such that the interval $a < t < b$ can be divided into a *finite* number of subintervals over each of which $h(t)$ is a constant. By applying the argument given in the preceding paragraph to each subinterval, we see that

$$\lim_{y \to \infty} \int_a^b h(t) \sin yt \, dt = 0 \tag{6}$$

Now, by a standard result of Lebesgue integral theory (Kestelman [1], p. 166), for any given $\varepsilon > 0$ the function $f(t)$ can be approximated by a step function $h(t)$ in such a way that

$$\int_a^b |f(t) - h(t)| \, dt < \varepsilon$$

Finally, for all real values of y, we can write

$$\left| \int_a^b f(t) \sin yt \, dt - \int_a^b h(t) \sin yt \, dt \right| \leq \int_a^b |f(t) - h(t)| \, dt < \varepsilon$$

This result combined with (6) implies the desired conclusion.

Lemma 2

If the locally integrable function $f(t)$ is absolutely integrable over $b \leq t < \infty$ and y is a real variable, then

$$\lim_{y \to \infty} \int_b^\infty f(t) \sin yt \, dt = 0 \tag{7}$$

Proof: Because of the absolute integrability of $f(t)$, we have that, for any given $\varepsilon > 0$, there exists a T such that

$$\left| \int_T^\infty f(t) \sin yt \, dt \right| \leq \int_T^\infty |f(t)| \, dt < \varepsilon \tag{8}$$

this inequality holding for all real values of y. Also, by Lemma 1, there exists a Y such that for $y > Y$

$$\left| \int_b^T f(t) \sin yt \, dt \right| < \varepsilon \tag{9}$$

(8) and (9) imply (7).

Lemma 3

Let the locally integrable function $f(t)$ be absolutely integrable over $-\infty < t < \infty$, let $\tilde{f}(\omega)$ be given by (2), and let

$$g_y(t) \triangleq \frac{1}{2\pi} \int_{-y}^{y} \tilde{f}(\omega) e^{i\omega t} \, d\omega \tag{10}$$

Then,

$$g_y(t) = \frac{1}{\pi} \int_{-\infty}^{\infty} f(t-x) \frac{\sin yx}{x} \, dx \tag{11}$$

Proof: By replacing t by x in (2) and then combining the result with (10), we get

$$g_y(t) = \frac{1}{2\pi} \int_{-y}^{y} d\omega \int_{-\infty}^{\infty} f(x) e^{i\omega(t-x)} \, dx \tag{12}$$

The integrand in (12) is absolutely integrable over the domain $-y \leq \omega \leq y$, $-\infty < x < \infty$. Therefore, the order of integration in (12) may be interchanged, according to Fubini's theorem (Kestelman [1], p. 206):

$$g_y(t) = \frac{1}{2\pi} \int_{-\infty}^{\infty} dx \int_{-y}^{y} f(x) e^{i\omega(t-x)} \, d\omega$$
$$= \frac{1}{\pi} \int_{-\infty}^{\infty} f(x) \frac{\sin y(t-x)}{t-x} \, dx$$

A change of variable will alter the last integral into the right-hand side of (11). Q.E.D.

Proof of Theorem 7.2-1: By comparing (3) and (10), we see that we need only establish the following limit for $a < t < b$:

$$\lim_{y \to \infty} g_y(t) = f(t) \tag{13}$$

176 **Distribution theory and transform analysis** *Chap. 7*

It is a fact that

$$\lim_{\substack{\alpha \to -\infty \\ \beta \to \infty}} \int_\alpha^\beta \frac{\sin \tau}{\tau} d\tau = \pi \tag{14}$$

Setting $\tau = yx$ ($y > 0$), we can write the improper integral,

$$f(t) = \frac{1}{\pi} \int_{-\infty}^\infty f(t) \frac{\sin yx}{x} dx$$

By invoking Lemma 3, we can also write

$$f(t) - g_y(t) = \frac{1}{\pi} \int_{-\infty}^\infty [f(t) - f(t-x)] \frac{\sin yx}{x} dx \tag{15}$$

Let us break up the improper integral on the right-hand side of (15) as follows:

$$\int_{-\infty}^\infty = \int_{-\infty}^{-1} + \int_{-1}^0 + \int_0^1 + \int_1^\infty \tag{16}$$

We shall first establish that the integral over $0 \leq x \leq 1$ approaches zero as $y \to \infty$. Let

$$h(x) \triangleq \begin{cases} \dfrac{f(t) - f(t-x)}{x} & x \neq 0 \\ f^{(1)}(t) & x = 0 \end{cases}$$

(Here, we are considering t as an arbitrary but fixed number.) Under the stated assumptions on f in the hypothesis of Theorem 7.2-1, $h(x)$ is an integrable function on $0 \leq x \leq 1$. Consequently, by Lemma 1

$$\int_0^1 \frac{f(t) - f(t-x)}{x} \sin yx \, dx = \int_0^1 h(x) \sin yx \, dx \to 0$$

as $y \to \infty$.

Next we shall show that

$$\int_1^\infty [f(t) - f(t-x)] \frac{\sin yx}{x} dx$$

also converges to zero as $y \to \infty$. Indeed, in view of the convergence of the improper integral

$$\int_1^\infty \frac{\sin v}{v} dv$$

we have that

$$\int_1^\infty f(t) \frac{\sin yx}{x} dx = f(t) \int_y^\infty \frac{\sin v}{v} dv \to 0$$

as $y \to \infty$. Furthermore, since $|f(t-x)/x| \leq |f(t-x)|$ for $1 \leq x < \infty$, $f(t-x)/x$ is absolutely integrable over this interval. Thus, by Lemma 2,

$$\int_1^\infty \frac{f(t-x)}{x} \sin yx \, dx \to 0$$

as $y \to \infty$.

Similar arguments show that the integrals in (16) over the intervals $-\infty < t \leq 1$ and $-1 \leq t \leq 0$ also approach zero as $y \to \infty$. Thus, (15) converges to zero, which proves the theorem.

Theorem 7.2-1 also asserts the *uniqueness property* of the Fourier transformation, which we explicitly display as

Corollary 7.2-1a

If $f(t)$ and $g(t)$ satisfy the hypotheses of Theorem 7.2-1 and if their Fourier transforms $\tilde{f}(\omega)$ and $\tilde{g}(\omega)$ are equal everywhere, then $f(t) = g(t)$ for $a < t < b$.

Proof: For $a < t < b$,

$$f(t) = \mathfrak{F}^{-1}\tilde{f}(\omega) = \mathfrak{F}^{-1}\tilde{g}(\omega) = g(t) \quad \text{Q.E.D.}$$

At this point it may be interesting to take note of a formal development for the inverse Fourier transformation, which one rather commonly finds in the engineering literature. Throwing caution to the wind, we proceed as follows. With $f(t)$ being an ordinary function, we write

$$\tilde{f}(\omega) = \int_{-\infty}^\infty f(x) e^{-i\omega x} \, dx$$

and

$$\frac{1}{2\pi} \int_{-\infty}^\infty \tilde{f}(\omega) e^{i\omega t} \, d\omega = \frac{1}{2\pi} \int_{-\infty}^\infty d\omega \int_{-\infty}^\infty f(x) e^{i\omega(t-x)} \, dx$$

After formal inversion of the order of integration, this becomes

$$\frac{1}{2\pi} \int_{-\infty}^\infty f(x) \left[\int_{-\infty}^\infty \cos \omega(t-x) \, d\omega \right] dx$$

According to Example 2.3-4, the inner integral is a symbolic expression for $2\pi \delta(t-x)$, so that we finally obtain the inverse Fourier transform

$$\frac{1}{2\pi} \int_{-\infty}^\infty \tilde{f}(\omega) e^{i\omega t} \, d\omega = \int_{-\infty}^\infty f(x) \delta(t-x) \, dx = f(t)$$

Needless to say, these symbolic manipulations by no means constitute a rigorous argument. Nevertheless, a proof along these lines can still be constructed if these symbolic expressions are treated as distributions. Indeed, we may state

178 *Distribution theory and transform analysis* Chap. 7

Theorem 7.2-2

Let the locally integrable function $f(t)$ be absolutely integrable over $-\infty < t < \infty$. Then

$$f(t) = \lim_{y \to \infty} g_y(t) \triangleq \lim_{y \to \infty} \frac{1}{2\pi} \int_{-y}^{y} \tilde{f}(\omega) e^{i\omega t}\, d\omega$$

where it is now understood that this is a limit in $\mathfrak{D}'$; that is, as $y \to \infty$ the regular distributions $g_y(t)$ converge in $\mathfrak{D}'$ to the regular distribution $f(t)$.

Proof: In view of Lemma 3, we wish to show that as $y \to \infty$

$$\frac{1}{\pi} \int_{-\infty}^{\infty} f(t - x) \frac{\sin yx}{x}\, dx \tag{17}$$

converges in $\mathfrak{D}'$ to the regular distribution $f(t)$. Let

$$k_y(x) \triangleq \begin{cases} \dfrac{\sin yx}{x} & |x| \leq 1 \\ 0 & |x| > 1 \end{cases}$$

We shall first show that, if

$$\frac{1}{\pi} \int_{-\infty}^{\infty} f(t - x) k_y(x)\, dx \tag{18}$$

converges in $\mathfrak{D}'$ to some limit as $y \to \infty$, then (17) also converges in $\mathfrak{D}'$ to the same limit. Indeed, for any ϕ in $\mathfrak{D}$

$$\left\langle \int_{-\infty}^{\infty} f(t - x) \left[\frac{\sin yx}{x} - k_y(x) \right] dx,\, \phi(t) \right\rangle$$
$$= \int_{-\infty}^{\infty} dt \int_{|x|>1} f(t - x) \frac{\sin yx}{x} \phi(t)\, dx$$

The last integrand is absolutely integrable over the entire (x, t) plane because $|(\sin yx)/x| \leq |y|$ for all x and y, $f(t)$ is absolutely integrable over $-\infty < t < \infty$, and $\phi(t)$ is continuous and of bounded support. Therefore, by Fubini's theorem we may interchange the order of integration to get

$$\int_{|x|>1} \frac{\sin yx}{x} \int_{-\infty}^{\infty} f(t - x) \phi(t)\, dt\, dx$$

where the inner integral is an absolutely integrable function of x over $-\infty < x < \infty$. By Lemma 2 the last expression approaches zero as $y \to \infty$. Thus, to establish the theorem, we need merely show that (18) converges in $\mathfrak{D}'$ to $f(t)$ as $y \to \infty$.

As was indicated in Example 5.4-1, (18) is equal (in the distributional sense) to the distributional convolution

$$\frac{1}{\pi} f(t) * k_y(t)$$

in view of the fact that $k_y(t)$ has a bounded support.

Furthermore, $k_y(t)$ converges in $\mathfrak{D}'$ to $\pi \delta(t)$. Indeed, for every ϕ in $\mathfrak{D}$

$$\lim_{y \to \infty} \langle k_y(t), \phi(t) \rangle = \lim_{y \to \infty} \int_{-1}^{1} \frac{\sin yt}{t} \phi(t)\, dt \tag{19}$$

Consider the function

$$\phi(t) 1_+(-t - 1) 1_+(t - 1)$$

and compare (3), (10), (11). It follows from Theorem 7.2-1 that the right-hand side of (19) converges to $\pi \phi(0) = \langle \pi \delta, \phi \rangle$.

Finally, for every value of y, the support of $k_y(t)$ is contained in the interval $-1 \leq t \leq 1$. Therefore, by the continuity of the convolution process (see Theorem 5.6-1),

$$\lim_{y \to \infty} \frac{1}{\pi} f(t) * k_y(t) = \frac{1}{\pi} f(t) * [\lim_{y \to \infty} k_y(t)]$$
$$= f(t) * \delta(t) = f(t)$$

where now the convergence is understood to be in $\mathfrak{D}'$. Q.E.D.

Since a regular distribution determines the function generating it almost everywhere, we may extend the uniqueness assertion for the Fourier transformation as follows:

Corollary 7.2-2a

If the locally integrable functions $f(t)$ and $g(t)$ are absolutely integrable over $-\infty < t < \infty$ and if their Fourier transforms $\tilde{f}(\omega)$ and $\tilde{g}(\omega)$ are equal everywhere, then $f(t) = g(t)$ almost everywhere.

Since integration is a linear operation, so is the Fourier transformation. That is, if $f(t)$ and $g(t)$ are locally integrable and absolutely integrable over $-\infty < t < \infty$ and if α and β are any two numbers, then

$$\mathfrak{F}(\alpha f + \beta g) = \alpha \mathfrak{F} f + \beta \mathfrak{F} g$$

The inverse Fourier transformation is also a linear operation.

Another relationship between f, $\tilde{f}$, g, and $\tilde{g}$, which we shall use in our definition of the distributional Fourier transformation, is known as *Parseval's equation*. One form of it is given by

Theorem 7.2-3

If the locally integrable functions $f(t)$ and $g(t)$ are absolutely integrable over $-\infty < t < \infty$, then

$$\int_{-\infty}^{\infty} f(x)\tilde{g}(x)\,dx = \int_{-\infty}^{\infty} \tilde{f}(x)g(x)\,dx \tag{20}$$

Proof: As was pointed out near the beginning of this section, $\tilde{f}(\omega)$ and $\tilde{g}(\omega)$ are bounded and continuous for all ω. Therefore, both sides of (20) converge. Moreover,

$$\int_{-\infty}^{\infty} f(x)\tilde{g}(x)\,dx = \int_{-\infty}^{\infty} dx \int_{-\infty}^{\infty} f(x)g(y)e^{-ixy}\,dy \tag{21}$$

Since the integrand of the last iterated integral is absolutely integrable over the entire (x, y) plane, Fubini's theorem allows us to interchange the order of integration. Thus, (21) becomes

$$\int_{-\infty}^{\infty} dy \int_{-\infty}^{\infty} f(x)g(y)e^{-ixy}\,dx = \int_{-\infty}^{\infty} \tilde{f}(y)g(y)\,dy$$

which proves this theorem.

PROBLEMS

1 Establish (14). HINT: One way to do this is to integrate e^{iz}/z around the upper half-plane, indenting the path of integration around the pole at the origin.

2 Assuming that $g(t)$ is twice differentiable in the ordinary sense and that $f(t)$, $g(t)$, $g^{(1)}(t)$, and $g^{(2)}(t)$ are continuous and absolutely integrable over $-\infty < t < \infty$, establish the following forms for Parseval's equations:

$$\int_{-\infty}^{\infty} \tilde{f}(x)\tilde{g}(x)\,dx = 2\pi \int_{-\infty}^{\infty} f(x)g(-x)\,dx$$

$$\int_{-\infty}^{\infty} \tilde{f}(x)\overline{\tilde{g}(x)}\,dx = 2\pi \int_{-\infty}^{\infty} f(x)\overline{g(x)}\,dx$$

3 Establish the following Fourier transform formulas:

a. $\mathfrak{F}[1_+(t-a) - 1_+(b-t)] = \dfrac{e^{-i\omega a} - e^{-i\omega b}}{i\omega} \qquad a < b$

b. $\mathfrak{F}e^{-a|t|} = \dfrac{2a}{a^2 + \omega^2} \qquad a > 0$

c. $\mathfrak{F}[t^k e^{-at} 1_+(t)] = \dfrac{k!}{(a+i\omega)^{k+1}} \qquad a > 0$

d. $\mathfrak{F}\left[\dfrac{1}{a^2 + t^2}\right] = \dfrac{\pi}{|a|} e^{-|a\omega|} \qquad a$ real

4 Show that

$$\mathfrak{F}e^{-at^2} = \sqrt{\dfrac{\pi}{a}} e^{-\omega^2/4a} \qquad a > 0$$

Sec. 7.3   *The Fourier transformation*   181

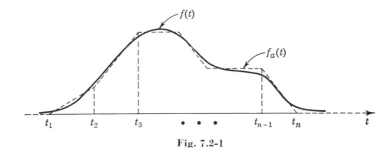

Fig. 7.2-1

HINT: Show that $\tilde{f}(\omega) \triangleq \mathfrak{F} \exp(-at^2)$ satisfies the differential equation

$$-2a\tilde{f}^{(1)}(\omega) = \omega\tilde{f}(\omega)$$

by integrating by parts in $\mathfrak{F} \exp(-at^2)$ and then noting that the result is a derivative under the integral sign. The constant of integration occurring in the solution to this differential equation can be evaluated by using the definite integral

$$\int_{-\infty}^{\infty} e^{-at^2}\,dt = \sqrt{\frac{\pi}{a}}$$

5 Let the locally integrable function $f(t)$ be absolutely integrable over $-\infty < t < \infty$ and let $f_a(t)$ be an approximation to it consisting of a finite number of straight line segments as shown in Fig. 7.2-1. Note that $f_a(t)$ is continuous and has a bounded support. Let A_k denote the change in slope in $f_a(t)$ at the point $t = t_k$, and let $F(i\omega) = \mathfrak{F}f(t)$ and $F_a(i\omega) = \mathfrak{F}f_a(t)$. Show that the approximation $F_a(i\omega)$ to $F(i\omega)$ is given by

$$F_a(i\omega) = -\frac{1}{\omega^2} \langle f_a^{(2)}(t), e^{-i\omega t}\rangle = -\frac{1}{\omega^2} \sum_{k=1}^{n} A_k e^{-i\omega t_k}$$

Also, show that a bound on the error of approximation is

$$|F(i\omega) - F_a(i\omega)| \leq \int_{-\infty}^{\infty} |f(t) - f_a(t)|\,dt$$

This technique for the approximate evaluation of the ordinary Fourier transformation can be applied to a variety of other integral transformations (Zemanian [1]).

6 Let the function $f(t)$ be such that $f(t)$ and $f^{(1)}(t)$ are continuous and absolutely integrable over $-\infty < t < \infty$. Also, let $f(t) = 0$ for $t < 0$. Set $\tilde{f}_R(\omega) = \operatorname{Re} \tilde{f}(\omega)$ and $\tilde{f}_X(\omega) = \operatorname{Im} \tilde{f}(\omega)$. Show that for $t > 0$

$$f(t) = \frac{2}{\pi}\int_0^\infty \tilde{f}_R(\omega)\cos\omega t\,d\omega = -\frac{2}{\pi}\int_0^\infty \tilde{f}_X(\omega)\sin\omega t\,d\omega$$

★7.3 THE FOURIER TRANSFORMS OF TESTING FUNCTIONS OF RAPID DESCENT

In this section we shall show that the Fourier transformation maps the space $\mathcal{S}$ *onto* itself; i.e., if $\phi(t)$ is in $\mathcal{S}_t$, then $\mathfrak{F}\phi(t) \triangleq \tilde{\phi}(\omega)$ is in $\mathcal{S}_\omega$ and, moreover, every function in $\mathcal{S}_\omega$ is the Fourier transform of some function in $\mathcal{S}_t$. We shall also show that the Fourier transformation is a *continuous linear* mapping of $\mathcal{S}$ onto itself, continuous in the sense that, if $\{\phi_\nu\}_{\nu=1}^\infty$ is a sequence that converges in $\mathcal{S}$ to ϕ, then $\{\tilde{\phi}_\nu\}_{\nu=1}^\infty$ is a sequence that converges in $\mathcal{S}$ to $\tilde{\phi} \triangleq \mathfrak{F}\phi$. These fundamental properties are essential for the definition of the Fourier transforms of distributions of slow growth, which will be discussed in the next section.

Theorem 7.3-1

If $\phi(t)$ is in $\mathcal{S}$, then its Fourier transform,

$$\tilde{\phi}(\omega) \triangleq \mathfrak{F}\phi(t) \triangleq \int_{-\infty}^{\infty} \phi(t) e^{-i\omega t}\, dt \tag{1}$$

is also in $\mathcal{S}$.

Proof: If $\phi(t)$ is in $\mathcal{S}$, its Fourier transform certainly exists. Moreover, we may differentiate (in the ordinary sense) the right-hand side of (1) under the integral sign with respect to ω to get

$$\tilde{\phi}^{(1)}(\omega) = \int_{-\infty}^{\infty} (-it)\phi(t) e^{-i\omega t}\, dt \tag{2}$$

This is because the right-hand side of (2) converges uniformly for $-\infty < \omega < \infty$ in view of the fact that $t\phi(t)$ is also in $\mathcal{S}$ (Titchmarsh [1], p. 59). (Note that the symbol $\tilde{\phi}^{(1)}$ denotes the derivative of the Fourier transform of ϕ and not the Fourier transform of the derivative of ϕ.) In fact, for any nonnegative integer k, $t^k \phi(t)$ is again in $\mathcal{S}$, so that $\tilde{\phi}(\omega)$ is infinitely smooth and

$$\tilde{\phi}^{(k)}(\omega) = \int_{-\infty}^{\infty} (-it)^k \phi(t) e^{-i\omega t}\, dt \tag{3}$$

Second, $\tilde{\phi}^{(k)}(\omega)$ is of rapid descent as $\omega \to \infty$. Indeed, integrating (3) by parts m times and using the fact that as $t \to \infty$

$$\frac{d^\nu}{dt^\nu}[t^k \phi(t)] \to 0 \qquad \nu = 0, 1, 2, \ldots$$

we may write

$$|(i\omega)^m \tilde{\phi}^{(k)}(\omega)| = \left| \int_{-\infty}^{\infty} e^{-i\omega t} \frac{d^m}{dt^m}[(-it)^k \phi(t)]\, dt \right|$$

$$\leq \int_{-\infty}^{\infty} \left| \frac{d^m}{dt^m}[t^k \phi(t)] \right| dt \tag{4}$$

The fact that $\phi(t)$ is in $\mathcal{S}$ also implies that the integral on the right-hand side of (4) is bounded (say by a constant C_{mk} that depends upon m and k). Thus,

$$|\omega^m \phi^{(k)}(\omega)| \leq C_{mk} \qquad -\infty < \omega < \infty$$

for every pair of nonnegative integers m and k. This completely establishes our assertion. Q.E.D.

We have just shown that the Fourier transformation maps the space $\mathfrak{S}$ into itself. However, even more is true: the Fourier transformation maps the space $\mathfrak{S}$ *onto* itself.

Corollary 7.3-1a

Every function in $\mathfrak{S}_\omega$ is the Fourier transform of a function in $\mathfrak{S}_t$. In fact, the Fourier transformation and its inverse provide a one-to-one correspondence between $\mathfrak{S}_\omega$ and $\mathfrak{S}_t$.

Proof: By applying the same proof as that of the preceding theorem to the inverse Fourier transform

$$\mathfrak{F}^{-1}\tilde{\phi}(\omega) = \frac{1}{2\pi} \int_{-\infty}^{\infty} \tilde{\phi}(\omega) e^{i\omega t}\, dt$$

of an arbitrary function $\tilde{\phi}(\omega)$ in $\mathfrak{S}_\omega$, we see that $\mathfrak{F}^{-1}\tilde{\phi}(\omega)$ is also in $\mathfrak{S}_t$. The uniqueness of the Fourier transformation establishes the one-to-one correspondence between $\mathfrak{S}_\omega$ and $\mathfrak{S}_t$.

Theorem 7.3-2

The Fourier transformation and its inverse are continuous linear mappings of $\mathfrak{S}$ onto itself.

Proof: Only the direct Fourier transformation will be considered, since the proof for its inverse is exactly the same.

The direct Fourier transformation is clearly linear. To show its continuity, assume that the sequence $\{\phi_\nu\}_{\nu=1}^\infty$ converges in $\mathfrak{S}$. Because of the afore-mentioned linearity, we may restrict the proof to the case where the limit function of this sequence is zero. We wish to show that $\{\tilde{\phi}_\nu\}_{\nu=1}^\infty$ also converges in $\mathfrak{S}$ to zero. With m and k being nonnegative integers, we may write

$$(i\omega)^m \tilde{\phi}_\nu^{(k)}(\omega) = \int_{-\infty}^{\infty} e^{-i\omega t} \frac{d^m}{dt^m}[(-it)^k \phi_\nu(t)]\, dt$$

$$= (-i)^k \sum_{\mu=0}^{m} \binom{m}{\mu} \int_{-\infty}^{\infty} e^{-i\omega t} \left(\frac{d^\mu}{dt^\mu} t^k\right) \left[\frac{d^{m-\mu}}{dt^{m-\mu}} \phi_\nu(t)\right] dt$$

where $\binom{m}{\mu}$ are the binomial coefficients

$$\binom{m}{\mu} \triangleq \frac{m!}{\mu!(m-\mu)!}$$

Hence,

$$|\omega^m \tilde{\phi}_\nu^{(k)}(\omega)| \leq \sum_{\mu=0}^{\inf(k,m)} \binom{m}{\mu} \frac{k!}{(k-\mu)!} \int_{-\infty}^{\infty} \frac{|(t^2+1)t^{k-\mu}\phi_\nu^{(m-\mu)}(t)|}{t^2+1} dt$$

By the definition of convergence in $\mathfrak{S}$, for each fixed choice of the nonnegative integers m, k, and μ there exists a sequence of constants $\{C_{mk\mu\nu}\}_{\nu=1}^{\infty}$ that converges to zero and is such that the quantities

$$|(t^2+1)t^{k-\mu}\phi_\nu^{(m-\mu)}(t)|$$

are bounded by the corresponding constants in the said sequence. Thus,

$$|\omega^m \tilde{\phi}_\nu^{(k)}(\omega)| \leq \pi \sum_{\mu=0}^{\inf(k,m)} \binom{m}{\mu} \frac{k!}{(k-\mu)!} C_{mk\mu\nu} \to 0 \qquad (5)$$

as $\nu \to \infty$. This shows that for each fixed pair of m and k the left-hand side of (5) converges uniformly to zero for $-\infty < \omega < \infty$. In other words, $\{\tilde{\phi}_\nu\}_{\nu=1}^{\infty}$ converges in $\mathfrak{S}$ to zero. This completes the proof.

★7.4 THE FOURIER TRANSFORMS OF DISTRIBUTIONS OF SLOW GROWTH

Parseval's equation provides a definition for the Fourier transforms of distributions of slow growth. If the locally integrable function $f(t)$ is absolutely integrable for $-\infty < t < \infty$ and if ϕ is a testing function of rapid descent, then their respective Fourier transforms $\tilde{f}$ and $\tilde{\phi}$ certainly exist and one form of Parseval's equation reads

$$\int_{-\infty}^{\infty} \tilde{f}(\omega)\phi(\omega)\,d\omega = \int_{-\infty}^{\infty} f(t)\tilde{\phi}(t)\,dt \qquad (1)$$

according to Theorem 7.2-3. In our customary notation, this becomes

$$\langle \tilde{f}, \phi \rangle = \langle f, \tilde{\phi} \rangle \qquad (2)$$

or

$$\langle \mathfrak{F}f, \phi \rangle = \langle f, \mathfrak{F}\phi \rangle$$

We may generalize relation (2) by letting f be any distribution of slow growth. As ϕ traverses $\mathfrak{S}$, (2) will then define $\tilde{f}$ as a functional on $\mathfrak{S}$. In words, the Fourier transform $\tilde{f}$ of a distribution f of slow growth is defined as that functional which assigns to each ϕ in $\mathfrak{S}$ the same number that f assigns to the Fourier transform $\tilde{\phi}$ of ϕ.

Other definitions of $\tilde{f}$ can be constructed by using different forms of Parseval's equation. For instance, (1) can be replaced by

$$\int_{-\infty}^{\infty} \tilde{f}(\omega)\tilde{\phi}(\omega)\,d\omega = 2\pi \int_{-\infty}^{\infty} f(t)\phi(-t)\,dt$$

which leads to the definition

$$\langle \tilde{f}, \tilde{\phi} \rangle \triangleq 2\pi \langle f, \check{\phi} \rangle$$

where, as usual, $\check{\phi}$ denotes the transpose of ϕ. Similarly, when we replace (1) by

$$\int_{-\infty}^{\infty} \tilde{f}(\omega) \overline{\tilde{\phi}(\omega)} \, d\omega = 2\pi \int_{-\infty}^{\infty} f(t) \overline{\phi(t)} \, dt$$

we get still another possible definition as follows:

$$\langle \tilde{f}, \bar{\tilde{\phi}} \rangle \triangleq 2\pi \langle f, \bar{\phi} \rangle$$

Since $\mathfrak{F}\check{\phi} = 2\pi\check{\tilde{\phi}}$ and $\mathfrak{F}\bar{\tilde{\phi}} = 2\pi\bar{\phi}$ and since ϕ, $\check{\phi}$, $\tilde{\phi}$, $\check{\tilde{\phi}}$, and $\bar{\tilde{\phi}}$ are all in $\mathfrak{S}$ whenever any one of them is, it follows that all these definitions are equivalent.

Theorem 7.4-1

If f is a distribution of slow growth, then its Fourier transform $\tilde{f}$ is also a distribution of slow growth.

Proof: The linearity of $\tilde{f}$ being clear, consider its continuity. If $\{\phi_\nu\}_{\nu=1}^{\infty}$ converges in $\mathfrak{S}$ to zero, then, as was shown in the preceding section, $\{\tilde{\phi}_\nu\}_{\nu=1}^{\infty}$ also converges in $\mathfrak{S}$ to zero. Now, as $\nu \to \infty$

$$\langle \tilde{f}, \phi_\nu \rangle = \langle f, \tilde{\phi}_\nu \rangle \to 0$$

and hence $\tilde{f}$ is a continuous linear functional on $\mathfrak{S}$, as was asserted.

Relation (2) also serves as a definition of the inverse Fourier transformation $\mathfrak{F}^{-1}$ of distributions of slow growth. If we set $\mathfrak{F}f = g$ and $\mathfrak{F}\phi = \psi$, we may rewrite (2) as

$$\langle \mathfrak{F}^{-1}g, \psi \rangle = \langle g, \mathfrak{F}^{-1}\psi \rangle \qquad (3)$$

where g is in $\mathfrak{S}'$ and ψ is in $\mathfrak{S}$. Thus, the inverse Fourier transform of an arbitrary distribution g in $\mathfrak{S}'$ is that functional which assigns to each ψ in $\mathfrak{S}$ the same number that g assigns to the inverse Fourier transform of ψ. $\mathfrak{F}^{-1}g$ can again be shown to be a distribution of slow growth by the same sort of argument as was used for Theorem 7.4-1.

Since, with f in $\mathfrak{S}'$ and ϕ in $\mathfrak{S}$

$$\langle \mathfrak{F}^{-1}\mathfrak{F}f, \phi \rangle = \langle \mathfrak{F}f, \mathfrak{F}^{-1}\phi \rangle = \langle f, \mathfrak{F}\mathfrak{F}^{-1}\phi \rangle = \langle f, \phi \rangle$$

it follows that $\mathfrak{F}^{-1}\mathfrak{F}f = f$. Similarly, $\mathfrak{F}\mathfrak{F}^{-1}f = f$. Thus, the Fourier transform and its inverse provide one-to-one mappings of the space $\mathfrak{S}'$ onto itself. It also follows that $\mathfrak{F}f = 0$ if and only if $f = 0$. (Here f is the zero *distribution* but, when f is taken to be a function, its values may be different from zero on a set of measure zero.)

It may be worth noting that our present definition of the Fourier transformation is not applicable when f is an arbitrary distribution in $\mathfrak{D}'$. This is because $\mathfrak{F}\phi$ will not be in $\mathfrak{D}$ when ϕ is in $\mathfrak{D}$ and $\phi \not\equiv 0$. Consequently, the right-hand side of (2) may not have a sense. (That $\mathfrak{F}\phi$ is not in $\mathfrak{D}$ when $\phi \not\equiv 0$ follows from the fact that $\mathfrak{F}\phi$ is an entire function and, therefore, can be zero only at isolated points. See Sec. 7.6 for a discussion of this point.) As was mentioned in the introduction to this chapter, by employing still another space of testing functions and its dual space of continuous linear functionals, it becomes possible to construct the Fourier transform of any distribution in $\mathfrak{D}'$. It will be constructed in Secs. 7.6 to 7.8.

It is not difficult to demonstrate that, as ϕ traverses $\mathfrak{S}$, (1) uniquely determines the regular distribution corresponding to the ordinary Fourier transform of a locally integrable function that is absolutely integrable over $-\infty < t < \infty$. (Show this.) Thus, the ordinary Fourier transformation is a special case of our distributional Fourier transformation.

Multiplication of $f(t)$ by $-it$ corresponds to the differentiation of $\tilde{f}(\omega)$. More generally, for any positive integer k,

$$\mathfrak{F}[(-it)^k f(t)] = \tilde{f}^{(k)}(\omega) \tag{4}$$

[We again remind the reader that $\tilde{f}^{(k)}(\omega)$ denotes $(\mathfrak{F}f)^{(k)}$ and not $\mathfrak{F}(f^{(k)})$.] Formula (4) can be proved as follows. For ϕ in $\mathfrak{S}$,

$$\langle \tilde{f}^{(1)}(\omega), \phi(\omega) \rangle = \langle \tilde{f}(\omega), -\phi^{(1)}(\omega) \rangle = \langle f(t), -\mathfrak{F}\phi^{(1)}(\omega) \rangle$$

Upon integrating by parts, we get

$$\left\langle f(t), -\int_{-\infty}^{\infty} \phi^{(1)}(\omega) e^{-i\omega t}\, d\omega \right\rangle = \left\langle -f(t), it \int_{-\infty}^{\infty} \phi(\omega) e^{-i\omega t}\, d\omega \right\rangle$$
$$= \langle -itf(t), \mathfrak{F}\phi(\omega) \rangle = \langle \mathfrak{F}[-itf(t)], \phi(\omega) \rangle$$

By repetition, (4) is obtained.

A similar development shows that differentiation of $f(t)$ corresponds to multiplication of $\tilde{f}(\omega)$ by $i\omega$. That is,

$$\mathfrak{F}f^{(k)}(t) = (i\omega)^k \tilde{f}(\omega) \tag{5}$$

It can just as easily be shown that the shifting of $f(t)$ by the real amount τ is converted by the Fourier transformation into the multiplication of $\tilde{f}(\omega)$ by $e^{-i\omega\tau}$:

$$\mathfrak{F}f(t-\tau) = \tilde{f}(\omega) e^{-i\omega\tau} \tag{6}$$

Analogously, we have

$$\mathfrak{F}[f(t)e^{-i\tau t}] = \tilde{f}(\omega + \tau) \tag{7}$$

Also, the formula for the change of scale of the independent variable is

$$\mathfrak{F}f(at) = \frac{1}{|a|}\tilde{f}\left(\frac{\omega}{a}\right) \tag{8}$$

where a is a real nonzero constant.

The Fourier transformation and its inverse are clearly *linear* mappings of $\mathcal{S}'$ onto itself. Furthermore, they are *continuous*. For instance, if $\{f_\nu\}_{\nu=1}^\infty$ converges in $\mathcal{S}'$ to f, then, by the fact that $\mathcal{S}'$ is closed under convergence, f is in $\mathcal{S}'$, it has a Fourier transform $\tilde{f}$, and

$$\langle \tilde{f}_\nu, \phi \rangle = \langle f_\nu, \tilde{\phi} \rangle \to \langle f, \tilde{\phi} \rangle = \langle \tilde{f}, \phi \rangle$$

for every ϕ in $\mathcal{S}$. Thus, we have the important

Theorem 7.4-2

The Fourier transformation and its inverse are continuous linear mappings of $\mathcal{S}'$ onto itself. Consequently, if a series $\sum_{\nu=1}^{\infty} g_\nu$ converges in $\mathcal{S}'$ to g, then the Fourier transformation may be applied to this series term by term to obtain

$$\tilde{g} = \sum_{\nu=1}^{\infty} \tilde{g}_\nu$$

where the last series again converges in $\mathcal{S}'$.

This theorem constitutes still another advantage of distribution theory, since such term-by-term transformation is not in general permissible in classical analysis. The situation here is quite analogous to that for the differentiation of distributions, which was discussed in Sec. 2.4.

Theorem 7.4-3

If f is a distribution with bounded support, then its Fourier transform $\tilde{f}$ is a function that is infinitely smooth. It is given by

$$\tilde{f}(\omega) = \langle f(t), e^{-i\omega t} \rangle$$

The right-hand side is understood to be the value

$$\langle f(t), \lambda(t) e^{-i\omega t} \rangle$$

where $\lambda(t)$ is any testing function in $\mathcal{D}$ that is equal to one over some neighborhood of the support of $f(t)$.

Proof: In view of the discussion of Sec. 1.8, the quantity $\tilde{f}(\omega)$ is certainly the same for all such $\lambda(t)$. For ϕ in $\mathcal{S}$,

$$\langle \tilde{f}(\omega), \phi(\omega) \rangle = \left\langle f(t), \int_{-\infty}^{\infty} \phi(\omega) e^{-i\omega t}\, d\omega \right\rangle$$
$$= \left\langle f(t), \int_{-\infty}^{\infty} \phi(\omega) \lambda(t) e^{-i\omega t}\, d\omega \right\rangle$$

But $f(t)$ is in S'_t and $\phi(\omega)\lambda(t)e^{-i\omega t}$ is in $S_{t,\omega}$. So, by Corollary 5.3-2b,

$$\langle \tilde{f}(\omega), \phi(\omega) \rangle = \int_{-\infty}^{\infty} \phi(\omega) \langle f(t), \lambda(t)e^{-i\omega t} \rangle \, d\omega$$
$$= \langle \langle f(t), e^{-i\omega t} \rangle, \phi(\omega) \rangle$$

Finally, by Theorem 2.7-2,

$$\langle f(t), \lambda(t)e^{-i\omega t} \rangle$$

is an infinitely smooth function. Q.E.D.

We shall conclude this section with some simple examples. Upon the establishing of a particular formula $\mathfrak{F}f = \tilde{f}$, the inverse formula,

$$\mathfrak{F}^{-1}[\overline{f(t)}] = \frac{\overline{\tilde{f}(\omega)}}{2\pi}$$

will also be obtained because

$$\overline{\mathfrak{F}^{-1}[2\pi \overline{f(t)}]} = \mathfrak{F}f(t)$$

Therefore, in the following examples we shall restrict our attention exclusively to the direct Fourier transformation.

Example 7.4-1 The delta functional and its derivatives possess exponential functions as their transforms. More specifically, for k a positive integer and τ a real constant,

$$\mathfrak{F}\delta = 1 \tag{9}$$
$$\mathfrak{F}\delta(t - \tau) = e^{-i\omega\tau} \tag{10}$$
$$\mathfrak{F}\delta^{(k)}(t) = (i\omega)^k \tag{11}$$
$$\mathfrak{F}\delta^{(k)}(t - \tau) = (i\omega)^k e^{-i\omega\tau} \tag{12}$$

We need merely prove relation (12), since it contains the preceding three relations as special cases. For ϕ in S,

$$\langle \mathfrak{F}\delta^{(k)}(t - \tau), \phi(\omega) \rangle = \langle \delta^{(k)}(t - \tau), \tilde{\phi}(t) \rangle = (-1)^k \tilde{\phi}^{(k)}(\tau)$$
$$= \left[(-1)^k \frac{d^k}{dy^k} \int_{-\infty}^{\infty} \phi(\omega)e^{-i\omega y} \, d\omega \right]_{y=\tau}$$
$$= \int_{-\infty}^{\infty} (i\omega)^k e^{-i\omega\tau} \phi(\omega) \, d\omega = \langle (i\omega)^k e^{-i\omega\tau}, \phi(\omega) \rangle$$

In a similar way, we can establish the relations

$$\mathfrak{F}1 = 2\pi\delta(\omega) \tag{13}$$
$$\mathfrak{F}e^{-it\tau} = 2\pi\delta(\omega + \tau) \tag{14}$$
$$\mathfrak{F}(it)^k = (-1)^k 2\pi\delta^{(k)}(\omega) \tag{15}$$
$$\mathfrak{F}[(it)^k e^{-it\tau}] = (-1)^k 2\pi\delta^{(k)}(\omega + \tau) \tag{16}$$

The Fourier transform of any polynomial is obtainable from (15):

$$\mathfrak{F}(a_n t^n + a_{n-1} t^{n-1} + \cdots + a_0)$$
$$= 2\pi[a_n i^n \delta^{(n)}(\omega) + a_{n-1} i^{n-1} \delta^{(n-1)}(\omega) + \cdots + a_0 \delta(\omega)]$$

Example 7.4-2 Distribution theory provides a very simple means of obtaining one form of *Poisson's summation formula*, which is an identity that equates the sum of certain values of a function to the sum of certain values of its Fourier transform. According to Example 2.4-4,

$$\sum_{\nu=-\infty}^{\infty} e^{i\nu t} = \sum_{\nu=-\infty}^{\infty} 2\pi \delta(t - 2\pi\nu) \tag{17}$$

Moreover, both sides of (17) converge in $\mathcal{S}'$, as is easily shown. By the continuity of the Fourier transformation as a mapping of $\mathcal{S}'$ onto $\mathcal{S}'$, we may transform the left-hand side of (17) term by term to get

$$\sum_{\nu=-\infty}^{\infty} \delta(\omega - \nu) = \mathfrak{F} \Big[\sum_{\nu=-\infty}^{\infty} \delta(t - 2\pi\nu) \Big] \tag{18}$$

(This remarkable formula is an example of a *self-reciprocal transform*; i.e., the transform of the distribution yields the distribution again, except for a change in scale of the independent variable.) By applying the two sides of (18) to any testing function ϕ in $\mathcal{S}$, we obtain the formula we seek. In particular,

$$\Big\langle \sum_{\nu=-\infty}^{\infty} \delta(\omega - \nu), \phi(\omega) \Big\rangle = \Big\langle \sum_{\nu=-\infty}^{\infty} \delta(t - 2\pi\nu), \tilde{\phi}(t) \Big\rangle$$

so that our special form of Poisson's summation formula is found to be

$$\sum_{\nu=-\infty}^{\infty} \phi(\nu) = \sum_{\nu=-\infty}^{\infty} \tilde{\phi}(2\pi\nu) \tag{19}$$

There are more complicated formulas of this nature (see, for instance, Titchmarsh [2], pp. 60–62). One of them is stated in Prob. 6.

PROBLEMS

1 Show that with f in $\mathcal{S}'$

$$\mathfrak{F}\mathfrak{F}f(t) = 2\pi f(-t)$$

2 Show that the ordinary Fourier transformation of locally integrable functions that are absolutely integrable over $-\infty < t < \infty$ is a special case of the distributional Fourier transformation.

3 Show that Theorem 2.8-2 is encompassed as a special case of the distributional Fourier transformation.

4 Establish formulas (5) to (8) and then verify relations (13) to (16).

5 Let the symbolism $\check{f} = (f)^{\vee}$ denote the transpose $f(-t)$ of $f(t)$ and let $\bar{f} = (f)^{-}$ denote the complex conjugate of f. For f in $\mathcal{S}'$, establish the following formulas:

$$\mathfrak{F}f = 2\pi(\mathfrak{F}^{-1}\bar{f})^{-} = 2\pi(\mathfrak{F}^{-1}f)^{\vee}$$
$$(\mathfrak{F}f)^{\vee} = \mathfrak{F}(\check{f})$$
$$(\mathfrak{F}f)^{-} = \mathfrak{F}[(\bar{f})^{\vee}] = \mathfrak{F}[(\check{f})^{-}]$$
$$\mathfrak{F}\bar{f} = [(\mathfrak{F}f)^{-}]^{\vee} = [(\mathfrak{F}f)^{\vee}]^{-}$$

6 Prove the following more general form of Poisson's summation formula:

$$\sum_{\nu=-\infty}^{\infty} \phi(t + \nu T) = \frac{1}{T} \sum_{\nu=-\infty}^{\infty} e^{i2\pi\nu t/T}\, \tilde{\phi}\left(\frac{2\pi\nu}{T}\right)$$

Here T is a real constant, t is a real variable, and ϕ is in $\mathcal{S}$.

7 Let

$$g_k(t) = \frac{1}{k} \sum_{\nu=-k^2}^{k^2} \exp\frac{i2\pi\nu t}{k}$$

Show that $\{g_k\}_{k=1}^{\infty}$ converges in $\mathcal{D}'$ to δ. Then let f be a distribution of bounded support. Construct a sequence of trigonometric polynomials that converges in $\mathcal{D}'$ to f.

8 By a direct computation from definition (2), show that

$$\mathfrak{F}1_+(t) = \pi\delta(\omega) + \mathrm{Pv}\,\frac{1}{i\omega} \tag{20}$$

HINT: Establish and use the following relation, wherein ε is a positive number.

$$\lim_{T\to\infty} \left(\int_{-\infty}^{-\varepsilon} + \int_{\varepsilon}^{\infty}\right) \frac{\phi(\omega)}{\omega} \cos\omega T\, d\omega = 0$$

9 Show that

$$\mathfrak{F}[t^n 1_+(t)] = i^n \pi \delta^{(n)}(\omega) + \mathrm{Pf}\,\frac{n!}{(i\omega)^{n+1}} \qquad n = 0, 1, 2, \ldots$$

in two ways. First, use relations (4) and (20). Second, use the results of Sec. 2.5, Prob. 6.

10 Establish the following formulas:

$$\mathfrak{F}(t^n \operatorname{sgn} t) = 2\,\mathrm{Pf}\,\frac{n!}{(i\omega)^{n+1}}$$

$$\mathfrak{F}\left(\mathrm{Pf}\,\frac{1}{t^n}\right) = \pi\frac{(-i)^n}{(n-1)!}\,\omega^{n-1}\operatorname{sgn}\omega$$

Here $n = 1, 2, 3, \ldots$ and $\operatorname{sgn} x = 1_+(x) - 1_+(-x)$.

11 Let k be a positive integer and let f be a distribution of slow growth that satisfies the equation $t^k f(t) = 1$, for all t. By using the Fourier transformation, find all possible solutions for f. Compare your answer with that of Sec. 3.2, Prob. 3.

12 Find the Fourier transform of

$$f(t) = \frac{\sin t - t \cos t}{t^3}$$

by using formulas (4) and (14). HINT: Note that, because $f(t)$ is continuous everywhere and absolutely integrable over $-\infty < t < \infty$, $\tilde{f}(\omega)$ approaches zero as $|\omega| \to \infty$.

★7.5 THE FOURIER TRANSFORMATION OF CONVOLUTIONS OF DISTRIBUTIONS HAVING BOUNDED SUPPORTS

It will now be shown that under a certain restriction the Fourier transformation converts the convolution of two distributions $f * g$ into the product of their Fourier transforms. The condition that will be imposed on f and g is that both their supports are bounded. However, this conversion of convolution into multiplication is valid under less restrictive conditions. For instance, only one of the distributions need have a bounded support, as we shall show in Sec. 7.9. In addition, it will be shown in Sec. 8.5 that the Laplace transformation behaves in a similar way: it converts the convolution of two Laplace-transformable distributions, whose supports are bounded on the same side, into the product of their Laplace transforms. It is furthermore true that there need be no restriction on the supports of $f(t)$ and $g(t)$ so long as their behavior as $t \to \pm \infty$ is sufficiently confined. (See Schwartz [1], vol. II, pp. 124–126.)

As we shall see subsequently, this is quite a useful property of the Fourier and Laplace transformations, since the convolution process is in general more difficult to perform than is ordinary multiplication. Moreover, these transformations may be used to convert convolution equations into equations involving ordinary multiplication, which may be more easily solved.

Theorem 7.5-1

Let f and g be distributions with bounded supports and let $\tilde{f}$ and $\tilde{g}$ be their respective Fourier transforms. Then,

$$\mathfrak{F}(f * g) = \tilde{f}\tilde{g} \tag{1}$$

Proof: $h \triangleq f * g$ has a bounded support (see Corollary 5.4-2a) and, by Theorem 7.4-3, its Fourier transform $\tilde{h}(\omega)$ equals $\langle h(t), \lambda(t)e^{-i\omega t}\rangle$, where $\lambda(t)$ is a testing function in $\mathfrak{D}$ that equals one over a neighborhood of the support of $h(t)$. Hence, we may write

$$h(\omega) = \langle h(t), \lambda(t)e^{-i\omega t}\rangle = \langle f(t), \langle g(\tau), \lambda(t+\tau)e^{-i\omega(t+\tau)}\rangle\rangle$$

We are free to replace $\lambda(t + \tau)$ by $\psi(t)\psi(\tau)$, where $\psi(t)$ is in $\mathfrak{D}$ and $\psi(t)\psi(\tau)$ equals one over a neighborhood of the support of $f(t) \times g(\tau)$. Thus,

$$\tilde{h}(\omega) = \langle f(t), \psi(t)e^{-i\omega t}\rangle\langle g(\tau), \psi(\tau)e^{-i\omega\tau}\rangle = \tilde{f}(\omega)\tilde{g}(\omega)$$

as was asserted.

Under the hypothesis of this theorem, we also have that

$$\mathfrak{F}^{-1}(f * g) = 2\pi(\mathfrak{F}^{-1}f)(\mathfrak{F}^{-1}g) \tag{2}$$

Example 7.5-1 Relation (6) of Sec. 7.4 can readily be established by employing Theorem 7.5-1 and the fact that

$$f(t - \tau) = f(t) * \delta(t - \tau)$$

Indeed,

$$\mathfrak{F}f(t - \tau) = \mathfrak{F}[f(t) * \delta(t - \tau)] = \tilde{f}(\omega)e^{-i\omega\tau}$$

Similarly, relation (5) of Sec. 7.4 is obtained through

$$\mathfrak{F}f^{(k)}(t) = \mathfrak{F}(\delta^{(k)} * f) = (i\omega)^k \tilde{f}(\omega)$$

In order to use Theorem 7.5-1 in this way, we have to assume that f has a bounded support; but this assumption is really unnecessary, as we shall see in Sec. 7.9.

PROBLEM

1 Let f and g be distributions with bounded supports. Establish (2). Also, for each of the following products, determine an equivalent expression in terms of a Fourier transform of a convolution.

$(\mathfrak{F}f)(\mathfrak{F}^{-1}g) \qquad (\mathfrak{F}f)^-(\mathfrak{F}^{-1}g) \qquad (\mathfrak{F}f)^\vee(\mathfrak{F}g)$

Here the notation used is the same as that of Sec. 7.4, Prob. 5.

7.6 THE SPACE Z OF TESTING FUNCTIONS WHOSE FOURIER TRANSFORMS ARE IN $\mathfrak{D}$

As was mentioned in Sec. 7.4, our definition of the Fourier transformation cannot be used for arbitrary distributions in $\mathfrak{D}'$. Let us be specific. If f is in $\mathfrak{D}'$ but not in $\mathcal{S}'$, then, as ϕ traverses $\mathfrak{D}$, the right-hand side of the relation

$$\langle \tilde{f}, \phi \rangle = \langle f, \tilde{\phi} \rangle \tag{1}$$

will certainly have a sense. However, the left-hand side must be interpreted in a new way. This is because ϕ will not be in $\mathfrak{D}$ but will instead traverse some new space of testing functions that we have not as yet discussed. Actually, the Fourier transform $\tilde{f}$ of an arbitrary distribution in $\mathfrak{D}'$ is not, in general, a distribution but is instead another kind of continuous linear functional which is defined over this new space of testing functions. Such a functional is, at times, called an *ultradistribution* (Sebastião e Silva [1]), and this is the terminology we shall adopt in this book.

Our purpose in this section is to discuss this new space of testing functions; and, in particular, we wish to answer the following question: What is the space Z of functions whose Fourier transforms are testing functions in $\mathfrak{D}$? The answer to this question will provide a means for defining the ultradistributions and, thereby, the Fourier transforms of the distributions in $\mathfrak{D}'$. (This extension of the Fourier transformation is due to Ehrenpreis [1] and to Gelfand and Shilov [1], vol. I, chap. 2, and vol. II, chap. 3.)

Let $\tilde{\phi}(t)$ be an arbitrary testing function in $\mathfrak{D}$ whose support is contained in the finite interval $-a \leq t \leq a$. Its inverse Fourier transform can be written as an integral over this finite interval as follows:

$$\phi(\omega) = \frac{1}{2\pi} \int_{-a}^{a} \tilde{\phi}(t) e^{i\omega t}\, dt \tag{2}$$

Because of this, $\phi(\omega)$ can be extended to an entire function over the complex z plane ($z = \omega + iy$). That is, the function

$$\phi(z) = \frac{1}{2\pi} \int_{-a}^{a} \tilde{\phi}(t) e^{izt}\, dt \tag{3}$$

is analytic for all finite z. This follows from the following facts (Titchmarsh [1], p. 99):

1. The integral (3) converges uniformly over every bounded domain of the z plane.

2. The integrand is a continuous function of (z, t) for every complex z and every real t.

3. The integrand is an analytic function of z for every real t.

Moreover, if (3) is integrated by parts k times, we obtain

$$(-iz)^k \phi(z) = \frac{1}{2\pi} \int_{-a}^{a} \tilde{\phi}^{(k)}(t) e^{izt}\, dt$$

so that for all z

$$|z^k \phi(z)| \leq C_k e^{a|y|} \qquad k = 0, 1, 2, \ldots \tag{4}$$

where

$$C_k = \frac{1}{2\pi} \int_{-a}^{a} |\tilde{\phi}^{(k)}(t)|\, dt$$

Thus, if $\tilde\phi(t)$ is in $\mathfrak{D}$ with its support in $-a \leq t \leq a$, $\phi(\omega)$ can be extended to an entire function and there exists a set of constants C_k ($k = 0, 1, 2, \ldots$) such that (4) is satisfied.

The converse is also true. Indeed,

$$\tilde\phi(t) = \int_{-\infty}^{\infty} \phi(\omega) e^{-i\omega t}\, d\omega$$

and, with $\phi(z)$ entire and satisfying (4), the path of integration may be shifted in the z plane onto any line that is parallel to the ω axis. This may be justified by Cauchy's theorem and the fact that, for all y in any fixed finite interval, $\phi(\omega + iy)$ goes to zero faster than any power of $1/|\omega|$ as $|\omega| \to \infty$, according to (4). Thus, by shifting the path of integration, we have that, for every y,

$$\tilde\phi(t) = \int_{-\infty}^{\infty} \phi(\omega + iy) e^{-i(\omega+iy)t}\, d\omega$$

$$= e^{yt} \int_{-\infty}^{\infty} \phi(\omega + iy) e^{-i\omega t}\, d\omega$$

Because of (4), the last integral converges uniformly for $-\infty < t < \infty$. By formally differentiating under the integral sign, we obtain

$$\tilde\phi^{(1)}(t) = \int_{-\infty}^{\infty} (-iz)\phi(z) e^{-izt}\, d\omega$$

$$= e^{yt} \int_{-\infty}^{\infty} (-iz)\phi(z) e^{-i\omega t}\, d\omega \qquad (5)$$

where the last integral is again uniformly convergent for $-\infty < t < \infty$. Hence, $\tilde\phi(t)$ possesses an ordinary first derivative given by (5) (Titchmarsh [1], p. 59). Continuing in this way, we see that $\tilde\phi(t)$ is infinitely smooth and that

$$\tilde\phi^{(k)}(t) = \int_{-\infty}^{\infty} (-iz)^k \phi(z) e^{-izt}\, d\omega \qquad (6)$$

We shall now show that $\tilde\phi(t)$ has its support contained in the interval $-a \leq t \leq a$. By using (4) for both $k = 0$ and $k = 2$, we may write

$$|\phi(z)| \leq e^{a|y|} \inf\left(C_0, \frac{C_2}{|z|^2}\right) \leq \frac{Ce^{a|y|}}{1 + \omega^2}$$

where C is still another constant. Consequently, for every y,

$$|\tilde\phi(t)| \leq e^{ty+a|y|} \int_{-\infty}^{\infty} \frac{C}{1 + \omega^2}\, d\omega = C\pi e^{ty+a|y|} \qquad (7)$$

Now, let $u < 0$ and $y = u|t|/t$. Then,

$$|\tilde\phi(t)| \leq C\pi e^{u|t|+a|u|} = C\pi e^{u(|t|-a)}$$

For $|t| > a$, the right-hand side of this inequality goes to zero as $u \to -\infty$. This implies that $|\tilde\phi(t)| = 0$ for $|t| > a$.

We can summarize our discussion so far with

Lemma 1

A necessary and sufficient condition for $\tilde{\phi}(t)$ to be in $\mathfrak{D}$, with its support contained in $-a \leq t \leq a$, is that its inverse Fourier transform can be extended to an entire function that satisfies the inequalities (4).

The symbol $\mathcal{Z}$ will denote the space of all functions whose Fourier transforms are elements of $\mathfrak{D}$. The functions of $\mathcal{Z}$ will also be called *testing functions*. Lemma 1 gives a direct characterization of such functions: $\mathcal{Z}$ is the space of all entire functions that satisfy a set of inequalities of the form (4) for some constants C_k and a. Clearly, $\mathcal{Z}$ is a linear space. (See Appendix A.) Also, the direct and inverse Fourier transformations are linear one-to-one mappings of $\mathcal{Z}$ onto $\mathfrak{D}$ and $\mathfrak{D}$ onto $\mathcal{Z}$, respectively.

By letting $\tilde{\phi}(t)$ be in $\mathfrak{D}$ and by denoting the extension over the z plane of the inverse Fourier transform of $\tilde{\phi}(t)$ by $\mathfrak{F}_c^{-1}\tilde{\phi}(t)$, we have the following standard formulas, which can be derived in the customary way. Here, τ is a real constant and α is a complex constant.

$$(-iz)^k \phi(z) = \mathfrak{F}_c^{-1}\tilde{\phi}^{(k)}(t) \tag{8}$$
$$\phi^{(k)}(z) = \mathfrak{F}_c^{-1}[(it)^k \tilde{\phi}(t)] \tag{9}$$
$$e^{iz\tau}\phi(z) = \mathfrak{F}_c^{-1}\tilde{\phi}(t - \tau) \tag{10}$$
$$\phi(z - i\alpha) = \mathfrak{F}_c^{-1}[e^{\alpha t}\tilde{\phi}(t)] \tag{11}$$

Let us now turn to the concept of convergence in the space $\mathcal{Z}$. The sequence $\{\phi_\nu\}_{\nu=1}^{\infty}$ is said *to converge in* $\mathcal{Z}$ if the following three conditions are fulfilled:

1. Each ϕ_ν is in $\mathcal{Z}$.
2. There exists a set of constants a and C_k ($k = 0, 1, 2, \ldots$), which do not depend upon ν, such that for all $z = \omega + iy$

$$|z^k \phi_\nu(z)| \leq C_k e^{a|y|} \quad k = 0, 1, 2, \ldots \tag{12}$$

3. $\{\phi_\nu(z)\}_{\nu=1}^{\infty}$ converges uniformly on every bounded domain of the z plane.

As a consequence of this definition, the limit function ϕ of $\{\phi_\nu\}_{\nu=1}^{\infty}$ is also in $\mathcal{Z}$; for it will obviously satisfy the inequalities (12) and the uniformity of the convergence in condition 3 ensures that $\phi(z)$ is analytic for all z (Titchmarsh [1], p. 95). Thus, the space $\mathcal{Z}$ *is closed under convergence*.

Note that conditions 2 and 3 together imply that $\{\omega^k \phi_\nu(\omega)\}_{\nu=1}^{\infty}$ converges to $\omega^k \phi(\omega)$ uniformly for $-\infty < \omega < \infty$.

Theorem 7.6-1

The sequence $\{\tilde{\phi}_\nu\}_{\nu=1}^{\infty}$ converges in $\mathfrak{D}$ to the limit $\tilde{\phi}$ if and only if the inverse Fourier transforms $\{\phi_\nu\}_{\nu=1}^{\infty}$ converge in $\mathcal{Z}$ to the limit $\phi = \mathfrak{F}^{-1}\tilde{\phi}$.

Proof: First, let $\{\tilde\phi_\nu\}_{\nu=1}^\infty$ converge in $\mathfrak{D}$ to $\tilde\phi$ and let the supports of all the $\tilde\phi_\nu$ be contained in $-a \leq t \leq a$. Then, the ϕ_ν and ϕ are in Z. Also,

$$\begin{aligned}|z^k \phi_\nu(z)| &= \left| \frac{1}{2\pi} \int_{-a}^{a} \tilde\phi_\nu^{(k)}(t) e^{izt}\, dt \right| \\ &\leq \frac{e^{a|y|}}{2\pi} \int_{-a}^{a} |\tilde\phi_\nu^{(k)}(t)|\, dt \\ &\leq \frac{a}{\pi} e^{a|y|} \sup_t |\tilde\phi_\nu^{(k)}(t)| \end{aligned} \qquad (13)$$

Since $\{\tilde\phi_\nu\}_{\nu=1}^\infty$ converges to $\tilde\phi$ in $\mathfrak{D}$, $\sup_t |\tilde\phi_\nu^{(k)}(t)|$ is uniformly bounded for all ν. Thus, the ϕ_ν satisfy the inequalities (12).

The third condition for convergence in Z is also fulfilled. Indeed, according to (13),

$$|\phi_\nu(z) - \phi(z)| \leq \frac{a}{\pi} e^{a|y|} \sup_t |\tilde\phi_\nu(t) - \tilde\phi(t)|$$

But, as $\nu \to \infty$, $\sup_t |\tilde\phi_\nu(t) - \tilde\phi(t)| \to 0$ and, on each bounded domain of the z plane, $e^{a|y|}$ is bounded. Thus, $|\phi_\nu(z) - \phi(z)| \to 0$ uniformly on each such domain.

Conversely, if $\{\phi_\nu\}_{\nu=1}^\infty$ converges in Z to ϕ, then by (12) and Lemma 1 all $\tilde\phi_\nu(t)$ and $\tilde\phi(t)$ are in $\mathfrak{D}$ and they have their supports contained in $-a \leq t \leq a$. Also, for each nonnegative integer k,

$$\begin{aligned}|\tilde\phi_\nu^{(k)}(t) - \tilde\phi^{(k)}(t)| &= \left| \int_{-\infty}^{\infty} (-i\omega)^k [\phi_\nu(\omega) - \phi(\omega)] e^{-i\omega t}\, d\omega \right| \\ &\leq \int_{-\infty}^{\infty} \left| \frac{(\omega^k + \omega^{k+2})[\phi_\nu(\omega) - \phi(\omega)]}{1 + \omega^2} \right| d\omega \\ &\leq \pi \sup_\omega |(\omega^k + \omega^{k+2})[\phi_\nu(\omega) - \phi(\omega)]| \end{aligned}$$

By conditions 2 and 3 of our definition of convergence in Z, the right-hand side of the last relation converges to zero. Thus, $\{\tilde\phi_\nu\}_{\nu=1}^\infty$ converges in $\mathfrak{D}$ to $\tilde\phi$. Q.E.D.

Another way to state Theorem 7.6-1 is to say that the direct and inverse Fourier transformations are continuous mappings of Z onto $\mathfrak{D}$ and $\mathfrak{D}$ onto Z, respectively. In view of what we have previously established, we can say that these transformations are continuous linear one-to-one mappings of the spaces $\mathfrak{D}$ and Z onto one another, where $\mathfrak{F}$ acts on Z and $\mathfrak{F}^{-1}$ acts on $\mathfrak{D}$.

Since ϕ is an entire function, it cannot be zero on any interval $a < t < b$ except when it is zero everywhere (Titchmarsh [1], p. 88). Thus, the spaces $\mathfrak{D}$ and Z do not intersect except in the identically zero testing function. On the other hand, in common with the space $\mathfrak{D}$, Z possesses the property that it is a proper subspace of $\mathfrak{S}$, as we shall now

The Fourier transformation

show. (Here it is understood that, when we speak of Z as a subspace of S, the independent variable for the functions of Z is a real variable.)

Theorem 7.6-2

Z *is a proper subspace of* S.

Proof: If ϕ is in Z, it is an entire function and hence $\phi(\omega)$ is infinitely smooth for all ω. Also, its Fourier transform $\tilde{\phi}(t)$ is in $\mathcal{D}$, which implies that $t^k \tilde{\phi}(t)$ is also in $\mathcal{D}$. Therefore, by (9), $\phi^{(k)}(\omega)$ is again in Z, so that for each pair of nonnegative integers m and k

$$|\omega^m \phi^{(k)}(\omega)| \leq C_{mk} \qquad -\infty < \omega < \infty$$

Hence, ϕ is in S.

Finally, $\mathcal{D}$ is part of S and $\mathcal{D}$ does not intersect Z except for the zero function. Therefore, Z is truly a proper subspace of S.

A similar argument shows that convergence in Z implies convergence in S.

Theorem 7.6-3

If the sequence $\{\phi_\nu\}_{\nu=1}^\infty$ *converges in* Z *to* ϕ, *then it also converges in* S *to* ϕ.

Proof: By Theorem 7.6-1, our hypothesis implies that $\{\tilde{\phi}_\nu(t)\}_{\nu=1}^\infty$ converges in $\mathcal{D}$ to $\tilde{\phi}(t)$. But then, for $k = 0, 1, 2, \ldots$, $\{t^k \tilde{\phi}_\nu(t)\}_{\nu=1}^\infty$ converges in $\mathcal{D}$ to $t^k \tilde{\phi}(t)$. Hence, $\{\phi_\nu^{(k)}(\omega)\}_{\nu=1}^\infty$ converges in Z to $\phi^{(k)}(\omega)$ and, by our definition of convergence in Z, for each pair of nonnegative integers m and k, $\{\omega^m \phi_\nu^{(k)}(\omega)\}_{\nu=1}^\infty$ converges to $\omega^m \phi^{(k)}(\omega)$ uniformly for $-\infty < \omega < \infty$. Q.E.D.

Theorem 7.6-4

Z *is dense in* S. *That is, for each ϕ in S there exists a sequence* $\{\phi_\nu\}_{\nu=1}^\infty$ *with elements exclusively in Z that converges in S to ϕ.*

Proof: Since the Fourier transformation maps S onto itself, $\tilde{\phi}$ is also in S. Moreover, $\mathcal{D}$ is dense in S. Therefore, choose a sequence $\{\tilde{\phi}_\nu(t)\}_{\nu=1}^\infty$ of functions in $\mathcal{D}$ which converges in S to $\tilde{\phi}$. By the continuity of the inverse Fourier transformation as a mapping of S onto itself, $\{\phi_\nu(\omega)\}_{\nu=1}^\infty$ is the sequence we seek.

We close this section with a listing of several operations under which the space Z is closed. In the following assume that ϕ and ϕ_1 are both in Z:

1. Addition: $\phi + \phi_1 \in Z$.
2. Multiplication by a constant α: $\alpha\phi \in Z$.

(It follows now that Z is a linear space as mentioned previously.)

3. **Shifting:** $\phi(\omega - \alpha) \in Z$, where α is a real or complex number. Note that $\phi(\omega - \alpha)$ is still defined when α is complex because ϕ is an entire function. This is in contrast to the situation for the space $\mathfrak{D}$; there the shift must be a real one.

4. **Transposition:** $\phi(-\omega) \in Z$.

5. **Multiplication of the independent variable by a positive constant** a: $\phi(a\omega) \in Z$.

6. **Differentiation:** $\phi^{(1)} \in Z$.

7. **Multiplication by an entire function** ψ that satisfies an inequality of the form

$$|\psi(z)| \leq C e^{b|y|}(1 + |z|^m) \tag{14}$$

where C and b are real constants and m is an integer: $\psi\phi \in Z$. Entire functions that satisfy (14) are called *multipliers in the space* Z.

PROBLEMS

1 Show that condition 3 of our definition for convergence in Z can be replaced by the requirement that $\{\phi_\nu(\omega)\}_{\nu=1}^\infty$ converge to $\phi(\omega)$ uniformly on every finite interval of the ω axis.

2 Show that the space Z is truly closed under the seven operations listed at the end of this section.

7.7 THE SPACE Z' OF ULTRADISTRIBUTIONS

Z' will denote the dual of Z. That is, Z' is the space of all ultradistributions (continuous linear functionals on the space Z). Still more specifically, f is in Z' if and only if it satisfies the following two conditions:

Linearity If ϕ_1 and ϕ_2 are testing functions in Z and α and β are two complex numbers, then

$$\langle f, \alpha\phi_1 + \beta\phi_2 \rangle = \alpha\langle f, \phi_1 \rangle + \beta\langle f, \phi_2 \rangle$$

Continuity If $\{\phi_\nu\}_{\nu=1}^\infty$ converges in Z to zero, then the sequence of numbers $\{\langle f, \phi_\nu \rangle\}_{\nu=1}^\infty$ converges to zero.

It follows from these two conditions that, if $\{\phi_\nu\}_{\nu=1}^\infty$ converges in Z to a limit function ϕ that is not identically zero, then $\{\langle f, \phi_\nu \rangle\}_{\nu=1}^\infty$ converges to $\langle f, \phi \rangle$.

Theorem 7.7-1

If $f \in S'$, then $f \in Z'$.

Proof: Z is a subspace of S, according to Theorem 7.6-2. Hence, each distribution f of slow growth is linear on Z. Furthermore, since convergence in Z implies convergence in S (see Theorem 7.6-3), f is also continuous on Z. Q.E.D.

We have thus shown that the delta functional and all its derivatives are in Z'.

Every ultradistribution f that can be related to a locally integrable function $f(\omega)$ through the relation

$$\langle f, \phi \rangle = \int_{-\infty}^{\infty} f(\omega)\phi(\omega)\, d\omega$$

where ϕ traverses Z, is said to be a regular ultradistribution.

If f is a distribution of slow growth, then a knowledge of $\langle f, \phi \rangle$ for every ϕ in Z determines $\langle f, \psi \rangle$ for every ψ in S. For Z is dense in S; and consequently, for each ψ in S, a sequence $\{\phi_\nu\}_{\nu=1}^{\infty}$ of testing functions in Z that converges in S to ψ can be chosen. Then the limit of $\{\langle f, \phi_\nu \rangle\}_{\nu=1}^{\infty}$ is the value $\langle f, \psi \rangle$ that we seek. (This is similar to the situation discussed in Sec. 4.3, where f was extended from $\mathfrak{D}$ onto S in a continuous fashion.)

Equality for ultradistributions is defined in the usual way. If f and g are ultradistributions and

$$\langle f, \phi \rangle = \langle g, \phi \rangle$$

for every ϕ in Z, then f is said to be equal to g. (In symbols, $f = g$.)

A number of operations on ultradistributions can be defined in the same way as they were for distributions. A summary of these definitions follows, where it is understood that f and g are ultradistributions and that each relation holds for every testing function ϕ in Z. The right-hand sides of the following equations are known values that are assigned by the definition to the left-hand-side expressions. The operations indicated on the testing functions in the right-hand sides were defined at the end of the preceding section, and the closure of the space Z under these operations is essential for the validity of the following definitions. The reader should convince himself that the space Z' is also closed under these operations.

1. The addition of ultradistributions:

$$\langle f + g, \phi \rangle \triangleq \langle f, \phi \rangle + \langle g, \phi \rangle \tag{1}$$

2. The multiplication of an ultradistribution by a constant α:

$$\langle \alpha f, \phi \rangle \triangleq \langle f, \alpha\phi \rangle \tag{2}$$

The closure of Z' under these first two operations implies that Z' is a linear space. (See Appendix A.)

3. The shifting of an ultradistribution: Let α be a complex constant.

$$\langle f(\omega - \alpha), \phi(\omega)\rangle \triangleq \langle f(\omega), \phi(\omega + \alpha)\rangle \tag{3}$$

Note that the argument of f in the left-hand side of this definition is, in general, complex. We shall say more about this in a moment.

4. The transposition of an ultradistribution:

$$\langle f(-\omega), \phi(\omega)\rangle \triangleq \langle f(\omega), \phi(-\omega)\rangle \tag{4}$$

5. The multiplication of the independent variable by a positive constant a:

$$\langle f(a\omega), \phi(\omega)\rangle \triangleq \left\langle f(\omega), \frac{1}{a}\phi\left(\frac{\omega}{a}\right)\right\rangle \tag{5}$$

6. The differentiation of an ultradistribution:

$$\langle f^{(1)}, \phi\rangle \triangleq \langle f, -\phi^{(1)}\rangle \tag{6}$$

7. The multiplication of an ultradistribution by a multiplier in the space Z: Let ψ be a multiplier in the space Z [i.e., an entire function that satisfies (14) of Sec. 7.6]. Then

$$\langle \psi f, \phi\rangle \triangleq \langle f, \psi\phi\rangle \tag{7}$$

Our definition of the shifting of an ultradistribution serves to give a meaning to the symbol $f(z)$, where z is the complex variable $\omega + iy$. In accordance with (3), $f(z)$ is defined as follows. For every $\phi(\omega)$ in Z,

$$\langle f(z), \phi(\omega)\rangle \triangleq \langle f(\omega + iy), \phi(\omega)\rangle \triangleq \langle f(\omega), \phi(\omega - iy)\rangle$$

In this way f can be extended from the real ω axis onto the entire z plane. In these expressions, y is held constant, so that the path along which the expression is evaluated is a line running parallel to the ω axis. This will always be understood whenever we write $f(z)$ or $\langle f(z), \phi(\omega)\rangle$. Thus, $f(z)$ exists as an ultradistribution on each and every line in the z plane that is parallel to the ω axis. As a simple example of this, we may write

$$\langle \delta^{(\nu)}(z), \phi(\omega)\rangle = \langle \delta^{(\nu)}(\omega + iy), \phi(\omega)\rangle = \langle \delta(\omega), (-1)^\nu \phi^{(\nu)}(\omega - iy)\rangle$$
$$= (-1)^\nu \phi^{(\nu)}(-iy)$$

Also, note that we are free to write

$$\langle f(\omega), \phi(\omega)\rangle = \langle f(z), \phi(z)\rangle$$

because

$$\langle f(z), \phi(z)\rangle = \langle f(\omega + iy), \phi(\omega + iy)\rangle = \langle f(\omega), \phi(\omega + iy - iy)\rangle$$
$$= \langle f(\omega), \phi(\omega)\rangle$$

A sequence $\{f_\nu\}_{\nu=1}^\infty$ is said *to converge in* $\mathcal{Z}'$ if every f_ν is in $\mathcal{Z}'$ and if, for each ϕ in $\mathcal{Z}$, the sequence $\{\langle f_\nu, \phi\rangle\}_{\nu=1}^\infty$ converges. As ϕ traverses $\mathcal{Z}$, the limits of $\{\langle f_\nu, \phi\rangle\}_{\nu=1}^\infty$ define a functional f on $\mathcal{Z}$, and f is said to be the limit of $\{f_\nu\}_{\nu=1}^\infty$. Once again, we may adapt the proof of Theorem 2.2-1 (which asserts that $\mathfrak{D}'$ is closed under convergence) to show that f is linear and continuous as a functional on $\mathcal{Z}$. In other words, $\mathcal{Z}'$ *is also closed under convergence.* (A shorter proof of this fact is indicated in Sec. 7.8, Prob. 6.)

As usual, an infinite series is said to converge in $\mathcal{Z}'$ if the sequence of its partial sums converges in $\mathcal{Z}'$.

With the customary identification of testing functions with their corresponding regular distributions, we have that

$$\mathcal{Z} \subset \mathcal{S} \subset \mathcal{S}' \subset \mathcal{Z}' \tag{8}$$

and that convergence in any one of these spaces implies convergence in the spaces appearing in (8) to the right of the original space. We have already established most of this assertion. [See Sec. 4.3, Eq. (5), and Theorems 7.6-2, 7.6-3, and 7.7-1.] The only thing that still needs verification is that convergence in $\mathcal{S}'$ implies convergence in $\mathcal{Z}'$. But this follows directly from the facts that $\mathcal{Z} \subset \mathcal{S}$ and $\mathcal{S}' \subset \mathcal{Z}'$.

Every ultradistribution can be decomposed into a series that has the same form as a Taylor's series. That is, if f is in $\mathcal{Z}'$ and if α is any complex constant, then the shifting of f by the amount $-\alpha$ is given by

$$f(\omega + \alpha) = \sum_{\nu=0}^\infty \frac{\alpha^\nu}{\nu!} f^{(\nu)}(\omega) \tag{9}$$

To show the validity of (9), let ϕ be in $\mathcal{Z}$. Since ϕ is an entire function, its Taylor's series expansion,

$$\phi(\omega - \alpha) = \sum_{\nu=0}^\infty \frac{(-\alpha)^\nu}{\nu!} \phi^{(\nu)}(\omega) \tag{10}$$

converges for all α. Furthermore, consider the Fourier transforms of the partial sums of this Taylor's series:

$$\sum_{\nu=0}^n \frac{(-\alpha)^\nu}{\nu!} (it)^\nu \tilde{\phi}(t)$$

As $n \to \infty$, these partial sums converge in $\mathfrak{D}$ to $e^{-i\alpha t}\tilde{\phi}(t)$, since $\tilde{\phi}(t)$ has a bounded support and the exponential series and every one of its derivatives converge uniformly on every finite interval of the t axis. Therefore, according to Theorem 7.6-1, the infinite series (10) converges in $\mathcal{Z}$. By the continuity of f on $\mathcal{Z}$, we may now write

$$\langle f(\omega + \alpha), \phi(\omega)\rangle = \langle f(\omega), \phi(\omega - \alpha)\rangle$$

$$= \lim_{n\to\infty} \left\langle f(\omega), \sum_{\nu=0}^{n} \frac{(-\alpha)^\nu}{\nu!} \phi^{(\nu)}(\omega)\right\rangle$$

$$= \lim_{n\to\infty} \left\langle \sum_{\nu=0}^{n} \frac{\alpha^\nu}{\nu!} f^{(\nu)}(\omega), \phi(\omega)\right\rangle$$

$$= \left\langle \sum_{\nu=0}^{\infty} \frac{\alpha^\nu}{\nu!} f^{(\nu)}(\omega), \phi(\omega)\right\rangle$$

So (9) is justified.

PROBLEMS

1 Verify that the space Z' is closed under the seven operations listed in this section, and then compare the definitions of these operations with the corresponding ones for the space $\mathfrak{D}'$. Explain any discrepancies.

2 Revise the proof of Theorem 2.2-1 to show that the space Z' is closed under convergence.

3 Show that differentiation is a continuous linear operation in the space Z'. This will imply that an infinite series of ultradistributions that converges in Z' can be differentiated term by term.

4 Show that $\exp t^2$ is a distribution but not an ultradistribution. Also, show that $\delta(t - i\tau)$, where τ is a positive constant, is an ultradistribution but not a distribution. Since $\delta(t)$ is both a distribution and an ultradistribution, these examples demonstrate that the spaces $\mathfrak{D}'$ and Z' intersect but that neither is contained in the other.

5 Every distribution f can always be decomposed into a sum of distributions all of which have bounded supports. This can be done by multiplying f by an appropriate partitioning of unity. (See Sec. 1.8.) Verify that ultradistributions cannot be decomposed in this fashion. Actually, ultradistributions have only a global character (they must be considered as single entities) in contrast to distributions, which have a local character.

6 Show that for any f in Z',

$$\frac{df}{dz} = \lim_{\Delta z \to 0} \frac{f(z + \Delta z) - f(z)}{\Delta z}$$

where the limit is understood in the sense of convergence in Z'.

7.8 THE FOURIER TRANSFORMS OF ARBITRARY DISTRIBUTIONS

The definition for the Fourier transforms of arbitrary distributions in $\mathfrak{D}'$ is also based upon Parseval's equation, just as it was for the distributions in $\mathcal{S}'$. As in Sec. 7.4, we generalize one of the forms for Parseval's equation to obtain

$$\langle \tilde{f}, \phi \rangle \triangleq \langle f, \tilde{\phi} \rangle \tag{1}$$

as the definition for the Fourier transform $\tilde{f}$ of any distribution f in $\mathfrak{D}'$. As $\tilde{\phi}$ traverses $\mathfrak{D}$, ϕ traverses $\mathcal{Z}$, so that $\tilde{f}$ is defined as that functional which assigns to each ϕ in $\mathcal{Z}$ the same number that f assigns to $\tilde{\phi}$.

The derivation of the various properties of the Fourier transformation, when applied to $\mathfrak{D}'$, proceeds in a fashion that is quite analogous to that of Sec. 7.4. By using Theorem 7.6-1, we can show that, *for f in $\mathfrak{D}'$, $\tilde{f}$ is a continuous linear functional on $\mathcal{Z}$ and thus an ultradistribution*. It should be emphasized here that the Fourier transform of a distribution in $\mathfrak{D}'$ is not, in general, also in $\mathfrak{D}'$ but is instead in $\mathcal{Z}'$. This is in contrast to the situation that we had when dealing with distributions of slow growth; there, f and $\tilde{f}$ were both in $\mathcal{S}'$.

Every ultradistribution $\tilde{f}$ defines through (1) a distribution f (this can be shown by again using Theorem 7.6-1), so that (1) also serves as a definition of the inverse Fourier transformation. It follows that the Fourier transformation is a mapping of $\mathfrak{D}'$ onto $\mathcal{Z}'$ and the inverse Fourier transformation is a mapping of $\mathcal{Z}'$ onto $\mathfrak{D}'$. Clearly, this correspondence is also one-to-one. The reader should convince himself that, if f is a distribution of slow growth, our present definition of $\tilde{f}$ is equivalent to that given in Sec. 7.4. Here we have merely extended the definition of the Fourier transformation to cover all distributions in $\mathfrak{D}'$ without altering the definition for distributions in $\mathcal{S}'$. It then follows that the ordinary Fourier transformation of locally integrable functions that are absolutely integrable over $-\infty < t < \infty$ is a special case of this generalized Fourier transformation (see Sec. 7.4, Prob. 2).

The analogue to Theorem 7.4-2 is

Theorem 7.8-1

The Fourier transformation is a continuous linear mapping of $\mathfrak{D}'$ onto $\mathcal{Z}'$. Hence, if $\sum_{\nu=1}^{\infty} g_\nu$ converges in $\mathfrak{D}'$ to g, then

$$\tilde{g} = \sum_{\nu=1}^{\infty} \tilde{g}_\nu$$

where the last series converges in $\mathcal{Z}'$. The inverse Fourier transformation has the same properties as a mapping of $\mathcal{Z}'$ onto $\mathfrak{D}'$.

Proof: The Fourier transformation is obviously linear. For continuity, let $\{f_\nu\}_{\nu=1}^{\infty}$ converge in $\mathfrak{D}'$ to f. By the fact that $\mathfrak{D}'$ is closed under convergence, f is in $\mathcal{D}'$ and therefore has a Fourier transform $\tilde{f}$. Thus, as $\nu \to \infty$,

$$\langle \tilde{f}_\nu, \phi \rangle = \langle f_\nu, \tilde{\phi} \rangle \to \langle f, \tilde{\phi} \rangle = \langle \tilde{f}, \phi \rangle$$

for each ϕ in Z. Consequently, $\{\tilde{f}_\nu\}_{\nu=1}^\infty$ converges in Z' to $\tilde{f}$. A similar argument can be used for the inverse Fourier transformation.

The following formulas can be established in the same way as they are for the Fourier transforms of distributions of slow growth. In the present case, f is an arbitrary distribution in $\mathcal{D}'$, k is a positive integer, τ is a real number, α is a complex number, and a is a nonzero real number.

$$\mathfrak{F}[(-it)^k f(t)] = \tilde{f}^{(k)}(\omega) \tag{2}$$
$$\mathfrak{F} f^{(k)}(t) = (i\omega)^k \tilde{f}(\omega) \tag{3}$$
$$\mathfrak{F} f(t - \tau) = e^{-i\tau\omega} \tilde{f}(\omega) \tag{4}$$
$$\mathfrak{F}[e^{-i\alpha t} f(t)] = \tilde{f}(\omega + \alpha) \tag{5}$$
$$\mathfrak{F} f(at) = \frac{1}{|a|} \tilde{f}\left(\frac{\omega}{a}\right) \tag{6}$$

Example 7.8-1 Let us find the Fourier transform of $e^{\alpha t}$, where α is any complex number. Since

$$e^{\alpha t} = \sum_{\nu=0}^\infty \frac{(\alpha t)^\nu}{\nu!}$$

and since the right-hand side converges in $\mathcal{D}'$ (according to Theorem 2.3-1), we may invoke the continuity of the Fourier transformation to write

$$\mathfrak{F} e^{\alpha t} = 2\pi \sum_{\nu=0}^\infty \frac{(i\alpha)^\nu}{\nu!} \delta^{(\nu)}(\omega) = 2\pi \delta(\omega + i\alpha) \tag{7}$$

Here we have used (9) of Sec. 7.7 and the previously established fact that $\mathfrak{F} t^\nu = 2\pi i^\nu \delta^{(\nu)}(\omega)$. If we set $\alpha = -i\tau$, where τ is real, $e^{\alpha t}$ will be of slow growth and (7) will agree with (14) of Sec. 7.4. It also follows from (7) that

$$\mathfrak{F} \sin \alpha t = -i\pi[\delta(\omega - \alpha) - \delta(\omega + \alpha)] \tag{8}$$
$$\mathfrak{F} \cos \alpha t = \pi[\delta(\omega - \alpha) + \delta(\omega + \alpha)] \tag{9}$$
$$\mathfrak{F} \sinh \alpha t = \pi[\delta(\omega + i\alpha) - \delta(\omega - i\alpha)] \tag{10}$$
$$\mathfrak{F} \cosh \alpha t = \pi[\delta(\omega + i\alpha) + \delta(\omega - i\alpha)] \tag{11}$$

Still another Fourier transform can be obtained by applying (2) to (7):

$$\mathfrak{F} t^k e^{\alpha t} = 2\pi i^k \delta^{(k)}(\omega + i\alpha) \tag{12}$$

PROBLEMS

1 Show that f is in $\mathfrak{D}'$ if and only if $\tilde{f}$ is in $\mathfrak{Z}'$.

2 Verify formulas (2) to (6).

3 Assuming that f is in $\mathfrak{S}$, show that the definition of $\tilde{f}$ given in this section is the same as the one given in Sec. 7.4.

4 Show that the relations stated in Sec. 7.4, Probs. 1 and 5, are still valid when the Fourier transformation is applied to an arbitrary distribution in $\mathfrak{D}'$.

5 Find a series expansion in terms of $\delta^{(\nu)}$ for the Fourier transform of an entire function.

6 The proof of Theorem 7.8-1 can be used to show that the fact that $\mathfrak{Z}'$ is closed under convergence is a consequence of the corresponding property of $\mathfrak{D}'$. Do this.

7.9 THE FOURIER TRANSFORMATION OF THE CONVOLUTION OF TWO DISTRIBUTIONS ONE OF WHICH HAS A BOUNDED SUPPORT

In this section we continue the discussion of the Fourier transformation of convolution, which was started in Sec. 7.5. Our purpose is to extend our previous result by showing that the Fourier transform of the convolution of two distributions is the product of the Fourier transforms of the distributions even when only one of the distributions has a bounded support. (In Sec. 7.5 both their supports were required to be bounded.)

Lemma 1

If f is a distribution of bounded support, then $\tilde{f}$ is a multiplier in $\mathfrak{Z}$ [that is, $\tilde{f}(z)$ is an entire function such that, for $z = \omega + iy$,

$$|\tilde{f}(z)| \leq Ce^{b|y|}(1 + |z|^m) \tag{1}$$

where C and b are real constants and m is an integer].

Proof: First note that

$$\tilde{f}(z) = \langle f(t), \lambda(t)e^{-izt} \rangle \tag{2}$$

where $\lambda(t)$ is a testing function in $\mathfrak{D}$ that equals one over a neighborhood of the support of f. The proof of this fact is almost the same as that of Theorem 7.4-3.

Next, consider

$$\frac{\tilde{f}(z + \Delta z) - \tilde{f}(z)}{\Delta z} = \langle f(t), \phi_{\Delta z}(t) \rangle$$

where $\Delta z \neq 0$ and

$$\phi_{\Delta z}(t) = \lambda(t)e^{-izt}\frac{e^{-i\Delta z t} - 1}{\Delta z} \tag{3}$$

Some computation shows that $\phi_{\Delta z}(t)$ converges in $\mathfrak{D}$ as $|\Delta z| \to 0$. Hence, $\tilde{f}(z)$ has an ordinary derivative everywhere in the z plane; that is, $\tilde{f}(z)$ is an entire function.

Finally, because $\lambda(t)$ is fixed, Theorem 3.3-1 shows that, for some constant K and some nonnegative integer r,

$$\begin{aligned}|\tilde{f}(z)| &= |\langle f(t), \lambda(t)e^{-izt}\rangle| \\ &\leq K \sup_t \left|\frac{d^r}{dt^r}[\lambda(t)e^{-izt}]\right| \\ &\leq K \sup_t \sum_{k=0}^r \binom{r}{k} |\lambda^{(r-k)}(t)| \, |z|^k |e^{-izt}|\end{aligned}$$

Choose b so large that the interval $-b < t < b$ contains the support of $\lambda(t)$. Then,

$$|\tilde{f}(z)| \leq K e^{b|y|} \sup_t \sum_{k=0}^r \binom{r}{k} |\lambda^{(r-k)}(t)| \, |z|^k$$

(1) follows directly from this inequality.   Q.E.D.

Theorem 7.9-1

If f and g are both in $\mathfrak{D}'$ and if f has a bounded support, then

$$\mathfrak{F}(f * g) = \tilde{f}\tilde{g}$$

Proof: $f * g$ exists as a distribution and, therefore, has a Fourier transform. Consequently, for any ϕ in Z, we may write

$$\begin{aligned}\langle \mathfrak{F}(f * g), \phi \rangle &= \langle f * g, \tilde{\phi} \rangle = \langle g(\tau), \langle f(t), \tilde{\phi}(t + \tau)\rangle\rangle \\ &= \langle g(\tau), \langle \check{f}(t), \tilde{\phi}(\tau - t)\rangle\rangle = \langle g, \check{f} * \tilde{\phi}\rangle \\ &= \langle \tilde{g}, \mathfrak{F}^{-1}(\check{f} * \tilde{\phi})\rangle\end{aligned}$$

But, from Sec. 7.5, Eq. (2),

$$\mathfrak{F}^{-1}(\check{f} * \tilde{\phi}) = (2\pi \mathfrak{F}^{-1}\check{f})\phi = \tilde{f}\phi$$

By Lemma 1, $\tilde{f}$ is a multiplier in Z, and therefore

$$\langle \mathfrak{F}(f * g), \phi \rangle = \langle \tilde{g}, \tilde{f}\phi \rangle = \langle \tilde{f}\tilde{g}, \phi \rangle \quad \text{Q.E.D.}$$

Example 7.9-1 Let

$$p(t) = \begin{cases} 1 & |t| < 1 \\ 0 & |t| \geq 1 \end{cases}$$

and let $f(t) = p(t) * e^{\alpha t}$, α being some complex number. Now, $\mathfrak{F}p(t) = (2 \sin \omega)/\omega$, and therefore

$$\tilde{f}(\omega) = \mathfrak{F}[p(t) * e^{\alpha t}] = 4\pi \frac{\sin \omega}{\omega} \delta(\omega + i\alpha)$$

according to Sec. 7.8, Eq. (7). Here we remind the reader that $\tilde{f}$ is an ultradistribution but not, in general, a distribution.

Example 7.9-2 Let

$$f(t) = \delta^{(1)}(t) * \sinh \alpha t$$

Then, by Sec. 7.4, Eq. (11), and Sec. 7.8, Eq. (10),

$$\tilde{f}(\omega) = \pi i \omega [\delta(\omega + i\alpha) - \delta(\omega - i\alpha)]$$

The following identity is easily verified:

$$i\omega[\delta(\omega + i\alpha) - \delta(\omega - i\alpha)] = \alpha[\delta(\omega + i\alpha) + \delta(\omega - i\alpha)]$$

Hence, by Sec. 7.8, Eq. (11),

$$\tilde{f}(\omega) = \alpha \mathfrak{F} \cosh \alpha t$$

This result agrees with the fact that

$$f(t) = \delta^{(1)}(t) * \sinh \alpha t = \frac{d}{dt} \sinh \alpha t = \alpha \cosh \alpha t$$

Example 7.9-3 We have previously seen that

$$f^{(\mu)}(t - t_\nu) = f(t) * \delta^{(\mu)}(t - t_\nu)$$

where μ is a nonnegative integer, t_ν is a real constant, and f is an arbitrary distribution. Therefore, the Fourier transform of a general difference-differential expression with constant coefficients $a_{\mu\nu}$ is

$$\mathfrak{F} \sum_{\mu=0}^{m} \sum_{\nu=0}^{n} a_{\mu\nu} f^{(\mu)}(t - t_\nu) = \tilde{f}(\omega) \sum_{\mu=0}^{m} \sum_{\nu=0}^{n} a_{\mu\nu} (i\omega)^\mu e^{-i\omega t_\nu}$$

PROBLEMS

1 Establish (2).

2 Show that (3) converges in $\mathfrak{D}$ as $|\Delta z| \to 0$.

3 Let α be a complex number and let f be a distribution of bounded support. Establish the following formula:

$$\mathfrak{F}\left[e^{\alpha t} \sum_{\nu=-\infty}^{\infty} f(t - \nu)\right] = 2\pi \sum_{\nu=-\infty}^{\infty} \tilde{f}(2\pi\nu) \delta(\omega + i\alpha - 2\pi\nu)$$

HINT: First, show that

$$\sum_{\nu=-\infty}^{\infty} \exp[-i\nu(\omega + i\alpha)] = 2\pi \sum_{\nu=-\infty}^{\infty} \delta(\omega + i\alpha - 2\pi\nu)$$

Then make use of Lemma 1.

7.10 THE GENERAL SOLUTION OF A HOMOGENEOUS LINEAR DIFFERENTIAL EQUATION WITH CONSTANT COEFFICIENTS

The ultimate objective of this section is to use ultradistributions for the determination of the general solution of a homogeneous linear differential equation with constant coefficients. In order to do this, however, we shall have to extend to ultradistributions the results of Sec. 3.2, where it was shown that the delta functional and its derivatives are the only distributions that satisfy equations of the form

$$t^k f(t) = 0$$

where k is some positive integer. Indeed, as will be shown below, this is still the case even when $f(t)$ is allowed to be an ultradistribution. Most of this section will be devoted to the derivation of a more general form of this result. The argument is almost the same as that given in Sec. 3.2, except that now some simplification arises from the fact that the testing functions in Z are entire functions.

We again proceed by means of two lemmas.

Lemma 1

Let $p(\omega)$ be the polynomial

$$p(\omega) \triangleq (\omega - \gamma_1)^{k_1}(\omega - \gamma_2)^{k_2} \cdots (\omega - \gamma_q)^{k_q} \tag{1}$$

where the γ_μ are complex numbers and the k_μ are positive integers. In order for a testing function $\chi(\omega)$ in Z to have the form

$$\chi(\omega) = p(\omega)\phi(\omega) \tag{2}$$

where $\phi(\omega)$ is also in Z, it is necessary and sufficient that

$$\chi^{(\nu)}(\gamma_\mu) = 0 \quad \nu = 0, 1, \ldots, k_\mu - 1; \mu = 1, 2, \ldots, q \tag{3}$$

Proof: (2) clearly implies (3). Conversely, assume that (3) holds with $\chi(\omega)$ being in Z. This means that $\chi(z)$ is an entire function and has zeros at the points $z = \gamma_\mu$ whose multiplicities are at least k_μ, respectively.

Therefore,
$$\phi(z) = \frac{\chi(z)}{p(z)}$$
is also an entire function. We have yet to show that $\phi(z)$ is appropriately bounded.

Let m denote a nonnegative integer and consider
$$z^m \phi(z) = \frac{z^m \chi(z)}{p(z)} \tag{4}$$

The right-hand side of (4) is analytic and therefore bounded over every bounded domain of the z plane. Let n denote the degree of the polynomial $p(z)$. As $|z| \to \infty$,
$$\left| \frac{z^m}{p(z)} \right| \sim |z|^{m-n}$$
Since
$$|z^k \chi(z)| \leq C_k e^{a|y|} \qquad k = 0, 1, 2, \ldots$$
where the C_k and a are constants, it follows that
$$|z^m \phi(z)| = \left| \frac{z^m}{p(z)} \chi(z) \right| \leq K_0 e^{a|y|} \qquad m = 0, 1, \ldots, n$$
and that
$$|z^m \phi(z)| = \left| \frac{z^m}{p(z)} \chi(z) \right| \leq K_{m-n} e^{a|y|} \qquad m = n+1, n+2, \ldots$$
where the K_ν are also constants. Thus, ϕ is truly in $\mathcal{Z}$. Q.E.D.

Lemma 2

Let γ_μ denote the roots of $p(z)$ and let k_μ be their respective multiplicities, as before. Let r be the largest of the k_μ. Finally, let $\lambda_\mu(z)$ ($\mu = 1, 2, \ldots, q$) be fixed testing functions in $\mathcal{Z}$ with the following properties. At every point $z = \gamma_\eta$ ($\eta \neq \mu$), $\lambda_\mu(z)$ has a zero whose multiplicity is at least r; at the point $z = \gamma_\mu$, $\lambda_\mu(\gamma_\mu) = 1$ and $\lambda_\mu^{(\nu)}(\gamma_\mu) = 0$ for $\nu = 1, 2, \ldots, r$. Then any testing function $\psi(z)$ in $\mathcal{Z}$ can be uniquely represented in the form
$$\psi(z) = \sum_{\mu=1}^{q} \sum_{\nu=0}^{k_\mu - 1} \lambda_\mu(z) \frac{1}{\nu!} \psi^{(\nu)}(\gamma_\mu)(z - \gamma_\mu)^\nu + \chi(z) \tag{5}$$
where $\chi(z)$ is in $\mathcal{Z}$ and satisfies (3).

Proof: $\chi(z)$ is uniquely determined by the given $\psi(z)$, $\lambda_\mu(z)$, and γ_μ. It is in Z because the $\psi(z)$ and $\lambda_\mu(z)$ are. That $\chi(z)$ satisfies (3) can be ascertained by appropriately differentiating (5). Q.E.D.

The property of ultradistributions that we are seeking is the following:

Theorem 7.10-1

Let $p(\omega)$ be given by (1). A necessary and sufficient condition for an ultradistribution $f(\omega)$ to satisfy the equation

$$p(\omega)f(\omega) = 0 \tag{6}$$

is that

$$f(\omega) = \sum_{\mu=1}^{q} \sum_{\nu=0}^{k_\mu - 1} b_{\mu\nu} \delta^{(\nu)}(\omega - \gamma_\mu) \tag{7}$$

where the $b_{\mu\nu}$ are arbitrary constants.

Proof: That (7) implies (6) is clear. To demonstrate the converse, let ϕ be an arbitrary testing function in Z. From (6), we have that

$$\langle pf, \phi \rangle = \langle f, p\phi \rangle = 0 \tag{8}$$

For any ψ in Z we may invoke Lemma 2 to write

$$\langle f, \psi \rangle = \sum_{\mu=1}^{q} \sum_{\nu=0}^{k_\mu - 1} \psi^{(\nu)}(\gamma_\mu) \left\langle f(z), \frac{1}{\nu!}(z - \gamma_\mu)^\nu \lambda_\mu(z) \right\rangle + \langle f(z), \chi(z) \rangle \tag{9}$$

where χ is in Z and satisfies (3). By Lemma 1, χ has the form $p\phi$, so that (8) may be used to get $\langle f, \chi \rangle = 0$. If we define the constants $b_{\mu\nu}$ by

$$b_{\mu\nu} \triangleq (-1)^\nu \left\langle f(z), \frac{1}{\nu!}(z - \gamma_\mu)^\nu \lambda_\mu(z) \right\rangle \tag{10}$$

we can now conclude from (9) that

$$f(z) = \sum_{\mu=1}^{q} \sum_{\nu=0}^{k_\mu - 1} b_{\mu\nu} \delta^{(\nu)}(z - \gamma_\mu) \tag{11}$$

This completes the proof.

As an application of ultradistributions, let us now take up the problem of finding the general solution to any homogeneous linear differential equation with constant coefficients. Let

$$a_n f^{(n)} + a_{n-1} f^{(n-1)} + \cdots + a_0 f = 0 \qquad a_n \neq 0;\, n \geq 1 \tag{12}$$

where the a_ν are constants and f is allowed to be any distribution in $\mathcal{D}'$.

The Fourier transform of (12) is
$$[a_n(i\omega)^n + a_{n-1}(i\omega)^{n-1} + \cdots + a_0]\tilde{f}(\omega) = 0$$
where $\tilde{f}(\omega)$ is an ultradistribution. If we factor the polynomial as a function of ω, we obtain
$$(\omega - \gamma_1)^{k_1}(\omega - \gamma_2)^{k_2} \cdots (\omega - \gamma_q)^{k_q}\tilde{f}(\omega) = 0$$
Therefore, by the last theorem,
$$\tilde{f}(\omega) = \sum_{\mu=1}^{q} \sum_{\nu=0}^{k_\mu - 1} b_{\mu\nu} \delta^{(\nu)}(\omega - \gamma_\mu)$$
By taking the inverse Fourier transform, using Sec. 7.8, Eq. (12), and setting
$$c_{\mu\nu} = \frac{b_{\mu\nu}}{2\pi i^\nu}$$
we finally obtain the general solution to (12):
$$f(t) = \sum_{\mu=1}^{q} \sum_{\nu=0}^{k_\mu - 1} c_{\mu\nu} t^\nu e^{i\gamma_\mu t} \tag{13}$$

We have shown that the distributional solutions to (12) are none other than the classical solutions. This concurs with the discussion given in Sec. 6.3.

PROBLEM

1   When we compare the exponential terms in (13) with those in Sec. 6.3, Eq. (6), we find an extra factor of i in the exponents of (13). Why?

2   Show that there exist λ_μ in Z of the form described in Lemma 2.

8 The Laplace Transformation

★8.1 INTRODUCTION

Before discussing the Laplace transformation for distributions, we shall first develop in the next section some classical properties of the ordinary Laplace transformation for functions. As in the preceding chapter, only those classical results which will be needed in the later exposition on the distributional transformation will be investigated. Because of this, we shall place in the next section some rather strong restrictions on the functions under consideration.

Sections 8.3 to 8.7 are devoted to a development of the right-sided Laplace transformation for distributions whose supports are bounded on the left. The results obtained there are extended in Sec. 8.8 to the left-sided Laplace transformation for distributions whose supports are bounded on the right. The two-sided Laplace transformation for distributions whose supports are in general unbounded is taken up in Sec. 8.9.

As we shall see, not all distributions have Laplace transforms according to the definitions that we shall use. However, it is the right-sided distributions that arise in many practical applications, and enough of these possess Laplace transforms to make our Laplace transformation a very useful tool.

The discussion of this chapter will remain within the framework of distribution theory because the Laplace transforms of the distributions considered here will again be distributions (in fact, analytic functions). This is in contrast to the extension of the Fourier transformation to all distributions, which gave rise to the ultradistributions. On the other hand, one can generalize the definition of the Laplace transformation so that it too will apply to all distributions (Ishihara [1]), but then the ultradistributions will once again be generated.

◆8.2 THE LAPLACE TRANSFORMS OF ORDINARY RIGHT-SIDED FUNCTIONS

When we defined the ordinary Fourier transform of the locally integrable function $f(t)$, we assumed that $f(t)$ was absolutely integrable over $-\infty < t < \infty$. This eliminated from consideration functions that grow as fast as $e^t 1_+(t)$ as $t \to \infty$.

The Laplace transformation is an adaptation of the Fourier transformation that allows some of these rapidly growing functions to be taken into account. However, in relaxing the requirement that $f(t)$ be absolutely integrable over $-\infty < t < \infty$, we shall impose in its place the condition that $f(t)$ is a right-sided function. More specifically, we shall concern ourselves in this section with locally integrable functions $f(t)$ that satisfy the following conditions.

Conditions A:

1. $f(t) = 0$ for $-\infty < t < T$.
2. There exists a real number c such that $f(t)e^{-ct}$ is absolutely integrable over $-\infty < t < \infty$.

Let s denote the complex variable $s \triangleq \sigma + i\omega$. The *Laplace transformation* is an operation $\mathfrak{L}$ that assigns a function $F(s)$ of the complex variable s to each locally integrable function $f(t)$ that satisfies conditions A. $\mathfrak{L}$ is defined by

$$\mathfrak{L}f(t) \triangleq F(s) \triangleq \int_{-\infty}^{\infty} f(t)e^{-st}\,dt \qquad (1)$$

We shall also refer to $\mathfrak{L}$ as the *direct* Laplace transformation to distinguish it from the inverse Laplace transformation, which will be introduced later on. The function $F(s)$ is called the *Laplace transform* of $f(t)$.

The fact that $f(t) = 0$ for $-\infty < t < T$ allows us to write

$$\mathfrak{L}f(t) \triangleq F(s) \triangleq \int_{T}^{\infty} f(t)e^{-st}\,dt \qquad (2)$$

We shall refer to (2) as a *right-sided* Laplace transform in order to indicate that the lower limit on the integral is a finite number.

The expression (2) may be broken up into the sum of two integrals:

$$\mathfrak{L}f(t) = \mathfrak{L}_B f(t) + \mathfrak{L}_+ f(t) \tag{3}$$

Here,

$$\mathfrak{L}_B f(t) \triangleq F_B(s) \triangleq \int_T^0 f(t)e^{-st}\,dt \tag{4}$$

$$\mathfrak{L}_+ f(t) \triangleq F_+(s) \triangleq \int_0^\infty f(t)e^{-st}\,dt \tag{5}$$

The second part of conditions A implies that (4) converges absolutely for every (finite) value of s. Furthermore, since

$$|f(t)e^{-st}| \leq |f(t)e^{-ct}|$$

whenever $t > 0$ and $\operatorname{Re} s \geq c$, (5) converges absolutely for all s in the half-plane $\operatorname{Re} s \geq c$. We can therefore conclude that (2) converges absolutely for all s in this half-plane.

The greatest lower bound σ_a on all possible values of c, for which the second of conditions A holds, is called the *abscissa of absolute convergence*, and the open half-plane $\operatorname{Re} s > \sigma_a$ is called the *half-plane* (or *region*) *of absolute convergence* for the Laplace transform (2). We shall refer to a half-plane that is bounded on the left but extends infinitely to the right as a right-sided half-plane. Note that the Laplace transform of a right-sided function that satisfies conditions A has as its region of convergence either a right-sided half-plane or [as is indicated by the example $f(t) = 1_+(t) \exp(-t^2)$] the entire s plane.

In certain cases (2) may also converge absolutely for all s whose real parts equal σ_a. An example of this is provided by the function

$$f(t) = \frac{1_+(t)}{1+t^2}$$

whose Laplace transform has the origin as its abscissa of absolute convergence $\sigma_a = 0$ and converges absolutely for all imaginary s. We shall never include this borderline, $s = \sigma_a + i\omega$, as part of the half-plane of absolute convergence. In other words, this half-plane is by definition an open one.

Moreover, there are certain functions for which (2) converges conditionally in a half-plane $\operatorname{Re} s > \sigma_c$, where $\sigma_c < \sigma_a$. (In line with this, see Prob. 1.) There are still other functions for which (2) does not converge absolutely for any value of s even though it converges conditionally for $\operatorname{Re} s > \sigma_c < \infty$. (See Widder [1], p. 47.) We shall not pursue this ramification; instead, we shall restrict our attention to functions that satisfy conditions A and to their corresponding half-planes of absolute convergence.

The relationship between the Fourier transform and the Laplace transform of a locally integrable function that satisfies conditions A is

$$\mathfrak{L}f(t) = \mathfrak{F}[f(t)e^{-\sigma t}] \qquad \sigma > \sigma_a \tag{6}$$

$\mathfrak{L}$ is a linear transformation in the following way. Let the locally integrable functions f and g satisfy conditions A and let their Laplace transforms have the abscissas of absolute convergence σ_{af} and σ_{ag}, respectively. Then, with α and β being any two constants, we have

$$\mathfrak{L}(\alpha f + \beta g) = \alpha \mathfrak{L} f + \beta \mathfrak{L} g \qquad \sigma > \sup{(\sigma_{af}, \sigma_{ag})} \tag{7}$$

In certain cases the abscissa of convergence for $\mathfrak{L}(\alpha f + \beta g)$ may actually be less than $\sup{(\sigma_{af}, \sigma_{ag})}$.

Theorem 8.2-1

Let $f(t)$ be a continuous function that satisfies conditions A and let σ_a be the abscissa of absolute convergence for $\mathfrak{L}f(t) \triangleq F(s)$. Then, $F(s)$ is an analytic function for Re $s > \sigma_a$ *and*

$$F^{(k)}(s) = \int_T^\infty (-t)^k f(t) e^{-st}\, dt \qquad \text{Re } s > \sigma_a \tag{8}$$

Here again we have hypothesized more than is needed. A considerably more general result will be obtained when we discuss the distributional Laplace transformation.

Proof: First, assume that $T \geq 0$. For Re $s \geq c > \sigma_a$,

$$\int_T^\infty |f(t)e^{-st}|\, dt \leq \int_T^\infty |f(t)|e^{-ct}\, dt < \infty$$

Consequently, (2) converges uniformly for Re $s \geq c$. Moreover, the integrand of (2) is a continuous function of (s, t) for all s and all t and an analytic function of s for every t. According to a standard theorem (Titchmarsh [1], pp. 99–100), $F(s)$ is analytic for Re $s \geq c > \sigma_a$ and we may differentiate (in the ordinary sense) under the integral sign to get

$$F^{(1)}(s) = -\int_T^\infty tf(t)e^{-st}\, dt$$

The same sort of argument shows that we may again differentiate the last integral under the integral sign. By continuing in this fashion, we obtain (8). Since c may be chosen as close to σ_a as we wish, these conclusions hold for Re $s > \sigma_a$.

On the other hand, if $T < 0$, then we may decompose $\mathfrak{L}f(t)$ according to (3). The argument of the preceding paragraph has already established the desired result for $\mathfrak{L}_+f(t)$. For $\mathfrak{L}_Bf(t)$, we need merely note that, over every bounded domain Ω of the s plane,

$$|e^{-st}| \leq e^{-bt} \quad T \leq t \leq 0$$

where b is any upper bound on Re s for all s in Ω. Thus, $\mathfrak{L}_Bf(t)$ converges uniformly over Ω and we may proceed as before. Q.E.D.

If the continuous function $f(t)$ has a bounded support, then its Laplace transform $F(s)$ will have the entire s plane as its region of absolute convergence and will therefore be an entire function.

Because of the relationship (6) between the ordinary Fourier and Laplace transforms, an inversion formula for the Laplace transformation can be obtained by adapting the corresponding formula for the Fourier transformation. We shall again assume stronger conditions on the functions $f(t)$ than we really need.

Theorem 8.2-2

Let the locally integrable function $f(t)$ satisfy conditions A, let its ordinary first derivative $f^{(1)}(t)$ exist and be continuous throughout some open interval $a < t < b$, and let σ_a be the abscissa of absolute convergence for $\mathfrak{L}f(t) \triangleq F(s)$. For each real constant c greater than σ_a and for $a < t < b$,

$$f(t) = \mathfrak{L}^{-1}F(s) \triangleq \frac{1}{2\pi i} \lim_{y \to \infty} \int_{c-iy}^{c+iy} F(s)e^{st}\,ds \tag{9}$$

where the path of integration is along the vertical line $s = c + i\omega$.

Here, $\mathfrak{L}^{-1}$ denotes the *inverse Laplace transformation* and $\mathfrak{L}^{-1}F(s)$ is the *inverse Laplace transform* of $F(s)$. It will always be understood that the path of integration for $\mathfrak{L}^{-1}F(s)$ is a vertical line inside the region of absolute convergence of the direct transform $\mathfrak{L}f$.

Proof: The right-hand side of (9) can be written as

$$\frac{e^{ct}}{2\pi} \lim_{y \to \infty} \int_{-y}^{y} F(c + i\omega)e^{i\omega t}\,d\omega \tag{10}$$

According to (6),

$$F(c + i\omega) = \mathfrak{F}[f(t)e^{-ct}]$$

Consequently, we may use the inverse Fourier transformation (see Theorem 7.2-1) to convert (10) into

$$e^{ct}f(t)e^{-ct} = f(t) \quad a < t < b \quad \text{Q.E.D.}$$

Note that any real value of c greater than σ_a may be used in (9) and that every such choice leads to the same value for $f(t)$ at any given point in the interval $a < t < b$.

Moreover, the uniqueness of the Laplace transformation is implicit in Theorem 8.2-2. In other words, we have

Corollary 8.2-2a

If both $f(t)$ and $g(t)$ satisfy the hypothesis of Theorem 8.2-2 and if their respective Laplace transforms $F(s)$ and $G(s)$ are equal on some vertical line $s = c + i\omega$ in their regions of absolute convergence, then $f(t) = g(t)$ for $a < t < b$.

Proof: For $a < t < b$,

$$f(t) = \mathfrak{L}^{-1}F(c + i\omega) = \mathfrak{L}^{-1}G(c + i\omega) = g(t) \qquad \text{Q.E.D.}$$

Another result that will be needed subsequently is given by the following theorem, which states a sufficient set of conditions for a function $F(s)$ to be a Laplace transform.

Theorem 8.2-3

Assume that over the half-plane $\operatorname{Re} s \geq a$, $F(s)$ is analytic and satisfies the inequality

$$|F(s)| \leq \frac{C}{|s|^2} \tag{11}$$

where C is a constant. If the integral $\mathfrak{L}^{-1}F(s)$ is taken over some vertical line in the half-plane $\operatorname{Re} s \geq a$, then $\mathfrak{L}^{-1}F(s) \triangleq f(t)$ exists and is a continuous function for all t. Moreover, $f(t) = 0$ for $t < 0$ and $\mathfrak{L}f(t) = F(s)$, at least for $\operatorname{Re} s > a$.

Proof: Because of (11),

$$\int_{-\infty}^{\infty} |F(a + i\omega)e^{i\omega t}|\, d\omega \leq \int_{-\infty}^{\infty} |F(a + i\omega)|\, d\omega < \infty \tag{12}$$

and we may write

$$e^{-at}f(t) = \frac{1}{2\pi} \int_{-\infty}^{\infty} F(a + i\omega)e^{i\omega t}\, d\omega$$

where the last integral converges absolutely and uniformly for $-\infty < t < \infty$. The inequality (12) also implies that $e^{-at}f(t)$ is a bounded function for all t. Furthermore, the same argument as that given at the beginning of Sec. 7.2 shows that $f(t)$ is continuous for all t.

Next, we shall prove that $f(t) = 0$ for $t < 0$. Consider the integral

$$\oint F(s)e^{st}\, ds \tag{13}$$

where the closed path of integration lies in the half-plane $\operatorname{Re} s \geq a$ and

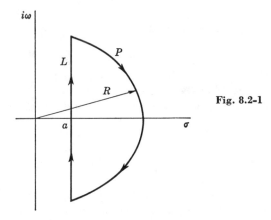

Fig. 8.2-1

consists of a segment L of the vertical line $s = a + i\omega$ and an arc P of a circle centered at the origin and of radius R, as shown in Fig. 8.2-1. By the analyticity of $F(s)$ for $\operatorname{Re} s \geq a$ and by Cauchy's integral theorem, (13) is zero for every $R > |a|$. Furthermore, when $t < 0$, the integral on the circular arc P goes to zero as $R \to \infty$ because

$$\left| \int_P F(s)e^{st}\,ds \right| \leq \int_P |F(s)||e^{\sigma t}|\,|ds| \leq e^{at} \int_P |F(s)|\,|ds|$$

and the right-hand side goes to zero as $R \to \infty$ in view of (11). Thus, we have that, for $t < 0$,

$$f(t) = \frac{1}{2\pi i} \int_{a-i\infty}^{a+i\infty} F(s)e^{st}\,ds = \lim_{R\to\infty} \frac{1}{2\pi i} \oint F(s)e^{st}\,ds = 0$$

Also, note that, because of (11) and Cauchy's integral theorem, we may shift the vertical path of integration to the right without affecting $f(t)$.

Since $f(t)$ is continuous for all t and zero for $t < 0$ and since $e^{-at}f(t)$ is bounded for all t, $e^{-ct}f(t)$ is absolutely integrable for each $c > a$. We have thus shown that $f(t)$ satisfies conditions A and therefore has a Laplace transform. We have yet to verify that this Laplace transform is $F(s)$.

On the vertical line $s = c + i\omega$ ($c > a$) the inverse Laplace transform $f(t) = \mathfrak{L}^{-1}F(s)$ can be written as an ordinary direct Fourier transform:

$$2\pi e^{-ct}\overline{f(t)} = \mathfrak{F}\overline{F(c + i\omega)} \tag{14}$$

Since $F(s)$ is analytic for $\operatorname{Re} s \geq a$, $F(c + i\omega)$ satisfies the hypotheses of Theorem 7.2-1. Therefore, by the uniqueness of the Fourier transformation (see Corollary 7.2-1a) there is only one function of s that is continuous on the line $s = c + i\omega$, and that produces $f(t)$ according to (14). By applying the ordinary inverse Fourier transformation to both sides of (14), we obtain

$$\frac{1}{2\pi}\int_{-\infty}^{\infty}[2\pi e^{-ct}\overline{f(t)}]e^{i\omega t}\,dt = \int_{0}^{\infty}\overline{f(t)}e^{(-c+i\omega)t}\,dt$$
$$= \overline{F(c+i\omega)}$$

We may take the complex conjugate of this equality to obtain $\mathfrak{L}f(t) = F(s)$. This result will hold for at least Re $s > a$ because c is arbitrary so long as it is greater than a. This completes the proof.

Example 8.2-1 Let us establish that

$$\mathfrak{L}\left[\frac{t^k}{k!}1_+(t)\right] = \frac{1}{s^{k+1}} \qquad \text{Re } s > 0;\, k = 0, 1, 2, \ldots \qquad (15)$$

through an inductive argument. For $k = 0$,

$$\mathfrak{L}1_+(t) = \int_0^\infty e^{-st}\,dt = \frac{1}{s} \qquad \text{Re } s > 0$$

Now, let k be a positive integer. Through an integration by parts, we may write

$$\int_0^\infty \frac{t^k}{k!}e^{-st}\,dt = \frac{1}{s}\int_0^\infty \frac{t^{k-1}}{(k-1)!}e^{-st}\,dt$$
$$= \frac{1}{s}\mathfrak{L}\left[\frac{t^{k-1}}{(k-1)!}1_+(t)\right] \qquad \text{Re } s > 0$$

According to the last expression, (15) will certainly hold so long as it holds when k is replaced by $k - 1$. This proves (15).

Example 8.2-2 Let the locally integrable function f satisfy conditions A. If $\mathfrak{L}f = F(s)$ for Re $s > \sigma_a$ and if α is some complex number, then

$$\mathfrak{L}[e^{\alpha t}f(t)] = F(s - \alpha) \qquad \text{Re } s > \sigma_a + \text{Re } \alpha \qquad (16)$$

This follows directly from definition (1) since

$$\int_{-\infty}^{\infty} e^{\alpha t}f(t)e^{-st}\,dt = \int_{-\infty}^{\infty} f(t)e^{-(s-\alpha)t}\,dt$$

As an application of (16), we may conclude from (15) that

$$\mathfrak{L}\left[\frac{1}{k!}t^k e^{\alpha t}1_+(t)\right] = \frac{1}{(s-\alpha)^{k+1}} \qquad \text{Re } s > \text{Re } \alpha \qquad (17)$$

Example 8.2-3 As a final example, we shall employ the inverse Laplace transformation to establish the formula

$$\mathfrak{L}^{-1}\frac{1}{\sqrt{s^2+1}} = J_0(t)1_+(t) \qquad (18)$$

where it is understood that the half-plane of convergence is a right-sided one and where $J_0(t)$ is the zero-order Bessel function of first kind,

$$J_0(t) \triangleq \sum_{\nu=0}^{\infty} \frac{(-1)^\nu}{(\nu!)^2} \left(\frac{t}{2}\right)^{2\nu} \qquad (19)$$

By using the binomial series expansion, we may write

$$\frac{1}{\sqrt{s^2+1}} - \frac{1}{s} = -\frac{1}{2} \times \frac{1}{s^3} + \frac{1 \times 3}{2 \times 4} \times \frac{1}{s^5} - \frac{1 \times 3 \times 5}{2 \times 4 \times 6} \times \frac{1}{s^7}$$
$$+ \cdots \qquad |s| > 1 \quad (20)$$

As $|s| \to \infty$, (20) is asymptotic to $\frac{1}{2}s^3$, and it therefore satisfies the hypothesis of Theorem 8.2-3. Since $\mathfrak{L}^{-1}1/s = 1_+(t)$ (Re $s > 0$), it follows that $\mathfrak{L}^{-1}1/\sqrt{s^2+1}$ is a function $f(t)$ that equals zero for $t < 0$, is continuous for $t > 0$, and has an ordinary discontinuity at $t = 0$ such that $f(0+) = 1$.

In order to obtain $f(t)$ for $t > 0$, we shall apply the inverse Laplace transformation to the right-hand side of (20) term by term. [This term-by-term integration is valid because the right-hand side of (20) is an inverse power series whose leading term is $-\frac{1}{2}s^3$ and because on the vertical line, $s = c + i\omega$, $|e^{st}|$ equals e^{ct}.] Thus, by using (15), we obtain for $t > 0$

$$\mathfrak{L}^{-1}\frac{1}{\sqrt{s^2+1}} - 1 = -\frac{1}{2} \times \frac{t^2}{2!} + \frac{1 \times 3}{2 \times 4} \times \frac{t^4}{4!}$$
$$- \frac{1 \times 3 \times 5}{2 \times 4 \times 6} \times \frac{t^6}{6!} + \cdots$$
$$= -\left(\frac{t}{2}\right)^2 + \frac{1}{(2!)^2}\left(\frac{t}{2}\right)^4 - \frac{1}{(3!)^2}\left(\frac{t}{2}\right)^6 + \cdots$$

Upon comparing this result with the series (19), we conclude (18).

Although s was restricted to the half-plane Re $s \geq c > 1$ in the above manipulation, the region of absolute convergence for $\mathfrak{L}[J_0(t)1_+(t)]$ is actually the half-plane Re $s > 0$. This follows from the fact that as $t \to \infty$

$$J_0(t) \approx \sqrt{\frac{2}{\pi t}} \cos\left(t - \frac{\pi}{4}\right)$$

All the results of this section can be extended to the ordinary left-sided Laplace transformation, which is the restriction of (1) to certain functions whose supports are bounded on the right. A more general extension of (1) can also be made, where now certain functions having supports that are unbounded on both sides are taken into account; in this latter case, (1) is called an ordinary two-sided Laplace transform. We shall not explicitly make these extensions here, since such extensions will be made in Secs. 8.8 and 8.9 for the more general distributional Laplace transformation.

PROBLEMS

1 Show that

$$\int_0^\infty e^{-st} \sin e^t \, dt \tag{21}$$

converges absolutely only for those s whose real parts are positive and conditionally for all real s that lie in the interval $-1 < s \leq 0$. (HINT: Make the change of variable $u = e^t$.) It is a fact that, if (21) converges at the real point $s = \sigma_0$, it also converges for all s in the half-plane Re $s > \sigma_0$. (Can you show this?) Hence, the fact that (21) converges conditionally for all real s in $-1 < s \leq 0$ implies that it converges at least conditionally for all s in the half-plane Re $s > -1$.

2 Verify the following Laplace transform pairs. Here, a is a real positive constant.

    a. $\mathfrak{L}[1_+(t) \sin at] = \dfrac{a}{s^2 + a^2}$ Re $s > 0$

    b. $\mathfrak{L}[1_+(t) \cos at] = \dfrac{s}{s^2 + a^2}$ Re $s > 0$

    c. $\mathfrak{L}[1_+(t) \sinh at] = \dfrac{a}{s^2 - a^2}$ Re $s > a$

    d. $\mathfrak{L}[1_+(t) \cosh at] = \dfrac{s}{s^2 - a^2}$ Re $s > a$

3 Assume that $f(t)$ is a function that satisfies conditions A and let $\mathfrak{L}f(t) \triangleq F(s)$ (Re $s > \sigma_a$). Establish the following formulas, wherein σ_b and τ are real constants and a is a real positive constant.

    a. $\mathfrak{L}f^{(1)}(t) = sF(s)$ Re $s > \sigma_b$

[Also assume here that $f(t)$ is absolutely continuous for all t and that $f^{(1)}(t)$ satisfies conditions A.]

    b. $\mathfrak{L}f(t - \tau) = e^{-s\tau}F(s)$ Re $s > \sigma_a$

    c. $\mathfrak{L}f(at) = \dfrac{1}{a}F\left(\dfrac{s}{a}\right)$ Re $s > a\sigma_a$

4 Find the inverse Laplace transforms of the following functions, where the region of convergence is taken as a right-sided half-plane. [That is, $\mathfrak{L}^{-1}F(s)$ is integrated over some vertical line lying to the right of all the poles of each function.] HINT: Make partial-fraction expansions.

    a. $\dfrac{1}{s(s + 1)}$

    b. $\dfrac{1}{s^3 + 2s^2 + 2s}$

    c. $\dfrac{2s + 1}{s^2(s + 1)^2}$

    d. $\dfrac{s}{s^4 + 1}$

5 Show that

$$\int_0^\infty e^{-st} \log t \, dt = -\frac{\gamma + \log s}{s} \qquad \text{Re } s > 0$$

where γ is Euler's constant,

$$\gamma \triangleq -\int_0^\infty e^{-x} \log x \, dx = 0.5772 \cdots$$

6 By an appropriate limiting process, relate the Fourier and Laplace transforms of $1_+(t)$.

7 Extend the results of this section to the ordinary left-sided Laplace transformation. In particular, assume that the locally integrable function $f(t)$ is zero for $T < t < \infty$ and that there exists a real number c such that $f(t)e^{-ct}$ is absolutely integrable over $-\infty < t < \infty$. Define the left-sided Laplace transformation by

$$\mathfrak{L}f(t) \triangleq F(s) \triangleq \int_{-\infty}^T f(t)e^{-st} \, dt$$

Then, describe the region of absolute convergence for this integral and alter all the theorems of this section so that they will apply to the present case. HINT: This can be done very simply by making some changes of variables.

8 Extend the results of this section to the ordinary two-sided Laplace transformation. That is, assume now that the support of the locally integrable function $f(t)$ is not bounded on either side and that, for every value of the real constant c in some (nonvoid) open interval, $f(t)e^{-ct}$ is absolutely integrable over $-\infty < t < \infty$. The two-sided Laplace transformation is defined by (1) of this section. Describe the general form of its region of absolute convergence. Then, alter all the theorems of this section for this more general case. HINT: The two-sided transform can be decomposed into a left-sided one and a right-sided one.

★8.3 THE LAPLACE TRANSFORMS OF RIGHT-SIDED DISTRIBUTIONS

Let us repeat the definition of the ordinary Laplace transform $G(s)$ of a locally integrable function $g(t)$ that satisfies conditions A:

$$G(s) \triangleq \mathfrak{L}g(t) \triangleq \int_T^\infty g(t)e^{-st} \, dt \qquad \text{Re } s > \sigma_a \qquad (1)$$

This relation may be rewritten in the form

$$G(s) = \langle g(t), e^{-st} \rangle \qquad (2)$$

It is natural to attempt to define the Laplace transform of a distribution $f(t)$ whose support is bounded on the left by formally extending (2) to obtain

$$F(s) \triangleq \mathfrak{L}f(t) \triangleq \langle f(t), e^{-st} \rangle \qquad (3)$$

Of course, we have yet to ascertain under what conditions (3) can have a sense.

Let us assume that there exists a real number c for which $e^{-ct}f(t)$ is a distribution of slow growth. We may convert (3) into

$$F(s) \triangleq \mathfrak{L}f(t) \triangleq \langle e^{-ct}f(t), \lambda(t)e^{-(s-c)t}\rangle \tag{4}$$

where $\lambda(t)$ is any infinitely smooth function with support bounded on the left, which equals one over a neighborhood of the support of $f(t)$. For Re $s > c$, $\lambda(t)e^{-(s-c)t}$ is a testing function in $\mathcal{S}_t$ and, therefore, the right-hand side of (4) possesses a sense as the application of a distribution in $\mathcal{S}'_t$ to a testing function in $\mathcal{S}_t$. *Thus if $f(t)$ is a right-sided distribution and if $f(t)$ is such that $e^{-ct}f(t)$ is in $\mathcal{S}'$ for some real c, the right-hand side of (4) is taken as the definition of the Laplace transform of $f(t)$.* Henceforth, it will be understood that the right-hand side of (3) possesses that sense which is assigned to it by the right-hand side of (4); (3) will exist only if (4) exists. In this case, we shall say that $f(t)$ is *Laplace-transformable.* Its Laplace transform will also be called right-sided to denote again that $f(t)$ has its support bounded on the left; *this should not be taken to mean that $F(s)$ is a right-sided function.*

Not all right-sided distributions are Laplace-transformable. An example is the regular distribution $1_+(t) \exp t^2$. There is no value of c for which $1_+(t) \exp (t^2 - ct)$ will be a distribution of slow growth.

Note that the ordinary Laplace transform discussed in the preceding section is a special case of our distributional Laplace transform (4).

Clearly, the Laplace transform $F(s)$ of a Laplace-transformable $f(t)$ in $\mathfrak{D}'_R$ is a function of s defined over the right-sided half-plane Re $s > c$. In analogy to the ordinary Laplace transformation, the greatest lower bound σ_1 of all real constants c for which $e^{-ct}f(t)$ is in $\mathcal{S}'$ is called the *abscissa of convergence* and the half-plane Re $s > \sigma_1$ is called the *region of convergence* or the *half-plane of convergence* for $\mathfrak{L}f(t)$. In certain cases, $\sigma_1 = -\infty$. Note that the half-plane of convergence is defined as an open one and the borderline $s = \sigma_1 + i\omega$ is not part of it.

A distinction should be made between the "abscissa of convergence σ_1" defined here and the "abscissa of absolute convergence σ_a" defined in Sec. 8.2. When $f(t)$ is a singular distribution, our use of the word "convergence" does not imply the convergence of some integral. Moreover, when $f(t)$ is a regular distribution (i.e., a locally integrable function) that satisfies conditions A of Sec. 8.2, its Laplace transform may have an abscissa of convergence σ_1 that is smaller than its abscissa of absolute convergence σ_a. (An example of this is indicated in Prob. 1.)

Another point should be made here. *The Laplace transform $F(s)$ is independent of the choice of c in the definition (4) so long as $c > \sigma_1$.* To see this, let $e^{-ct}f(t)$ be in $\mathfrak{D}'_R$ and in $\mathcal{S}'$ and let Re $s > a > c$. Then,

$$\langle e^{-ct}f(t), \lambda(t)e^{-(s-c)t}\rangle = \langle e^{(c-a)t}e^{-ct}f(t), e^{(a-c)t}\lambda(t)e^{-(s-c)t}\rangle$$
$$= \langle e^{-at}f(t), \lambda(t)e^{-(s-a)t}\rangle$$

The right-hand side possesses a sense, since $e^{-at}f(t)$ is certainly in $\mathcal{S}'$ and $\lambda(t)e^{-(s-a)t}$ is in $\mathcal{S}$. This means that, if c is replaced by a larger real number a, the definition of $\mathfrak{L}f$ yields precisely the same function $F(s)$ so long as Re $s > a$. But this in turn implies that $F(s)$ is independent of all choices of c greater than σ_1, since c may be chosen arbitrarily close to σ_1.

The distributional right-sided Laplace transformation is linear in the same way that the ordinary right-sided Laplace transformation is. That is, if f and g are Laplace-transformable distributions in $\mathfrak{D}'_R$ and if

$$\mathfrak{L}f = F(s) \qquad \text{Re } s > \sigma_f$$

and

$$\mathfrak{L}g = G(s) \qquad \text{Re } s > \sigma_g$$

then, with α and β being any two constants, we have that

$$\mathfrak{L}(\alpha f + \beta g) = \alpha F(s) + \beta G(s) \qquad \text{Re } s > \sup(\sigma_f, \sigma_g), \text{ at least}$$

The Laplace transform and the Fourier transform of a distribution can be related in the following way: Let $\sigma_1 < c < \sigma$, let $\phi(\omega)$ be in $\mathcal{S}_\omega$, and consider

$$\langle e^{-ct}f(t), \lambda(t)e^{-(\sigma-c+i\omega)t}\phi(\omega)\rangle$$

The testing function in this expression is in $\mathcal{S}_{\omega,t}$. According to Theorem 4.5-1, the expression itself is an element of $\mathcal{S}_\omega$. We can conclude, therefore, that the expression can be integrated over $-\infty < \omega < \infty$ and that $F(\sigma + i\omega)$ is a continuous function. Moreover, from the boundedness property of distributions of slow growth [see (1) of Sec. 4.4], it follows that, as a function of ω, $F(\sigma + i\omega)$ is of slow growth. Consequently, we may write

$$\int_{-\infty}^{\infty} \langle e^{-ct}f(t), \lambda(t)e^{-(\sigma-c+i\omega)t}\phi(\omega)\rangle \, d\omega = \langle F(\sigma + i\omega), \phi(\omega)\rangle = \langle \mathfrak{L}f, \phi\rangle$$

Upon invoking Corollary 5.3-2b, we see that the left-hand side of the last equation equals

$$\left\langle e^{-ct}f(t), \lambda(t)e^{(c-\sigma)t}\int_{-\infty}^{\infty}\phi(\omega)e^{-i\omega t}\,d\omega\right\rangle$$
$$= \langle e^{-\sigma t}f(t), \tilde{\phi}(t)\rangle = \langle \mathfrak{F}[e^{-\sigma t}f(t)], \phi(\omega)\rangle$$

Thus, we see that

$$\mathfrak{L}f(t) = \mathfrak{F}[e^{-\sigma t}f(t)] \qquad \sigma > \sigma_1 \tag{5}$$

Actually, (5) can be used as an alternative means of defining the Laplace transform of any $f(t)$ in $\mathfrak{D}'_R$ so long as $e^{-\sigma t}f(t)$ is in $\mathcal{S}'_t$ for all $\sigma > \sigma_1$.

By using (5) the uniqueness theorem for our distributional Laplace transformation can be established. In particular, assume that $f(t)$ is a Laplace-transformable distribution in $\mathcal{D}'_R$ and that $F(s) \triangleq \mathfrak{L}f(t) = 0$ over some vertical line $s = c + i\omega$ in the region of convergence. We shall show that $f(t)$ must be the zero distribution. Indeed, let ϕ and, therefore, $\tilde{\phi}$ traverse $\mathcal{S}$. Then,

$$0 = \langle F(c + i\omega), \phi(\omega)\rangle = \langle \mathfrak{F}[e^{-ct}f(t)], \phi(\omega)\rangle$$
$$= \langle e^{-ct}f(t), \tilde{\phi}(t)\rangle$$

Hence, $e^{-ct}f(t)$ and, therefore, $f(t)$ are the zero distribution. In view of the linearity of $\mathfrak{L}$, this result implies the following uniqueness theorem.

Theorem 8.3-1 (*The uniqueness theorem*)

Let f and g be Laplace-transformable distributions in $\mathcal{D}'_R$. If $\mathfrak{L}f = \mathfrak{L}g$ on some vertical line $s = c + i\omega$ in their regions of convergence, then $f = g$.

We shall now establish the analyticity property of Laplace transforms.

Theorem 8.3-2 (*The analyticity theorem*)

Let $f(t)$ be a Laplace-transformable distribution in $\mathcal{D}'_R$. Then, $F(s) \triangleq \mathfrak{L}f$ is an analytic function in its region of convergence $\operatorname{Re} s > \sigma_1$ and

$$\frac{dF}{ds} = \left\langle f(t), \frac{\partial}{\partial s} e^{-st}\right\rangle = \langle -tf(t), e^{-st}\rangle$$
$$\triangleq \langle e^{-ct}f(t), -t\lambda(t)e^{-(s-c)t}\rangle \qquad \operatorname{Re} s > c > \sigma_1 \quad (6)$$

where c is an arbitrary real number greater than σ_1.

Our proof of this theorem will make use of the following lemma.

Lemma 1

Let $\lambda(t)$ be the function specified in the definition of the Laplace transform of $f(t)$. Also, let s be fixed with $\operatorname{Re} s > 0$ and let all Δs be so restricted that $|\Delta s| \leq \operatorname{Re} s - a > 0$, a being a real positive number. Finally, let

$$\phi_{\Delta s}(t) \triangleq \lambda(t) \frac{e^{-(s+\Delta s)t} - e^{-st}}{\Delta s} \qquad |\Delta s| \neq 0$$

Then, as $\Delta s \to 0$, $\phi_{\Delta s}(t)$ converges in $\mathcal{S}$ to

$$\lambda(t) \frac{\partial}{\partial s} e^{-st} = -t\lambda(t)e^{-st}$$

Proof: We shall merely sketch out the proof of this lemma and will leave the details to the reader. First of all,

$$\left| \frac{e^{-\Delta st} - 1}{\Delta s} \right| \leq |t| e^{|\Delta st|}$$

so that for any nonnegative integer m

$$|t^m \phi_{\Delta s}(t)| \leq |t^{m+1} \lambda(t) e^{-st + |\Delta st|}|$$

The right-hand side is bounded uniformly for all t and for all Δs that satisfy $|\Delta s| \leq \operatorname{Re} s - a > 0$. A similar argument shows that for each nonnegative integer k

$$|t^m \phi_{\Delta s}{}^{(k)}(t)|$$

is also bounded uniformly for all t and for all Δs that satisfy the same restriction.

Also, as $\Delta s \to 0$, $e^{-\Delta st} \to 1$ and

$$\frac{e^{-\Delta st} - 1}{\Delta s} \to -t$$

uniformly on every bounded t interval. Because of this, for each k, $\phi_{\Delta s}{}^{(k)}(t)$ tends uniformly to a limit on each bounded t interval. Thus, $\phi_{\Delta s}(t)$ converges in $\mathcal{S}$ to $-t\lambda(t)e^{-st}$, as was asserted.

Proof of Theorem 8.3-2: Let c be an arbitrary real number greater than σ_1 and let

$$H(s) \triangleq \langle e^{-ct}f(t), \lambda(t)e^{-st}\rangle = F(s+c)$$

$H(s)$ certainly exists for $\operatorname{Re} s > 0$. By the linearity of $e^{-ct}f(t)$ as a functional on $\mathcal{S}_t$, we have, for $|\Delta s| \leq \operatorname{Re} s - a > 0$ $(a > 0)$

$$\frac{H(s + \Delta s) - H(s)}{\Delta s} = \langle e^{-ct}f(t), \phi_{\Delta s}(t)\rangle \qquad |\Delta s| \neq 0$$

By the continuity of $e^{-ct}f(t)$ as a functional on $\mathcal{S}_t$ and by Lemma 1, this yields in the limit as $\Delta s \to 0$

$$\frac{dF(s+c)}{ds} = \frac{dH}{ds} = \langle e^{-ct}f(t), -t\lambda(t)e^{-st}\rangle$$

Finally,

$$F(s) = H(s - c) = \langle e^{-ct}f(t), \lambda(t)e^{-(s-c)t}\rangle$$

Since a and c are arbitrary (except that $c > \sigma_1$ and $a > 0$), $F(s)$ is analytic for $\operatorname{Re} s > \sigma_1$ and

$$\frac{dF}{ds} = \langle e^{-ct}f(t), -t\lambda(t)e^{-(s-c)t}\rangle \triangleq \langle f(t), -te^{-st}\rangle \qquad \operatorname{Re} s > c > \sigma_1$$

Q.E.D.

Previously, we showed (see Theorem 7.4-3) that if $f(t)$ is a distribution with bounded support, then its Fourier transform $\tilde{f}(\omega)$ is an infinitely smooth function given by

$$\tilde{f}(\omega) = \langle f(t), e^{-i\omega t}\rangle \triangleq \langle f(t), \lambda(t)e^{-i\omega t}\rangle$$

where now $\lambda(t)$ is a testing function in $\mathfrak{D}$ that equals one over a neighborhood of the support of $f(t)$. In this case, $e^{-\sigma t}f(t)$ will be in $\mathcal{S}'_t$ for every σ so that

$$F(s) = \mathfrak{L}f = \langle f(t), \lambda(t)e^{-st}\rangle$$

Here $F(s)$ is an analytic function for all s. Thus, when f has a bounded support, its Laplace transform $F(s)$ can be obtained from its Fourier transform $\tilde{f}(\omega) = F(i\omega)$ by extending $F(i\omega)$ from the $s = i\omega$ axis to an entire function on the s plane.

A similar possibility presents itself when the support of f is bounded only on the left. According to (5), we may obtain $\mathfrak{L}f(t)$ over its region of convergence by extending the values of $\mathfrak{F}[e^{-ct}f(t)]$, where $c > \sigma_1$, from the vertical line $s = c + i\omega$ to an analytic function on that region of convergence.

Example 8.3-1 Since δ has its support concentrated on a single point, the following Laplace transforms can be obtained by extending the results of Example 7.4-1. They may also be computed directly from the definition (4). Here, τ is a real constant and k is a positive integer.

$$\mathfrak{L}\delta = 1 \qquad -\infty < \operatorname{Re} s < \infty$$
$$\mathfrak{L}\delta(t - \tau) = e^{-s\tau} \qquad -\infty < \operatorname{Re} s < \infty$$
$$\mathfrak{L}\delta^{(k)}(t) = s^k \qquad -\infty < \operatorname{Re} s < \infty$$
$$\mathfrak{L}\delta^{(k)}(t - \tau) = s^k e^{-s\tau} \qquad -\infty < \operatorname{Re} s < \infty$$

Example 8.3-2 Let us compute the Laplace transform $F(s)$ of the pseudofunction

$$f(t) = \operatorname{Pf} \frac{1_+(t)1_+(1 - t)}{1 - t}$$

Since f has a bounded support, its transform exists for all finite s. To obtain $F(s)$, we need merely refer to (21) of Sec. 2.5, in which we set $\phi(t) = e^{-st}$, $k = 1$, $a = 0$, and $b = 1$. This yields

$$F(s) = \lim_{\varepsilon \to 0+} \left[\int_0^{1-\varepsilon} \frac{e^{-st}}{1 - t} dt + e^{-s} \log \varepsilon \right]$$
$$= \lim_{\varepsilon \to 0+} e^{-s} \int_0^{1-\varepsilon} \frac{e^{s(1-t)} - 1}{1 - t} dt$$
$$= e^{-s} \int_0^1 \frac{e^{s(1-t)} - 1}{1 - t} dt = e^{-s} \int_0^s \frac{e^z - 1}{z} dz$$

The integrand of the last expression is entire and can therefore be expanded into a Maclaurin's series. This series may be integrated term by term to obtain a series for $F(s)$ that converges for all s:

$$F(s) = e^{-s} \int_0^s \sum_{\nu=1}^\infty \frac{z^{\nu-1}}{\nu!} dz = e^{-s} \sum_{\nu=1}^\infty \frac{s^\nu}{\nu!\nu}$$

The last series is related to the exponential integral

$$\text{Ei}^*(s) \triangleq i\pi - \int_s^\infty \frac{e^z}{z} dz$$

where, as $z \to \infty$, $\arg z \to \beta$ and $\pi/2 \leq \beta \leq 3\pi/2$. (See Jahnke, Emde, and Losch [1], p. 17.) Indeed, we have

$$F(s) = e^{-s}[\text{Ei}^*(s) - \gamma - \log s]$$

where γ denotes Euler's constant ($\gamma = 0.5772 \cdots$) and the principal branches of $\text{Ei}^*(s)$ and $\log s$ are understood. This expression takes on an indeterminate form when $s = 0$, whereas the series expansion for $F(s)$ holds for all finite s.

A number of standard formulas concerning the ordinary Laplace transform are preserved for the distributional Laplace transform. As usual, let $f(t)$ be a Laplace-transformable distribution in $\mathfrak{D}'_R$ so that

$$\mathfrak{L}f(t) = F(s) \qquad \text{Re } s > \sigma_1$$

If k is a positive integer, τ a real number, α a complex number, and a a real positive number, then

$$\mathfrak{L}[t^k f(t)] = (-1)^k F^{(k)}(s) \qquad \text{Re } s > \sigma_1 \tag{7}$$
$$\mathfrak{L}f^{(k)}(t) = s^k F(s) \qquad \text{Re } s > \sigma_1 \tag{8}$$
$$\mathfrak{L}f(t - \tau) = e^{-s\tau} F(s) \qquad \text{Re } s > \sigma_1 \tag{9}$$
$$\mathfrak{L}[e^{-\alpha t} f(t)] = F(s + \alpha) \qquad \text{Re } s > \sigma_1 - \text{Re } \alpha \tag{10}$$
$$\mathfrak{L}f(at) = \frac{1}{a} F\left(\frac{s}{a}\right) \qquad \text{Re } s > a\sigma_1 \tag{11}$$

Formula (7) is an immediate consequence of Theorem 8.3-2. To establish (8), first note that, if $e^{-ct} f(t)$ is in $\mathcal{S}'$ for a certain c, then its first derivative is also in $\mathcal{S}'$. Hence, the same may be said of

$$e^{-ct} f^{(1)}(t) = \frac{d}{dt}[e^{-ct} f(t)] + c e^{-ct} f(t)$$

Consequently, for Re $s > c > \sigma_1$,

$$\mathfrak{L}f^{(1)}(t) = \left\langle \frac{d}{dt}[e^{-ct}f(t)] + ce^{-ct}f(t), \lambda(t)e^{-(s-c)t} \right\rangle$$
$$= \langle e^{-ct}f(t), \lambda(t)(s-c)e^{-(s-c)t} + c\lambda(t)e^{-(s-c)t} \rangle$$
$$= \langle e^{-ct}f(t), \lambda(t)se^{-(s-c)t} \rangle$$
$$= sF(s) \qquad (12)$$

[In the second line of this manipulation, we have used the fact that $\lambda(t)$ is constant over a neighborhood of the support of $f(t)$.] (8) now follows from (12).

Formulas (9) to (11) can be established even more readily.

Example 8.3-3 As was indicated in Sec. 8.2, Prob. 5, it is a fact that

$$\mathfrak{L}[1_+(t) \log t] = \int_0^\infty (\log t)e^{-st}\, dt = -\frac{\gamma + \log s}{s} \qquad \text{Re } s > 0 \qquad (13)$$

where γ again denotes Euler's constant ($\gamma = 0.5772 \cdots$) and the principal branch of $\log s$ is understood. Starting from this formula, we can compute the Laplace transform of

$$\operatorname{Pf} \frac{1_+(t)}{t^k} \qquad k = 1, 2, 3, \ldots$$

by using (8). First, we note from Sec. 2.5, Prob. 5, that

$$\frac{d}{dt}[1_+(t) \log t] = \operatorname{Pf} \frac{1_+(t)}{t}$$

and hence

$$\mathfrak{L} \operatorname{Pf} \frac{1_+(t)}{t} = s\mathfrak{L}[1_+(t) \log t] = -\log s - \gamma \qquad \text{Re } s > 0 \qquad (14)$$

Also, Sec. 2.5, formula (23), states that

$$\frac{d}{dt} \operatorname{Pf} \frac{1_+(t)}{t^k} = \operatorname{Pf} \frac{-k1_+(t)}{t^{k+1}} + \frac{(-1)^k}{k!} \delta^{(k)}(t) \qquad k = 1, 2, 3, \ldots$$

Consequently,

$$\mathfrak{L} \operatorname{Pf} \frac{1_+(t)}{t^2} = \mathfrak{L}\left[-\frac{d}{dt} \operatorname{Pf} \frac{1_+(t)}{t} - \delta^{(1)}(t)\right]$$
$$= s(\log s + \gamma - 1) \qquad \text{Re } s > 0$$
$$\mathfrak{L} \operatorname{Pf} \frac{1_+(t)}{t^3} = \frac{1}{2}\mathfrak{L}\left[-\frac{d}{dt} \operatorname{Pf} \frac{1_+(t)}{t^2} + \frac{\delta^{(2)}(t)}{2!}\right]$$
$$= -\frac{s^2}{2}\left(\log s + \gamma - \frac{3}{2}\right) \qquad \text{Re } s > 0$$

By continuing in this fashion, we can show in general that

$$\mathfrak{L}\,\operatorname{Pf}\frac{1_+(t)}{t^k} = -\frac{(-s)^{k-1}}{(k-1)!}\left(\log s + \gamma - \sum_{\nu=1}^{k-1}\frac{1}{\nu}\right)$$
$$\operatorname{Re} s > 0;\ k = 2, 3, 4, \ldots \quad (15)$$

Example 8.3-4 Classical Laplace transform analysis is for the most part based on the ordinary transformation

$$\mathfrak{L}_+ f(t) \triangleq F(s) \triangleq \int_0^\infty f(t)e^{-st}\,dt$$

Assuming that $f^{(1)}(t)$ exists and is continuous for $0 < t < \infty$, that $f(0+)$ exists, and that $f(t)e^{-\sigma t}$ and $f^{(1)}(t)e^{-\sigma t}$ are absolutely integrable over $0 < t < \infty$ for all $\sigma > \sigma_a$, we can integrate by parts to get

$$\mathfrak{L}_+ f^{(1)}(t) = sF(s) - f(0+) \qquad \operatorname{Re} s > \sigma_a \qquad (16)$$

This classical formula can also be obtained in a distributional way as follows. First, consider the function $f_e(t)$ that is zero for $-\infty < t < 0$ and equals $f(t)$ for $0 < t < \infty$. We are free to write

$$f_e(t) \triangleq f(t)1_+(t) \qquad -\infty < t < \infty \qquad (17)$$

Clearly, $\mathfrak{L}f_e = \mathfrak{L}_+ f \triangleq F(s)$. Then, a distributional differentiation yields

$$f_e^{(1)}(t) = f^{(1)}(t)1_+(t) + f(0+)\delta(t)$$

[Here, $f^{(1)}(t)1_+(t)$ is understood to be an ordinary function undefined at $t = 0$.] According to (8), $\mathfrak{L}f_e^{(1)} = sF(s)$. Therefore, the Laplace transformation applied to the last equation produces (16).

More generally, assuming that $f^{(k)}(t)$ exists and is continuous for all t, we may differentiate (17) k times to get

$$f_e^{(k)}(t) = f^{(k)}(t)1_+(t) + f(0+)\delta^{(k-1)}(t) + f^{(1)}(0+)\delta^{(k-2)}(t)$$
$$+ \cdots + f^{(k-1)}(0+)\delta(t)$$

where $f^{(k)}(t)1_+(t)$ is taken to be undefined at $t = 0$. Then, if $f_e(t)$ is Laplace-transformable for $\operatorname{Re} s > \sigma_1$, an application of the Laplace transformation and the use of (8) yield the classical formula

$$\mathfrak{L}_+ f^{(k)}(t) = s^k F(s) - f(0+)s^{k-1} - f^{(1)}(0+)s^{k-2}$$
$$- \cdots - f^{(k-1)}(0+) \qquad \operatorname{Re} s > \sigma_1$$

We have already mentioned that the distributional Laplace transformation is a linear operation. We shall now show that it is a continuous one.

Theorem 8.3-3
Let $\{f_\nu\}_{\nu=1}^\infty$ be a sequence of distributions with the following two properties:
a. All the f_ν have their supports contained in a fixed semi-infinite interval $T \leq t < \infty$.
b. There is a real number c such that the sequence $\{e^{-ct}f_\nu(t)\}_{\nu=1}^\infty$ converges in $\mathcal{S}'$ to the limit $e^{-ct}f(t)$.
Then, the sequence $\{F_\nu(s)\}_{\nu=1}^\infty = \{\mathfrak{L}f_\nu\}_{\nu=1}^\infty$ converges to $F(s) \triangleq \mathfrak{L}f$ for $\text{Re } s > c$.

Proof: First of all, by the fact that $\mathcal{S}'$ is closed under convergence, $e^{-ct}f(t)$ is in $\mathcal{S}'$. Also, f will have its support bounded on the left at $t = T$. Therefore, f is Laplace-transformable and $F(s) \triangleq \mathfrak{L}f$ exists for $\text{Re } s > c$. Second, since $\lambda(t)e^{-(s-c)t}$ is in $\mathcal{S}$ for $\text{Re } s > c$,

$$\mathfrak{L}f_\nu = \langle e^{-ct}f_\nu(t), \lambda(t)e^{-(s-c)t}\rangle \rightarrow \langle e^{-ct}f(t), \lambda(t)e^{-(s-c)t}\rangle = \mathfrak{L}f$$

as was asserted.

A useful criterion that may be used to ascertain whether a sequence converges in $\mathcal{S}'$ (as is required by condition b of the above hypothesis) is given by Theorem 4.3-1.

PROBLEMS

1 Show that the distributional Laplace transform of $1_+(t)e^t \cos e^t$ has an abscissa of convergence σ_1 that is less than the abscissa of absolute convergence σ_a for the corresponding ordinary Laplace transform.

2 Fill in the details of the proof of Lemma 1.

3 Establish formulas (9) to (11).

4 The hypothesis of Theorem 8.3-1 may be weakened by merely requiring that $\mathfrak{L}f_1 = \mathfrak{L}f_2$ over some segment of an arbitrarily oriented line in their common regions of convergence. Why is this so?

5 Which of the following distributions are Laplace-transformable? For those that are, find their Laplace transforms and specify the regions of convergence. Here, a and θ are real constants and b is a real positive constant.

a. $\displaystyle\sum_{\nu=0}^\infty (t-\nu)^\nu 1_+(t-\nu)$

b. $\displaystyle\sum_{\nu=0}^\infty \delta(t-\nu)$

c. $\displaystyle\sum_{\nu=0}^\infty \delta^{(\nu)}(t-\nu)$

d. $1_+(t) \sin(at + \theta)$

e. $f(t) = \begin{cases} 0 & t < 0 \\ \sin at & 0 < t < T \\ 0 & T < t \end{cases}$

f. $f(t) = \begin{cases} 0 & t < 0 \\ t & 0 < t < 1 \\ 2 - t & 1 < t < 2 \\ 0 & 2 < t \end{cases}$

g. $b^t 1_+(t)$

h. $t^t 1_+(t)$

6 Show that

$$\mathcal{L} \, \text{Pf}\,[t^\eta 1_+(t)] = \frac{\Gamma(\eta + 1)}{s^{\eta+1}} \qquad \text{Re } s > 0; \, \eta \neq -1, -2, -3, \ldots \tag{18}$$

where

$$\Gamma(z) \triangleq \int_0^\infty e^{-x} x^{z-1}\, dx \qquad \text{Re } z > 0$$

$$\Gamma(z + 1) = z\Gamma(z)$$

and η is any real number other than a negative integer. When η is not a positive integer, the principal branches of t^η and $s^{\eta+1}$ are understood. When $\eta > -1$, the symbol Pf may be discarded.

7 Establish the following Laplace transform formulas. Here, a, b, c, and τ are real constants with $b > 0$ and $c \neq 0$; α is a complex constant. The principal branches of the multivalued functions are again understood.

a. $\mathcal{L}\delta(ct - \tau) = \dfrac{1}{|c|} e^{-s\tau/c} \qquad -\infty < \text{Re } s < \infty$

b. $\mathcal{L} \, \text{Pf} \, \dfrac{1_+(t) \cos \alpha t}{t} = -\gamma - \log \sqrt{s^2 + \alpha^2} \qquad \text{Re } s > |\text{Im } \alpha|$

c. $\mathcal{L} \dfrac{1_+(t) \sinh \alpha t}{t} = \dfrac{1}{2} \log \dfrac{s + \alpha}{s - \alpha} \qquad \text{Re } s > |\text{Re } \alpha|$

d. $\mathcal{L} \dfrac{1_+(t)(\sin \alpha t)^2}{t} = \dfrac{1}{4} \log \left(1 + \dfrac{4\alpha^2}{s^2}\right) \qquad \text{Re } s > 2|\text{Im } \alpha|$

e. $\mathcal{L} \left[\delta(t) + 2 \sum_{\nu=1}^\infty (-1)^\nu \delta(t - 2a\nu)\right] = \tanh as \qquad \text{Re } s > 0$

f. $\mathcal{L} \, \text{Pf} \, \dfrac{1_+(t) 1_+(b - t)}{t} = \text{Ei}(-bs) - \log s - \gamma \qquad -\infty < \text{Re } s < \infty$

Here you may use the fact that

$$\text{Ei}(-z) = \int_{-\infty}^{-z} \frac{e^\zeta}{\zeta}\, d\zeta \qquad \text{Re } z > 0$$

g. $\mathcal{L} \dfrac{1_+(t)}{t + b} = -e^{bs} \, \text{Ei}(-bs) \qquad \text{Re } s > 0$

h. $\mathcal{L}t \, \text{Pf} \, \dfrac{1_+(t - 1)}{(t - 1)^{\frac{3}{2}}} = 2\sqrt{\pi s} \, e^{-s} \left(\dfrac{1}{2s} - 1\right) \qquad \text{Re } s > 0$

8 According to (14),

$$\mathcal{L} \, \text{Pf} \, \frac{1_+(t)}{t} = -\log s - \gamma \qquad \text{Re } s > 0 \tag{19}$$

Now let a denote a real positive number. Let us replace t by at in the left-hand side of (19) and let us alter the right-hand side according to (11). This yields

$$\mathfrak{L} \operatorname{Pf} \frac{1_+(at)}{at} = \frac{1}{a} \mathfrak{L} \operatorname{Pf} \frac{1_+(t)}{t} = -\frac{1}{a} (\log s + \gamma - \log a) \qquad \operatorname{Re} s > 0$$

The last expression contradicts (19). Where is the mistake?

9 The Bessel function of first kind and order p ($p = 0, 1, 2, \ldots$) is defined for all t by the series

$$J_p(t) \triangleq \sum_{\nu=0}^{\infty} \frac{(-1)^\nu}{\nu!(\nu+p)!} \left(\frac{t}{2}\right)^{p+2\nu}$$

Use the results of Examples 8.2-3 and 8.3-4 and the fact that

$$J_0^{(1)}(t) = -J_1(t)$$

to show that

$$\mathfrak{L}[J_1(t)1_+(t)] = \frac{\sqrt{s^2+1} - s}{\sqrt{s^2+1}} \qquad \operatorname{Re} s > 0$$

More generally, use the recurrence relation

$$J_{p+1}(t) = J_{p-1}(t) - 2J_p^{(1)}(t) \qquad p = 1, 2, 3, \ldots$$

to obtain

$$\mathfrak{L}[J_p(t)1_+(t)] = \frac{(\sqrt{s^2+1} - s)^p}{\sqrt{s^2+1}} \qquad \operatorname{Re} s > 0$$

10 Show that

$$\mathfrak{L}[t^n J_n(t)] = \frac{1 \times 3 \times 5 \times \cdots \times (2n-1)}{(s^2+1)^{n+\frac{1}{2}}} \qquad n = 1, 2, 3, \ldots; \operatorname{Re} s > 0$$

HINT: By using the recurrence relationship

$$\frac{d}{dt}[t^n J_n(t)1_+(t)] = t^n J_{n-1}(t)1_+(t)$$

first show that

$$\mathfrak{L}[t^n J_n(t)] = -\frac{1}{s}\frac{d}{ds} \mathfrak{L}[t^{n-1} J_{n-1}(t)]$$

11 Let f and g be Laplace-transformable right-sided distributions and let their Laplace transforms be $F(s)$ and $G(s)$, respectively. If

$$F(s) = \sum_{\nu=1}^{q} \frac{c_\nu}{s - s_\nu}$$

where c_ν and s_ν are constants, show that

$$\mathfrak{L}[f(t)g(t)] = \sum_{\nu=1}^{q} c_\nu G(s - s_\nu)$$

12 Let f be a Laplace-transformable distribution in $\mathfrak{D}'_R$ and let $\mathfrak{L}f = F(s)$ (Re $s > \sigma_1$). Show that, for every s in the region of convergence, $F(\sigma + i\omega)$ satisfies the Cauchy-Riemann equation

$$\left(\frac{\partial}{\partial \sigma} + i\frac{\partial}{\partial \omega}\right) F(\sigma + i\omega) = 0$$

without invoking the fact that $F(s)$ is analytic there. HINT: Use (5) and the equation

$$\frac{\partial}{\partial \sigma} \mathfrak{F}[e^{-\sigma t}f(t)] = \mathfrak{F}[-te^{-\sigma t}f(t)]$$

which can be proved by using ultradistributions.

13 Let $f(t)$ be a distribution of bounded support and let $\mathfrak{L}f(t) = F(s)(-\infty < $ Re $s < \infty$). Also, let

$$g(t) = \sum_{\nu=0}^{\infty} f(t - \nu b) \tag{20}$$

where b is a positive number. In Sec. 4.3, Prob. 11, we asked the reader to show that the series (20) converges in $\mathcal{S}'$. Now, show that

$$\mathfrak{L}g(t) = \frac{F(s)}{1 - e^{-bs}} \quad \text{Re } s > 0$$

14 By interpreting (5) in terms of ultradistributions, we can extend the Laplace transformation onto all of $\mathfrak{D}'$. For example, let f be an arbitrary distribution. (Its support need not be bounded on the left.) We can define the Laplace transform of $f(t)$ as an ultradistribution, namely, as the Fourier transform of $e^{-\sigma t}f(t)$. According to Sec. 7.8, Eq. (5), if $\mathfrak{F}f(t) = \hat{f}(\omega)$, then $\mathfrak{L}f(t) = \hat{f}(\omega - i\sigma)$; this Laplace transform always exists for all values of σ, in contrast to our previous definition.

As an example of this, show that

$$\mathfrak{L}1_+(t) = \pi\delta(\omega - i\sigma) + \text{Pv}\,\frac{1}{\sigma + i\omega}$$

where ω is the independent variable for this ultradistribution, whereas σ is a parameter. Here, Pv $1/(\sigma + i\omega)$ denotes the ultradistribution that assigns to each ϕ in $\mathcal{Z}$ the number

$$\left\langle \text{Pv}\,\frac{1}{\omega},\, \phi(\omega - i\sigma) \right\rangle$$

Verify that, if $\sigma > 0$, this result coincides with $1/(\sigma + i\omega)$, which is the classical Laplace transform of $1_+(t)$. Note that this coincidence fails to exist if $\sigma \leq 0$.

More generally, show that, if $f(t)$ is in $\mathfrak{D}'_R$ and if $f(t)e^{-ct}$ is in $\mathcal{S}'$ for $c > \sigma_1$, then our ultradistributional definition of $\mathfrak{L}f$ coincides with our previous definition (4) for Re $s > \sigma_1$.

15 By using the ultradistributional definition of the Laplace transformation given in the preceding problem, find the Laplace transforms of the following distributions:

a. $e^{-bt^2} \quad b > 0$

b. $\dfrac{t^2}{1 + t^2}$

Sec. 8.4   **The Laplace transformation**   235

   c. $\displaystyle\sum_{\nu=-\infty}^{\infty} \delta(t-\nu)$

   d. $\displaystyle\sum_{\nu=-\infty}^{\infty} \delta^{(\nu)}(t-\nu)$

16 Let f be a distribution whose support is contained in $0 < t < \infty$. Furthermore, assume that there is a real number c such that $t^c f(t)$ is a bounded distribution. We can define a distributional Mellin transformation $\mathfrak{M}$ for f as follows:

$$F(s) \triangleq \mathfrak{M}f \triangleq \langle f(t), t^{s-1}\rangle$$
$$\triangleq \langle t^c f(t), \lambda(t) t^{s-c-1}\rangle \qquad \text{Re } s < c < \sigma_2 \tag{21}$$

Here the function $\lambda(t)$ is infinitely smooth; it equals zero for $-\infty < t \leq \varepsilon$ ($\varepsilon > 0$) and equals one over a neighborhood of the support of $f(t)$. Clearly, $\lambda(t) t^{s-c-1}$ is in $\mathfrak{D}_{L_1}$ for Re $s < c$, so that the right-hand side of (21) has a sense. σ_2 is the least upper bound of all possible values for c.

   a. Show that $F(s) \triangleq \mathfrak{M}f$ exists and is analytic in the half-plane Re $s < \sigma_2$.
HINT: First, show that

$$\lambda(t) t^{s-c-1} \frac{t^{\Delta s} - 1}{\Delta s}$$

converges in $\mathfrak{D}_{L_1}$ as $|\Delta s| \to 0$.

   b. Show that

$$\mathfrak{M}[tf^{(1)}(t)] = -sF(s) \qquad \text{Re } s < \sigma_2$$

   c. More generally, show that for each positive integer p

$$\mathfrak{M}[t^p f^{(p)}(t)] = (-1)^p s(s+1) \cdots (s+p-1) F(s) \qquad \text{Re } s < \sigma_2$$

$$\mathfrak{M}\left[\left(t\frac{d}{dt}\right)^p f\right] \triangleq t\frac{d}{dt}\left[t\frac{d}{dt}\left(\cdots t\frac{df}{dt}\right)\right] = (-s)^p F(s) \qquad \text{Re } s < \sigma_2$$

   d. Find the Mellin transform of $\delta(t-1)$ and of

$$\frac{1_+(t-1)}{t^2}$$

An application of the Mellin transformation for the solution of an ordinary linear differential equation with variable coefficients is indicated in Sec. 9.7, Prob. 2.

Actually, this Mellin transformation can be obtained quite simply by applying a change of variable to the left-sided distributional Laplace transformation. This is indicated in Sec. 8.8, Prob. 3.

★8.4 THE INVERSION OF THE LAPLACE TRANSFORMATION FOR RIGHT - SIDED DISTRIBUTIONS

Under what conditions is a function $F(s)$ the Laplace transform of a distribution in $\mathfrak{D}'_R$? A partial answer to this question is given by

Theorem 8.4-1

Let $F(s)$ be a function that is analytic over a half-plane $\operatorname{Re} s \geq c$ and is bounded according to

$$|F(s)| \leq P(|s|) \qquad \operatorname{Re} s \geq c \tag{1}$$

where $P(|s|)$ is some polynomial in $|s|$. Then, $F(s)$ is the Laplace transform of a distribution $f(t)$ whose support is bounded on the left at $t = 0$.

Proof: First, consider a function $G(s)$ that is analytic for $\operatorname{Re} s \geq a > 0$ and satisfies

$$|G(s)| \leq \frac{C}{|s|^2} \qquad \operatorname{Re} s \geq a > 0 \tag{2}$$

where C is a constant. Then, according to Theorem 8.2-3, the ordinary inverse Laplace transform $g(t)$ exists, is continuous for all t, and equals zero for $t < 0$. Moreover, $\mathfrak{L}g(t) = G(s)$ for at least $\operatorname{Re} s > a$.

Now, according to (1), there certainly exist two real positive constants C and a and a positive integer m for which

$$|F(s)| \leq C|s|^m \qquad \operatorname{Re} s \geq a \geq c$$

Set $F(s) = s^{m+2} G(s)$ so that $G(s)$ satisfies (2). By Sec. 8.3, Eq. (8), $F(s)$ is the Laplace transform of the distribution $g^{(m+2)}(t)$. Clearly, $g^{(m+2)}(t)$ has its support bounded on the left at $t = 0$. Q.E.D.

We now see that every rational function of s is the Laplace transform of a right-sided distribution so long as the region of convergence is taken as the largest open half-plane lying to the right of all the poles of the rational function.

According to Sec. 8.3, Eq. (9),

$$\mathfrak{L}f(t + T) = e^{sT}\mathfrak{L}f(t) = e^{sT}F(s)$$

By replacing $F(s)$ by $e^{sT}F(s)$ in Theorem 8.4-1, we obtain a more general result.

Corollary 8.4-1a

Let $F(s)$ be a function that is analytic in some half-plane $\operatorname{Re} s \geq c$ and is bounded according to

$$|F(s)| \leq e^{-\operatorname{Re} sT} P(|s|) \qquad \operatorname{Re} s \geq c \tag{3}$$

where T is a real constant and $P(|s|)$ is some polynomial in $|s|$. Then $F(s)$ is the Laplace transform of a distribution whose support is bounded on the left at $t = T$.

Actually, the converse of this corollary is also true. (The proof is based on the fact that distributions of slow growth are of finite order. See Prob. 1.) *Thus, a necessary and sufficient condition for a function $F(s)$ to be the Laplace transform of a distribution whose support is bounded on the left at $t = T$ is that it be analytic in some half-plane $\operatorname{Re} s \geq c$ and satisfy* (3). This, then, is the complete answer to the question posed at the beginning of this section.

Two simple examples concerned with the actual computation of inverse Laplace transforms follow.

Example 8.4-1 We shall compute the inverse Laplace transform of

$$F(s) = \frac{s^3 + 2s + 1}{s^2 + 1} \tag{4}$$

in two different ways, assuming that the region of convergence is $\operatorname{Re} s > 0$.

First method: A partial-fraction expansion of $F(s)$ will be made and then the inverse Laplace transform will be taken term by term. We have that

$$F(s) = s + \frac{1}{\sqrt{2}} \left[\frac{e^{-i\pi/4}}{s - i} + \frac{e^{i\pi/4}}{s + i} \right] \tag{5}$$

By using Sec. 8.2, Eq. (17), with $k = 0$ and invoking the uniqueness theorem 8.3-1, we can write down the inverse Laplace transform immediately:

$$\begin{aligned} f(t) &= \delta^{(1)}(t) + \frac{1_+(t)}{\sqrt{2}} \left(e^{i(t - \pi/4)} + e^{-i(t - \pi/4)} \right) \\ &= \delta^{(1)}(t) + \sqrt{2}\, 1_+(t) \cos\left(t - \frac{\pi}{4}\right) \end{aligned} \tag{6}$$

Second method: Since

$$\mathcal{L}^{-1} \frac{1}{s^2 + 1} = 1_+(t) \sin t \tag{7}$$

and since $\mathcal{L}^{-1}$ converts multiplication by s into differentiation with respect to t, we can get $f(t)$ by differentiating (7) in accordance with the numerator of $F(s)$. Thus,

$$\begin{aligned} f(t) &= \left(\frac{d^3}{dt^3} + 2\frac{d}{dt} + 1 \right) [1_+(t) \sin t] \\ &= \delta^{(1)}(t) + 1_+(t) \cos t + 1_+(t) \sin t \\ &= \delta^{(1)}(t) + \sqrt{2}\, 1_+(t) \cos\left(t - \frac{\pi}{4}\right) \end{aligned}$$

Let us be a little more precise here. Our uniqueness theorem merely states that (6) is the *distribution* whose Laplace transform is (4). Hence,

$$1_+(t) \cos\left(t - \frac{\pi}{4}\right) \tag{8}$$

should be interpreted as a regular distribution or, equivalently, as the class of functions that differ from the *function* (8) on no more than a set of measure zero. Thus, over $-\infty < t < 0$ and over $0 < t < \infty$ the distributional inverse Laplace transform of (4) is the regular distribution corresponding to a continuous function. Actually, we could invoke Theorem 8.2-2 and conclude that the ordinary inverse Laplace transform of the bracketed portion of (5) equals (8) for all nonzero values of t.

Example 8.4-2 Let us establish the general form of the right-sided distribution whose Laplace transform is the arbitrary rational function

$$F(s) = \frac{\alpha_n s^n + \alpha_{n-1} s^{n-1} + \cdots + \alpha_0}{\beta_m s^m + \beta_{m-1} s^{m-1} + \cdots + \beta_0} \tag{9}$$

where the α_ν and β_ν are complex constants and m and n are positive integers. If $n > m$, we can divide the denominator into the numerator to obtain

$$F(s) = \sum_{\nu=0}^{n-m} \xi_\nu s^\nu + \frac{\rho_p s^p + \rho_{p-1} s^{p-1} + \cdots + \rho_0}{\beta_m s^m + \beta_{m-1} s^{m-1} + \cdots + \beta_0}$$

where the ξ_ν and ρ_ν are also complex constants and $p < m$. If $m > n$, the ξ_ν will all be zero. By carrying out a partial-fraction expansion, we obtain

$$F(s) = \sum_{\nu=0}^{n-m} \xi_\nu s^\nu + \sum_{\mu=1}^{q} \sum_{\nu=1}^{k_\mu} \frac{\zeta_{\mu\nu}}{(s - \gamma_\mu)^\nu}$$

where

$$\sum_{\mu=1}^{q} k_\mu = m$$

the γ_μ are the finite poles of $F(s)$, the k_μ are the corresponding multiplicities, and the $\zeta_{\mu\nu}$ are the coefficients of the expansion. We now assume that the half-plane of convergence is Re $s > \sigma_1$, where σ_1 is the largest of the real parts of the γ_μ. By Sec. 8.2, Eq. (17), and the uniqueness theorem,

$$f(t) = \sum_{\nu=0}^{n-m} \xi_\nu \delta^{(\nu)}(t) + \sum_{\mu=1}^{q} \sum_{\nu=1}^{k_\mu} \frac{\zeta_{\mu\nu}}{(\nu-1)!} t^{\nu-1} e^{\gamma_\mu t} 1_+(t) \tag{10}$$

If $n < m$, the delta functional and its derivatives do not appear. Over $-\infty < t < 0$ and over $0 < t < \infty$, this distributional inverse Laplace transform of $F(s)$ is a regular distribution corresponding to a continuous function. By referring to Theorem 8.2-3, we see that, if $n \leq m - 2$, $f(t)$ corresponds to a function that is continuous everywhere. Finally, the form of (10) indicates that when $n = m - 1$, $f(t)$ corresponds to a function that has at most an ordinary discontinuity at $t = 0$. We shall show in Sec. 8.6 that this discontinuity has the magnitude $|\alpha_n/\beta_m|$.

PROBLEMS

1 Use the boundedness property of distributions of slow growth, given by Sec. 4.4, Eq. (1), to prove the converse of Corollary 8.4-1a.

2 Compute the inverse Laplace transforms of each of the following functions by assuming that the region of convergence is in each case the largest half-plane lying to the right of all the poles.

a. $\dfrac{s}{s+1}$

b. $\dfrac{s^2 + i}{s^2 - 3s + 2}$

c. $\dfrac{(s+1)^3}{s^3 + 1}$

d. $\dfrac{s^4 + s}{(s^2 + 4)^2}$

e. $\dfrac{s^5}{s^2 + (1-i)s - i}$

3 Starting from the fact that

$$\mathfrak{L} \operatorname{Pf} \frac{1_+(t)}{t} = -\log Cs \qquad \operatorname{Re} s > 0$$

where $\log C$ is Euler's constant, find the inverse Laplace transforms of the following functions, always assuming that the region of convergence is a right-sided half-plane. In the following, α and β denote complex numbers and a denotes a real number. The principal branches of all multivalued functions are, as usual, understood.

a. $s^2 \log Cs$

b. $\log \dfrac{s + \alpha}{s + \beta}$

c. $(\cosh s) \log \dfrac{s + a}{s - a}$

d. $\dfrac{\log s}{s - 1}$

★8.5 THE LAPLACE TRANSFORMATION OF CONVOLUTIONS OF RIGHT-SIDED DISTRIBUTIONS

We saw in Secs. 7.5 and 7.9 how the Fourier transformation converts the convolution of two distributions into the product of their transforms, this conversion being valid under certain restrictions on the supports of the distributions. The Laplace transformation acts in the same way. The conditions that will now be imposed on the distributions are the same as those of the two preceding sections; i.e., the distributions are assumed to be Laplace-transformable elements of $\mathfrak{D}'_R$.

Theorem 8.5-1

Let f and g be Laplace-transformable distributions in $\mathfrak{D}'_R$ and let

$$\mathfrak{L}f = F(s) \qquad \text{Re } s > \sigma_f$$

and

$$\mathfrak{L}g = G(s) \qquad \text{Re } s > \sigma_g$$

Then, $f * g$ is also a Laplace-transformable distribution in $\mathfrak{D}'_R$ and

$$\mathfrak{L}(f * g) = F(s)G(s) \qquad \text{Re } s > \sup (\sigma_f, \sigma_g) \tag{1}$$

Proof: $f * g$ certainly exists and has a support bounded on the left. Let the constant c be greater than $\sup (\sigma_f, \sigma_g)$. Then,

$$[e^{-ct}f(t)] * [e^{-ct}g(t)] = e^{-ct}[f(t) * g(t)] \tag{2}$$

since, for any ϕ in $\mathfrak{D}$,

$$\begin{aligned}
\langle [e^{-ct}f(t)] * [e^{-ct}g(t)], \phi(t) \rangle &= \langle e^{-ct}f(t), \langle e^{-c\tau}g(\tau), \phi(t + \tau) \rangle \rangle \\
&= \langle f(t), \langle g(\tau), e^{-c(t+\tau)}\phi(t + \tau) \rangle \rangle \\
&= \langle f(t) * g(t), e^{-ct}\phi(t) \rangle \\
&= \langle e^{-ct}[f(t) * g(t)], \phi(t) \rangle
\end{aligned}$$

As was pointed out in Sec. 5.4 (see the paragraph following Corollary 5.4-2a), (2) is a distribution of slow growth.

Now, let λ be an infinitely smooth function whose support is bounded on the left. Also, assume that $\lambda(t) = 1$ over neighborhoods of the supports of f, g, and $f * g$. Then, by using (2), we get

$$\mathfrak{L}(f * g) = \langle e^{-ct}f(t) \times e^{-c\tau}g(\tau), \lambda(t + \tau)e^{-(s-c)(t+\tau)} \rangle \qquad \text{Re } s > c$$

Both $\lambda(t + \tau)$ and $\lambda(t)\lambda(\tau)$ equal one over some neighborhood of the support of $f(t) \times g(\tau)$. Consequently, $\lambda(t + \tau)$ may be replaced by $\lambda(t)\lambda(\tau)$

in the last equation. Thus,

$$\mathcal{L}(f * g) = \langle e^{-ct}f(t), \lambda(t)e^{-(s-c)t}\rangle\langle e^{-c\tau}g(\tau), \lambda(\tau)e^{-(s-c)\tau}\rangle$$
$$= F(s)G(s) \qquad \text{Re } s > c$$

This result holds for all $c > \sup(\sigma_f, \sigma_g)$, so that (1) is established.

Example 8.5-1 Let

$$f(t) = \sum_{\nu=0}^{\infty} e^{\nu}\delta(t - \nu) \tag{3}$$

By a direct computation, it can be shown that

$$f(t) * f(t) = \sum_{\nu=0}^{\infty} (\nu + 1)e^{\nu}\delta(t - \nu) \tag{4}$$

This result can also be obtained through the use of Theorem 8.5-1. For $c = 1$,

$$e^{-ct}f(t) = \sum_{\nu=0}^{\infty} \delta(t - \nu)$$

which is a distribution of slow growth whose support is bounded on the left. Thus, the Laplace transform of (3) is

$$\mathcal{L}f = \langle e^{-t}f(t), \lambda(t)e^{-(s-1)t}\rangle$$
$$= \sum_{\nu=0}^{\infty} e^{-(s-1)\nu} \qquad \text{Re } s > 1$$

By Theorem 8.5-1,

$$\mathcal{L}(f * f) = (\mathcal{L}f)(\mathcal{L}f) = \sum_{\nu=0}^{\infty} (\nu + 1)e^{-(s-1)\nu}$$

By taking the inverse Laplace transform term by term, we obtain (4), as is to be expected. (This term-by-term inversion can be justified by invoking Theorems 8.3-1 and 8.3-3.)

Example 8.5-2 A classical example is provided by the zero-order Bessel function of first kind, $J_0(t)$. It was shown in Example 8.2-3 that

$$\mathcal{L}[J_0(t)1_+(t)] = \frac{1}{\sqrt{s^2 + 1}} \qquad \text{Re } s > 0$$

Hence,

$$\mathcal{L}\{[J_0(t)1_+(t)] * [J_0(t)1_+(t)]\} = \frac{1}{s^2 + 1} \qquad \text{Re } s > 0$$

By taking the inverse Laplace transform, we obtain a standard formula:

$$1_+(t) \int_0^t J_0(\tau) J_0(t-\tau)\, d\tau = 1_+(t) \sin t \tag{5}$$

We can delete the factors of $1_+(t)$ on both sides of (5), since $J_0(t)$ is an even function and $\sin t$ is an odd function.

PROBLEMS

1 Use the uniqueness theorem of the Laplace transformation and the theorem of this section to show that the space of all Laplace-transformable right-sided distributions does not possess divisors of zero when multiplication is replaced by convolution. That is, if f and g are such distributions, then the fact that $f * g = 0$ implies that either $f = 0$ and/or $g = 0$.

2 By using the Laplace transformation and perhaps the Laplace transform table in Appendix B, determine the following convolutions:

a. $\operatorname{Pf} \dfrac{1_+(t)}{t} * \operatorname{Pf} \dfrac{1_+(t)}{t}$

b. $[\delta^{(2)}(t) - \delta(t)] * [1_+(t) e^{-t}]$

c. $[\operatorname{Pf} t^\eta 1_+(t)] * [\operatorname{Pf} t^\zeta 1_+(t)]$

Here η and ζ are real numbers that are not equal to negative integers.

d. $\dfrac{1_+(t) \sin \sqrt{bt}}{t} * \dfrac{1_+(t) \cosh \sqrt{bt}}{\sqrt{t}}$

e. $\operatorname{Pf} \dfrac{1_+(t)}{t^{\frac{3}{2}}} * \dfrac{1_+(t) e^{-b^2/4t}}{t^{\frac{1}{2}}}$

f. $J_k(t) 1_+(t) * \dfrac{J_n(t) 1_+(t)}{t} \qquad k = 0, 1, 2, \ldots\,;\, n = 1, 2, 3, \ldots$

g. $\dfrac{J_m(t) 1_+(t)}{t} * \dfrac{J_n(t) 1_+(t)}{t} \qquad m, n = 1, 2, 3, \ldots$

3 A useful result for estimating the inverse Laplace transformations of rational functions is the following. Let $F(s)$ be the rational function with real coefficients,

$$F(s) = \frac{a_n s^n + a_{n-1} s^{n-1} + \cdots + a_0}{s^m + b_{m-1} s^{m-1} + \cdots + b_0}$$

whose poles all have nonpositive real parts. If $0 \leq 2n \leq m \geq 1$, then for $t \geq 0$

$$|f(t)| \leq \frac{|a_n| t^{m-n-1}}{(m-n-1)!} + \frac{|a_{n-1}| t^{m-n}}{(m-n)!} + \cdots + \frac{|a_1| t^{m-2}}{(m-2)!} + \frac{|a_0| t^{m-1}}{(m-1)!}$$

Prove this assertion. HINT: First show that it is true for all of the following forms:

$$\frac{1}{s+\rho} \qquad \frac{1}{(s+\alpha)^2 + \beta^2} \qquad \frac{s}{(s+\alpha)^2 + \beta^2} \qquad \frac{s}{(s+\rho)(s+\eta)}$$

wherein α, β, η, and ρ are real nonnegative constants. (Actually, this assertion is a special case of a more general result. See Zemanian [2 and 3].)

8.6 SOME ABELIAN THEOREMS OF THE INITIAL - VALUE TYPE

The theorems with which we shall be concerned in this and the next section relate the behavior of a Laplace transform $F(s) \triangleq \mathfrak{L}f(t)$ as s approaches zero or infinity to the behavior of $f(t)$ as t approaches infinity or zero, respectively. Theorems of this nature are called Abelian and Tauberian theorems, and there is quite a variety of them. (See Widder [1], chap. V, and Doetsch [1], part V.) We shall discuss two classical Abelian theorems and some generalizations to distributions.

The first of these is taken up in this section. It relates the asymptotic behavior of $f(t)$ as $t \to 0+$ to the asymptotic behavior of $F(\sigma)$ as $\sigma \to \infty$. This result is also referred to as an "initial-value theorem," since it is the initial behavior of $f(t)$ that is considered. The second classical theorem, discussed in the next section, is called a "final-value theorem," since it connects the behavior of $f(t)$ as $t \to \infty$ to the behavior of $F(\sigma)$ as $\sigma \to 0+$.

It will be assumed at times that $f(t)$ is a locally integrable function and satisfies conditions A of Sec. 8.2, which we repeat here.

Conditions A:

1. $f(t) = 0$ for $-\infty < t < T$.
2. There exists a real number c such that $f(t)e^{-ct}$ is absolutely integrable over $-\infty < t < \infty$.

We turn now to the classical initial-value theorem.

Theorem 8.6-1

If the locally integrable function $f(t)$ satisfies conditions A with $T = 0$ [i.e., the support of $f(t)$ is in $0 \leq t < \infty$] and if there exist a complex number α and a real number η ($\eta > -1$) such that

$$\lim_{t \to 0+} \frac{f(t)}{t^\eta} = \alpha \tag{1}$$

then

$$\lim_{\sigma \to \infty} \frac{\sigma^{\eta+1} F(\sigma)}{\Gamma(\eta + 1)} = \alpha \tag{2}$$

where $F(s) \triangleq \mathfrak{L}f(t)$, σ is real, and Γ is the gamma function

$$\Gamma(x) \triangleq \int_0^\infty u^{x-1} e^{-u} \, du \qquad x > 0$$

Proof: First note that for $\eta > -1$ and $\sigma > 0$

$$\int_0^\infty t^\eta e^{-\sigma t} \, dt = \frac{1}{\sigma^{\eta+1}} \int_0^\infty (\sigma t)^\eta e^{-\sigma t} \, d(\sigma t)$$

$$= \frac{\Gamma(\eta + 1)}{\sigma^{\eta+1}}$$

By using this equation and assuming that $\sigma > 0$ and $y > 0$, we may write

$$|\sigma^{\eta+1}F(\sigma) - \alpha\Gamma(\eta+1)|$$
$$= \sigma^{\eta+1}\left|\int_0^\infty f(t)e^{-\sigma t}\,dt - \alpha\int_0^\infty t^\eta e^{-\sigma t}\,dt\right|$$
$$\leq \sigma^{\eta+1}\int_0^y e^{-\sigma t}|f(t) - \alpha t^\eta|\,dt + \sigma^{\eta+1}\int_y^\infty e^{-\sigma t}|f(t) - \alpha t^\eta|\,dt \quad (3)$$

An estimate for the first term on the right-hand side of (3) can be constructed as follows:

$$\sigma^{\eta+1}\int_0^y e^{-\sigma t}|f(t) - \alpha t^\eta|\,dt \leq \int_0^{\sigma y} e^{-\sigma t}(\sigma t)^\eta\,d(\sigma t)\sup_{0\leq t\leq y}\left|\frac{f(t)}{t^\eta} - \alpha\right|$$
$$\leq \Gamma(\eta+1)\sup_{0\leq t\leq y}\left|\frac{f(t)}{t^\eta} - \alpha\right| \quad (4)$$

To obtain an estimate for the second term on the right-hand side of (3), first note that, according to the second of conditions A and the assumption that $c > 0$ (we are free to choose c as large as we wish),

$$e^{-ct}|f(t) - \alpha t^\eta|$$

is absolutely integrable over $0 < t < \infty$. Then, for $\sigma > c$

$$\sigma^{\eta+1}\int_y^\infty e^{-\sigma t}|f(t) - \alpha t^\eta|\,dt = \sigma^{\eta+1}\int_y^\infty e^{-(\sigma-c)t}e^{-ct}|f(t) - \alpha t^\eta|\,dt$$
$$\leq K\sigma^{\eta+1}e^{-(\sigma-c)y} \quad (5)$$

where K is the constant

$$K \triangleq \int_0^\infty e^{-ct}|f(t) - \alpha t^\eta|\,dt$$

Now, let $\varepsilon > 0$. Because of (1), we can choose y so small that the right-hand side of (4), which is independent of σ, becomes less than ε. By fixing y in this way, we can then choose σ so large that the right-hand side of (5) also becomes less than ε. Since ε is arbitrary, the left-hand side of (3) can be made arbitrarily small for all sufficiently large σ, which, in turn, proves (2).

There are two ways to use Theorem 8.6-1. In the first case the function $f(t)$ is given and its asymptotic behavior (1) is known. From this, the asymptotic behavior (2) of $F(\sigma)$ can be determined. In the second case $F(\sigma)$ and its asymptotic behavior (2) are given and we wish to determine the behavior (1) of $f(t)$. Before we can do this we must know whether $f(t)$ is locally integrable and satisfies conditions A and whether its limit (1) exists. This can at times be ascertained from the character of $F(s)$.

For instance, it was shown in Example 8.4-2 that, if $F(s)$ is a rational function having more poles than zeros in the finite s plane, its distributional inverse Laplace transform is the regular distribution corresponding to the function

$$f(t) = \sum_{\mu=1}^{q} \sum_{\nu=1}^{k_\mu} \frac{\zeta_{\mu\nu}}{(\nu-1)!} t^{\nu-1} e^{\gamma_\mu t} 1_+(t) \tag{6}$$

(The notation used here is defined in Example 8.4-2.) This piecewise-continuous function certainly satisfies conditions A and possesses $F(s)$ as its ordinary Laplace transform. Moreover, (1) clearly exists with η being some nonnegative integer. Thus, in this case (2) implies (1) and we see from (2) that we may set $\eta = m - n - 1$. More specifically, we may state

Corollary 8.6-1a

Let $F(s)$ be a rational function having m poles and n zeros in the finite s plane, where $m > n$. Let $f(t)$ denote the right-sided function which is continuous for all $t > 0$ and whose ordinary Laplace transform is $F(s)$. Then,

$$\lim_{\sigma \to \infty} \frac{\sigma^{m-n} F(\sigma)}{\Gamma(m-n)} = \lim_{t \to 0+} \frac{f(t)}{t^{m-n-1}}$$

Example 8.6-1 The function

$$F(s) = \frac{s-1}{s^2 + 2s + 2}$$

satisfies the hypothesis of Corollary 8.6-1a. The corresponding right-sided function $f(t)$ has the "initial value"

$$f(0+) = \lim_{\sigma \to \infty} \sigma F(\sigma) = 1$$

Moreover, the first derivative of $f(t)$ has the Laplace transform $sF(s)$ and its initial value $f^{(1)}(0+)$ can be obtained by expanding $sF(s)$ into

$$sF(s) = 1 - \frac{3s+2}{s^2 + 2s + 2} \triangleq 1 + F_1(s)$$

and noting that $\mathfrak{L}^{-1} 1 = \delta(t)$. Thus,

$$f^{(1)}(0+) = \lim_{\sigma \to \infty} \sigma F_1(\sigma) = -3$$

Similarly, to obtain $f^{(2)}(0+)$, we write

$$s^2 F(s) = s - 3 + \frac{4s+6}{s^2 + 2s + 2} \triangleq s - 3 + F_2(s)$$

and then obtain
$$f^{(2)}(0+) = \lim_{\sigma \to \infty} \sigma F_2(\sigma) = 4$$

The quantities $f^{(\nu)}(0+)$ ($\nu = 3, 4, \ldots$) may be computed in the same fashion.

Under certain circumstances, Theorem 8.6-1 can be extended to a distribution that in some neighborhood of the origin corresponds to an ordinary function. In order to make this extension, we shall need

Lemma 1

Let $f(t)$ be a Laplace-transformable distribution with its support in $y \leq t < \infty$, where $y > 0$. Let the real number c be such that $e^{-ct}f(t)$ is in $\mathcal{S}'$. Then, for $c + 1 < \sigma < \infty$,

$$|F(\sigma)| \leq Me^{-\sigma\tau}$$

where M is a constant and τ is any real number satisfying $0 < \tau < y$.

Proof: For $\sigma > c$,

$$F(\sigma) = \langle f(t)e^{-ct}, \lambda(t)e^{-(\sigma-c)t} \rangle$$

We are free to choose $\lambda(t)$ such that its support is contained in $x \leq t < \infty$, where x is a fixed number satisfying $0 < x < y$. Moreover, a distribution of slow growth satisfies a boundedness condition of the type given in Theorem 4.4-1. Thus,

$$|F(\sigma)| \leq C \sup_{x \leq t < \infty} \left| (1 + t^2)^r \frac{d^r}{dt^r} [\lambda(t)e^{-(\sigma-c)t}] \right|$$
$$= C \sup_{x \leq t < \infty} \left| (1 + t^2)^r \sum_{\nu=0}^{r} \binom{r}{\nu} \lambda^{(\nu)}(t)(\sigma - c)^{r-\nu} e^{-(\sigma-c)(t-x)} e^{-(\sigma-c)x} \right| \quad (7)$$

where the constant C and the integer r depend only on f. Moreover, if τ is a fixed number such that $0 < \tau < x$, then

$$(\sigma - c)^{r-\nu} e^{-(\sigma-c)x} \leq Be^{-(\sigma-c)\tau} \qquad \sigma > c$$

for each $\nu = 0, 1, \ldots, r$, where B is a constant independent of ν. Furthermore, with $\sigma > c + 1$

$$e^{-(\sigma-c)(t-x)} \leq e^{-(t-x)} \qquad t \geq x$$

so that the right-hand side of (7) is dominated by

$$CBe^{-(\sigma-c)\tau} \sup_{x \leq t < \infty} \left| (1 + t^2)^r \sum_{\nu=0}^{r} \binom{r}{\nu} \lambda^{(\nu)}(t)e^{-(t-x)} \right|$$

But the function of t inside the last sup symbol is bounded for all $t \geq x$. Consequently, the last expression is less than $Me^{-\sigma\tau}$, where M is a sufficiently large constant. Note that in the foregoing argument τ can be chosen anywhere in the interval $0 < \tau < y$ by first choosing x appropriately. This completes the proof.

Theorem 8.6-2

Let $f(t)$ be a Laplace-transformable distribution having its support in $0 \leq t < \infty$ and assume that over some neighborhood of the origin $f(t)$ is a regular distribution corresponding to a (Lebesgue) integrable function $h(t)$. Also, assume that there exist a complex number α and a real number $\eta > -1$ such that

$$\lim_{t \to 0+} \frac{h(t)}{t^\eta} = \alpha$$

Then,

$$\lim_{\sigma \to \infty} \frac{\sigma^{\eta+1} F(\sigma)}{\Gamma(\eta + 1)} = \alpha \tag{8}$$

where

$$F(s) \triangleq \mathfrak{L}f(t)$$

Proof: Let the neighborhood over which $f(t)$ is a regular distribution be $-\infty < t < T$ ($T > 0$) and let $0 < y < T$. Then, $f(t)$ can be decomposed into

$$f(t) = f_1(t) + f_2(t)$$

where the supports of $f_1(t)$ and $f_2(t)$ are contained in the respective intervals $0 \leq t \leq y$ and $y \leq t < \infty$. Thus, $f_1(t)$ is a regular distribution that corresponds to $h(t)$ for $0 < t < y$. Let $F_1(s) \triangleq \mathfrak{L} f_1(t)$ and $F_2(s) \triangleq \mathfrak{L} f_2(t)$. It follows from Lemma 1 that

$$\lim_{\sigma \to \infty} \frac{\sigma^{\eta+1} F_2(\sigma)}{\Gamma(\eta + 1)} = 0 \tag{9}$$

Since the ordinary Laplace transform of a (Lebesgue) integrable function is identical to the distributional Laplace transform of the corresponding regular distribution, we have from Theorem 8.6-1 that

$$\lim_{\sigma \to \infty} \frac{\sigma^{\eta+1} F_1(\sigma)}{\Gamma(\eta + 1)} = \alpha \tag{10}$$

Now, $F(s) = F_1(s) + F_2(s)$ and, thus, (9) and (10) prove (8).

Example 8.6-2 It was established in Example 8.3-2 that the Laplace transform of

$$f(t) = \operatorname{Pf} \frac{1_+(t)1_+(1-t)}{1-t} \tag{11}$$

is

$$F(s) = e^{-s} \sum_{\nu=1}^{\infty} \frac{s^\nu}{\nu!\nu} = e^{-s}[\operatorname{Ei}^*(s) - \gamma - \log s]$$

Moreover, (11) satisfies the hypothesis of Theorem 8.6-2 with $\eta = 0$ and $\alpha = 1$. It follows, then, that $\lim_{\sigma \to \infty} \sigma F(\sigma)$ must equal one. This can be verified by using the following asymptotic expansion for $\operatorname{Ei}^*(\sigma)$, which is valid for large σ (Jahnke, Emde, and Losch [1], p. 18):

$$\operatorname{Ei}^*(\sigma) \sim \frac{e^\sigma}{\sigma}\left(1 + \frac{1!}{\sigma} + \frac{2!}{\sigma^2} + \frac{3!}{\sigma^3} + \cdots\right)$$

Hence,

$$\sigma F(\sigma) \sim \left(1 + \frac{1!}{\sigma} + \frac{2!}{\sigma^2} + \cdots\right) - \sigma e^{-\sigma}(\gamma + \log \sigma) \to 1$$

when $\sigma \to \infty$, as was expected.

PROBLEMS

1 Verify that the "initial value" of each of the following expressions is determined correctly by a theorem of this section:
 a. e^{-t}
 b. $1_+(t)(\cos t)^2$
 c. $1_+(t)1_+(1-t)$
 d. $\operatorname{Pf} \dfrac{1_+(t-1)}{(t-1)^{\frac{1}{2}}}$

2 Let $F(s)$ be a rational function having m poles and n zeros ($m > n$) in the finite s plane and let $f(t)$ be the right-sided function which is continuous for $t > 0$ and whose ordinary Laplace transform is $F(s)$. Show that, for $\nu = 1, 2, 3, \ldots,$

$$f^{(\nu)}(0+) = \lim_{\sigma \to \infty} [\sigma^{\nu+1}F(\sigma) - \sigma^\nu f(0+) - \sigma^{\nu-1}f^{(1)}(0+) - \cdots - \sigma f^{(\nu-1)}(0+)]$$

where $f^{(\mu)}(0+) = 0$ for $\mu = 0, 1, \ldots, m - n - 2$, if $m - n > 1$.

3 Compute the quantities $f(0+)$ and $f^{(1)}(0+)$ corresponding to the following Laplace transforms of $f(t)$. Assume in each case that $f(t)$ equals zero for $t < 0$ and is a continuous function for $t > 0$.

 a. $\dfrac{1}{s(s+1)}$

b. $\dfrac{3s^2 + 4s + 2}{s^3 + 2s^2 + 1}$

c. $\dfrac{s^3 - 3s^2 + 2}{s + 1}$

d. $\dfrac{1}{\sqrt{s^2 + 1}}$

Here $\sqrt{s^2 + 1}$ is understood to be real and positive when s is real and positive.

4 Find the Laplace transform $F(s)$ of

$$f(t) = \frac{e^{-t} - e^{-2t}}{t} 1_+(t)$$

in the following way. Set $g(t) = tf(t)$ and $G(s) = \mathfrak{L}g(t)$. Therefore, $F^{(1)}(s) = -G(s)$. Solve this differential equation for $F(s)$ and evaluate the constant of integration by using an initial-value theorem.

8.7 SOME ABELIAN THEOREMS OF THE FINAL-VALUE TYPE

Now, we shall relate the behavior of $f(t)$ as $t \to \infty$ to the behavior of $F(\sigma)$ as $\sigma \to 0+$. The development of this section is quite similar to that given in the preceding one. First, we shall establish a classical theorem that is analogous to Theorem 8.6-1.

Theorem 8.7-1

If the locally integrable function $f(t)$ satisfies conditions A and if there exist a complex number α and a real number η ($\eta > -1$) such that

$$\lim_{t \to \infty} \frac{f(t)}{t^\eta} = \alpha \tag{1}$$

then

$$\lim_{\sigma \to 0+} \frac{\sigma^{\eta+1} F(\sigma)}{\Gamma(\eta + 1)} = \alpha \tag{2}$$

Proof: First of all, note that (1) indicates that $f(t)$ is a function of slow growth and that its Laplace transform therefore has a region of convergence that contains at least the half-plane $\operatorname{Re} s > 0$. For the moment, assume that the support of $f(t)$ is bounded on the left at $t = T$, where $T \geq 0$. Restricting σ and y to being real positive quantities, we proceed as in the proof of Theorem 8.6-1 to obtain

$$\begin{aligned}|\sigma^{\eta+1} F(\sigma) - \alpha \Gamma(\eta + 1)| &\leq \sigma^{\eta+1} \int_0^\infty e^{-\sigma t} |f(t) - \alpha t^\eta|\, dt \\ &\leq \sigma^{\eta+1} \int_0^y |f(t) - \alpha t^\eta|\, dt + \sigma^{\eta+1} \int_y^\infty e^{-\sigma t} |f(t) - \alpha t^\eta|\, dt\end{aligned} \tag{3}$$

Moreover, the second term on the right-hand side of (3) is dominated by

$$\int_{\sigma y}^{\infty} (\sigma t)^\eta e^{-\sigma t}\, d(\sigma t) \sup_{y \leq t < \infty} \left| \frac{f(t)}{t^\eta} - \alpha \right| \leq \Gamma(\eta + 1) \sup_{y \leq t < \infty} \left| \frac{f(t)}{t^\eta} - \alpha \right| \quad (4)$$

Now let ε be a fixed but arbitrary positive number. In view of (1), y can certainly be chosen so large that the right-hand side of (4), which is independent of σ, can be made less than ε. Having fixed y in this way, we can then choose σ so small that the first term on the right-hand side of (3) is also less than ε. This proves (2) for the case where $T \geq 0$.

On the other hand, if $T < 0$, the additional term

$$\sigma^{\eta+1} \int_T^0 e^{-\sigma t} |f(t)|\, dt$$

appears on the right-hand side of (3). But this term is dominated by

$$\sigma^{\eta+1} e^{-\sigma T} \int_T^0 |f(t)|\, dt$$

a quantity that approaches zero as $\sigma \to 0+$. Hence, (2) holds for any value of T. Q.E.D.

We can extend the preceding theorem to certain distributions as follows:

Theorem 8.7-2

Let $f(t)$ be a Laplace-transformable distribution in $\mathfrak{D}'_R$ and assume that, over some semi-infinite interval $\tau < t < \infty$, $f(t)$ is a regular distribution corresponding to a locally integrable function $h(t)$ that satisfies the second of conditions A. Also, assume that there exist a complex number α and a real number η ($\eta > -1$) such that

$$\lim_{t \to \infty} \frac{h(t)}{t^\eta} = \alpha \quad (5)$$

Then,

$$\lim_{\sigma \to 0+} \frac{\sigma^{\eta+1} F(\sigma)}{\Gamma(\eta + 1)} = \alpha \quad (6)$$

Proof: $f(t)$ can be decomposed into

$$f(t) = f_1(t) + f_2(t)$$

where $f_2(t)$ has its support contained in $y \leq t < \infty$ ($y > \tau$) and $f_1(t)$ has a bounded support that does not extend to the right of $t = y$. $F_1(s) \triangleq \mathfrak{L}f_1(t)$ is then an entire function of s, so that

$$\lim_{\sigma \to 0+} \sigma^{\eta+1} F_1(\sigma) = 0$$

Therefore, with

$$F_2(s) \triangleq \mathfrak{L} f_2(t) = \mathfrak{L}[h(t)1_+(t-y)]$$

we have that

$$\lim_{\sigma \to 0+} \sigma^{\eta+1} F(\sigma) = \lim_{\sigma \to 0+} \sigma^{\eta+1} F_2(\sigma)$$

The right-hand side equals $\alpha \Gamma(\eta + 1)$, according to Theorem 8.7-1. This proves (6).

In certain practical cases, it is the function $F(s)$ that is given and the limit (5) that is to be determined. This can be done by computing (6) if it is known that $F(s)$ is truly the Laplace transform of a distribution that satisfies the hypothesis of Theorem 8.7-2. This will certainly be the case if $F(s)$ is a rational function such that, for some nonnegative integer η, $s^{\eta+1}F(s)$ is analytic over the half-plane Re $s > 0$ and on the finite imaginary axis.

To see this, first note that the poles of $F(s)$ can occur only at $s = 0$, at $s = \infty$, and in the half-plane Re $s < 0$. Expand $F(s)$ into a polynomial in s and a sum of partial fractions as explained in Example 8.4-2. By taking the inverse Laplace transform, we see that $f(t)$ will be of the general form of Sec. 8.4, Eq. (10). Here, all the γ_μ will have negative real parts except possibly for one of them, and this possible γ_μ will equal zero. Thus, as $t \to \infty$, all terms in Sec. 8.4, Eq. (10), will disappear except for the terms due to the possible pole at the origin, and among these the term αt^η will dominate, where α is given by (5).

Similar comments hold for a rational function that is multiplied by the exponential factor e^{as}, where a is a real constant, because this factor merely serves to shift the distribution over the t axis. Thus, we have

Corollary 8.7-2a

Let $F(s) = G(s)e^{as}$, where a is a real constant and $G(s)$ is a rational function. Also, for some nonnegative integer η, let $s^{\eta+1}F(s)$ be analytic over the half-plane Re $s > 0$ and on the imaginary axis. If $f(t)$ denotes that right-sided function which is continuous for $t > -a$ and whose Laplace transform is $F(s)$, then

$$\lim_{t \to \infty} \frac{f(t)}{t^\eta} = \lim_{\sigma \to 0+} \frac{\sigma^{\eta+1} F(\sigma)}{\Gamma(\eta+1)}$$

Example 8.7-1 Let

$$F(s) = \frac{s^4 + 2}{s^2(s+1)}$$

With $\eta = 1$, this satisfies the hypothesis of our corollary. Hence,

$$\lim_{t \to \infty} \frac{f(t)}{t} = \lim_{\sigma \to 0+} \frac{\sigma^4 + 2}{\sigma + 1} = 2$$

PROBLEMS

1 Verify that the behavior for $t \to \infty$ of each of the following expressions is determined correctly by one of the theorems of this section:

   a. $(1 - e^{-t})1_+(t)$

   b. $\displaystyle\sum_{\nu=1}^{\infty} \frac{1}{\nu^2} 1_+(t - \nu)$

   c. $\mathrm{Pf}\,\dfrac{t 1_+(t - 1)}{(t - 1)^{\frac{3}{2}}}$

2 Determine the behavior for $t \to \infty$ of $f(t) \triangleq \mathfrak{L}^{-1} F(s)$ for each of the following expressions for $F(s)$. Assume in each case that $f(t)$ is right-sided and continuous for $t > 0$.

   a. $\dfrac{1}{s(s + 1)}$

   b. $\dfrac{3s^2 + 4s + 2}{s^3 + 2s^2 + s}$

   c. $\dfrac{3s^2 + 2}{s^3 + 2^{\frac{3}{2}}} e^{-s}$

8.8 THE LAPLACE TRANSFORMS OF LEFT-SIDED DISTRIBUTIONS

If the distribution $f(t)$ is in $\mathfrak{D}'_L$ and such that $e^{-ct}f(t)$ is in $\mathcal{S}'$ for some real c, then $f(t)$ is again said to be Laplace-transformable and its Laplace transform (which we now call a left-sided Laplace transform) is defined as follows:

$$F(s) \triangleq \mathfrak{L}f(t) \triangleq \langle e^{-ct}f(t), \mu(t)e^{-(s-c)t}\rangle \qquad \mathrm{Re}\, s < \sigma_2 \qquad (1)$$

Here, $\mu(t)$ is an infinitely smooth function, with support bounded on the right, that equals one over a neighborhood of the support of $f(t)$. The right-hand side of (1) possesses a sense for all s in the half-plane $\mathrm{Re}\, s < c$, and σ_2 denotes the least upper bound for all c such that $e^{-ct}f(t)$ is of slow growth. Thus, in this case the *region of convergence* is a *left-sided* open half-plane that extends infinitely to the left and possesses the vertical line $s = \sigma_2 + i\omega$ as its right-hand boundary. The *abscissa of convergence* is now σ_2. For certain distributions, $\sigma_2 = +\infty$. As before, $F(s)$ is independent of all choices of c in the range $-\infty < c < \sigma_2$.

Sec. 8.8   *The Laplace transformation*   253

Note the analogy between the right-sided and left-sided transforms; the right-sidedness or left-sidedness of the half-plane of convergence (if it is not the whole plane) conforms with the right-sidedness or left-sidedness of the transform.

We could continue our discussion of (1) along the lines of the preceding five sections, but it is simpler to merely point out that a left-sided transform can be converted into a right-sided one by making some changes of variable. Let $\tau = -t$, $p = -s$, and $c = -b$. Then, (1) can be written as

$$\begin{aligned} F(s) &\triangleq \langle e^{-ct}f(t),\ \mu(t)e^{-(s-c)t}\rangle \\ &= \langle e^{-b\tau}f(-\tau),\ \mu(-\tau)e^{-(p-b)\tau}\rangle \\ &\triangleq G(p) \qquad \operatorname{Re} p > -\sigma_2 \end{aligned}$$

Note that $f(-\tau)$ is a right-sided Laplace-transformable distribution and that the region of convergence has been converted into a right-sided one, $\operatorname{Re} p > -\sigma_2$. Indeed, all the theorems and results of the preceding five sections can be made applicable to the left-sided transformation merely by making these changes of variable in the appropriate places. Because of this, we shall merely state in this section a number of properties for this transformation without repeating their proofs.

The left-sided transformation is linear in the following sense. If f and g are Laplace-transformable distributions in $\mathfrak{D}'_L$ having, as their corresponding abscissas of convergence, σ_f and σ_g, respectively, and if α and β are two constants, then

$$\mathfrak{L}(\alpha f + \beta g) = \alpha \mathfrak{L} f + \beta \mathfrak{L} g \qquad \operatorname{Re} s < \inf(\sigma_f, \sigma_g), \text{ at least} \tag{2}$$

Its relationship to the Fourier transformation is

$$\mathfrak{L}f = \mathfrak{F}[e^{-\sigma t}f(t)] \qquad \sigma < \sigma_f \tag{3}$$

Theorem 8.8-1 (*The uniqueness theorem*)

If f and g are Laplace-transformable distributions in $\mathfrak{D}'_L$ and if $\mathfrak{L}f = \mathfrak{L}g$ on some vertical line $s = c + i\omega$ in their common regions of convergence, then $f = g$.

Theorem 8.8-2 (*The analyticity theorem*)

If $f(t)$ is a Laplace-transformable distribution in $\mathfrak{D}'_L$ with the half-plane of convergence $\operatorname{Re} s < \sigma_2$, then $F(s) \triangleq \mathfrak{L}f$ is an analytic function for $\operatorname{Re} s < \sigma_2$ and

$$\begin{aligned} \frac{dF}{ds} &= \langle -tf(t),\ e^{-st}\rangle \\ &\triangleq \langle e^{-ct}f(t),\ -t\mu(t)e^{-(s-c)t}\rangle \qquad \operatorname{Re} s < c < \sigma_2 \end{aligned} \tag{4}$$

Theorem 8.8-3 (*The continuity theorem*)
Let $\{f_\nu\}_{\nu=1}^\infty$ be a sequence of distributions having the following two properties:
a. All the f_ν have their supports contained in a fixed semi-infinite interval $-\infty < t \leq T$.
b. There is a real number c such that the sequence $\{e^{-ct}f_\nu(t)\}_{\nu=1}^\infty$ converges in $\mathcal{S}'$ to the limit $e^{-ct}f(t)$.
Then, the sequence $\{F_\nu(s)\}_{\nu=1}^\infty \triangleq \{\mathfrak{L}f_\nu\}_{\nu=1}^\infty$ converges to $F(s) \triangleq \mathfrak{L}f$ for Re $s < c$.

With k a positive integer, τ a real number, α a complex number, and a a real positive number, the following standard formulas again hold. If $\mathfrak{L}f = F(s)$ for Re $s < \sigma_2$, then

$$\mathfrak{L}[t^k f(t)] = (-1)^k F^{(k)}(s) \qquad \text{Re } s < \sigma_2 \qquad (5)$$

$$\mathfrak{L}f^{(k)}(t) = s^k F(s) \qquad \text{Re } s < \sigma_2 \qquad (6)$$

$$\mathfrak{L}f(t - \tau) = e^{-s\tau} F(s) \qquad \text{Re } s < \sigma_2 \qquad (7)$$

$$\mathfrak{L}[e^{-\alpha t} f(t)] = F(s + \alpha) \qquad \text{Re } s < \sigma_2 - \text{Re } \alpha \qquad (8)$$

$$\mathfrak{L}f(at) = \frac{1}{a} F\left(\frac{s}{a}\right) \qquad \text{Re } s < a\sigma_2 \qquad (9)$$

PROBLEMS

1 Alter Corollary 8.4-1a and Theorems 8.5-1, 8.6-2, and 8.7-2 so that they will apply to Laplace-transformable distributions in $\mathcal{D}'_L$. Prove your statements.

2 Determine the inverse Laplace transform of

$$\frac{1}{s(s-1)}$$

by assuming that the half-plane of convergence is Re $s < 0$. Compare your answer with the result that would have been obtained had the half-plane of convergence been Re $s > 1$.

3 By using an appropriate change of variable, apply formula (7) of Sec. 1.7, Prob. 10, to the left-sided Laplace transform (1) to obtain the Mellin transform given by Sec. 8.3, Eq. (21).

8.9 THE LAPLACE TRANSFORMS OF DISTRIBUTIONS HAVING, IN GENERAL, UNBOUNDED SUPPORTS

The Laplace transform of a distribution having an unbounded support can be defined as the sum of a left-sided transform and a right-sided one. In this case it is called a *two-sided transform*. First of all, let us note that every distribution can be decomposed into the following sum:

$$f = f_L + f_R \qquad (1)$$

Here, f_L and f_R denote distributions in $\mathfrak{D}'_L$ and $\mathfrak{D}'_R$, respectively. Indeed, by assuming that $\theta(t)$ is an infinitely smooth function that equals one for $T \leq t < \infty$ and equals zero for $-\infty < t \leq -T$, where $T > 0$, we may set up the decomposition (1) by letting $f_R = \theta f$ and $f_L = (1 - \theta)f$.

If, in addition, f_L and f_R are both Laplace-transformable and

$$\mathfrak{L}f_L = F_L(s) \qquad \text{Re } s < \sigma_2$$
$$\mathfrak{L}f_R = F_R(s) \qquad \text{Re } s > \sigma_1$$

then the Laplace transform of f will be defined by

$$\mathfrak{L}f \triangleq \mathfrak{L}f_L + \mathfrak{L}f_R = F_L(s) + F_R(s) \tag{2}$$

so long as there is an open region of the s plane for which (2) exists. Three different possibilities arise.

First of all, if the abscissa of convergence σ_2 of $\mathfrak{L}f_L$ is smaller than the abscissa of convergence σ_1 of $\mathfrak{L}f_R$, then there are no common values of s for which $\mathfrak{L}f_L$ and $\mathfrak{L}f_R$ exist. In this case, (2) cannot be used as a definition for $\mathfrak{L}f$. One may expect, at first, that a different decomposition (say $f = f'_L + f'_R$, where f'_L is in $\mathfrak{D}'_L$ and f'_R is in $\mathfrak{D}'_R$) may overcome this difficulty. But this is not the case, since the abscissas of convergence, σ_1 and σ_2, remain the same for all such decompositions of f. (Indeed, the abscissas of convergence of f_R and of f'_R are determined by their behavior as $t \to \infty$. Moreover, there will be some semi-infinite interval extending to the right in which f_R and f'_R are equal. A similar statement may be made for f_L and f'_L.) Thus, when $\sigma_2 < \sigma_1$, $\mathfrak{L}f$ does not exist.

Second, if $\sigma_1 = \sigma_2$, then there is no *open* region of the s plane for which (2) exists. Here again, f does not have a Laplace transform. In this case, it may happen that $e^{-\sigma_1 t}f(t)$ is Fourier-transformable in the sense of Sec. 7.4. One such distribution is

$$f(t) = \frac{e^{\sigma_1 t}}{1 + t^2}$$

One might then say that $\mathfrak{L}f$ exists only on the vertical line $s = \sigma_1 + i\omega$. However, we shall not use this terminology. By our definition a Laplace transform must exist in an *open* region of the s plane if it is to exist at all.

Finally, if $\sigma_1 < \sigma_2$, $\mathfrak{L}f$ does exist and we have the definition

$$\mathfrak{L}f \triangleq F(s) \triangleq F_L(s) + F_R(s) \qquad \sigma_1 < \text{Re } s < \sigma_2 \tag{3}$$

It has already been shown that σ_1 and σ_2 are independent of the decomposition of f. We have yet to show that the transform $F(s)$ itself is also independent of this decomposition. We shall do this shortly.

The open region $\sigma_1 < \text{Re } s < \sigma_2$ is called the *region of convergence* and also the *strip of convergence* for $\mathfrak{L}f$. For this more general two-sided Laplace transform there are two *abscissas of convergence*, σ_1 and σ_2. σ_1 may equal $-\infty$, and σ_2 may equal ∞.

The two-sided Laplace transformation is related to the Fourier transformation, just as are the left-sided and right-sided ones. Assume, once again, that f is Laplace-transformable and has the strip of convergence $\sigma_1 < \text{Re } s < \sigma_2$. Also, let σ be a real number satisfying $\sigma_1 < \sigma < \sigma_2$. With the decomposition $f = f_L + f_R$ we have that $e^{-\sigma t}f_L(t)$ and $e^{-\sigma t}f_R(t)$ are both in $\mathcal{S}'_t$, so that $e^{-\sigma t}f(t)$ is also in $\mathcal{S}'_t$. By employing Sec. 8.3, Eq. (5), and Sec. 8.8, Eq. (3), we now obtain

$$\mathfrak{L}f = \mathfrak{L}f_L + \mathfrak{L}f_R = \mathfrak{F}[e^{-\sigma t}f_L(t)] + \mathfrak{F}[e^{-\sigma t}f_R(t)]$$
$$= \mathfrak{F}[e^{-\sigma t}f(t)] \qquad \sigma_1 < \sigma < \sigma_2 \qquad (4)$$

This relationship can be used as an alternative means of defining the distributional two-sided Laplace transform in terms of the Fourier transform, so long as there exists a nonvoid open interval $\sigma_1 < \sigma < \sigma_2$ for which $e^{-\sigma t}f(t)$ is of slow growth. All the properties of the two-sided Laplace transformation that we shall develop from definition (3) can also be obtained directly from (4) (Schwartz [3]).

Since $\mathfrak{F}[e^{-\sigma t}f(t)]$ is independent of any decomposition of f in the form (1), it follows from (4) that $\mathfrak{L}f$ *is also invariant with respect to all such decompositions of f.* Thus, the decomposition merely serves as a tool for defining and computing $\mathfrak{L}f$ and the abscissas of convergence. It does not affect these quantities.

The linearity of the two-sided Laplace transformation is expressed as follows. If f and g are Laplace-transformable distributions in $\mathcal{D}'$ having as their corresponding strips of convergence $\sigma_{f1} < \text{Re } s < \sigma_{f2}$ and $\sigma_{g1} < \text{Re } s < \sigma_{g2}$, respectively, if these strips intersect, and if α and β are two constants, then

$$\mathfrak{L}(\alpha f + \beta g) = \alpha \mathfrak{L}f + \beta \mathfrak{L}g$$
$$\sup(\sigma_{f1}, \sigma_{g1}) < \text{Re } s < \inf(\sigma_{f2}, \sigma_{g2}), \text{ at least}$$

Through the uniqueness of the Fourier transformation, (4) establishes the uniqueness of the two-sided Laplace transformation. More specifically, we have

Theorem 8.9-1 (*The uniqueness theorem*)

Let f and g be Laplace-transformable distributions in $\mathcal{D}'$. If $\mathfrak{L}f = \mathfrak{L}g$ on some vertical line $s = c + i\omega$ in their common strips of convergence, then $f = g$.

Since $F_L(s)$ and $F_R(s)$ are analytic for $\sigma_1 < \text{Re } s$ and for $\text{Re } s < \sigma_2$, respectively, $F(s)$ is analytic for $\sigma_1 < \text{Re } s < \sigma_2$. Theorems 8.3-2 and 8.8-2 together yield

Theorem 8.9-2 (*The analyticity theorem*)

If $f(t)$ is a Laplace-transformable distribution in $\mathfrak{D}'$ with the strip of convergence $\sigma_1 < \operatorname{Re} s < \sigma_2$, then $F(s) \triangleq \mathfrak{L}f$ is analytic for $\sigma_1 < \operatorname{Re} s < \sigma_2$ and

$$\frac{dF}{ds} = \langle -tf(t), e^{-st} \rangle \tag{5}$$

NOTE: As usual, it is understood that the right-hand side of (5) equals

$$\langle e^{-c_2 t} f_L(t), -t\mu(t) e^{-(s-c_2)t} \rangle + \langle e^{-c_1 t} f_R(t), -t\lambda(t) e^{-(s-c_1)t} \rangle$$

where $f_L + f_R$ is any decomposition of f and $\sigma_1 < c_1 < \operatorname{Re} s < c_2 < \sigma_2$.

The distributional two-sided Laplace transformation is also a continuous operation.

Theorem 8.9-3 (*The continuity theorem*)

Let $\{f_\nu\}_{\nu=1}^\infty$ be a sequence of distributions and let all the f_ν be decomposed into $f_{L\nu} + f_{R\nu}$, where the $f_{L\nu}$ have their supports contained in the fixed semi-infinite interval $-\infty > t \geq T_1$ and the $f_{R\nu}$ have their supports contained in the fixed semi-infinite interval $T_2 \leq t < \infty$. If there exist two distinct real numbers c_1 and c_2 ($c_1 < c_2$) for which $\{e^{-c_2 t} f_{L\nu}(t)\}_{\nu=1}^\infty$ converges in $\mathcal{S}'$ to $e^{-c_2 t} f_L(t)$ and $\{e^{-c_1 t} f_{R\nu}(t)\}_{\nu=1}^\infty$ converges in $\mathcal{S}'$ to $e^{-c_1 t} f_R(t)$, then $\{F_\nu(s)\}_{\nu=1}^\infty \triangleq \{\mathfrak{L}f_\nu\}_{\nu=1}^\infty$ converges to $F(s) \triangleq \mathfrak{L}(f_L + f_R)$ for $c_1 < \operatorname{Re} s < c_2$.

NOTE: If this hypothesis holds for some particular decomposition of the f_ν, it will hold for all such decompositions.

Proof: The fact that $c_1 < c_2$ ensures that all the $\mathfrak{L}f_\nu$ exist for $c_1 < \operatorname{Re} s < c_2$. Then, Theorems 8.3-3 and 8.8-3 can be invoked to obtain

$$\mathfrak{L}f_\nu = \mathfrak{L}f_{L\nu} + \mathfrak{L}f_{R\nu} \to \mathfrak{L}f_L + \mathfrak{L}f_R = \mathfrak{L}(f_L + f_R) \qquad c_1 < \operatorname{Re} s < c_2$$

Q.E.D.

For easy reference, we shall list some standard formulas that follow readily from the corresponding ones for the right-sided and left-sided Laplace transforms. Once again, k is a positive integer, τ a real number, α a complex number, and a a real positive number. If $\mathfrak{L}f = F(s)$ for $\sigma_1 < \operatorname{Re} s < \sigma_2$, then

$$\mathfrak{L}[t^k f(t)] = (-1)^k F^{(k)}(s) \qquad \sigma_1 < \operatorname{Re} s < \sigma_2 \tag{6}$$

$$\mathfrak{L}f^{(k)}(t) = s^k F(s) \qquad \sigma_1 < \operatorname{Re} s < \sigma_2 \tag{7}$$

$$\mathfrak{L}f(t - \tau) = e^{-s\tau} F(s) \qquad \sigma_1 < \operatorname{Re} s < \sigma_2 \tag{8}$$

$$\mathfrak{L}[e^{-\alpha t} f(t)] = F(s + \alpha) \qquad \sigma_1 - \operatorname{Re} \alpha < \operatorname{Re} s < \sigma_2 - \operatorname{Re} \alpha \tag{9}$$

$$\mathfrak{L}f(at) = \frac{1}{a} F\left(\frac{s}{a}\right) \qquad a\sigma_1 < \operatorname{Re} s < a\sigma_2 \tag{10}$$

By combining Corollary 8.4-1a with the analogous result for the Laplace transforms of distributions in $\mathfrak{D}'_L$, we immediately obtain the following inversion theorem:

Theorem 8.9-4

Let $F(s)$ be a function that is analytic in a strip $\sigma_1 < \mathrm{Re}\, s < \sigma_2$ and assume that $F(s)$ can be decomposed into

$$F(s) = F_L(s) + F_R(s)$$

where $F_L(s)$ and $F_R(s)$ possess the following properties. $F_R(s)$ is analytic for $\mathrm{Re}\, s > \sigma_1$, and in some half-plane $\mathrm{Re}\, s \geq c_1 > \sigma_1$

$$|F_R(s)| \leq P_1(|s|)e^{-\mathrm{Re}\, s T_1}$$

where $P_1(|s|)$ is a polynomial in $|s|$ and T_1 is a real constant. $F_L(s)$ is analytic for $\mathrm{Re}\, s < \sigma_2$, and in some half-plane $\mathrm{Re}\, s \leq c_2 < \sigma_2$

$$|F_L(s)| \leq P_2(|s|)e^{-\mathrm{Re}\, s T_2}$$

where $P_2(|s|)$ is also a polynomial in $|s|$ and T_2 is a real constant. If $c_1 < c_2$, then $F(s)$ is the Laplace transform of a distribution f that can be decomposed into $f = f_L + f_R$, where $f_L = \mathfrak{L}^{-1} F_L(s)$ and $f_R = \mathfrak{L}^{-1} F_R(s)$. Moreover, f_R has its support bounded on the left at T_1 and f_L has its support bounded on the right at T_2.

Actually, an inversion theorem for a two-sided Laplace transform that does not depend upon a decomposition for $F(s)$ can be stated. In fact, if $F(s)$ is analytic in a strip $\sigma_1 < \mathrm{Re}\, s < \sigma_2$ and if, in each substrip $\sigma_1 < c_1 \leq \mathrm{Re}\, s \leq c_2 < \sigma_2$, $F(s)$ is bounded by a polynomial in $|s|$ (this polynomial depending upon the choice of the substrip), then $F(s)$ is the two-sided Laplace transform of a distribution. Moreover, these conditions on $F(s)$ are necessary, as well as sufficient, for $F(s)$ to be such a Laplace transform (Schwartz [3], proposition 6).

PROBLEMS

1 We have mentioned that the relationship

$$\mathfrak{L}f = \mathfrak{F}[e^{-\sigma t}f(t)] \qquad \sigma_1 < \sigma < \sigma_2$$

can be used as an alternative definition of the distributional two-sided Laplace transform. Without decomposing f, show that, if there exist two distinct real numbers c_1 and c_2 ($c_1 < c_2$) for which $e^{-c_1 t}f(t)$ and $e^{-c_2 t}f(t)$ are both in $\mathcal{S}'_t$, then $e^{-\sigma t}f(t)$ is also in $\mathcal{S}'_t$ if $c_1 \leq \sigma \leq c_2$. This implies that the set of values of σ for which $e^{-\sigma t}f(t)$ is in $\mathcal{S}'_t$ is a single interval (i.e., a convex set in $\mathfrak{R}^1$). The largest open interval $\sigma_1 <$

$\sigma < \sigma_2$ in this convex set determines the strip of convergence for $\mathfrak{L}f$. HINT: Use the relationship

$$e^{-\sigma t}f(t) = \theta(t)e^{-c_1 t}f(t) + \theta(t)e^{-c_2 t}f(t)$$

where

$$\theta(t) = \frac{e^{-\sigma t}}{e^{-c_1 t} + e^{-c_2 t}}$$

2 Develop formulas (6) to (10) first by using the results for the one-sided Laplace transforms and then by employing (4) and the corresponding relations for the Fourier transforms.

3 Show that, when a is a real negative number, formula (10) must be replaced by

$$\mathfrak{L}f(at) = \frac{1}{|a|} F\left(\frac{s}{a}\right) \qquad a\sigma_2 < \text{Re } s < a\sigma_1$$

4 Determine the inverse Laplace transform of $1/s(s-1)$ by assuming that the strip of convergence is $0 < \text{Re } s < 1$. Compare your answer with the results of Sec. 8.8, Prob. 2.

5 Which of the following distributions are Laplace-transformable? For those that are, find their Laplace transforms and specify the regions of convergence. Here a and b are real constants with $b > 0$.

   a. e^t
   b. $\delta(t-a) + \delta(t+a)$
   c. $\displaystyle\sum_{\nu=-\infty}^{\infty} \delta(t-\nu)$
   d. $\displaystyle\sum_{\nu=-\infty}^{\infty} e^{-|\nu|}\delta(t-\nu)$
   e. $\text{Pf}\,\dfrac{1}{\sin t}$
   f. $\text{Pf}\,\dfrac{e^{-b|t|}}{|t|}$
   g. e^{-bt^2}

HINT: Set

$$z = t\sqrt{b} + \frac{s}{2\sqrt{b}}$$

and use the fact that

$$\int_{-\infty}^{\infty} e^{-z^2}\,dx = \sqrt{\pi}$$

6 Prove Theorem 3.2-1 by using the Laplace transformation.

9 The Solution of Differential and Difference Equations by Transform Analysis

★9.1 INTRODUCTION

This chapter is devoted to a study of how the Laplace transformation may be used to solve certain types of equations, namely, ordinary linear differential and integrodifferential equations with constant coefficients and also ordinary linear difference equations with constant coefficients of both the continuous- and discrete-variable types. As was mentioned in our discussion of distributional convolution, all these equations are special types of convolution equations. Our particular concern here will be to solve them in the convolution algebra $\mathfrak{D}'_R$; thus we shall be restricting ourselves to the initial-condition type of problem.

In the last section we briefly consider ordinary linear differential equations with polynomial coefficients. These are not representable as convolution equations. Nevertheless, they can at times be solved by using the Laplace transformation.

★9.2 THE USE OF THE LAPLACE TRANSFORMATION IN SOLVING CONVOLUTION EQUATIONS IN THE ALGEBRA $\mathcal{D}'_R$

Consider the convolution equation

$$f * u = g \tag{1}$$

where f, u, and g are distributions in $\mathcal{D}'_R$, f and g being given and u being unknown. As was discussed in Sec. 6.2, we can find u by first determining an inverse in $\mathcal{D}'_R$ for f. That is, we first seek in $\mathcal{D}'_R$ a distribution f^{*-1} that satisfies the equation

$$f^{*-1} * f = \delta \tag{2}$$

When such an f^{*-1} exists, it is unique. Moreover, we may then write

$$u = f^{*-1} * g \tag{3}$$

and (3) is the one and only solution in $\mathcal{D}'_R$ for (1).

The Laplace transformation may be employed to facilitate this procedure in the following way. Let f be a Laplace-transformable distribution in $\mathcal{D}'_R$ and let $\mathfrak{L}f \triangleq F(s)$ with $\operatorname{Re} s > \sigma_f$ as its half-plane of convergence. Since the Laplace transformation converts convolution into multiplication, (2) may be formally transformed into

$$(\mathfrak{L}f^{*-1})F(s) = 1$$

If $1/F(s)$ is itself a Laplace transform (i.e., if it possesses the properties specified in Theorem 8.4-1 or in Corollary 8.4-1a in some half-plane $\operatorname{Re} s > \sigma_I$), we may then write

$$f^{*-1} = \mathfrak{L}^{-1} \frac{1}{F(s)} \qquad \operatorname{Re} s > \sigma_I \tag{4}$$

One may check the validity of these manipulations merely by computing $1/F(s)$ and seeing whether it is a Laplace transform according to Corollary 8.4-1a. If so, it follows from Theorems 8.3-1 and 8.5-1 that the desired inverse f^{*-1} is truly given by (4).

If g is also Laplace-transformable and if $\mathfrak{L}g \triangleq G(s)$ with $\operatorname{Re} s > \sigma_g$ as its half-plane of convergence, then the Laplace transform may again be used to compute u. Indeed, from (3) we get

$$u = \mathfrak{L}^{-1} \frac{G(s)}{F(s)} \qquad \operatorname{Re} s > \sup(\sigma_I, \sigma_g) \tag{5}$$

However, g need not be Laplace-transformable, and in that case we can determine u (at least in principle) by directly evaluating (3).

Example 9.2-1 Let us determine a u in $\mathfrak{D}'_R$ that satisfies

$$[1_+(t) \cos t] * u(t) = (e^{t^2} - 1)1_+(t) \tag{6}$$

Now,

$$\mathfrak{L}[1_+(t) \cos t] = \frac{s}{s^2 + 1} \qquad \text{Re } s > 0 \tag{7}$$

The reciprocal of the right-hand side of (7) is analytic for Re $s > 0$ and is bounded by a polynomial in $|s|$ for Re $s \geq b > 0$. Hence, it is a Laplace transform and we have

$$[1_+(t) \cos t]^{*-1} = \mathfrak{L}^{-1} \frac{s^2 + 1}{s} \qquad \text{Re } s > 0$$
$$= \delta^{(1)}(t) + 1_+(t)$$

By convolving this with both sides of (6), we get

$$u(t) = 2te^{t^2}1_+(t) + 1_+(t) \int_0^t (e^{x^2} - 1)\, dx$$

Here $u(t)$ turns out to be a locally integrable function. Hence, (6) involves only locally integrable functions, and for $t > 0$ it can be written as

$$\int_0^t \cos(t - \tau)\, u(\tau)\, d\tau = e^{t^2} - 1$$

We have, in effect, solved this integral equation.

However, if we replace the right-hand side of (6) by $1_+(t) \exp t^2$, the solution of the resulting equation will be

$$u(t) = \delta(t) + 2te^{t^2}1_+(t) + 1_+(t) \int_0^t e^{x^2}\, dx$$

The solution is now a singular distribution even though the other factors in (6) are locally integrable functions.

The Laplace transform is also useful for solving sets of simultaneous convolution equations in the algebra $\mathfrak{D}'_R$. By using matrix notation to represent such a set, we may write

$$\mathbf{f} * \mathbf{u} = \mathbf{g} \tag{8}$$

where

$$\mathbf{f} = \begin{bmatrix} f_{11} & f_{12} & \cdots & f_{1n} \\ f_{21} & f_{22} & \cdots & f_{2n} \\ \vdots & & & \vdots \\ f_{n1} & f_{n2} & \cdots & f_{nn} \end{bmatrix}$$

Sec. 9.2 Differential and difference equations

$$\mathbf{u} = \begin{bmatrix} u_1 \\ u_2 \\ \cdot \\ \cdot \\ \cdot \\ u_n \end{bmatrix} \qquad \mathbf{g} = \begin{bmatrix} g_1 \\ g_2 \\ \cdot \\ \cdot \\ \cdot \\ g_n \end{bmatrix}$$

All the elements of $\mathbf{f}$, $\mathbf{u}$, and $\mathbf{g}$ are restricted to being in $\mathcal{D}'_R$ (that is, $\mathbf{f} \in \mathcal{D}'_{R,n \times n}$, $\mathbf{u} \in \mathcal{D}'_{R,n \times 1}$, $\mathbf{g} \in \mathcal{D}'_{R,n \times 1}$); $\mathbf{f}$ and $\mathbf{g}$ are assumed known and $\mathbf{u}$ is to be determined. As was also discussed in Sec. 6.2, $\mathbf{u}$ may be found by first determining in $\mathcal{D}'_{R,n \times n}$ an inverse $\mathbf{f}^{*-1}$ to $\mathbf{f}$ so that

$$\mathbf{f}^{*-1} * \mathbf{f} = \boldsymbol{\delta}_{n \times n} \tag{9}$$

When such an inverse exists, it is unique in $\mathcal{D}'_{R,n \times n}$ and there is precisely one solution for (8) in $\mathcal{D}'_{R,n \times 1}$. It is given by

$$\mathbf{u} = \mathbf{f}^{*-1} * \mathbf{g} \tag{10}$$

The Laplace transform $\mathbf{F}(s)$ of a matrix $\mathbf{f}(t)$ is defined by taking the Laplace transform of each element in $\mathbf{f}(t)$. $\mathbf{f}(t)$ is said to be Laplace-transformable if each of its elements is Laplace-transformable and if there is a nonvoid open intersection between the regions of convergence of all its elements. This intersection is now called the region of convergence for the Laplace transform of $\mathbf{f}$.

In addition to the condition that $\mathbf{f}$ is in $\mathcal{D}'_{R,n \times n}$ let us now assume that $\mathbf{f}$ is Laplace-transformable. Thus, its transform $\mathfrak{L}\mathbf{f} \triangleq \mathbf{F}(s)$ will possess some half-plane of convergence, Re $s > \sigma_f$. Proceeding as before, we formally transform (9) into

$$(\mathfrak{L}\mathbf{f}^{*-1}) \mathbf{F}(s) = \mathbf{1}_{n \times n}$$

where $\mathbf{1}_{n \times n}$ denotes the $n \times n$ unit matrix. If $\mathbf{F}(s)$ possesses a matrix inverse $\mathbf{F}^{-1}(s)$ in the ordinary sense, then we shall also write

$$\mathfrak{L}\mathbf{f}^{*-1} = \mathbf{F}^{-1}(s) \tag{11}$$

Moreover, if $\mathbf{F}^{-1}(s)$ is the Laplace transform of some matrix in $\mathcal{D}'_{R,n \times n}$, then (11) possesses a sense and we may take the inverse Laplace transform to get

$$\mathbf{f}^{*-1} = \mathfrak{L}^{-1}\mathbf{F}^{-1}(s) \tag{12}$$

We have yet to verify that (12) is truly the inverse of f. But this follows immediately, since

$$\mathbf{f}^{*-1} * \mathbf{f} = \mathfrak{L}^{-1}[\mathbf{F}^{-1}(s) \mathbf{F}(s)]$$
$$= \mathfrak{L}^{-1} \mathbf{1}_{n \times n} = \boldsymbol{\delta}_{n \times n}$$

We emphasize that this method works so long as **f** is Laplace-transformable and its transform $\mathbf{F}(s)$ has an inverse $\mathbf{F}^{-1}(s)$, which is also the Laplace transform of some matrix in $\mathfrak{D}'_{R,n\times n}$. This means that $\mathbf{F}^{-1}(s)$ also has a half-plane of convergence, say $\operatorname{Re} s > \sigma_I$.

Let us review the steps for computing $\mathbf{F}^{-1}(s)$ from $\mathbf{F}(s)$. The cofactor $C_{ik}(s)$ of each element $F_{ik}(s)$ in $\mathbf{F}(s)$ is the product of $(-1)^{i+k}$ with the determinant of the $(n-1)\times(n-1)$ matrix obtained by striking out the ith row and kth column of $\mathbf{F}(s)$. Then, the adjoint $\mathbf{A}(s)$ of $\mathbf{F}(s)$ is the matrix transpose of the matrix of cofactors. (This terminology is not uniformly used; some writers use the phrase "adjoint matrix" to mean the complex-conjugate matrix transpose of a given matrix.)

$$\mathbf{A}(s) \triangleq \begin{bmatrix} C_{11} & C_{21} & \cdots & C_{n1} \\ C_{12} & C_{22} & \cdots & C_{n2} \\ \cdots & \cdots & \cdots & \cdots \\ C_{1n} & C_{2n} & \cdots & C_{nn} \end{bmatrix}$$

Finally, if $\det \mathbf{F}(s)$ is not identically zero, then $\mathbf{F}(s)$ has an inverse given by

$$\mathbf{F}^{-1}(s) = \frac{\mathbf{A}(s)}{\det \mathbf{F}(s)}$$

In this case, it can be shown that

$$\mathbf{F}^{-1}(s)\mathbf{F}(s) = \mathbf{F}(s)\mathbf{F}^{-1}(s) = \mathbf{1}_{n\times n}$$

If at each step of this procedure we take the inverse Laplace transform, we shall obtain precisely the procedure developed in Sec. 6.2 for obtaining $\mathbf{f}^{*-1}$ from $\mathbf{f}$.

Finally, if **g** is also Laplace-transformable and if $\mathfrak{L}\mathbf{g} = \mathbf{G}(s)$ with $\operatorname{Re} s > \sigma_g$ as its region of convergence, then

$$\mathbf{u} = \mathfrak{L}^{-1}[\mathbf{F}^{-1}(s)\mathbf{G}(s)] \qquad \operatorname{Re} s > \sup(\sigma_I, \sigma_g) \tag{13}$$

and this expression is quite often easier to evaluate than (10).

Example 9.2-2 Let us solve

$$\begin{aligned} \left(\frac{d}{dt}+1\right)u_1(t) - u_2(t) &= 1_+(t)e^t \\ \left(\frac{d}{dt}+1\right)u_1(t) + \frac{d}{dt}u_2(t) &= 0 \end{aligned} \tag{14}$$

in the algebra $\mathfrak{D}'_R$. By using matrix notation, we may write $\mathbf{f} * \mathbf{u} = \mathbf{g}$, where

$$\mathbf{f} = \begin{bmatrix} \delta^{(1)}+\delta & -\delta \\ \delta^{(1)}+\delta & \delta^{(1)} \end{bmatrix}$$

$$\mathbf{u} = \begin{bmatrix} u_1 \\ u_2 \end{bmatrix} \qquad \mathbf{g} = \begin{bmatrix} 1_+(t)e^t \\ 0 \end{bmatrix}$$

Upon applying the Laplace transformation, we get

$$\mathbf{F}(s) = \begin{bmatrix} s+1 & -1 \\ s+1 & s \end{bmatrix} \qquad -\infty < \operatorname{Re} s < \infty$$

$$\mathbf{G}(s) = \begin{bmatrix} \dfrac{1}{s-1} \\ 0 \end{bmatrix} \qquad \operatorname{Re} s > 1$$

The inverse of $\mathbf{F}(s)$ is easily computed to be

$$\mathbf{F}^{-1}(s) = \frac{1}{(s+1)^2} \begin{bmatrix} s & 1 \\ -s-1 & s+1 \end{bmatrix}$$

and in the half-plane $\operatorname{Re} s > -1$ its elements are all Laplace transforms. Hence, with $\mathbf{U}(s) \triangleq \mathfrak{L}\mathbf{u}(t)$ we have

$$\mathbf{U}(s) = \mathbf{F}^{-1}(s)\mathbf{G}(s) = \begin{bmatrix} \dfrac{s}{(s-1)(s+1)^2} \\ \dfrac{-1}{(s-1)(s+1)} \end{bmatrix}$$

so that

$$u_1(t) = 1_+(t)(\tfrac{1}{2}te^{-t} - \tfrac{1}{4}e^{-t} + \tfrac{1}{4}e^t)$$
$$u_2(t) = 1_+(t)(\tfrac{1}{2}e^{-t} - \tfrac{1}{2}e^t)$$

The bulk of this chapter is devoted to applications of this simple technique for solving convolution equations of various types. As we shall see, it is a very useful technique, since many different types of convolution equations can be attacked in this way.

PROBLEMS

1 $J_0(t)$ denotes the zero-order Bessel function of first kind. Use the fact that

$$\mathfrak{L}[1_+(t)J_0(t)] = \frac{1}{\sqrt{s^2+1}} \qquad \operatorname{Re} s > 0$$

to determine the inverse in $\mathfrak{D}'_R$ of $1_+(t)J_0(t)$. Then, solve the equation

$$[1_+(t)J_0(t)] * \mathbf{u}(t) = 1_+(t)\cos t$$

in the algebra $\mathfrak{D}'_R$.

2 Solve the following equations in the algebra $\mathfrak{D}'_R$. You may use the table in Appendix B.

a. $1_+(t-1) * u(t) = \dfrac{1_+(t)}{t+1}$

b. $[1_+(t) - 1_+(t-1)] * u(t) = 1_+(t) \sin \pi t$

c. $[1_+(t) \sin t] * u(t) = \delta^{(2)}(t-\pi)$

d. $[te^t 1_+(t)] * u(t) = \dfrac{1_+(t)}{1+t^2}$

e. $[e^t 1_+(t)] * u(t) = \operatorname{Pf} \dfrac{1_+(t)}{t}$

f. $\left[\operatorname{Pf} \dfrac{1_+(t)}{t^2} \right] * u(t) = 1_+(t)$

g. $\left[\operatorname{Pf} \dfrac{1_+(t) \log t}{t} \right] * u(t) = 1_+(t) (\log t)^2$

3 Solve the following simultaneous equations in the algebra $\mathfrak{D}'_R$:

a. $[1_+(t) \cos t] * u_1(t) - [1_+(t) \sin t] * u_2(t) = \delta(t)$
$[1_+(t) \sin t] * u_1(t) + [1_+(t) \cos t] * u_2(t) = 1_+(t)$

b. $[\operatorname{Pf} t^{-\frac{3}{2}} 1_+(t)] * u_1(t) + \delta(t+1) * u_2(t) = \delta^{(1)}(t)$
$\delta(t-1) * u_1(t) + [\operatorname{Pf} t^{-\frac{3}{2}} 1_+(t)] * u_2(t) = 0$

c. $1_+(t) \displaystyle\int_0^t u_1(x)\,dx + \dfrac{d}{dt} u_3(t) = 1_+(t-1)$

$u_1(t) + 1_+(t) \displaystyle\int_0^t e^{x-t} u_2(x)\,dx + \dfrac{d}{dt} u_3(t) = 0$

$u_2(t) + \dfrac{d}{dt} u_3(t) = 1_+(t) e^{-t}$

4 Repeat Sec. 6.2, Prob. 6, this time using the Laplace transformation. In addition to the restrictions imposed on $k(t)$ in that problem, you may also assume that $k^{(1)}(t)$ is continuous for $t > 0$ and that $k(t)e^{-ct}$ is absolutely integrable over $0 < t < \infty$ for some real c.

5 Find a function $u(t)$ that satisfies the following integral equation for all t:

$$u(t) - e^t \int_{-\infty}^t e^{-\tau} u^{(1)}(\tau)\,d\tau = t^2 1_+(t)$$

Would this equation have an ordinary sense if the right-hand side were $t 1_+(t)$?

★9.3 ORDINARY LINEAR DIFFERENTIAL EQUATIONS WITH CONSTANT COEFFICIENTS

We saw in Sec. 6.3 how the convolution algebra $\mathfrak{D}'_R$ may be used to solve the initial-condition problem for ordinary linear differential equations with constant coefficients. We shall now enhance this method with the Laplace transformation.

Once again, let

$$L \triangleq a_n \frac{d^n}{dt^n} + a_{n-1} \frac{d^{n-1}}{dt^{n-1}} + \cdots + a_0 \tag{1}$$

where the a_ν ($\nu = 0, 1, \ldots, n$) are complex constants, $a_n \neq 0$, and $n \geq 1$. The differential equation

$$Lu = g \tag{2}$$

can be written as the convolution equation

$$(L\delta) * u = g \tag{3}$$

Now, $L\delta$ is certainly Laplace-transformable and

$$F(s) \triangleq \mathfrak{L}(L\delta) = a_n s^n + a_{n-1} s^{n-1} + \cdots + a_0 \tag{4}$$

where the region of convergence is the entire s plane. Hence, the inverse in $\mathfrak{D}'_R$ of $L\delta$ is

$$\begin{aligned}(L\delta)^{*-1} &= \mathfrak{L}^{-1} \frac{1}{F(s)} \\ &= \mathfrak{L}^{-1} \frac{1}{a_n s^n + a_{n-1} s^{n-1} + \cdots + a_0} \quad \operatorname{Re} s > \sigma_I\end{aligned} \tag{5}$$

[The abscissa of convergence σ_I equals the largest of the real parts of the poles of $F(s)$.] Upon making a partial-fraction expansion of $1/F(s)$ and then taking the inverse Laplace transform of each fraction, we find that

$$(L\delta)^{*-1} = 1_+(t) h(t)$$

where

$$h(t) = \sum_{\mu=1}^{q} \sum_{\nu=1}^{k_\mu} c_{\mu\nu} t^{\nu-1} e^{\gamma_\mu t} \tag{6}$$

$$\sum_{\mu=1}^{q} k_\mu = n$$

Here, the γ_μ are the distinct poles of $1/F(s)$, the k_μ are the corresponding multiplicities, and the $c_{\mu\nu}$ are the coefficients of the partial-fraction expansion.

This inverse is precisely the same as that specified in Theorem 6.3-1. Indeed, by applying the initial-value theorem to (5) (see Corollary 8.6-1a and Sec. 8.6, Prob. 2), we obtain

$$h(0) = h^{(1)}(0) = \cdots = h^{(n-2)}(0) = 0$$

$$h^{(n-1)}(0) = \frac{1}{a_n}$$

which are the correct initial conditions for the Green's function $h(t)1_+(t)$.

It now follows that, whenever g is a right-sided distribution, (3) will have a unique solution in $\mathfrak{D}'_R$ given by

$$u(t) = [1_+(t)h(t)] * g(t) \tag{7}$$

Moreover, we can account for all possible solutions to (3) by adding to (7) a complementary solution (i.e., a solution of the homogeneous equation $Lu = 0$). If $g(t)$ has its support contained in $t_0 \leq t < \infty$, then we may specify the complementary solution by introducing the values of $u(t_0 - \varepsilon)$, $u^{(1)}(t_0 - \varepsilon)$, ..., $u^{(n-1)}(t_0 - \varepsilon)$, where ε is some (perhaps arbitrarily small) positive number. As was shown in Sec. 6.3, for $t > t_0 - \varepsilon$ this is equivalent to solving in $\mathcal{D}'_R$ the equation

$$Lu(t) = g(t) + \sum_{\nu=0}^{n-1} b_\nu \delta^{(\nu)}(t - t_0 + \varepsilon) \tag{8}$$

where

$$b_\nu = a_{\nu+1} u(t_0 - \varepsilon) + a_{\nu+2} u^{(1)}(t_0 - \varepsilon) + \cdots + a_n u^{(n-1-\nu)}(t_0 - \varepsilon)$$

This may again be done by using the Laplace transformation. Assuming that $\mathfrak{L}g \triangleq G(s)$ exists with $\operatorname{Re} s > \sigma_g$ as its half-plane of convergence, the solution becomes

$$u(t) = \mathfrak{L}^{-1}\left\{\frac{1}{F(s)}\left[G(s) + \sum_{\nu=0}^{n-1} b_\nu s^\nu e^{-(t_0-\varepsilon)s}\right]\right\} \qquad \operatorname{Re} s > \sup(\sigma_I, \sigma_g) \tag{9}$$

or

$$u(t) = [1_+(t) h(t)] * g(t) + 1_+(t - t_0 + \varepsilon) \sum_{\nu=0}^{n-1} b_\nu h^{(\nu)}(t - t_0 + \varepsilon) \tag{10}$$

If we drop the factor $1_+(t - t_0 + \varepsilon)$ multiplying the summation in (10), we shall have

$$u(t) = [1_+(t) h(t)] * g(t) + \sum_{\nu=0}^{n-1} b_\nu h^{(\nu)}(t - t_0 + \varepsilon) \tag{11}$$

This is the solution of $Lu = g$, which now holds for all t. It is not in $\mathcal{D}'_R$ but satisfies the given initial conditions at $t = t_0 - \varepsilon$.

When g is a locally integrable function, $u(t)$ is a continuous function and (11) is precisely the classical solution for $Lu = g$. In this case $(1_+ h) * g$ is zero at $t = t_0$, so that we may set $\varepsilon = 0$ in order to introduce the initial conditions at $t = t_0$.

Example 9.3-1 We shall solve

$$u^{(2)} - u = \delta^{(2)}$$

under the initial conditions $u(-\varepsilon) = u^{(1)}(-\varepsilon) = 1$. According to (8), we may consider the new equation

$$u^{(2)}(t) - u(t) = \delta^{(2)}(t) + u^{(1)}(-\varepsilon) \delta(t + \varepsilon) + u(-\varepsilon) \delta^{(1)}(t + \varepsilon)$$

By using the Laplace transformation, we get
$$(s^2 - 1)U(s) = s^2 + e^{\varepsilon s} + se^{\varepsilon s}$$
or
$$U(s) = \frac{s^2}{s^2 - 1} + \frac{e^{\varepsilon s}}{s - 1} \qquad \text{Re } s > 1$$

This yields
$$u(t) = \delta(t) + \tfrac{1}{2} 1_+(t)(e^t - e^{-t}) + 1_+(t + \varepsilon)e^{t+\varepsilon}$$

As $\varepsilon \to 0-$, this solution becomes
$$u(t) = \delta(t) + 1_+(t)(\tfrac{3}{2}e^t - \tfrac{1}{2}e^{-t})$$

Simultaneous ordinary linear differential equations with constant coefficients can be written in the form

$$\mathbf{Lu} = (\mathbf{L\delta}) * \mathbf{u} = \mathbf{g} \tag{12}$$

where $\mathbf{u}$ and $\mathbf{g}$ are the customary unknown and known $n \times 1$ matrices of distributions, respectively. By $\mathbf{L\delta}$, we mean an $n \times n$ matrix whose general element has the form

$$L_{ik}\delta = a_m \delta^{(m)} + a_{m-1}\delta^{(m-1)} + \cdots + a_0$$

Here, the integer m and the constant coefficients a_ν ($\nu = 0, 1, \ldots, m$) depend upon the choice of i and k, of course. The Laplace transform of $\mathbf{L\delta}$ is an $n \times n$ matrix $\mathbf{F}(s)$ all of whose elements are polynomials in s.

Concerning the determinant of $\mathbf{F}(s)$, two possibilities now arise. First of all, det $\mathbf{F}(s)$ may be identically zero for all values of s. In this case, neither will $\mathbf{F}(s)$ have an ordinary inverse nor will $\mathbf{f}(t)$ have an inverse in $\mathfrak{D}'_{R,n\times n}$. Thus, by Theorem 6.2-3, (12) will not have a solution in $\mathfrak{D}'_{R,n\times 1}$ for every $\mathbf{g}$ in $\mathfrak{D}'_{R,n\times 1}$. The other possibility is that det $\mathbf{F}(s) \not\equiv 0$ is a polynomial in s. [The case where det $\mathbf{F}(s)$ is a nonzero constant is allowed here as a polynomial of zero degree.] Since a polynomial has only a finite number of roots, there will be some half-plane Re $s > \sigma_I$ in which det $\mathbf{F}(s)$ is never zero. Consequently, $\mathbf{F}(s)$ possesses an inverse $\mathbf{F}^{-1}(s)$ for Re $s > \sigma_I$. $\mathbf{F}^{-1}(s)$ will be the Laplace transform of some matrix in $\mathfrak{D}'_{R,n\times n}$, according to Theorem 8.4-1. Hence,

$$(\mathbf{L\delta})^{*-1} = \mathfrak{L}^{-1}\mathbf{F}^{-1}(s) \qquad \text{Re } s > \sigma_I \tag{13}$$

and the solution in $\mathfrak{D}'_{R,n\times 1}$ of (12) is

$$\mathbf{u} = (\mathbf{L\delta})^{*-1} * \mathbf{g} \tag{14}$$

As before, if $\mathfrak{L}\mathbf{g} \triangleq \mathbf{G}(s)$ (Re $s > \sigma_g$) exists, we may use the following expression to evaluate **u**:

$$\mathbf{u} = \mathcal{L}^{-1}[\mathbf{F}^{-1}(s)\mathbf{G}(s)] \qquad \text{Re } s > \sup(\sigma_I, \sigma_g) \tag{15}$$

It is worthwhile to see how these conclusions conform with the classical solutions of our differential equations and how initial conditions can be introduced. Our discussion here will be a concise version of that given in Sec. 6.3 for a single differential equation. We shall restrict ourselves to the case where there are only two simultaneous equations in two unknown variables. Essentially the same ideas apply to any number of simultaneous equations.

Consider the 2×2 matrix of differential operators with constant coefficients

$$\mathbf{L} = \begin{bmatrix} L_{11} & L_{12} \\ L_{21} & L_{22} \end{bmatrix} \tag{16}$$

where

$$L_{11} = a_m D^m + a_{m-1} D^{m-1} + \cdots + a_0$$
$$L_{12} = b_p D^p + b_{p-1} D^{p-1} + \cdots + b_0$$
$$L_{21} = c_q D^q + c_{q-1} D^{q-1} + \cdots + c_0$$
$$L_{22} = e_r D^r + e_{r-1} D^{r-1} + \cdots + e_0$$
$$D \triangleq \frac{d}{dt}$$

and assume that det $(\mathbf{L}\delta)$ has an inverse in $\mathfrak{D}'_R$. If we apply $\mathbf{L}$ to the vector

$$\mathbf{u}(t) 1_+(t - t_0) = \begin{bmatrix} u_1(t) 1_+(t - t_0) \\ u_2(t) 1_+(t - t_0) \end{bmatrix} \tag{17}$$

where u_1 and u_2 are functions that are infinitely smooth, we shall obtain

$$\mathbf{L}[\mathbf{u}(t) 1_+(t - t_0)]$$
$$= \begin{bmatrix} 1_+(t-t_0)L_{11}u_1(t) + \sum_{\nu=0}^{m-1} \alpha_\nu \delta^{(\nu)}(t-t_0) + 1_+(t-t_0)L_{12}u_2(t) + \sum_{\nu=0}^{p-1} \beta_\nu \delta^{(\nu)}(t-t_0) \\ 1_+(t-t_0)L_{21}u_1(t) + \sum_{\nu=0}^{q-1} \gamma_\nu \delta^{(\nu)}(t-t_0) + 1_+(t-t_0)L_{22}u_2(t) + \sum_{\nu=0}^{r-1} \varepsilon_\nu \delta^{(\nu)}(t-t_0) \end{bmatrix}$$
$$\tag{18}$$

where

$$\begin{aligned} \alpha_\nu &= a_{\nu+1} u_1(t_0) + a_{\nu+2} u_1^{(1)}(t_0) + \cdots + a_m u_1^{(m-1-\nu)}(t_0) \\ \beta_\nu &= b_{\nu+1} u_2(t_0) + b_{\nu+2} u_2^{(1)}(t_0) + \cdots + b_p u_2^{(p-1-\nu)}(t_0) \\ \gamma_\nu &= c_{\nu+1} u_1(t_0) + c_{\nu+2} u_1^{(1)}(t_0) + \cdots + c_q u_1^{(q-1-\nu)}(t_0) \\ \varepsilon_\nu &= e_{\nu+1} u_2(t_0) + e_{\nu+2} u_2^{(1)}(t_0) + \cdots + e_r u_2^{(r-1-\nu)}(t_0) \end{aligned} \tag{19}$$

If $m = 0$ (that is, if $L_{11} = a_0$), the summation $\sum_{\nu=0}^{m-1}$ is replaced by zero on the right-hand side of (18); the same situation holds for the other summations on the right-hand side of (18).

Now, if **u** is a solution of the homogeneous system $\mathbf{Lu} = \mathbf{0}$ (such a solution is infinitely smooth), then (18) simplifies into

$$\mathbf{L}[\mathbf{u}(t)1_+(t-t_0)] = \begin{bmatrix} \sum_{\nu=0}^{m-1} \alpha_\nu \delta^{(\nu)}(t-t_0) + \sum_{\nu=0}^{p-1} \beta_\nu \delta^{(\nu)}(t-t_0) \\ \sum_{\nu=0}^{q-1} \gamma_\nu \delta^{(\nu)}(t-t_0) + \sum_{\nu=0}^{r-1} \varepsilon_\nu \delta^{(\nu)}(t-t_0) \end{bmatrix} \quad (20)$$

Moreover, $\mathbf{u}(t)$ may be determined over the interval $t_0 < t < \infty$ by solving (20) in the convolution algebra $\mathfrak{D}'_R$ since det $(\mathbf{L}\delta)$ is assumed to have an inverse in $\mathfrak{D}'_R$. Note that the right-hand side of (20) is completely determined by the values of u_1 and u_2 and their derivatives at $t = t_0$ that appear in (19). However, one important fact must be noted here. *It is usually impossible to choose these values independently of one another.* Only some of them may be chosen arbitrarily, and the rest will be determined by the system $\mathbf{Lu} = \mathbf{0}$. Thus, (20) indicates how a solution of the homogeneous system $\mathbf{Lu} = \mathbf{0}$ may be specified by its initial conditions at $t = t_0$. It does not indicate which of these initial conditions one is free to choose arbitrarily. Actually, the number of initial conditions that may be so chosen is equal to the highest order among the derivatives of δ appearing in det $(\mathbf{L}\delta)$. (A discussion of this is given in Frazer, Duncan, and Collar [1], chaps. V and VI.)

Example 9.3-2 Let us solve

$$\begin{aligned} (D^2 + 1)u_1 - (3D + 1)u_2 &= 0 \\ u_1 + u_2 &= 0 \end{aligned} \quad (21)$$

under some initial conditions. (These equations can certainly be solved by eliminating u_2 and analyzing the resulting ordinary differential equation. We shall not solve them in that way, however, because our purpose here is to illustrate the preceding discussion.) Since

$$\det (\mathbf{L}\delta) = \delta^{(2)} + 3\delta^{(1)} + 2\delta$$

we may assign two initial conditions at, say, $t = 0$. Clearly, these initial conditions must satisfy $u_1(0) = -u_2(0)$. Thus, we may choose

$$u_1(0) = u_1^{(1)}(0) = 1$$

and conclude that $u_2(0) = -1$. By substituting these into (20) and applying the Laplace transformation, we get

$$\begin{bmatrix} s^2 + 1 & -3s - 1 \\ 1 & 1 \end{bmatrix} \begin{bmatrix} U_1(s) \\ U_2(s) \end{bmatrix} = \begin{bmatrix} s + 4 \\ 0 \end{bmatrix}$$

which when solved yields

$$\begin{bmatrix} u_1(t) \\ u_2(t) \end{bmatrix} = \mathfrak{L}^{-1} \begin{bmatrix} \dfrac{s + 4}{(s + 1)(s + 2)} \\ \dfrac{-s - 4}{(s + 1)(s + 2)} \end{bmatrix} = 1_+(t) \begin{bmatrix} 3e^{-t} - 2e^{-2t} \\ -3e^{-t} + 2e^{-2t} \end{bmatrix}$$

This solution satisfies (21) for $t > 0$ and our initial conditions.

On the other hand, if we had chosen (incorrectly)

$$u_1(0) = u_1^{(1)}(0) = u_2(0) = 1$$

then the solution of (20) would turn out to be

$$\begin{bmatrix} u_1(t) \\ u_2(t) \end{bmatrix} = 1_+(t) \begin{bmatrix} -3e^{-t} + 4e^{-2t} \\ 3e^{-t} - 4e^{-2t} \end{bmatrix}$$

This satisfies (21) for $t > 0$. However, its initial values are not the incorrectly chosen values but rather $u_1(0) = 1$, $u_1^{(1)}(0) = -5$, and $u_2(0) = -1$, which incidentally satisfy the requirement that

$$u_1(0) = -u_2(0)$$

Consider now the nonhomogeneous equations

$$\begin{aligned} L_{11}u_1 + L_{12}u_2 &= g_1 \\ L_{21}u_1 + L_{22}u_2 &= g_2 \end{aligned} \tag{22}$$

where g_1 and g_2 are in $\mathfrak{D}'_R$. If **g** has its support in $t_0 \leq t < \infty$ and if **Lu** = **g** is solvable in $\mathfrak{D}'_R$, then its solution **u** will also have its support in $t_0 \leq t < \infty$. We may augment this solution by adding any complementary solution (i.e., any solution to **Lu** = **0**). One such complementary solution may be specified by appropriately assigning initial conditions at the point $t = t_0 - \varepsilon$ ($\varepsilon > 0$). Moreover, if the equation **Lu** = **g** is required to hold only over the interval $t_0 - \varepsilon < t < \infty$, these initial conditions may be incorporated into the convolution equations by replacing **Lu** = **g** by

$$\begin{aligned} L_{11}u_1 + L_{12}u_2 &= g_1 + \sum_{\nu=0}^{m-1} \alpha_\nu \delta^{(\nu)}(t - t_0 + \varepsilon) + \sum_{\nu=0}^{p-1} \beta_\nu \delta^{(\nu)}(t - t_0 + \varepsilon) \\ L_{21}u_1 + L_{22}u_2 &= g_2 + \sum_{\nu=0}^{q-1} \gamma_\nu \delta^{(\nu)}(t - t_0 + \varepsilon) + \sum_{\nu=0}^{r-1} \varepsilon_\nu \delta^{(\nu)}(t - t_0 + \varepsilon) \end{aligned} \tag{23}$$

where the α_ν, β_ν, γ_ν, and ε_ν are given by (19) (with t_0 replaced by $t_0 - \varepsilon$). As usual, if g_1 and g_2 are locally integrable functions, ε may be set equal to zero in (23). The procedure for solving (23) may again be facilitated by applying the Laplace transformation in the same way we did for (12). Let us repeat the warning that the initial conditions in (19) must be chosen consistently with the differential equations $\mathbf{L}u = 0$.

PROBLEMS

1 Solve the following differential equations with the indicated initial conditions by using the Laplace transformation:
 a. $3u^{(2)} + 2u^{(1)} + u = 1_+(t)$
   $u(0) = 1 \qquad u^{(1)}(0) = 2$
 b. $u^{(3)} + 3u^{(2)} + 3u^{(1)} + u = \delta^{(4)}(t-1)$
   $u(0) = u^{(2)}(0) = 1 \qquad u^{(1)}(0) = 0$
 c. $u^{(4)} + 1 = 1_+(t+2)\sinh t$
   $u(-2) = u^{(1)}(-2) = u^{(2)}(-2) = 0 \qquad u^{(3)}(-2) = -1$

2 For three simultaneous ordinary linear differential equations (with constant coefficients) in three unknowns, develop the general expressions corresponding to (20) and (23).

3 Solve the following simultaneous differential equations in $\mathfrak{D}'_R$:
 a. $(D+1)u_1 + Du_2 = 1_+(t)\sin t$
   $Du_1 + (D+1)u_2 = 0$
 b. $(D+1)u_1 + u_2 = \mathrm{Pf}\,\dfrac{1_+(t)}{t}$
   $Du_1 + u_2 = \delta^{(1)}(t)$
 c. $(D^2 + 2D + 1)u_1 + u_2 + Du_3 = \delta(t)$
   $Du_1 + Du_2 + u_3 = \delta(t)$
   $D^2 u_2 + u_3 = \delta(t)$

4 For each set of equations in Prob. 3 verify that the following initial values at $t = -1$ may be assigned arbitrarily:
 a. $u_1(-1)$
 b. None
 c. $u_1(-1)$, $u_2(-1)$, $u_3(-1)$

Then determine the other initial values at $t = -1$ that, in turn, are needed to determine the complementary solution by means of (20) or the result developed in Prob. 2.

★9.4 ORDINARY LINEAR INTEGRODIFFERENTIAL EQUATIONS WITH CONSTANT COEFFICIENTS

An ordinary linear integrodifferential equation with constant coefficients, which relates an unknown function $u(t)$ to a given continuous function $g(t)$, has the form

$$Lu = a_n u^{(n)} + \cdots + a_0 u + a_{-1} u^{(-1)} + \cdots + a_{-m} u^{(-m)} = g \quad (1)$$

Here $u^{(-k)}$ denotes as usual a primitive of kth order of u. We shall always assume that $a_n \neq 0$, $a_{-m} \neq 0$, $n \geq 0$, $m > 0$. If the change of variable

$$v = u^{(-m)} \qquad u = v^{(m)}$$

is made, (1) becomes the ordinary differential equation

$$a_n v^{(n+m)} + \cdots + a_0 v^{(m)} + a_{-1} v^{(m-1)} + \cdots + a_{-m} v = g$$

which may be solved by our previously established techniques. A unique solution becomes determinable upon the specification at $t = t_0$ of v (or equivalently of $u^{(-m)}$) and its first $n + m - 1$ derivatives. Thus, in comparison with the problem of solving an ordinary differential equation, that of solving an ordinary linear integrodifferential equation with constant coefficients presents no essentially new difficulties.

Now let us suppose that g is a distribution in $\mathfrak{D}'_R$ rather than just a function and let us seek in $\mathfrak{D}'_R$ a solution to (1). We shall do this by working with the primitives of u that are in $\mathfrak{D}'_R$. (Since for any given order k the kth-order primitives of u differ by polynomials, only one of them can be in $\mathfrak{D}'_R$.) Thus, we may write

$$u^{(-k)} = 1_+^{*k} * u = \left[1_+(t) \frac{t^{k-1}}{(k-1)!} \right] * u(t) \qquad k = 1, 2, 3, \ldots \quad (2)$$

Here, 1_+^{*k} denotes the convolution of 1_+ with itself $k - 1$ times ($k > 1$) and $1_+^{*1} \triangleq 1_+$. Thus, (1) becomes

$$(L\delta) * u = (a_n \delta^{(n)} + \cdots + a_0 \delta + a_{-1} 1_+ + \cdots + a_{-m} 1_+^{*m}) * u = g$$

The Laplace transform of $L\delta$ is

$$F(s) \triangleq \mathfrak{L}(L\delta) = a_n s^n + \cdots + a_0 + \frac{a_{-1}}{s} + \cdots + \frac{a_{-m}}{s^m}$$
$$\text{Re } s > 0 \quad (3)$$

so that the inverse in $\mathfrak{D}'_R$ of $L\delta$ is

$$(L\delta)^{*-1} = \mathfrak{L}^{-1} \frac{1}{F(s)}$$
$$= \mathfrak{L}^{-1} \frac{s^m}{a_n s^{m+n} + \cdots + a_0 s^m + a_{-1} s^{m-1} + \cdots + a_{-m}}$$
$$\text{Re } s > \sigma_I \quad (4)$$

where σ_I is the largest of the real parts of all the poles of $1/F(s)$. The solution we seek is now computable from

$$u = (L\delta)^{*-1} * g$$

If g is also Laplace-transformable and in $\mathfrak{D}'_R$, then, as usual, we may write

$$G(s) \triangleq \mathfrak{L}g \qquad \text{Re } s > \sigma_g$$

$$u = \mathfrak{L}^{-1} \frac{G(s)}{F(s)} \qquad \text{Re } s > \sup(\sigma_I, \sigma_g) \tag{5}$$

If g is zero for $-\infty < t < t_0$, then (5) is that solution which is also zero for $-\infty < t < t_0$. Any other solution will not be zero in this interval and may be obtained by adding to (5) a solution of $Lu = 0$. This allows us to specify initial conditions on u at $t = t_0 - \varepsilon$ ($\varepsilon > 0$). More specifically, we are free to state initial values on $u^{(-m)}$ and its first $n + m - 1$ derivatives as $t \to (t_0 - \varepsilon)+$. That is, as t approaches $t_0 - \varepsilon$ through values larger than $t_0 - \varepsilon$, we specify that the $u^{(k)}(t)$ ($k = -m, -m+1, \ldots, n-1$) approach the following values, respectively:

$$u_0^{(-m)}, \ldots, u_0, u_0^{(1)}, \ldots, u_0^{(n-1)} \tag{6}$$

In this case the procedure for getting u is the same as that used for Sec. 9.3, Eqs. (8) to (10). Over the interval $t_0 - \varepsilon < t < \infty$ the solution of $Lu = 0$ is found from

$$u = \mathfrak{L}^{-1} \left[\frac{1}{F(s)} \sum_{\nu=0}^{m+n-1} b_\nu s^\nu e^{-(t_0-\varepsilon)s} \right] \qquad \text{Re } s > \sigma_I \tag{7}$$

where now $F(s)$ is given by (3) and

$$b_\nu = a_{\nu-m+1} u_0^{(-m)} + a_{\nu-m+2} u_0^{(-m+1)} + \cdots + a_n u_0^{(n-1-\nu)} \tag{8}$$

By adding the right-hand sides of (5) and (7), we finally obtain

$$u = \mathfrak{L}^{-1} \left\{ \frac{1}{F(s)} \left[G(s) + \sum_{\nu=0}^{m+n-1} b_\nu s^\nu e^{-(t_0-\varepsilon)s} \right] \right\} \qquad \text{Re } s > \sup(\sigma_I, \sigma_g) \tag{9}$$

which is the solution that satisfies $Lu = g$ over the interval $t_0 - \varepsilon < t < \infty$ and converges to the initial conditions (6) as $t \to (t_0 - \varepsilon)+$.

The reader should carefully note that (7) and therefore (9) will in general contain the delta functional and some of its derivatives at $t = t_0 - \varepsilon$. When computing $u^{(-k)}$ after u has been found, the expression (2) should be used. The delta functional and its derivatives will then produce in the interval $t_0 - \varepsilon < t < \infty$ just those primitives of u which satisfy the given initial conditions.

Example 9.4-1 Let us solve in $\mathfrak{D}'_R$ the equation

$$u^{(1)} - 2u + u^{(-1)} = \text{Pf} \frac{1_+(t)}{t^{\frac{1}{2}}} \tag{10}$$

The Laplace transformation yields

$$\left(s - 2 + \frac{1}{s}\right) U(s) = -2\sqrt{\pi s}$$

So the solution in $\mathcal{D}'_R$ is

$$u(t) = \mathcal{L}^{-1}\left(-2\sqrt{\pi s}\,\frac{s}{s^2 - 2s + 1}\right) \qquad \text{Re } s > 1$$

$$= \left[\operatorname{Pf} \frac{1_+(t)}{t^{\frac{3}{2}}}\right] * [(t+1)e^t 1_+(t)] \tag{11}$$

This solution is zero over $-\infty < t < 0$.

If we also impose the initial conditions that, as $t \to (-1)+$, $u^{(-1)} \to u_0{}^{(-1)} = 2$ and $u \to u_0 = 3$, then (9) yields

$$u(t) = \mathcal{L}^{-1}\left[\frac{s}{s^2 - 2s + 1}\left(-2\sqrt{\pi s} + \sum_{\nu=0}^{1} b_\nu s^\nu e^s\right)\right]$$

where

$$b_0 = a_0 u_0{}^{(-1)} + a_1 u_0 = -1$$
$$b_1 = a_1 u_0{}^{(-1)} = 2$$

Hence,

$$u(t) = \left[\operatorname{Pf} \frac{1_+(t)}{t^{\frac{3}{2}}}\right] * [(t+1)e^t 1_+(t)]$$
$$+ (t+4)e^{t+1} 1_+(t+1) + 2\delta(t+1) \tag{12}$$

This solution satisfies (10) only over $-1 < t < \infty$ and over $-\infty < t < -1$, whereas (11) satisfies (10) for all t. However, from (12) we have that, as $t \to (-1)+$, $u \to 3$ and $u^{(-1)} = 1_+ * u \to 2$, as was required.

The analysis for simultaneous ordinary linear integrodifferential equations with constant coefficients proceeds without difficulty, and we shall merely summarize the pertinent formulas. Assuming that $\mathbf{u}$ and $\mathbf{g}$ are in $\mathcal{D}'_{R,n\times 1}$, such a system may be written in matrix notation as

$$\mathbf{L}\mathbf{u} = (\mathbf{L}\delta) * \mathbf{u} = \mathbf{g} \tag{13}$$

where the general element of $\mathbf{L}\delta$ is

$$L_{ik}\delta = a_n \delta^{(n)} + \cdots + a_0 \delta + a_{-1} 1_+ + \cdots + a_{-m}(1_+)^{*m}$$

By taking the Laplace transform, we obtain

$$\mathcal{L}(\mathbf{L}\delta) = \mathbf{F}(s) \qquad \text{Re } s > 0 \tag{14}$$

and
$$\mathfrak{L}(L_{ik}\delta) = F_{ik}(s) = a_n s^n + \cdots + a_0 + \frac{a_{-1}}{s} + \cdots + \frac{a_{-m}}{s^m}$$
$$\text{Re } s > 0$$

Hence,
$$(\mathbf{L}\delta)^{*-1} = \mathfrak{L}^{-1}\mathbf{F}^{-1}(s) \qquad \text{Re } s > \sigma_I$$
and
$$\mathbf{u} = (\mathbf{L}\delta)^{*-1} * \mathbf{g} \tag{15}$$

If $\mathfrak{L}\mathbf{g} = \mathbf{G}(s)$ (Re $s > \sigma_g$), then
$$\mathbf{u} = \mathfrak{L}^{-1}[\mathbf{F}^{-1}(s)\mathbf{G}(s)] \qquad \text{Re } s > \sup{(\sigma_I, \sigma_g)} \tag{16}$$

For these expressions to hold, $\mathbf{F}^{-1}(s)$ must exist, which means that det $\mathbf{F}(s)$ must not be identically zero for all s.

When $\mathbf{g} = 0$ for $-\infty < t < t_0$, the solutions (15) and (16) will also be zero over this interval. On the other hand, initial conditions on $\mathbf{u}$ for $t \to (t_0 - \varepsilon)+$ ($\varepsilon > 0$) may be introduced by adding to $\mathbf{u}$ over the interval $t_0 - \varepsilon < t < \infty$ some solution of $\mathbf{Lu} = 0$. For simplicity, let us again restrict ourselves to two equations in two unknowns. Consider

$$\mathbf{Lu} = \begin{bmatrix} L_{11} & L_{12} \\ L_{21} & L_{22} \end{bmatrix} \begin{bmatrix} u_1 \\ u_2 \end{bmatrix} = \begin{bmatrix} g_1 \\ g_2 \end{bmatrix} = \mathbf{g}$$

where
$$L_{11}\delta = a_{m_1}\delta^{(m_1)} + \cdots + a_0\delta + a_{-1}1_+ + \cdots + a_{-m_2}(1_+)^{*m_2}$$
$$L_{12}\delta = b_{p_1}\delta^{(p_1)} + \cdots + b_0\delta + b_{-1}1_+ + \cdots + b_{-p_2}(1_+)^{*p_2}$$
$$L_{21}\delta = c_{q_1}\delta^{(q_1)} + \cdots + c_0\delta + c_{-1}1_+ + \cdots + c_{-q_2}(1_+)^{*q_2}$$
$$L_{22}\delta = e_{r_1}\delta^{(r_1)} + \cdots + e_0\delta + e_{-1}1_+ + \cdots + e_{-r_2}(1_+)^{*r_2}$$

The solution of $\mathbf{Lu} = \mathbf{g}$, which holds for $t_0 - \varepsilon < t < \infty$ and satisfies initial conditions on u_1, u_2, and on some of their derivatives and primitives as $t \to (t_0 - \varepsilon)+$, is obtained by solving

$$L_{11}u_1 + L_{12}u_2 = g_1 + \sum_{\nu=0}^{m_1+m_2-1} \alpha_\nu \delta^{(\nu)}(t - t_0 + \varepsilon)$$
$$+ \sum_{\nu=0}^{p_1+p_2-1} \beta_\nu \delta^{(\nu)}(t - t_0 + \varepsilon)$$
$$L_{21}u_1 + L_{22}u_2 = g_2 + \sum_{\nu=0}^{q_1+q_2-1} \gamma_\nu \delta^{(\nu)}(t - t_0 + \varepsilon) \tag{17}$$
$$+ \sum_{\nu=0}^{r_1+r_2-1} \varepsilon_\nu \delta^{(\nu)}(t - t_0 + \varepsilon)$$

where

$$\begin{aligned}
\alpha_\nu &= a_{\nu-m_2+1} u_{10}{}^{(-m_2)} + a_{\nu-m_2+2} u_{10}{}^{(-m_2+1)} + \cdots + a_{m_1} u_{10}{}^{(m_1-1-\nu)} \\
\beta_\nu &= b_{\nu-p_2+1} u_{20}{}^{(-p_2)} + b_{\nu-p_2+2} u_{20}{}^{(-p_2+1)} + \cdots + b_{p_1} u_{20}{}^{(p_1-1-\nu)} \\
\gamma_\nu &= c_{\nu-q_2+1} u_{10}{}^{(-q_2)} + c_{\nu-q_2+2} u_{10}{}^{(-q_2+1)} + \cdots + c_{q_1} u_{10}{}^{(q_1-1-\nu)} \\
\varepsilon_\nu &= e_{\nu-r_2+1} u_{20}{}^{(-r_2)} + e_{\nu-r_2+2} u_{20}{}^{(-r_2+1)} + \cdots + e_{r_1} u_{20}{}^{(r_1-1-\nu)}
\end{aligned} \quad (18)$$

it being understood that in (18) $u_{10}{}^{(k)}$ and $u_{20}{}^{(k)}$ are the initial values approached by $u_1{}^{(k)}$ and $u_2{}^{(k)}$ as $t \to (t_0 - \varepsilon)+$. As with simultaneous differential equations, we are not free to choose arbitrarily all the initial conditions indicated in (18). Instead, they must be consistent with the equations

$$L_{11} u_1 + L_{12} u_2 = 0$$
$$L_{21} u_1 + L_{22} u_2 = 0$$

PROBLEMS

1 Solve the following equations under the indicated initial conditions. [Here, $u^{(k)}(-1+)$ indicates the limit of $u^{(k)}$ as $t \to -1$ through values larger than -1.]

  a. $u - u^{(-4)} = \delta(t - 1)$
$$u^{(-1)}(0+) = u^{(-2)}(0+) = u^{(-3)}(0+) = u^{(-4)}(0+) = 1$$

  b. $u^{(1)} - 2u + u^{(-1)} = e^t \operatorname{Pf} \dfrac{1_+(t)}{t^{\frac{3}{2}}}$

$$u(-1+) = 2 \qquad u^{(-1)}(-1+) = -1$$

  c. $u^{(2)} - 8u + 16 u^{(-2)} = e^{t^2} 1_+(t)$
$$u^{(1)}(-1+) = u^{(-1)}(-1+) = 1 \qquad u(-1+) = u^{(-2)}(-1+) = -1$$

2 Develop the expressions that correspond to (17) and (18) in the case where there are three simultaneous equations in three unknowns.

3 Solve in $\mathcal{D}'_R$ the following simultaneous equations. In these equations $D^{-k} u$ denotes, as usual, $u^{(-k)}$.

  a. $(1 + D^{-2}) u_1 - (2 D^{-1} + D^{-2}) u_2 = t 1_+(t)$
$$D^{-1} u_1 + D^{-2} u_2 = \delta^{(1)}\left(t + \frac{1}{2}\right)$$

  b. $(1 + D^{-1}) u_1 + u_2 - D^{-1} u_3 = \delta^{(1)}(t)$
$$D^{-1} u_1 + u_2 + 2 D^{-1} u_3 = 0$$
$$D^{-1} u_1 - D^{-1} u_2 + u_3 = 0$$

4 For each set of equations in Prob. 3, state a set of initial conditions at $t = -1+$ that may be chosen arbitrarily. Set every such initial value equal to one. Then obtain a new solution that satisfies the integrodifferential equations for $-1 < t < \infty$ and approaches these initial values as $t \to -1+$.

★9.5 ORDINARY LINEAR DIFFERENCE EQUATIONS WITH CONSTANT COEFFICIENTS: THE CONTINUOUS - VARIABLE CASE

An ordinary linear difference equation with constant coefficients is an equation that relates a given distribution g to a finite linear combination of certain translates of an unknown distribution u, as follows:

$$Ju(t) \triangleq a_n u(t + nT) + a_{n-1} u(t + nT - T) \\ + \cdots + a_1 u(t + T) + a_0 u(t) = g(t) \quad (1)$$

The adjective "ordinary" again signifies that the independent variable is one-dimensional. Here the a_ν ($\nu = 0, 1, \ldots, n$) are complex constants and T is a positive constant. We shall always assume that $n \geq 1$, $a_n \neq 0$, and $a_0 \neq 0$.

When u and g are restricted to being ordinary functions, we may consider t as a discrete variable taking on the values $t = \mu T$, where μ is a varying integer. In this type of problem, the solution $u(t)$ to (1) is defined only at the points $t = \mu T$. On the other hand, we may consider t as a continuous variable and seek a solution $u(t)$ that satisfies (1) for all t in some interval. Any solution to this more general problem will also be a solution to the discrete type of problem. However, if u and g are allowed to be distributions, then t must be interpreted as a continuous variable. In this section we shall take up the continuous-variable problem. In the next one we shall show how the discrete-variable problem can be solved by converting it into a continuous-variable one.

As in the case of ordinary differential equations, it is possible to impose initial conditions on the solution. However, in this case these initial conditions will take the form of a function or a distribution, as we shall see below. Before considering the technique for introducing these initial conditions, we shall first discuss how (1) may be solved, assuming that u and g have their supports bounded on the left.

First of all, let us take note of a step-by-step procedure. When t is replaced by $t - nT$, (1) becomes

$$u(t) = \frac{1}{a_n} [g(t - nT) - a_{n-1} u(t - T) - a_{n-2} u(t - 2T) \\ - \cdots - a_0 u(t - nT)] \quad (2)$$

This is a *recursion formula*, determining the distribution $u(t)$ over some interval $c < t < c + T$ in terms of its behavior over the n preceding intervals $c - T < t < c$, $c - 2T < t < c - T$, ..., $c - nT < t < c - (n-1)T$ and the behavior of $g(t)$ over $c - nT < t < c - (n-1)T$. Thus, if we assume, in particular, that $g(t)$ is zero over $-\infty < t < d$, then (2), combined with the condition that $u(t)$ is zero for all sufficiently small

t, determines $u(t)$ over the successive intervals $-\infty < t < d + nT$, $d + nT < t < d + (n+1)T, d + (n+1)T < t < d + (n+2)T, \ldots$ in the following way:

$$u(t) = 0 \qquad -\infty < t < d + nT$$

$$u(t) = \frac{1}{a_n} g(t - nT) \qquad d + nT < t < d + nT + T$$

$$u(t) = \frac{1}{a_n}[g(t - nT) - a_{n-1}u(t - T)]$$

$$= \frac{1}{a_n} g(t - nT) - \frac{a_{n-1}}{a_n{}^2} g(t - nT - T)$$

$$d + nT + T < t < d + nT + 2T$$

$$u(t) = \frac{1}{a_n}[g(t - nT) - a_{n-1}u(t - T) - a_{n-2}u(t - 2T)]$$

$$= \frac{1}{a_n} g(t - nT) - \frac{a_{n-1}}{a_n{}^2} g(t - nT - T)$$

$$+ \left(\frac{a_{n-1}^2}{a_n{}^3} - \frac{a_{n-2}}{a_n{}^2}\right) g(t - nT - 2T)$$

$$d + nT + 2T < t < d + nT + 3T$$

. . . .

Since the real number d may be decreased by any amount, this procedure determines $u(t)$ over any open interval of length T, at least in principle; for arbitrarily large t, we would have to perform an arbitrarily large number of steps of this recursion technique. We proved in Sec. 1.8 that the local behavior of a distribution uniquely determines the distribution for all t. Consequently, this procedure indirectly determines $u(t)$ as a distribution over $\mathcal{R}^1$.

We shall not pursue this approach any further; instead we shall discuss a method that uses the Laplace transformation to solve (1) in the convolution algebra $\mathcal{D}'_R$. This will lead to a solution that directly specifies $u(t)$ as a distribution over all of $\mathcal{R}^1$ rather than merely over arbitrary open intervals of length T.

Note that (1) may be written in the form

$$(J\delta) * u = [a_n \delta(t + nT) + a_{n-1}\delta(t + nT - T)$$
$$+ \cdots + a_0 \delta(t)] * u(t)$$
$$= g(t) \qquad (3)$$

As usual, the problem is essentially that of finding $(J\delta)^{*-1}$. The reciprocal of the Laplace transform of $J\delta$ is

$$\frac{1}{\mathfrak{L}(J\delta)} = \frac{1}{a_n e^{nTs} + a_{n-1} e^{(n-1)Ts} + \cdots + a_0}$$

$$= \frac{e^{-nTs}}{a_n + a_{n-1}e^{-Ts} + \cdots + a_0 e^{-nTs}} \qquad (4)$$

Since $T > 0$, there exists a half-plane $\operatorname{Re} s \geq b$ such that

$$|a_{n-1}e^{-Ts} + a_{n-2}e^{-2Ts} + \cdots + a_0 e^{-nTs}| < |a_n| - \varepsilon \tag{5}$$

where $\varepsilon > 0$. Therefore, $\mathfrak{L}(J\delta)$ has no zeros over this half-plane, which implies that its reciprocal is analytic there. The inequality (5) also implies that

$$\frac{1}{|\mathfrak{L}(J\delta)|} < \frac{e^{-nT \operatorname{Re} s}}{\varepsilon} \qquad \operatorname{Re} s \geq b$$

Hence, $[\mathfrak{L}(J\delta)]^{-1}$ satisfies the hypothesis of Corollary 8.4-1a and is therefore the Laplace transform of a distribution in $\mathfrak{D}'_R$.

An expression for $(J\delta)^{*-1}$ can be obtained as follows. It is permissible to divide $\mathfrak{L}(J\delta)$ into unity, starting with the term $a_n e^{nTs}$, in order to get a series expansion for $[\mathfrak{L}(J\delta)]^{-1}$ that is convergent (in the ordinary sense) over the half-plane $\operatorname{Re} s \geq b$. Thus,

$$\frac{1}{\mathfrak{L}(J\delta)} = \sum_{\nu=n}^{\infty} c_\nu e^{-\nu Ts} \tag{6}$$

By formally taking the inverse Laplace transform term by term, we get

$$(J\delta)^{*-1} = \sum_{\nu=n}^{\infty} c_\nu \delta(t - \nu T) \tag{7}$$

Since the right-hand side of (6) converges for $s = b$, it follows that $|c_\nu| e^{-\nu Tb}$ is bounded for all ν. Moreover,

$$e^{-tb} \sum_{\nu=n}^{\infty} c_\nu \delta(t - \nu T) = \sum_{\nu=n}^{\infty} e^{-\nu Tb} c_\nu \delta(t - \nu T)$$

Because of the boundedness of the $e^{-\nu Tb}|c_\nu|$, the last series converges in $\mathcal{S}'$. This means that the partial sums of the right-hand side of (7) satisfy the hypothesis of Theorem 8.3-3 and, therefore, the Laplace transform of the right-hand side of (7) truly is the right-hand side of (6). By the uniqueness of the Laplace transformation, we can conclude that (7) is a valid expression for $(J\delta)^{*-1}$.

Consequently, (1) always possesses a unique solution in $\mathfrak{D}'_R$ given by

$$u(t) = g * (J\delta)^{*-1} = g(t) * \sum_{\nu=n}^{\infty} c_\nu \delta(t - \nu T)$$

$$= \sum_{\nu=n}^{\infty} c_\nu g(t - \nu T) \tag{8}$$

where the coefficients c_ν are determined by the expansion (6). Note that, if t is restricted to some bounded open interval, (8) is actually a finite series because $g(t - \nu T)$ will be zero for all sufficiently large ν. Moreover,

(8) has essentially the same form as the solution obtained by the step-by-step procedure. It also possesses the same disadvantage; namely, we do not have an explicit expression for the constants c_ν for arbitrarily large ν.

However, if (4) is expanded into partial fractions before the inverse Laplace transformation is applied, it becomes possible to determine explicit expressions for the constants c_ν. Letting $\zeta = e^{Ts}$, we can write (4) as

$$\frac{1}{\mathfrak{L}(J\delta)} = \frac{1}{a_n \zeta^n + a_{n-1} \zeta^{n-1} + \cdots + a_0}$$

$$= \frac{1}{a_n (\zeta - \gamma_1)^{k_1} (\zeta - \gamma_2)^{k_2} \cdots (\zeta - \gamma_q)^{k_q}}$$

$$= \sum_{\mu=1}^{q} \sum_{\nu=1}^{k_\mu} \frac{b_{\mu\nu}}{(\zeta - \gamma_\mu)^\nu}$$

where

$$\sum_{\mu=1}^{q} k_\mu = n$$

and where the γ_μ are the poles of $1/\mathfrak{L}(J\delta)$, the k_μ are the corresponding multiplicities, and the $b_{\mu\nu}$ are the coefficients of the partial-fraction expansion. Moreover, each fraction can be expanded into a series of inverse powers of ζ. This yields

$$\frac{1}{\mathfrak{L}(J\delta)} = \sum_{\mu=1}^{q} \sum_{\nu=1}^{k_\mu} \sum_{\eta=0}^{\infty} \frac{(\nu + \eta - 1)! b_{\mu\nu} \gamma_\mu^\eta}{(\nu - 1)! \eta! \zeta^{\nu+\eta}}$$

[This series converges absolutely for all ζ outside some sufficiently large circle in the ζ plane and can therefore be rearranged according to similar powers of $1/\zeta$. If this is done, the coefficients of all terms up to the $(n-1)$st term will be found to be zero; that this must be so is indicated by (6) and the fact that a power-series expansion is unique.] Finally, the substitution of e^{Ts} for ζ and the application of the inverse Laplace transformation for a right-sided half-plane of convergence yield

$$(J\delta)^{*-1} = \sum_{\mu=1}^{q} \sum_{\nu=1}^{k_\mu} \sum_{\eta=0}^{\infty} \frac{(\nu + \eta - 1)!}{(\nu - 1)! \eta!} b_{\mu\nu} \gamma_\mu^\eta \delta(t - \nu T - \eta T) \qquad (9)$$

Thus, the solution in $\mathfrak{D}'_R$ of (1) is

$$u(t) = g * (J\delta)^{*-1}$$

$$= \sum_{\mu=1}^{q} \sum_{\nu=1}^{k_\mu} \sum_{\eta=0}^{\infty} \frac{(\nu + \eta - 1)!}{(\nu - 1)! \eta!} b_{\mu\nu} \gamma_\mu^\eta g(t - \nu T - \eta T) \qquad (10)$$

Although (9) and (10) are cumbersome, their coefficients are given explicitly, in contrast to (7) and (8). Here again, when t is restricted to a finite open interval, $g(t - \nu T - \eta T)$ will be zero for all sufficiently large ν, so that (10) will have only a finite number of terms.

Example 9.5-1 Let us solve in $\mathfrak{D}'_R$ the difference equation

$$u(t + 2) + 2u(t + 1) + u(t) = \delta(t + 2) + 3\delta(t + 1) \tag{11}$$

By taking the Laplace transform, we obtain

$$U(s) = \frac{e^{2s} + 3e^s}{e^{2s} + 2e^s + 1}$$

The series expansion for $U(s)$ in increasing powers of e^{-s} is

$$U(s) = 1 + \sum_{\nu=1}^{\infty} (-1)^{\nu-1}(2\nu - 1)e^{-\nu s}$$

Upon taking the inverse Laplace transform term by term, we find the solution to be

$$u(t) = \delta(t) + \sum_{\nu=1}^{\infty} (-1)^{\nu-1}(2\nu - 1)\delta(t - \nu) \tag{12}$$

We turn now to the problem of introducing initial conditions on the solution $u(t)$. Instead of specifying the right-hand member of (1) for all t, we could specify $u(t)$ for some initial period of time. In particular, assume that the difference equation $Ju = 0$ holds for $d < t < \infty$. Let us try to find in $\mathfrak{D}'_R$ a solution that satisfies the initial condition

$$u(t) = v(t) \qquad -\infty < t < d + nT \tag{13}$$

where $v(t)$ is a given right-sided distribution. [Actually, $v(t)$ need be specified only for $-\infty < t < d + nT$.]

By applying the difference operator J to $u(t)$ and using (13), we obtain

$$Ju(t) = a_n v(t + nT) + a_{n-1} v(t + nT - T) + \cdots + a_0 v(t)$$
$$-\infty < t < d \tag{14}$$

Now, if $v(t)$ is an integrable function in some sufficiently small neighborhood of each of the points $t_\nu = d + \nu T$ ($\nu = 0, 1, \ldots, n$), then the quantity $v(t)1_+(t_\nu - t)$ will possess a sense. In that case we may extend (14) into a difference equation that holds for all t by writing

$$Ju(t) = (J\delta) * u$$
$$= [a_n v(t + nT) + a_{n-1} v(t + nT - T)$$
$$+ \cdots + a_0 v(t)]1_+(d - t) \tag{15}$$

Thus, we have converted our initial-condition type of problem into a convolution equation in the algebra $\mathfrak{D}'_R$. The solution of (15) will be unique in $\mathfrak{D}'_R$. Therefore, it follows from the way we have set up (15) that $u(t)$ will satisfy the initial condition (13). Moreover, since the right-hand side of (15) has its support contained in $-\infty < t \leq d$, $u(t)$ will satisfy $Ju = 0$ for $d < t < \infty$.

It is worth emphasizing here that our method of setting up (15) has restricted somewhat the type of initial condition that may be imposed on $u(t)$, namely, that $v(t)$ is assumed to be an integrable function over some neighborhood of each of the points $t_\nu = d + \nu T$ ($\nu = 0, 1, \ldots, n$). In many instances, the latter requirement can be satisfied by choosing the constant d appropriately.

Example 9.5-2 Let us consider the same difference operator as that of Example 9.5-1, but this time we shall specify an initial condition on $u(t)$. In particular, we shall solve the difference equation

$$Ju(t) = u(t+2) + 2u(t+1) + u(t) = 0 \qquad 0 < t < \infty \qquad (16)$$

under the initial condition $u(t) = v(t)$ for $-\infty < t < 2$, where

$$v(t) = \delta(t) + \delta(t-1)$$

Note that $n = 2$ and that $T = 1$ in this case. If we choose $d = -\tfrac{1}{2}$, then $v(t)$ is an integrable function (in fact, it is zero) in certain neighborhoods of the points $t_\nu = d + \nu T$ ($\nu = 0, 1, 2$). By substituting the appropriate quantities into (15), we get

$$(J\delta) * u = [\delta(t+2) + 3\delta(t+1) + 3\delta(t) + \delta(t-1)]1_+(-t - \tfrac{1}{2})$$
$$= \delta(t+2) + 3\delta(t+1)$$

This equation is precisely the same as (11). Its solution (12) clearly satisfies the given initial conditions. Moreover, $Ju = 0$ for $t > -\tfrac{1}{2}$, so that (16) is satisfied.

Nonhomogeneous difference equations, whose solutions are required to satisfy certain initial conditions, can also be solved if we combine our previous techniques. Let $g(t)$ and $v(t)$ be distributions in $\mathfrak{D}'_R$. Moreover, let the support of $g(t)$ be contained in $d \leq t < \infty$. Once again, let $v(t)$ be an integrable function in certain neighborhoods of the points

$$t_\nu = d + \nu T \ (\nu = 0, 1, \ldots, n)$$

Then the solution in $\mathfrak{D}'_R$ of the difference equation

$$Ju(t) \triangleq a_n u(t + nT) + a_{n-1} u(t + nT - T) + \cdots + a_0 u(t)$$
$$= g(t) + [a_n v(t + nT) + a_{n-1} v(t + nT - T)$$
$$+ \cdots + a_0 v(t)]1_+(d - t)$$

will satisfy $Ju(t) = g(t)$ for $d < t < \infty$ and $u(t) = v(t)$ for $-\infty < t < d + nT$.

The extension of these results to simultaneous difference equations proceeds in the usual way. Consider

$$\mathbf{Ju} = (\mathbf{J}\delta) * \mathbf{u} = \mathbf{g} \tag{17}$$

where again $\mathbf{u}$ and $\mathbf{g}$ are, respectively, unknown and known $n \times 1$ matrices of distributions in $\mathcal{D}'_R$, and $\mathbf{J}\delta$ is an $n \times n$ matrix whose elements have the general form

$$J_{ik}\delta(t) = a_m\delta(t + mT) + a_{m-1}\delta(t + mT - T) + \cdots + a_0\delta(t)$$

Here the integer m and the complex constants a_ν depend upon i and k, and they may all be zero for some (but not all) of the elements. The Laplace transform $\mathbf{F}(s)$ of $\mathbf{J}\delta$ is an $n \times n$ matrix whose elements are polynomials in e^{sT}. If det $\mathbf{F}(s)$ is not zero for all s, then $\mathbf{F}(s)$ possesses an inverse $\mathbf{F}^{-1}(s)$ every element of which is the ratio of two polynomials in e^{sT}. By the same reasoning as that which was applied to (4), we can conclude that every such element is the Laplace transform of a distribution in $\mathcal{D}'_R$. Thus,

$$(\mathbf{J}\delta)^{*-1} = \mathcal{L}^{-1}\mathbf{F}^{-1}(s) \qquad \mathrm{Re}\, s > b$$

Every element of this matrix is either zero or can be expanded into a (finite or infinite) series of shifted delta functionals.

The solution in $\mathcal{D}'_{R,n\times 1}$ of (17) is

$$\mathbf{u} = (\mathbf{J}\delta)^{*-1} * \mathbf{g}$$

If $\mathbf{g}$ is Laplace-transformable and if $\mathbf{G}(s) = \mathcal{L}\mathbf{g}$, this can also be written as

$$\mathbf{u} = \mathcal{L}^{-1}[\mathbf{F}^{-1}(s)\mathbf{G}(s)] \qquad \mathrm{Re}\, s > a$$

The initial-condition problem for simultaneous difference equations can be converted into a simultaneous set of convolution equations by using the same procedure as that employed for a single equation. However, the difficulty that arose with simultaneous differential equations arises here also. These initial conditions on the various unknown distributions cannot be chosen independently of one another; for if they were so chosen, they would, in general, be inconsistent with the given difference equations.

PROBLEMS

1 When converting the initial-condition problem into a convolution equation in $\mathcal{D}'_R$, we assumed that $Ju(t) = 0$ for $d < t < \infty$ and that $u(t)$ is specified only over $-\infty < t < d + nT$. What will happen to the solution $u(t)$ for $d + nT < t < \infty$ if the specification of $u(t)$ for $-\infty < t < d$ is altered arbitrarily (without violating the condition that its support is bounded on the left)? Can we maintain the assumption that $Ju(t) = 0$ for $d < t < \infty$ and at the same time specify $u(t)$ in an arbitrary manner over the entire interval $-\infty < t < b$, where $b > d + nT$?

2 Solve the following difference equations with the indicated initial conditions. Give explicit expressions for all the coefficients occurring in your answers.

a. $u(t + 2) - 3u(t + 1) + u(t) = \text{Pf } \dfrac{1_+(t)}{t} \quad -1 < t < \infty$

$u(t) = 0 \quad -\infty < t < 1$

b. $u(t + 2) + 2u(t + 1) + 2u(t) = 0 \quad 0 < t < \infty$

$u(t) = \begin{cases} 0 & -\infty < t \leq 0 \\ 1 & 0 < t < 2 \end{cases}$

c. $u(t + 3) + u(t + 2) - u(t + 1) - u(t) = t \quad 0 < t < \infty$

$u(t) = \delta(t) + \delta(t - 1) + \delta(t - 2) \quad 0 < t < 3$

3 Solve in $\mathfrak{D}'_R$ the following pair of simultaneous equations:

$$u_1(t + 1) - u_1(t) + u_2(t + 1) - 2u_2(t) = \delta(t)$$

$$u_1(t + 1) + u_1(t) + u_2(t + 1) - u_2(t) = \text{Pf } \dfrac{1_+(t)}{t}$$

4 A *difference-differential equation* is one that contains both the translates and derivatives of an unknown distribution $u(t)$. A simple example of such an equation is

$$u^{(1)}(t + 2) + u(t) = g(t) \tag{18}$$

More precisely, $Mu(t) = g(t)$ is said to be an *ordinary linear difference-differential equation with constant coefficients* if t is one-dimensional, $u(t)$ and $g(t)$ are unknown and known distributions, respectively, and Mu is of the form

$$Mu = \sum_{\mu=1}^{q} \sum_{\nu=1}^{k_\mu} a_{\mu\nu} u^{(\nu)}(t - t_\mu) \tag{19}$$

where the $a_{\mu\nu}$ are complex constants and the t_μ are real constants. Our technique for solving convolution equations in $\mathfrak{D}'_R$ can also be applied to equations of this type, but now it is, in general, more difficult to determine whether or not $M\delta$ has an inverse in $\mathfrak{D}'_R$.

a. Solve (18) in $\mathfrak{D}'_R$ under the assumption that $g(t) = 1_+(t) \sin t$.

b. Assume that $g(t) = 0$ for $0 < t < \infty$ and that $u(t) = 1_+(t) - 1_+(t - 1)$ for $-\infty < t < 2$. Find a solution to (18) that holds for $0 < t < \infty$.

c. How much freedom does one have in choosing initial conditions for general equations of the form $Mu = 0$? Illustrate your answer with some examples.

★9.6 ORDINARY LINEAR DIFFERENCE EQUATIONS WITH CONSTANT COEFFICIENTS: THE DISCRETE-VARIABLE CASE

The type of equation that will be considered in this section can be written in the form

$$Ku_\mu \triangleq a_n u_{\mu+n} + a_{n-1} u_{\mu+n-1} + \cdots + a_0 u_\mu = g_\mu$$
$$\mu = \ldots, -1, 0, 1, \ldots; n \geq 1; a_n \neq 0; a_0 \neq 0 \tag{1}$$

Sec. 9.6 Differential and difference equations 287

where the a_ν ($\nu = 0, 1, \ldots, n$) are fixed complex constants, the g_μ comprise a given sequence of complex numbers, and the u_μ comprise an unknown sequence of complex numbers, which we seek. Here, the discrete variable μ replaces the continuous variable t in the difference equations of the preceding section. As before, (1) is called an "ordinary" equation because μ is one-dimensional.

This discrete-variable type of problem can be solved by converting it into a continuous-variable one, as follows. First of all, let us assume that the g_μ and u_μ are zero for all sufficiently small μ in order to use the convolution algebra $\mathfrak{D}'_R$. Then, let $p(t)$ be the pulse function

$$p(t) \triangleq 1_+(t) - 1_+(t - 1)$$

Let us replace the numbers g_μ by the pulse functions $g_\mu p(t - \mu)$ and the unknown discrete variable $u_{\mu+\nu}$ by the unknown distribution $u(t + \nu)$. Then the continuous-variable analogue to (1) is

$$Ju(t) \triangleq a_n u(t + n) + a_{n-1} u(t + n - 1) + \cdots + a_0 u(t)$$
$$= \sum_{\mu = -\infty}^{\infty} g_\mu p(t - \mu) \tag{2}$$

As usual, we shall discuss two distinct cases of this general problem. In the first case the g_μ are given and the initial conditions on the u_μ are zero; in the second one (2) holds only for positive μ, the g_μ are zero, and some nonzero initial conditions on the u_μ are specified.

Consider the first case. Here, the g_μ are given for all μ and are zero for those μ that are less than some fixed integer, which we can always take to be zero by simply renumbering the g_μ. By using the results of the preceding section, we can write down the following solution to (2):

$$u(t) = (J\delta)^{*-1} * \sum_{\mu=0}^{\infty} g_\mu p(t - \mu)$$
$$= \Big[\sum_{\nu=n}^{\infty} c_\nu \delta(t - \nu) \Big] * \Big[\sum_{\mu=0}^{\infty} g_\mu p(t - \mu) \Big]$$
$$= c_n g_0 p(t - n) + (c_n g_1 + c_{n+1} g_0) p(t - n - 1)$$
$$\quad + (c_n g_2 + c_{n+1} g_1 + c_{n+2} g_0) p(t - n - 2) + \cdots \tag{3}$$

Here, the c_ν are the coefficients of the series expansion of Sec. 9.5, Eq. (7). [As before, explicit expressions for the c_ν can be obtained by comparing (7) and (9) of that section.] We see from (3) that the solution $u(t)$ is the regular distribution corresponding to a step function that is constant over the intervals $\mu < t < \mu + 1$ ($\mu = \ldots, -1, 0, 1, \ldots$) and is zero for $-\infty < t < n$.

Now, let $\phi(t)$ be a testing function in $\mathfrak{D}$ whose support is contained in the arbitrary interval $\mu < t < \mu + 1$ and let

$$\int_{-\infty}^{\infty} \phi(t) \, dt = 1$$

If we apply the regular distribution corresponding to each side of (2) to $\phi(t)$, (2) becomes (1). Moreover, the application of the right-hand side of (3) to $\phi(t)$ yields the coefficient of the pulse $p(t - \mu)$. It follows that the sequence of coefficients in the right-hand side of (3) is the solution to (1). This latter sequence is

$$u_\mu = 0 \quad \mu = \ldots, n-3, n-2, n-1$$
$$u_\mu = c_n g_{\mu-n} + c_{n+1} g_{\mu-n-1} + \cdots + c_\mu g_0$$
$$\mu = n, n+1, n+2, \ldots \quad (4)$$

Moreover, (4) is the only solution to (1) whose values are all zero for sufficiently small μ. This is because, for every sequence of numbers that satisfies (1), there will be one and only one corresponding sequence of pulses that satisfies (2) and because (2) has only one solution in $\mathfrak{D}'_R$.

Now for the case when the initial conditions are not zero. Here the g_μ are unspecified for $\mu = \ldots, -3, -2, -1$ and are assumed to be zero for $\mu = 0, 1, 2, \ldots$ and the u_μ are specified for $\mu = \ldots, n-3, n-2, n-1$. Actually, each value of u_μ depends only on the n preceding values for u_μ. Thus, once u_μ is fixed for $\mu = 0, 1, \ldots, n-1$, we may set $u_\mu = 0$ for $\mu = \ldots, -3, -2, -1$ without affecting its unknown values for $\mu = n, n+1, n+2, \ldots$. Assuming these initial conditions, let us find the unknown values of u_μ that satisfy the equation $Ku_\mu = 0$ for $\mu = 0, 1, 2, \ldots$.

Turning to the continuous-variable analogue of this problem, we can set

$$v(t) = u_0 p(t) + u_1 p(t-1) + \cdots + u_{n-1} p(t-n+1)$$

and then apply Sec. 9.5, Eq. (15), to obtain

$$Ju(t) = \sum_{\mu=0}^{n-1} b_\mu p(t + \mu + 1) \tag{5}$$

where

$$b_\mu = a_{\mu+1} u_0 + a_{\mu+2} u_1 + \cdots + a_n u_{n-\mu-1}$$

[Note the similarity between these coefficients b_μ and those which were given by Sec. 6.3, Eq. (13), and which served an analogous role for differential equations.] By solving (5) in $\mathfrak{D}'_R$ in the usual way, we obtain

$$u(t) = \left[\sum_{\nu=n}^{\infty} c_\nu \delta(t - \nu) \right] * \left[\sum_{\mu=0}^{n-1} b_\mu p(t + \mu + 1) \right]$$

We can convolve these series term by term and rearrange the result according to the pulse functions $p(t - \mu)$. Their coefficients are found to be

$$u_\mu = 0 \quad \mu = \ldots, -3, -2, -1$$
$$u_\mu = c_n b_{n-\mu-1} + c_{n+1} b_{n-\mu} + \cdots + c_{n+\mu} b_{n-1}$$
$$\mu = 0, 1, \ldots, n-1 \quad (6)$$
$$u_\mu = c_{\mu+1} b_0 + c_{\mu+2} b_1 + \cdots + c_{n+\mu} b_{n-1}$$
$$\mu = n, n+1, n+2, \ldots$$

By the same reasoning as that given for (4), these coefficients comprise the one and only solution to this discrete-variable initial-condition problem.

Finally, solutions (4) and (6) can be added to obtain the solution to the general equation (1), assuming that it holds only for $\mu = 0, 1, 2, \ldots$ and that the u_μ are specified for $\mu = 0, 1, \ldots, n-1$.

PROBLEMS

1 By a direct computation show that expressions (6) become the identity $u_\mu = u_\mu$ for $\mu = 0, 1, \ldots, n-1$. This will verify that (6) truly satisfies the given initial conditions.

2 The compound-interest problem is the following. An amount of capital u_0 is invested, and it gains interest at the rate r per interest period. That is, if the accumulated capital at the beginning of any such period is u_μ ($\mu = 0, 1, 2, \ldots$), then the accumulated capital at the end of that period is $u_{\mu+1} = u_\mu(1 + r)$. Clearly, u_μ is related to u_0 through

$$u_\mu = u_0(1 + r)^\mu$$

Verify this formula by using the Laplace transformation to solve this initial-condition type of problem.

3 Solve the following discrete-type difference equations under the indicated initial conditions. Give explicit expressions for all coefficients appearing in your answers.
    a. $u_{\mu+1} - u_\mu = 1 \quad \mu = 0, 1, 2, \ldots$
       $u_0 = 0$
    b. $u_{\mu+2} - 2u_{\mu+1} + u_\mu = 0 \quad \mu = 0, 1, 2, \ldots$
       $u_0 = 1 \quad u_1 = 2$
    c. $u_{\mu+2} + 3u_{\mu+1} + u_\mu = 1 \quad \mu = 0, 1, 2, \ldots$
       $u_0 = -1 \quad u_1 = 1$

4 Extend the technique described in this section into a method for solving simultaneous ordinary linear difference equations of the discrete-variable type with constant coefficients in the case where the initial conditions are all zero.

9.7 ORDINARY LINEAR DIFFERENTIAL EQUATIONS WITH POLYNOMIAL COEFFICIENTS

In Sec. 5.4, Prob. 5, we indicated that linear differential equations with nonconstant polynomial coefficients could not be represented as convolution equations. Nevertheless, it may still be possible to solve such differential equations by using the Laplace transformation. In this section we shall illustrate this by some examples, without attempting a comprehensive discussion.

Consider

$$a_n(t)u^{(n)}(t) + a_{n-1}(t)u^{(n-1)}(t) + \cdots + a_0(t)u(t) = g(t) \tag{1}$$

where the coefficients $a_\nu(t)$ are polynomials. Assuming that u and g are Laplace-transformable, we may apply the formula

$$\mathfrak{L}[t^\mu u^{(\nu)}(t)] = (-1)^\mu \frac{d^\mu}{ds^\mu}[s^\nu U(s)] \tag{2}$$

to (1) to transform it into another ordinary linear differential equation with polynomial coefficients in the independent variable s. If m is the highest power of t in the coefficients of (1) and if $m < n$, then the transformed equation will be of lower order and probably easier to solve than the original equation. However, it may happen that the solutions of (1) are not all Laplace-transformable or that the transformed equation has solutions which are not Laplace transforms. Nevertheless, if the solution $U(s)$ of the transformed equation is a Laplace transform, then its inverse transform $u(t)$ will be a solution of the original equation.

Example 9.7-1 A special case of *Bessel's differential equation* is

$$tu^{(2)}(t) + u^{(1)}(t) + tu(t) = 0 \tag{3}$$

By applying (2), we convert this second-order differential equation into the following first-order one:

$$(s^2 + 1)U^{(1)}(s) + sU(s) = 0$$

This can be solved by separating the variables. We get

$$U(s) = \frac{C}{\sqrt{s^2 + 1}}$$

where C is a constant of integration. If we take as the region of convergence the half-plane $\operatorname{Re} s > 0$, we see that $U(s)$ is the Laplace transform of a right-sided function. In Example 8.2-3 this was found to be $CJ_0(t)1_+(t)$, where $J_0(t)$ is Bessel's function of first kind and zero order.

Actually, $J_0(t)$ satisfies (3) for all values of t, but since it does not have a Laplace transform, it is not generated by our transform method of solution. There is another linearly independent solution of (3), the Bessel function of second kind and zero order. It too is not generated by this method.

Example 9.7-2 Consider *Laguerre's differential equation*

$$tu^{(2)}(t) + (1 - t)u^{(1)}(t) + nu(t) = 0 \qquad n = 0, 1, 2, \ldots \tag{4}$$

This can be transformed into

$$(s - s^2)U^{(1)}(s) + (n + 1 - s)U(s) = 0$$

whose solution is

$$U(s) = \frac{C}{s}\left(1 - \frac{1}{s}\right)^n \tag{5}$$

where C is again a constant of integration. Clearly, (5) is a Laplace transform if we take the region of convergence to be the right-sided half-plane Re $s > 0$. By applying the binomial expansion to the right-hand side of (5) and taking the inverse Laplace transform, we get as a solution to (4)

$$u(t) = CL_n(t)1_+(t)$$

where

$$L_n(t) = \sum_{k=0}^{n} \binom{n}{k} \frac{(-t)^k}{k!}$$

The $L_n(t)$ are the *Laguerre polynomials*. They also satisfy (4) for all t but do not possess Laplace transforms. Here again, there is still another solution to (4) but it too is not Laplace-transformable.

Example 9.7-3 Let us now consider a differential equation that produces a singular distribution as a solution:

$$tu^{(2)}(t) + (t + 3)u^{(1)}(t) + u(t) = 0$$

The transformed equation is

$$(s^2 + s)U^{(1)}(s) - sU(s) = 0$$

which has the solution

$$U(s) = C(s + 1)$$

C again being the constant of integration. This is a Laplace transform over the entire s plane. Thus,

$$u(t) = C[\delta^{(1)}(t) + \delta(t)]$$

PROBLEMS

1 Find solutions to the following differential equations by using the Laplace transformation:
  a. $tu^{(1)}(t) + 2u(t) = 0$
  b. $tu^{(2)}(t) - tu^{(1)}(t) - u(t) = 0$
  c. $t^2 u^{(2)}(t) + 4tu^{(1)}(t) + 2u(t) = \delta(t)$

2 The Mellin transformation can be used to solve certain ordinary linear differential equations with polynomial coefficients. This transformation was described in Sec. 8.3, Prob. 16. As an indication of this, apply the Mellin transformation to

$$tu^{(1)}(t) + 2u(t) = \delta(t - 1) \tag{6}$$

Assume that $u(t)$ has its support contained in $0 < t < \infty$ and has a Mellin transform. Then use the result of part d in Sec. 8.3, Prob. 16, to formally obtain the solution $u(t)$. Substitute this solution into (6) in order to check it.

10 Passive Systems

10.1 INTRODUCTION

When analyzing a physical system, one may adopt either of two distinct points of view that may be called the "microscopic" and the "macroscopic" attitudes. In the microscopic approach, every part of the system is analyzed, which requires a knowledge of the internal structure of the system and of the mathematical relationships between all of its parts. For example, it is the microscopic attitude that is adopted when one analyzes an electrical network by means of Kirchhoff's laws or a mechanical device by means of Newton's laws. (See, for example, Dolezal and Vorel [1] and Dolezal [1].)

On the other hand, in many practical applications the complete internal behavior of the system is unimportant. Instead, the relationships between only a few of the variables is all that is required. It is then natural to adopt the macroscopic attitude by replacing the system by a mathematical characterization that relates only those variables which are of interest. It may happen that the development of this characterization may require a microscopic analysis; but once it has been achieved, the internal behavior of the system can be ignored. This is what is accomplished when a system is specified by means of its transfer function.

It is the macroscopic point of view that we shall adopt in this chapter. Our purpose is to discuss the fundamental properties of the mathematical characterizations of certain physical systems. Although a number of relationships between these properties will be developed, our primary conclusion will be the connection between the physical property of passivity and the mathematical property of positive-reality. A number of analyses achieve this conclusion without being restricted to special classes of systems such as those having lumped elements. (See Raisbeck [1], Youla, Castriota, and Carlin [1 and 2], König and Meixner [1], Wohlers and Beltrami [1], and Zemanian [4 and 5].) The one presented here is based on distribution theory and encompasses the usual simplifications that arise from it.

Furthermore, it will be assumed that there are only two quantities of interest: a driving variable $f(t)$ and a responding variable $v(t)$. The essential ideas of this analysis can be extended to the case where there are n driving and n responding variables (Zemanian [4]). This leads to a matrix analysis, in contrast to the development presented here.

Finally, it should be pointed out that the "physical properties," which we shall define in the next two sections, are in fact mathematical idealizations of certain types of physical behavior. Moreover, the physical interpretations that will be assigned to these definitions are really unessential for the subsequent development. Indeed, the reader may skip over the interpretive discussions; he need only take note of the definitions for "one-port," "convolution operator," "passivity," and "causality" and of the lemmas of Sec. 10.3 before proceeding with Sec. 10.4.

10.2 ONE-PORTS HAVING CONVOLUTION REPRESENTATIONS

Consider a physical system that sets up a correspondence between some driving variable $f(t)$ and some responding variable $v(t)$. (For the sake of definiteness, we may assume that t represents time.) By ignoring all other variables, we may characterize the system as an *operator* $\mathfrak{N}$ *in* $\mathfrak{D}'$. We have already introduced such operators in Sec. 5.8; let us now review some notation and terminology associated with them. We write

$$v = \mathfrak{N}f \quad f \in D(\mathfrak{N}) \quad v \in R(\mathfrak{N})$$

to indicate that $\mathfrak{N}$ assigns the distribution v to the distribution f. The set of all f, on which $\mathfrak{N}$ is defined, is the domain $D(\mathfrak{N})$ of $\mathfrak{N}$, and the set of all v corresponding to those in $D(\mathfrak{N})$ is the range $R(\mathfrak{N})$ of $\mathfrak{N}$. In general, $\mathfrak{N}$ may assign more than one v to each f.

A *real distribution* g is one that assigns a real number $\langle g, \phi \rangle$ to each real-valued testing function ϕ in $\mathfrak{D}$. An example is the delta functional. Moreover, an operator $\mathfrak{N}$ in $\mathfrak{D}'$ is said to be *real* if $v = \mathfrak{N}f$ is a real distribution whenever f is a real distribution. Since this is a characteristic of physical systems, we shall restrict our attention to such operators and, borrowing a terminology that is common in electrical network theory, we shall refer to them as "one-ports." More specifically, *by a one-port we shall mean a real operator in* $\mathfrak{D}'$.

Let us give two practical examples of one-ports.

Example 10.2-1 Figure 10.2-1 shows an electrical network having only two external terminals to which electrical connections can be made. It is understood here that the only way one may transfer energy from the outside into the network is through these connections. Thus, magnetic coupling between an external and an internal inductive coil is prohibited. The quantity $v(t)$ represents the voltage drop across the two terminals, and $f(t)$ is the current entering one terminal and leaving the other. In our generalized point of view, both $v(t)$ and $f(t)$ may be distributions. Depending upon the internal configuration, the network may or may not be able to absorb a particular distributional current $f(t)$. Moreover, if it can absorb this $f(t)$, there may be precisely one voltage response $v(t)$ or perhaps many (in fact, an infinite number of) different possible voltage responses. In any case, we can describe this network by a mathematical operator. The set of all currents $f(t)$ that can be absorbed by the network comprises the domain of the operator, and the set of all corresponding voltage responses constitutes its range. The actual correspondence between these currents and voltages defines the operator itself.

Here we have assumed that the current $f(t)$ is the driving variable and the voltage $v(t)$ is the responding one. If we switch the roles, the network will define a (usually) different operator which assigns one or perhaps many currents $f(t)$ to each voltage $v(t)$ that can be applied to the network.

We have also assumed that the current enters and leaves through

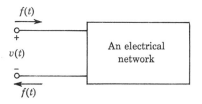

Fig. 10.2-1 An electrical network acting as a one-port.

the same two terminals across which the voltage is applied. This need not be the case, since the network will generally set up a correspondence between any voltage drop and any current occurring in it. However, the quantity Re $\bar{v}f$ (assuming that it exists as a distribution) will represent an electrical power transfer only if v and f occur at the same two terminals. Subsequently, when defining "passivity" for the network, we shall be concerned with the power transferred into the network. For this reason we shall restrict ourselves to the situation indicated in Fig. 10.2-1, where v and f occur at the same place. Indeed, in electrical network theory, the word "one-port" implies precisely this restriction.

Example 10.2-2 An example of a mechanical one-port is illustrated in Fig. 10.2-2. Here, the mechanical device is enclosed in a box from which only one piston protrudes. The only way one can externally apply power to the system is to apply a force $f(t)$ to the piston. The responding velocity of the piston is $v(t)$ and, because f and v occur at the same place, the power absorbed is Re $\bar{v}f$. Here again the correspondence between f and v, which is determined by the internal mechanical system, defines a mathematical operator.

In this chapter we shall concern ourselves only with one-ports that are *convolution operators*. In agreement with the definition of such operators given in Sec. 5.8, we shall say that *the one-port $\mathfrak{N}$ is a convolution operator $w *$ when*

$$v = \mathfrak{N}f = w * f \tag{1}$$

where w is a fixed distribution in $\mathfrak{D}'$, and if $D(\mathfrak{N})$ contains the set of all f for which $w * f$ exists in the sense of the distributional convolutions described in Theorems 5.4-1 and 5.7-1.

If $f = \delta$, then $v = w$. Consequently, the fact that $\mathfrak{N} \triangleq w *$ is a real operator implies that w is a real distribution. Moreover, w is called the

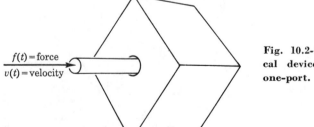

Fig. 10.2-2 A mechanical device acting as a one-port.

Sec. 10.2 Passive systems 297

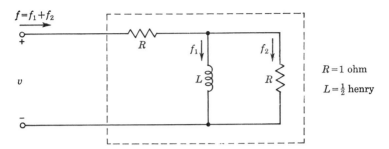

Fig. 10.2-3 An electrical one-port consisting of two resistors R and an inductor L.

unit impulse response of the one-port, the *unit impulse* being another name for δ commonly used in the technical literature.

From Theorem 5.4-1 we see that, for every w in $\mathfrak{D}'$, $D(w *)$ always contains $\mathcal{E}'$. Furthermore, if w is in $\mathfrak{D}'_R$ (or in $\mathfrak{D}'_L$), then $D(w *)$ contains $\mathfrak{D}'_R$ (or, respectively, $\mathfrak{D}'_L$). If w is in $\mathcal{E}'$, then $D(w *)$ is identical with $\mathfrak{D}'$. Finally, if w is in $\mathcal{S}'$, then $D(w *)$ contains $\mathcal{S}$, as we see from Sec. 5.7.

It was shown in Sec. 5.8 that an operator in $\mathfrak{D}'$ is a convolution operator if and only if it possesses the properties of single-valuedness, linearity, commutativity with the shifting operator, and continuity. The definitions of these properties are given in Sec. 5.8. Let us say a few words about their physical implications.

Single-valuedness   This is a rather common property of physical systems. Many systems can respond in only one way to a given driving force. However, systems that are multivalued (i.e., that do not possess single-valuedness) are not rare. A very simple example of a multivalued one-port is the electrical open circuit. The only current f that can be sent through an open circuit is the zero current, whereas any voltage v may occur across it. Thus, if we take current as the driving variable, the domain of the open circuit contains only one distribution, the zero distribution. Its range, however, is all of $\mathfrak{D}'$. If we take into consideration systems that are more complicated than one-ports, the occurrence of multivaluedness becomes even more frequent.

Furthermore, it is also possible for a particular physical system to act as a multivalued one-port if some arbitrary amount of energy had been placed within it at the time of its creation. For example, consider the electrical network of Fig. 10.2-3. With the indicated voltage v and currents f_1 and f_2 we may write

$$\frac{1}{2}\frac{df_1}{dt} = f_2 \tag{2}$$

and
$$v = f_1 + 2f_2 \tag{3}$$

If $f_1 = -f_2 \neq 0$, the driving current f will be zero but there will be a circulating current in the parallel circuit of L and R. Upon replacing f_1 by $-f_2$ in (2) we obtain

$$\frac{1}{2}\frac{df_2}{dt} + f_2 = 0$$

which implies that $f_2 = Be^{-2t}$, where B is an arbitrary constant. Since there is no current, and hence no voltage drop, in the series resistor, $v = f_2 = Be^{-2t}$. Thus, it is possible for the one-port to sustain this voltage v even though no driving current is applied. It follows that this one-port must be multivalued. For if v is a possible response to a given f, then $v + Be^{-2t}$ will also be a possible response for every value of B.

On the other hand, if we assume that the one-port had no initial energy stored in it at the time of its creation, then this one-port will be single-valued and, in fact, a convolution operator. Its unit impulse response will be

$$w(t) = 2\delta(t) - 2e^{-2t}1_+(t)$$

and its domain will contain the space $\mathcal{D}'_R$.

Linearity This too is a recurring property among physical systems, but most systems possess it only if their variables do not become too large in magnitude.

It should also be pointed out that our definition of linearity (see Sec. 5.8) requires that the response v of the one-port be zero when the driving variable f is zero. Many systems that one would ordinarily consider linear (such as electronic amplifiers) have constant nonzero responses when their driving variables are zero. For example, in the electrical one-port of Fig. 10.2-4, a driving current f is applied to a parallel connection of a resistor R and a constant-current source f_c. When $f = 0$,

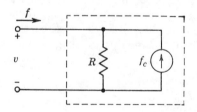

Fig. 10.2-4 An electrical one-port having a constant response $v = f_c R$ when its driving variable f is zero.

the constant voltage $f_c R$ appears at the port terminals. We can easily convert this one-port into a linear one (according to our definition) by defining the responding variable as being $v - f_c R$.

Commutativity with the shifting operator In the technical literature this property is more commonly called *time invariance* whenever the independent variable t denotes time. If the structure of the system and the values of its parameters remain fixed with time, it will not matter at what particular instant a certain driving variable is first applied; the system will always respond in exactly the same way. This is the customary interpretation of time invariance.

Continuity This property is more difficult to interpret physically. We may state it crudely by saying that the system responds in almost the same way to two different driving variables that are almost the same. To make this description precise, we would, of course, have to define what we mean by "almost the same." Since convergence is taking place in the space $\mathfrak{D}'$, this would lead naturally to a discussion of "neighborhoods" in the space $\mathfrak{D}'$. Two distributions would then be considered almost the same when they both occur in some sufficiently small neighborhood in the space $\mathfrak{D}'$. (The reader can find a discussion of such neighborhoods in Schwartz [1], vol. I, chap. III, and in Gelfand and Shilov [1], vol. 2, chaps. 1 and 2.) Let us simply emphasize at this point that this property is essential. Without it, the inevitable approximations that one makes when idealizing a physical system in order to analyze it might produce large errors.

The above four properties (or, more precisely, the physical traits that are idealized by these properties) quite often occur together in physical systems. Moreover, the satisfaction of these four properties is a necessary and sufficient condition for a one-port to be a convolution operator (see Theorem 5.8-2). Therefore, physical systems whose behavior can be approximated accurately enough by a convolution process are also not uncommon. Nevertheless, it is only fair to point out that most physical systems possess these four traits only over certain restricted ranges of their operation, if at all. For instance, there are many linear, time-invariant systems that possess neither single-valuedness nor continuity. A naturally more complicated nondistributional analysis that encompasses such degenerate systems is given by Youla, Castriota, and Carlin [1].

PROBLEM

1 Figure 10.2-5 shows a mechanical one-port consisting of a flywheel having a moment of inertia J. The driving variable is the torque f, and the responding variable is the angular velocity v. The flywheel's motion is retarded by a brake that applies

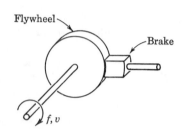

Fig. 10.2-5 A flywheel acting as a one-port.

a viscous friction. That is, the brake applies a retarding torque f_r to the flywheel that is proportional to the angular velocity v:

$$f_r = Fv$$

(F is a constant.) Find the convolution operator that characterizes this one-port. Is there any $v \neq 0$ for which $f = 0$?

10.3 CAUSALITY AND PASSIVITY

Two other fundamental properties of one-ports will be discussed in this section. The first one, called "causality," is defined as follows:

Causality *Let both f_1 and f_2 be arbitrary distributions in the domain of the one-port $\mathfrak{N}$ and let $v_1 \triangleq \mathfrak{N} f_1$ and $v_2 \triangleq \mathfrak{N} f_2$. The one-port $\mathfrak{N}$ is said to be causal (or to satisfy causality) if the condition $f_1(t) = f_2(t)$ for $t < t_0$ always implies that $v_1(t) = v_2(t)$ for $t < t_0$ and if this property holds for all real values of t_0.*

This means that the response v of the one-port over the interval $-\infty < t < t_0$ cannot depend on the future form of f (that is, for $t > t_0$); it depends only on the form of f over the interval $-\infty < t < t_0$. For if this were not the case, we could choose two driving variables f_1 and f_2 that are identical for $-\infty < t < t_0$ and are different for $t_0 < t < \infty$ in such a way that $\mathfrak{N} f_1$ would not equal $\mathfrak{N} f_2$ over some portion of the interval $-\infty < t < t_0$. This would violate our definition of causality.

One should not interpret the word "causality" too literally. It means only what is defined above and does not necessarily imply that f causes v. (The cock always crows before the sun rises, but his crowing does not cause the sun to rise.) Actually, the same comment applies to our use of the phrases "driving variable" and "responding variable."

Later on, we shall need the following two lemmas:

Lemma 1

Let the one-port $\mathfrak{N}$ be a convolution operator. A necessary and sufficient condition for $\mathfrak{N}$ to be causal is that its unit impulse response w be equal to zero for $t < 0$.

Proof: Necessity follows from the facts that $w = \Re\delta$ and that $\delta(t) = 0$ for $t < 0$.

For sufficiency, assume that $w(t) = 0$ for $t < 0$, that $v = \Re f$, and that ϕ is in $\mathfrak{D}$. We shall first show that $\langle v(t), \phi(t)\rangle$ is zero whenever $f(t) = 0$ for $t < t_0$ and the support of $\phi(t)$ is in $-\infty < t < t_0$. Consider

$$\langle v, \phi\rangle = \langle w * f, \phi\rangle = \langle w(t), \langle f(\tau), \phi(t + \tau)\rangle\rangle \tag{1}$$

If $\phi(t + \tau)$ (as a function of τ) has its support contained in $\tau < t_0 - t$ and if the support of $f(\tau)$ is contained in $t_0 \leq \tau$, then $\langle f(\tau), \phi(t + \tau)\rangle$ is certainly zero over some neighborhood of $0 \leq t < \infty$. Since the support of $w(t)$ is contained in $0 \leq t < \infty$, the right-hand side of (1) is zero. Thus, we have established that $v(t) = 0$ for $t < t_0$ whenever $f(t) = 0$ for $t < t_0$.

Now let f_1 and f_2 be in $D(\Re)$ and let $f_1(t) = f_2(t)$ for $t < t_0$. Since $\Re$ is both single-valued and linear, we may write

$$v_1 \triangleq \Re f_1 \quad v_2 \triangleq \Re f_2$$
$$v_1 - v_2 = \Re f_1 - \Re f_2 = \Re(f_1 - f_2) = 0 \quad t < t_0$$

Hence, $v_1(t) = v_2(t)$ for $t < t_0$, which shows that $\Re$ is causal. Q.E.D.

Lemma 2

If the one-port $\Re$ is a convolution operator and if $v(t) \triangleq \Re f(t) = 0$ for $t < t_0$ whenever $f(t)$ is a testing function in $\mathfrak{D}$ that equals zero for $t < t_0$, then $\Re$ is causal.

In other words, we need establish the causality property only for the subset $\mathfrak{D}$ of $D(\Re)$ in order for it to hold for all of $D(\Re)$.

Proof: For every $f(t)$ in $\mathfrak{D}$

$$v(t) = w(t) * f(t) = \langle w(\tau), f(t - \tau)\rangle \tag{2}$$

where $v(t)$ is everywhere a continuous function according to Theorem 5.5-1. Now, by definition, $w(\tau)$ will equal zero over an open interval if $\langle w(\tau), \phi(\tau)\rangle$ is zero for all ϕ in $\mathfrak{D}$ whose supports are contained in the open interval. But, if $f(t)$ has its support contained in $t_0 \leq t < \infty$, then on the τ axis the support of $f(t - \tau)$ will be in $-\infty < \tau \leq t - t_0$. Since by hypothesis $v(t) = 0$ for $t < t_0$, we can conclude from (2) that $w(\tau) = 0$ for $\tau < 0$. Lemma 1 completes the proof.

Let us now turn to the concept of passivity. Many physical systems are such that they can absorb energy but are unable to generate energy within themselves. If such a system imparts energy to its surroundings during some interval of time, it must have absorbed this energy during some previous period. This property is called *passivity*. In a one-port, energy can be absorbed or delivered by the system only through the action of the variables v and f. If $\bar{v}$ and f are ordinary functions whose product $\bar{v}f$ is integrable for all time, the total energy $\mathfrak{e}(x)$ absorbed by a one-port from its creation to the arbitrary instant of time x is

$$\begin{aligned}\mathfrak{e}(x) &= \operatorname{Re} \int_{-\infty}^{x} \overline{v(t)} f(t) \, dt \\ &= \operatorname{Re} \int_{-\infty}^{x} v(t) \overline{f(t)} \, dt\end{aligned} \qquad (3)$$

(Here we are again assuming that no initial energy was stored in the system at the time of its creation.) The passivity of the one-port can be simply described by stating that $\mathfrak{e}(x) \geq 0$ for all x. In so describing it, however, one must be careful to assign the correct polarities to the variables v and f in order for energy absorption to be a positive quantity. That is, for the electrical one-port of Fig. 10.2-1, positive current f enters the high-potential terminal for the voltage v. Similarly, in the mechanical one-port of Fig. 10.2-2, the positive orientations of the applied force f and the responding velocity v are taken in the same direction.

Another difficulty arises when $\mathfrak{e}(x)$ is defined according to (3). When $\bar{v}$ and f are integrable over $-\infty < t < x$, $\operatorname{Re} \bar{v}f$ may not be integrable, in which case $\mathfrak{e}(x)$ will not exist. Similarly, if we allow v and f to be distributions, the product symbol $\bar{v}f$ may be meaningless. To overcome this difficulty, $f(t)$ will be restricted to being a testing function in $\mathfrak{D}$ when we define passivity for a one-port. Then, if the one-port is a convolution operator, $\overline{v(t)}$ will be a continuous function and consequently $\mathfrak{e}(x)$ will exist for all x.

We are finally ready to state our precise definition of passivity.

Passivity *Let the one-port $\mathfrak{N}$ be a convolution operator. $\mathfrak{N}$ is said to be passive (or to satisfy passivity) if, for every $f(t)$ in $\mathfrak{D}$ and for $v(t) \triangleq \mathfrak{N}f(t)$, the energy function $\mathfrak{e}(x)$, given by (3), is nonnegative for all x.*

Actually, in establishing the passivity of a one-port, one need only ascertain the nonnegativity of (3) for real-valued $f(t)$ in $\mathfrak{D}$. This is because the one-port is a *real* operator, so that, if $f = f_R + if_X$, where f_R and f_X are real, then $v = v_R + iv_X$, where $v_R = \mathfrak{N}f_R$ and $v_X = \mathfrak{N}f_X$, and (3) becomes

$$\mathfrak{e}(x) = \int_{-\infty}^{x} v_R(t) f_R(t) \, dt + \int_{-\infty}^{x} v_X(t) f_X(t) \, dt \qquad (4)$$

Thus, if (3) is nonnegative for real-valued f in $\mathfrak{D}$, it is certainly nonnegative for complex-valued f in $\mathfrak{D}$.

Let us also take note of a remarkable fact that was first established by Youla, Castriota, and Carlin. (See Youla, Castriota, and Carlin [1], theorem 1.) Linearity and passivity imply causality. In our case, where we are restricting ourselves to one-ports that are convolution operators, this may be stated as follows:

Lemma 3

If a one-port $\mathfrak{N}$ is a convolution operator and is passive, then it is also causal.

Proof: Let f and f_1 be two real-valued testing functions in $\mathfrak{D}$ and let v and v_1 be their respective responses

$$v \triangleq \mathfrak{N}f \qquad v_1 \triangleq \mathfrak{N}f_1$$

Again, according to Theorem 5.5-1, v and v_1 are continuous functions. Furthermore, assume that $f(t) = 0$ for $t < t_0$. We shall show that $v(t) = 0$ for $t < t_0$ too.

Let a be an arbitrary real number and let $f_2 \triangleq f_1 + af$. Then $f_2(t) = f_1(t)$ for $t < t_0$. Moreover, if $v_2 \triangleq v_1 + av$, then, by the linearity of $\mathfrak{N}$, $v_2 = \mathfrak{N}f_2$. Also, by the passivity of $\mathfrak{N}$,

$$\int_{-\infty}^{x} v_2(t) f_2(t) \, dt \geq 0$$

For $x < t_0$, this becomes

$$\int_{-\infty}^{x} v_1(t) f_1(t) \, dt + a \int_{-\infty}^{x} v(t) f_1(t) \, dt \geq 0$$

Since this must hold for all real a, the second integral must be zero for all $x < t_0$, which in turn implies that $v(t) f_1(t) = 0$ for $t < t_0$. This means that $v(t) = 0$ whenever $f_1(t) \neq 0$ in this interval. Since $f_1(t)$ can be *any* real element in $\mathfrak{D}$, it follows that $v(t) = 0$ for $t < t_0$. By Lemma 2, it now follows that $\mathfrak{N}$ is causal.

Throughout the rest of this chapter, we shall assume that the one-port under consideration possesses the fundamental properties of single-valuedness, linearity, time invariance, continuity, passivity, and causality. In view of Theorem 5.8-2, the first four properties can be replaced by the equivalent assumption that the one-port is a convolution operator. Also, by Lemma 3, causality is a consequence of passivity for any one-port that is a convolution operator. Thus, the above six properties are subsumed by the following two postulates:

P1. *The one-port is a convolution operator.*
P2. *The one-port is passive.*

Taken together, the afore-mentioned six properties imply and are implied by these two postulates.

We shall also employ one more postulate:

P3. *The unit impulse response w of the one-port is a distribution of slow growth.*

Actually, this assumption is superfluous; it is a consequence of the two preceding postulates (Zemanian [4], sec. 4). The proof of this fact requires a knowledge of some concepts which we have not discussed. For the sake of brevity, we shall simply invoke Postulate P3, but the reader should bear in mind that it can be discarded. (See Prob. 2.)

Henceforth, in order to specify that a one-port satisfies Postulates P1, P2, and P3, we shall simply say that it is a *real passive convolution operator of slow growth.* Note that Postulates P1 and P3 together imply that the domain of the one-port contains the space $\mathcal{S}$. This is because the domain of a convolution operator $w *$ is by definition the set of all f for which $w * f$ exists in the sense of distributional convolution.

PROBLEMS

1 Demonstrate Lemma 3 in the following alternative way. With a and b arbitrary real numbers, with f_1 and f_2 arbitrary real testing functions in $\mathfrak{D}$, and with $v_1 \triangleq \mathfrak{N} f_1$ and $v_2 \triangleq \mathfrak{N} f_2$, show that

$$a^2 \int_{-\infty}^{x} v_1 f_1 \, dt + ab \int_{-\infty}^{x} \frac{v_1 f_2 + v_2 f_1}{2} dt \\ + ba \int_{-\infty}^{x} \frac{v_1 f_2 + v_2 f_1}{2} dt + b^2 \int_{-\infty}^{x} v_2 f_2 \, dt \geq 0$$

This is a symmetric nonnegative-definite quadratic form in the variables a and b. The fact that the determinant of its matrix must be nonnegative (see Mirsky [1], pp. 395, 403–405) may now be used to establish Lemma 3. (This proof is also due to Youla, Castriota, and Carlin [2], pp. 13–14.)

2 A distribution $g(t)$ is said to be nonnegative-definite if for every testing function $\phi(t)$ in $\mathfrak{D}$ the quantity

$$\int_{-\infty}^{\infty} \overline{\phi(t)} \langle g(\tau), \phi(t-\tau) \rangle \, dt = \left\langle g(\tau), \int_{-\infty}^{\infty} \overline{\phi(t)} \phi(t-\tau) \, dt \right\rangle$$

is nonnegative. Set

$$w_h(t) \triangleq \tfrac{1}{2}[w(t) + w(-t)]$$

Assuming that $w(t)$ is the unit impulse response of a one-port that satisfies Postulates P1 and P2, show that $w_h(t)$ is nonnegative-definite. Furthermore, it is a fact that every nonnegative-definite distribution is a distribution of slow growth (Schwartz [1], vol. II, p. 132). By using this fact, show that $w(t)$ is also of slow growth.

3 Let the one-port $\mathfrak{N}$ be a convolution operator whose unit impulse response $w(t)$ is Laplace-transformable. Show that necessary and sufficient conditions for $\mathfrak{N}$ to be causal are that $W(s) \triangleq \mathfrak{L} w(t)$ has a right-sided region of convergence (i.e., this region extends infinitely to the right) and that

$$|W(s)| \leq P(|s|) \qquad \text{Re } s \geq c$$

where $P(|s|)$ is some polynomial in $|s|$. HINT: The converse of Corollary 8.4-1a is also true. Use this fact.

10.4 THE POSITIVE - REALITY OF THE IMMITTANCE FUNCTION

In the technical literature the Laplace transform $W(s)$ of the unit impulse response $w(t)$ of some system is at times called the *immittance function*. This is the terminology we shall employ here. The object of this section is to show that the immittance function of any real passive convolution operator of slow growth is of a very special type known as a *positive-real function*. It is defined as follows:

A function $W(s)$ of the complex variable s is said to be positive-real if the following conditions are satisfied in the open right-half s plane (i.e., for Re $s > 0$):

1. *$W(s)$ is analytic.*
2. *$W(s)$ is real whenever s is real.*
3. *Re $W(s) \geq 0$.*

The fulfillment of the first two conditions is easily verified. According to Sec. 10.3, Lemma 3, Postulates P1 and P2 imply that the one-port is causal and, by Sec. 10.3, Lemma 1, its unit impulse response $w(t)$ is zero for $t < 0$. Postulate P3, which asserts that $w(t)$ is of slow growth, now ensures that w is Laplace-transformable and that the corresponding region of convergence contains at least the open right-half s plane. Consequently, $W(s)$ is analytic there. Finally, the one-port is a real operator by definition and w is therefore a real distribution. Hence, $W(s)$ is real for all real s in the region of convergence. We are left with condition 3, whose proof requires a bit more doing. We shall need

Lemma 1

If the one-port $\mathfrak{N}$ is a real passive convolution operator of slow growth, then the energy expression

$$\mathfrak{e}(x) \triangleq \text{Re} \int_{-\infty}^{x} v(t)\overline{f(t)}\, dt \qquad v = \mathfrak{N}f$$

exists and is nonnegative for all x even when f is an arbitrary testing function of rapid descent (that is, $f \in \mathcal{S}$).

Proof: Let $\{f_\nu\}_{\nu=1}^\infty$ be a sequence of testing functions in $\mathfrak{D}$ that converges in $\mathfrak{S}$ to the testing function f in $\mathfrak{S}$. Also, let $v_\nu = w * f_\nu$. According to Postulate P3, f is in $D(\mathfrak{N})$ and we may set $v \triangleq \mathfrak{N}f$. Moreover, by Theorem 5.7-1, there exists a positive integer p such that the sequence

$$\left\{\frac{v_\nu(t)}{(1+t^2)^{p-1}}\right\}_{\nu=1}^\infty \tag{1}$$

converges uniformly to the bounded function

$$\frac{v(t)}{(1+t^2)^{p-1}}$$

Thus, all the elements of the sequence (1) are uniformly bounded in magnitude by, say, the constant B. Also, given any $\varepsilon > 0$, there exists an integer N such that for all $\nu \geq N$

$$\frac{v(t)}{(1+t^2)^p} = \frac{v_\nu(t)}{(1+t^2)^p} + g_\nu(t)$$

and

$$(1+t^2)^p f(t) = (1+t^2)^p f_\nu(t) + h_\nu(t)$$

where the functions $g_\nu(t)$ and $h_\nu(t)$ ($\nu = N, N+1, N+2, \ldots$) are all bounded in magnitude by ε. Then

$$\left| \int_{-\infty}^x v(t)\overline{f(t)}\, dt - \int_{-\infty}^x v_\nu(t)\overline{f_\nu(t)}\, dt \right|$$

$$= \left| \int_{-\infty}^x g_\nu(t)(1+t^2)^p \overline{f(t)}\, dt + \int_{-\infty}^x \frac{v_\nu(t)\overline{h_\nu(t)}}{(1+t^2)^p}\, dt \right|$$

$$\leq \varepsilon \int_{-\infty}^x (1+t^2)^p |f(t)|\, dt + \varepsilon B \int_{-\infty}^x \frac{dt}{1+t^2}$$

Both the last two integrals are convergent, and hence the right-hand side of this inequality can be made arbitrarily small by choosing ε small enough. Since

$$\operatorname{Re} \int_{-\infty}^x v_\nu(t)\overline{f_\nu(t)}\, dt \geq 0$$

it follows that

$$\operatorname{Re} \int_{-\infty}^x v(t)\overline{f(t)}\, dt \geq 0 \quad \text{Q.E.D.}$$

We are now ready to show that $\operatorname{Re} W(s) \geq 0$ for $\operatorname{Re} s > 0$. When $f(t)$ is an arbitrary testing function in $\mathfrak{S}$, we see from Theorem 5.7-1 that

$$v(t) = \langle w(\tau), f(t-\tau)\rangle$$

By the preceding lemma,

$$0 \leq \operatorname{Re} \int_{-\infty}^{x} \overline{f(t)} \langle w(\tau), f(t-\tau) \rangle \, dt \qquad (2)$$

We may choose $f(t)$ equal to e^{st} over the interval $-\infty < t < a$, where $x < a < \infty$, so long as $\operatorname{Re} s > 0$. Since $w(\tau) = 0$ for $\tau < 0$, (2) may be converted into

$$0 \leq \operatorname{Re} \int_{-\infty}^{x} e^{\bar{s}t} \langle w(\tau), e^{s(t-\tau)} \rangle \, dt$$
$$= \int_{-\infty}^{x} e^{2 \operatorname{Re} st} \, dt \, \operatorname{Re} \langle w(\tau), e^{-s\tau} \rangle \qquad \operatorname{Re} s > 0$$

But $\langle w(\tau), e^{-s\tau} \rangle$ is simply the Laplace transform $W(s)$ of $w(\tau)$, and since

$$\int_{-\infty}^{x} e^{2 \operatorname{Re} st} \, dt > 0 \qquad \operatorname{Re} s > 0$$

it follows that $\operatorname{Re} W(s) \geq 0$ for $\operatorname{Re} s > 0$. We have thus achieved

Theorem 10.4-1

If a one-port is a real passive convolution operator of slow growth, then its immittance function $W(s)$ [i.e., the Laplace transform of its unit impulse response $w(t)$] exists and is a positive-real function.

As will be shown in Sec. 10.6, the converse of this theorem is also true. *Every positive-real function is the immittance function of some real passive convolution operator of slow growth.* This fact can be established by using a certain representation that is possessed by the unit impulse response of every such one-port. The next section is devoted to a development of this representation.

10.5 A REPRESENTATION FOR THE UNIT IMPULSE RESPONSE CORRESPONDING TO A POSITIVE-REAL IMMITTANCE FUNCTION

The desired representation for the unit impulse response w can be obtained by taking the inverse Laplace transform of Cauer's representation for a positive-real function. (See Cauer [1].)

Cauer's representation *The function $W(s)$ is positive-real if and only if, for $\operatorname{Re} s > 0$,*

$$W(s) = As + \int_{-\infty}^{\infty} \frac{s}{s^2 + \eta^2} (1 + \eta^2) \, dH(\eta) \qquad (1)$$

where A is a real nonnegative constant and $H(\eta)$ is a real odd nondecreasing bounded function.

(A concise proof of this is given by Loomis and Widder [1]. Actually, they develop a representation for a function that is harmonic and nonnegative in the upper half-plane, but their result can readily be converted into Cauer's representation.)

Let us now determine $w(t) \triangleq \mathfrak{L}^{-1}W(s)$ (Re $s > \sigma_1$). First note that (1) truly does possess an inverse Laplace transform in the distributional sense (but not, in general, in the classical sense) when the half-plane Re $s > 0$ is taken as the region of convergence or, perhaps, as part of it. For, over the fixed half-plane Re $s \geq b > 0$,

$$\left| \frac{1+\eta^2}{s^2+\eta^2} \right| < B(b)|s| \tag{2}$$

where the bound $B(b)$ is independent of η. Hence, from (1) we have that

$$|W(s)| \leq |s| \left[A + B(b)|s| \int_{-\infty}^{\infty} dH(\eta) \right]$$

and since $H(\eta)$ is a bounded function, it follows that, over any fixed half-plane Re $s \geq b > 0$, $|W(s)|$ is bounded by a quadratic function of $|s|$. By Theorem 8.4-1, $W(s)$ is truly a Laplace transform and, in addition, its inverse transform $w(t)$ has its support contained in $0 \leq t < \infty$.

To compute $w(t)$, we may use Sec. 8.3, Eq. (5), namely,

$$\mathfrak{L}w(t) = \mathfrak{F}[e^{-ct}w(t)] \tag{3}$$

where c is a real number greater than b. For ϕ in $\mathcal{S}$ and $\tilde{\phi} \triangleq \mathfrak{F}\phi$, the definition of the distributional Fourier transformation yields

$$\langle e^{-ct}w(t), \tilde{\phi}(t) \rangle = \langle W(c+i\omega), \phi(\omega) \rangle \tag{4}$$

Now, we already know that the inverse Laplace transform of As is simply $A\delta^{(1)}(t)$; so let us replace $W(c+i\omega)$ by merely the second term on the right-hand side of (1). The right-hand side of (4) then becomes

$$\int_{-\infty}^{\infty} d\omega \int_{-\infty}^{\infty} \phi(\omega) \frac{c+i\omega}{(c+i\omega)^2+\eta^2} (1+\eta^2) \, dH(\eta) \tag{5}$$

Because of our estimation (2) and the facts that $\phi(\omega)$ is in $\mathcal{S}$ and $H(\eta)$ is a bounded nondecreasing function, (5) converges absolutely and we may therefore apply Fubini's theorem (see Widder [1], theorem 15d, p. 26) to interchange the order of integration in (5). This yields

$$\int_{-\infty}^{\infty} dH(\eta) \int_{-\infty}^{\infty} \phi(\omega) \frac{c+i\omega}{(c+i\omega)^2+\eta^2} (1+\eta^2) \, d\omega \tag{6}$$

By using formula (4) to replace the inner integral of (6), we can convert (6) into

$$\lim_{Y \to \infty} \int_{-Y}^{Y} dH(\eta) \int_{-\infty}^{\infty} \tilde{\phi}(t)(1 + \eta^2)e^{-ct}1_+(t) \cos \eta t \, dt \tag{7}$$

Upon another application of Fubini's theorem, this becomes

$$\lim_{Y \to \infty} \int_{-\infty}^{\infty} dt \int_{-Y}^{Y} \tilde{\phi}(t)(1 + \eta^2)e^{-ct}1_+(t) \cos \eta t \, dH(\eta) \tag{8}$$

which can be represented symbolically by

$$\left\langle e^{-ct}1_+(t) \int_{-\infty}^{\infty} (1 + \eta^2) \cos \eta t \, dH(\eta), \tilde{\phi}(t) \right\rangle \tag{9}$$

In view of (3) and (4), we may drop the exponential factor to obtain the inverse Laplace transform of the second term on the right-hand side of (1); it is the symbolic expression

$$1_+(t) \int_{-\infty}^{\infty} (1 + \eta^2) \cos \eta t \, dH(\eta) \tag{10}$$

which is understood to be the distribution defined by (7) or, equivalently, by (8) after the factor e^{-ct} has been deleted.

This distribution is in S' because, with the factor e^{-ct} deleted, (7) converges for all $\tilde{\phi}(t)$ in S and determines a continuous linear functional on S. This can be seen from the following expression, which can be obtained from (7) through the deletion of e^{-ct} and some integrations by parts,

$$\int_0^{\infty} \left[\tilde{\phi}(t) \int_{-\infty}^{\infty} \cos \eta t \, dH(\eta) + \tilde{\phi}^{(2)}(t) \int_{-\infty}^{\infty} (1 - \cos \eta t) \, dH(\eta) \right] dt \tag{11}$$

By the reversibility of this development and the uniqueness of the Laplace transformation, we may conclude with

Theorem 10.5-1

The function $W(s)$ is positive-real if and only if its inverse Laplace transform for the half-plane $\operatorname{Re} s > 0$ is the distribution of slow growth

$$w(t) = A\delta^{(1)}(t) + 1_+(t) \int_{-\infty}^{\infty} (1 + \eta^2) \cos \eta t \, dH(\eta) \tag{12}$$

where A and $H(\eta)$ are restricted as in Cauer's representation. This symbolic expression designates the distribution that assigns to each testing function $\phi(t)$ in S the number

$$\langle w(t), \phi(t) \rangle = -A\phi^{(1)}(0)$$
$$+ \lim_{Y \to \infty} \int_0^\infty dt \int_{-Y}^Y \phi(t)(1+\eta^2) \cos \eta t \, dH(\eta)$$
$$= \langle A\delta^{(1)}(t), \phi(t) \rangle + \left\langle 1_+(t) \int_{-\infty}^\infty \cos \eta t \, dH(\eta), \phi(t) \right\rangle$$
$$+ \left\langle 1_+(t) \int_{-\infty}^\infty (1 - \cos \eta t) \, dH(\eta), \phi^{(2)}(t) \right\rangle$$

It is perhaps worth emphasizing that the integral in (12) must be interpreted in the distributional sense, since it does not converge, in general, as an ordinary integral.

PROBLEMS

1 Derive (11). Use it to show that (10) is truly a distribution of slow growth.

2 By constructing an appropriate example for $H(\eta)$, show that the integral in (12) does not always converge in the classical sense.

3 Show that, if $W(s)$ is a positive-real function, then its inverse Laplace transform $w(t)$ (for Re $s > 0$) has the form

$$w(t) = A\delta^{(1)}(t) + w_0(t)$$

where A is a nonnegative constant and $w_0(t)$ is a distribution of zero order. [Here the order of $w_0(t)$ is taken in the sense of Sec. 3.4. If we add the requirement that the even part of $w_0(t)$ be a nonnegative-definite distribution, then these conditions become both necessary and sufficient for $W(s)$ to be positive-real.]

10.6 THE REALIZABILITY OF EVERY POSITIVE-REAL FUNCTION

Following a development presented by König and Meixner [1], we shall show in this section that every positive-real function can be realized as the immittance function of some real passive convolution operator of slow growth. This is the converse of Theorem 10.4-1.

Every positive-real function generates through the inverse Laplace transformation (for the half-plane Re $s > 0$) a distribution of slow growth having a representation as specified in Theorem 10.5-1. This distribution may be taken as the unit impulse response of a one-port that is a convolution operator. Since Postulates P1 and P3 are automatically satisfied by this one-port, we need merely demonstrate its passivity. In doing this, we can restrict ourselves to real-valued elements f in $\mathfrak{D}$ without losing any generality. For each such testing function f the corresponding energy function is

$$e(x) = \int_{-\infty}^{x} f(t)\langle w(t - \tau), f(\tau)\rangle \, dt$$

By using Theorem 10.5-1, we may convert this into

$$e(x) = \int_{-\infty}^{x} f(t)\langle A\delta^{(1)}(t - \tau), f(\tau)\rangle \, dt$$
$$+ \int_{-\infty}^{x} f(t) \left[\lim_{Y \to \infty} \int_{-\infty}^{\infty} d\tau \int_{-Y}^{Y} f(\tau)(1 + \eta^2) 1_+(t - \tau) \right.$$
$$\left. \times \cos (\eta t - \eta \tau) \, dH(\eta) \right] dt \quad (1)$$

The first term on the right-hand side of (1) equals

$$A \int_{-\infty}^{x} f(t) f^{(1)}(t) \, dt = \frac{A}{2} [f(x)]^2$$

a nonnegative quantity.

Furthermore, because f is in $\mathfrak{D}$ and $H(\eta)$ is a bounded nondecreasing function, we may use the fact that for each t

$$\int_{-\infty}^{t} f(\tau) \cos (\eta t - \eta \tau) \, d\tau = O(\eta^{-2}) \qquad |\eta| \to \infty$$

and interchange the order of integration to convert the second term on the right-hand side of (1) into

$$\int_{-\infty}^{\infty} dH(\eta) \int_{-\infty}^{x} dt \int_{-\infty}^{t} (1 + \eta^2) f(t) f(\tau) \cos (\eta t - \eta \tau) \, d\tau \quad (2)$$

The integrand of (2) is symmetric in t and τ, and consequently (2) is equal to

$$\tfrac{1}{2} \int_{-\infty}^{\infty} dH(\eta) \int_{-\infty}^{x} dt \int_{-\infty}^{x} (1 + \eta^2) f(t) f(\tau) \cos (\eta t - \eta \tau) \, d\tau \quad (3)$$

By using

$$\cos (\eta t - \eta \tau) = \cos \eta t \cos \eta \tau + \sin \eta t \sin \eta \tau$$

we may convert (3) into

$$\tfrac{1}{2} \int_{-\infty}^{\infty} \left\{ \left[\int_{-\infty}^{x} f(t) \cos \eta t \, dt \right]^2 + \left[\int_{-\infty}^{x} f(t) \sin \eta t \, dt \right]^2 \right\} (1 + \eta^2) \, dH(\eta)$$

clearly another nonnegative quantity. Thus, $e(x) \geq 0$ and the one-port is therefore passive.

We have arrived at

Theorem 10.6-1

For every positive-real function there exists a real passive convolution operator of slow growth that possesses this function as its immittance function.

By combining Theorems 10.4-1, 10.5-1, and 10.6-1, we can state still another conclusion to this theory of idealized passive systems. *The representation given in Theorem* 10.5-1 *characterizes the unit impulse response of every real passive convolution operator of slow growth. Conversely, for every distribution of the form specified in Theorem* 10.5-1 *there exists a real passive convolution operator of slow growth which possesses this distribution as its unit impulse response.*

PROBLEM

1 Which of the following distributions can be the unit impulse responses of real passive convolution operators of slow growth? Justify your answers.

a. $-\text{Pf}\,\dfrac{1_+(t)}{t^{\frac{3}{2}}}$ Take the positive root when $t > 0$.

b. $\text{Pf}\,\dfrac{1_+(t)}{t}\,(e^{-t} - e^{-2t})$

c. $1_+(t)t^2$

d. $\text{Pf}\,\dfrac{1_+(t)}{t^2}$

11 Periodic Distributions

11.1 INTRODUCTION

Periodic distributions constitute a generalization of periodic functions. This is the subject we now take up. The development of this last chapter follows the broad outline of the preceding part of this book, and in a loose sense it acts as a summary of what we have done so far.

In the next two sections we introduce a space $\mathcal{P}_T$ of periodic testing functions of period T and a space $\mathcal{P}'_T$ of periodic distributions with period T. The elements of $\mathcal{P}'_T$ are then identified as continuous linear functionals on the space $\mathcal{P}_T$. Thus, $\mathcal{P}'_T$ is truly the dual of $\mathcal{P}_T$. Because of this identity between the periodic distributions and the continuous linear functionals on $\mathcal{P}_T$, many of their properties can be developed simply by adapting the corresponding properties of the distributions in $\mathcal{D}'$. This permits some brevity in the subsequent sections.

Since periodic distributions have unbounded supports, they cannot be convolved. However, a different type of convolution, T-convolution, can be defined for them, and it plays the same sort of role for the space $\mathcal{P}'_T$ as does our previous convolution for the space $\mathcal{D}'_R$. In particular, certain

314   *Distribution theory and transform analysis*   *Chap. 11*

types of equations can be represented as T-convolution equations and an algebra for solving them can be constructed. All this is discussed in Secs. 11.4 and 11.5.

Finally, Fourier series and the finite Fourier transformation are discussed in Secs. 11.6 and 11.7. These provide a useful method for solving T-convolution equations. Applications of this method for the solving of linear differential and difference equations (with constant coefficients) in the T-convolution algebra $\mathcal{P}'_T$ are presented in the last section.

As usual, all independent variables, such as t, τ, and x, are understood to be one-dimensional.

11.2 THE SPACE $\mathcal{P}_T$ OF PERIODIC TESTING FUNCTIONS

An ordinary function $f(t)$ is said to be *periodic* if there exists some real positive number T such that $f(t) = f(t - T)$ for all values of t. T is called a *period* of $f(t)$. It follows that, if a function is periodic, it will possess an infinity of periods, since nT ($n = 1, 2, 3, \ldots$) will be a period whenever T is a period. Note that by our definition all periods are positive numbers.

A function $\theta(t)$ will be called a *periodic testing function* if it is periodic and infinitely smooth. The space of all such periodic testing functions having the *common* period T (T being a *fixed* positive number) will be denoted by $\mathcal{P}_T$. $\mathcal{P}_T$ is a linear space.

Convergence in the space $\mathcal{P}_T$ is defined as follows. A sequence $\{\theta_\nu(t)\}_{\nu=1}^\infty$ is said to *converge in* $\mathcal{P}_T$ to a limit function θ if every θ_ν is in $\mathcal{P}_T$ and if for each nonnegative integer k the sequence $\{\theta_\nu^{(k)}(t)\}_{\nu=1}^\infty$ converges to $\theta(t)$ uniformly for all t. (As usual, the uniformity of the convergence must hold for each fixed k, but it need not hold over all k.) It follows that the limit function θ will also be in $\mathcal{P}_T$, and consequently $\mathcal{P}_T$ is closed under convergence.

Any testing function ϕ in $\mathfrak{D}$ generates a unique testing function θ in $\mathcal{P}_T$ through the expression

$$\theta(t) = \sum_{n=-\infty}^{\infty} \phi(t - nT) \tag{1}$$

Actually, over any bounded t interval there are only a finite number of nonzero terms in this summation because ϕ has a bounded support. Thus, we may differentiate term by term to get

$$\theta^{(k)}(t) = \sum_{n=-\infty}^{\infty} \phi^{(k)}(t - nT) \qquad k = 1, 2, 3, \ldots \tag{2}$$

Moreover, if $\{\phi_\nu\}_{\nu=1}^\infty$ converges in $\mathfrak{D}$ to ϕ and if we relate each ϕ_ν to a θ_ν in $\mathcal{P}_T$ by using (1), then $\{\theta_\nu\}_{\nu=1}^\infty$ clearly converges in $\mathcal{P}_T$ to θ.

Another type of function that we shall make use of is the so-called *unitary function* (Lighthill [1], p. 61). A function $\xi(t)$ is said to be unitary if it is an element of $\mathfrak{D}$ and if there exists a real number T for which

$$\sum_{n=-\infty}^{\infty} \xi(t - nT) = 1 \tag{3}$$

for all t. Many such functions exist. An example of one of them (for the case where $T = 1$) is

$$\xi(t) = \frac{\int_{|t|}^1 \exp \frac{-1}{x(1-x)} \, dx}{\int_0^1 \exp \frac{-1}{x(1-x)} \, dx} \qquad -1 < t < 1 \tag{4}$$
$$\xi(t) = 0 \qquad |t| \geq 1$$

The space of all functions that are unitary with respect to some fixed real number T will be denoted by $\mathfrak{U}_T$.

Here again, over any bounded t interval the number of nonzero terms on the left-hand side of (3) is finite. Consequently, we may differentiate term by term to get

$$\sum_{n=-\infty}^{\infty} \xi^{(k)}(t - nT) = 0 \qquad k = 1, 2, 3, \ldots \tag{5}$$

Clearly, if θ is in $\mathcal{P}_T$ and ξ is in $\mathfrak{U}_T$, then $\xi\theta$ is in $\mathfrak{D}$. Also, if $\{\theta_\nu\}_{\nu=1}^\infty$ converges in $\mathcal{P}_T$ to θ, then $\{\xi\theta_\nu\}_{\nu=1}^\infty$ converges in $\mathfrak{D}$ to $\xi\theta$. Each θ can be related to $\xi\theta$ according to (1) because the periodicity of θ may be employed to write

$$\sum_{n=-\infty}^{\infty} \xi(t-nT)\theta(t-nT) = \theta(t) \sum_{n=-\infty}^{\infty} \xi(t-nT) = \theta(t) \tag{6}$$

This shows that *every* θ in $\mathcal{P}_T$ can be generated through (1) from some ϕ in $\mathfrak{D}$.

PROBLEMS

1 Show that (4) is truly a unitary function.

2 Let $\xi(t)$ be in $\mathfrak{U}_T$. Show that its Fourier transform

$$\hat{\xi}(\omega) \triangleq \int_{-\infty}^{\infty} \xi(t) e^{-i\omega t} \, dt$$

satisfies

$$\xi\left(\frac{2\pi m}{T}\right) = \begin{cases} T & m = 0 \\ 0 & m = \pm 1, \pm 2, \pm 3, \ldots \end{cases}$$

11.3 THE SPACE $\mathcal{P}'_T$ OF PERIODIC DISTRIBUTIONS

A periodic distribution is defined in the same way as is a periodic function. In particular, the distribution f is said to be *periodic* if there exists a real positive number T such that

$$f(t) = f(t - T) \tag{1}$$

for all t. This means, of course, that for every ϕ in $\mathfrak{D}$

$$\langle f(t), \phi(t) \rangle = \langle f(t - T), \phi(t) \rangle \tag{2}$$

T is called a *period* of $f(t)$. As with ordinary functions, a distribution will have an infinity of periods nT ($n = 1, 2, 3, \ldots$) so long as it has at least one period T. Obviously, every constant distribution is a periodic distribution and each positive number is one of its periods.

Now, let T be a given (fixed) positive number. The space of all periodic distributions possessing T as one of its periods will be denoted by $\mathcal{P}'_T$.

Every element f of $\mathcal{P}'_T$ can also be identified as a continuous linear functional on the space $\mathcal{P}_T$. The (complex) number that f assigns to any θ in $\mathcal{P}_T$ will be denoted by the "dot product" $f \cdot \theta$ in order to avoid confusion with the number $\langle f, \phi \rangle$ that f assigns to any ϕ in $\mathfrak{D}$. This number $f \cdot \theta$ is defined by

$$f \cdot \theta \triangleq \langle f, \xi\theta \rangle \tag{3}$$

where ξ is any unitary function in $\mathfrak{U}_T$.

Before discussing the linearity and the continuity of f, we shall first show that (3) is independent of the choice of ξ. That is, *the right-hand side of* (3) *remains invariant if we replace ξ by any other unitary function in $\mathfrak{U}_T$*. It is a fact that

$$f(t) = \sum_{n=-\infty}^{\infty} f(t)\xi(t - nT) \tag{4}$$

Indeed, since this summation has only a finite number of nonzero terms over any bounded open t interval, we may write for any ϕ in $\mathfrak{D}$

$$\left\langle \sum_n f(t)\xi(t - nT), \phi(t) \right\rangle = \left\langle f(t), \phi(t) \sum_n \xi(t - nT) \right\rangle$$
$$= \langle f(t), \phi(t) \rangle$$

Now let ξ and κ be any two elements in $\mathcal{U}_T$. By invoking (4) and the fact that $f(t) = f(t + nT)$, we may write for every θ in $\mathcal{P}_T$

$$\langle f, \xi\theta \rangle = \left\langle \sum_n f(t)\kappa(t - nT), \xi(t)\theta(t) \right\rangle$$
$$= \sum_n \langle f(t)\kappa(t - nT)\xi(t), \theta(t) \rangle = \sum_n \langle f(t + nT)\kappa(t)\xi(t + nT), \theta(t) \rangle$$
$$= \left\langle \sum_n f(t)\xi(t + nT), \kappa(t)\theta(t) \right\rangle = \langle f, \kappa\theta \rangle$$

which is what was asserted.

As usual, f is said to be a *linear* functional on the space $\mathcal{P}_T$ if, for any θ_1 and θ_2 in $\mathcal{P}_T$ and for any two complex numbers α and β, we have

$$f \cdot (\alpha\theta_1 + \beta\theta_2) = \alpha(f \cdot \theta_1) + \beta(f \cdot \theta_2)$$

Similarly, f is said to be a *continuous* functional on $\mathcal{P}_T$ if, for any sequence $\{\theta_\nu\}_{\nu=1}^\infty$ that converges in $\mathcal{P}_T$ to θ, the sequence of numbers $\{f \cdot \theta_\nu\}_{\nu=1}^\infty$ converges to the number $f \cdot \theta$.

Theorem 11.3-1

If f is a periodic distribution with period T, then (3) defines it as a continuous linear functional on $\mathcal{P}_T$.

Proof: With θ_1 and θ_2 in $\mathcal{P}_T$ and α and β any two numbers, we have

$$f \cdot (\alpha\theta_1 + \beta\theta_2) = \langle f, \xi(\alpha\theta_1 + \beta\theta_2) \rangle = \alpha\langle f, \xi\theta_1 \rangle + \beta\langle f, \xi\theta_2 \rangle$$
$$= \alpha(f \cdot \theta_1) + \beta(f \cdot \theta_2)$$

Also, if $\{\theta_\nu\}_{\nu=1}^\infty$ converges in $\mathcal{P}_T$ to θ, then $\{\xi\theta_\nu\}_{\nu=1}^\infty$ converges in $\mathcal{D}$ to $\xi\theta$, whence, as $\nu \to \infty$,

$$f \cdot \theta_\nu = \langle f, \xi\theta_\nu \rangle \to \langle f, \xi\theta \rangle = f \cdot \theta$$

Thus, f is a continuous linear functional on $\mathcal{P}_T$.

Thus far, we have derived a continuous linear functional on $\mathcal{P}_T$ from a periodic distribution of period T. We can go in the other direction too. That is, from a given continuous linear functional on $\mathcal{P}_T$ we can derive a periodic distribution of period T in such a way that (3) still holds true. This can be done as follows.

Any ϕ in $\mathcal{D}$ will generate a θ in $\mathcal{P}_T$ through the expression

$$\theta(t) \triangleq \sum_{n=-\infty}^{\infty} \phi(t - nT) \tag{5}$$

Then, by knowing f as a functional on $\mathcal{P}_T$, we define the number $\langle f, \phi \rangle$ by

$$\langle f, \phi \rangle \triangleq f \cdot \theta \tag{6}$$

This defines f as a functional on $\mathfrak{D}$. Moreover, since both $\phi(t)$ and $\phi(t + T)$ generate through (5) the same $\theta(t)$, we have that

$$\langle f(t), \phi(t) \rangle = f(t) \cdot \theta(t) = \langle f(t), \phi(t + T) \rangle \tag{7}$$

We are now ready to prove

Theorem 11.3-2

If f is a continuous linear functional on $\mathcal{P}_T$ and if θ and ϕ are related by (5), then (6) defines f as a periodic distribution with period T.

Proof: Let θ_ν in $\mathcal{P}_T$ be related to ϕ_ν in $\mathfrak{D}$ in accordance with (5). The linearity of f on $\mathfrak{D}$ is established by

$$\langle f, \alpha\phi_1 + \beta\phi_2 \rangle = f \cdot (\alpha\theta_1 + \beta\theta_2) = \alpha(f \cdot \theta_1) + \beta(f \cdot \theta_2)$$
$$= \alpha\langle f, \phi_1 \rangle + \beta\langle f, \phi_2 \rangle$$

To show its continuity on $\mathfrak{D}$, assume that $\{\phi_\nu\}_{\nu=1}^\infty$ converges in $\mathfrak{D}$ to ϕ. As was pointed out in the preceding section, it follows that $\{\theta_\nu\}_{\nu=1}^\infty$ converges in $\mathcal{P}_T$ to the testing function (5). Since f is a continuous functional on $\mathcal{P}_T$,

$$\langle f, \phi_\nu \rangle = f \cdot \theta_\nu \to f \cdot \theta = \langle f, \theta \rangle$$

Finally, (7) shows that f is periodic with a period T. This completes the proof.

There is one more thing that we should demonstrate, namely, that *the definitions (3) and (6) are consistent.* By using (5) and the fact that $f(t) = f(t + nT)$, we may write for any ξ in $\mathfrak{U}_T$

$$\langle f, \xi\theta \rangle = \Big\langle f(t), \xi(t) \sum_n \phi(t - nT) \Big\rangle$$
$$= \sum_n \langle f(t), \xi(t)\phi(t - nT) \rangle = \sum_n \langle f(t + nT), \xi(t + nT)\phi(t) \rangle$$
$$= \Big\langle f(t), \phi(t) \sum_n \xi(t + nT) \Big\rangle = \langle f, \phi \rangle$$

Thus, (3) and (6) are really the same expression so long as θ and ϕ are related by (5). This is what we needed to show.

This result leads to

Corollary 11.3-2a

Two distributions f and g in $\mathcal{P}'_T$ are equal if $f \cdot \theta = g \cdot \theta$ for every θ in $\mathcal{P}_T$.

Proof: By using (5) and (6), we obtain for every ϕ in $\mathfrak{D}$

$$\langle f, \phi \rangle = f \cdot \theta = g \cdot \theta = \langle g, \phi \rangle$$

so that $f = g$.

Some examples of periodic distributions follow.

Example 11.3-1 When f is a regular distribution corresponding to a locally integrable periodic function of period T, we can interpret its dot product with any θ in $\mathcal{P}_T$ in a neat way. Indeed, with ξ being in $\mathcal{U}_T$ and a being any real number, we may write

$$\begin{aligned}
f \cdot \theta = \langle f, \xi\theta \rangle &= \int_{-\infty}^{\infty} f(t)\xi(t)\theta(t)\,dt \\
&= \sum_{n=-\infty}^{\infty} \int_{a+nT}^{a+nT+T} f(t)\xi(t)\theta(t)\,dt \\
&= \sum_{n} \int_{a}^{a+T} f(t+nT)\xi(t+nT)\theta(t+nT)\,dt \\
&= \int_{a}^{a+T} f(t)\theta(t) \sum_{n} \xi(t+nT)\,dt = \int_{a}^{a+T} f(t)\theta(t)\,dt
\end{aligned}$$

(The interchange of the summation and the integration is again justified by the fact that there are only a finite number of nonzero terms in the summation.) Thus, in this case the dot product $f \cdot \theta$ simply denotes the integration of the function $f(t)\theta(t)$ over an arbitrary *finite* interval of length T.

Example 11.3-2 Let

$$\delta_T(t) \triangleq \sum_{n=-\infty}^{\infty} \delta(t - nT) \qquad T > 0$$

Clearly, δ_T is in $\mathcal{P}_T'$. For θ in $\mathcal{P}_T$ and ξ in $\mathcal{U}_T$, we have

$$\begin{aligned}
\delta_T \cdot \theta = \langle \delta_T, \xi\theta \rangle &= \Big\langle \sum_n \delta(t-nT), \xi(t)\theta(t) \Big\rangle \\
&= \sum_n \xi(nT)\theta(nT) = \theta(0) \sum_n \xi(nT)
\end{aligned}$$

Since $\Sigma_n \, \xi(nT) = 1$, we have established that

$$\delta_T(t) \cdot \theta(t) = \theta(0)$$

A similar analysis shows that

$$\delta_T(t - \tau) \cdot \theta(t) = \theta(\tau)$$

Example 11.3-3 More generally, consider

$$\delta_T{}^{(k)}(t) = \sum_{n=-\infty}^{\infty} \delta^{(k)}(t - nT) \qquad k = 0, 1, 2, \ldots$$

This is also in $\mathcal{P}'_T$ and, as before, we have

$$\delta_{T^{(k)}} \cdot \theta = \sum_{n=-\infty}^{\infty} (-1)^k \frac{d^k}{dt^k} [\xi(t)\theta(t)]\Big|_{t=nT}$$

$$= \sum_{n=-\infty}^{\infty} (-1)^k \sum_{p=0}^{k} \binom{k}{p} \xi^{(p)}(nT)\theta^{(k-p)}(nT)$$

$$= (-1)^k \sum_{p=0}^{k} \binom{k}{p} \theta^{(k-p)}(0) \sum_{n=-\infty}^{\infty} \xi^{(p)}(nT)$$

In view of (3) and (5) of the preceding section, this equation yields

$$\delta_{T^{(k)}}(t) \cdot \theta(t) = (-1)^k \theta^{(k)}(0) \qquad k = 0, 1, 2, \ldots \tag{8}$$

Convergence in the space $\mathcal{P}'_T$ is defined in the usual way. A sequence $\{f_\nu\}_{\nu=1}^{\infty}$ is said to *converge in* $\mathcal{P}'_T$ *to a limit* f if each f_ν is in $\mathcal{P}'_T$ and if for each θ in $\mathcal{P}_T$ the sequence of numbers $f_\nu \cdot \theta$ converges to the number $f \cdot \theta$. This defines the limit f as a functional on the space $\mathcal{P}_T$. We can assert still further

Theorem 11.3-3

The space $\mathcal{P}'_T$ is closed under convergence. In other words, the limit f of every sequence $\{f_\nu\}_{\nu=1}^{\infty}$ that converges in $\mathcal{P}'_T$ is also in $\mathcal{P}'_T$.

Proof: For every ϕ in $\mathcal{D}$ there is a θ in $\mathcal{P}_T$ related to it through (5). Using (6), we have under our hypothesis that

$$\langle f_\nu, \phi \rangle = f_\nu \cdot \theta \to f \cdot \theta$$

Thus, $\{f_\nu\}_{\nu=1}^{\infty}$ converges in $\mathcal{D}'$ and, since $\mathcal{D}'$ is closed under convergence, its limit g is also in $\mathcal{D}'$. Moreover, g must also be periodic with period T, since every f_ν is such a distribution. Hence, g is in $\mathcal{P}'_T$ and $\langle g, \phi \rangle = f \cdot \theta$. By (6) again, $\langle g, \phi \rangle = g \cdot \theta$, so that $g \cdot \theta = f \cdot \theta$. This holds for every θ in $\mathcal{P}_T$, since each such θ can be generated by (5) from some ϕ in $\mathcal{D}$. Therefore, by Corollary 11.3-2a, $f = g$. Q.E.D.

Theorem 11.3-4

The sequence $\{f_\nu\}_{\nu=1}^{\infty}$ converges in $\mathcal{P}'_T$ to f if and only if it converges in $\mathcal{D}'$ to f and each f_ν is in $\mathcal{P}'_T$.

Proof: The preceding proof has already shown that convergence in $\mathcal{P}'_T$ implies convergence in $\mathcal{D}'$. Conversely, assume that the convergence is in $\mathcal{D}'$ and each f_ν is in $\mathcal{P}'_T$. It follows that the limit f must also be in $\mathcal{P}'_T$.

Then, for ξ in $\mathfrak{U}_T$ and θ in $\mathcal{P}_T$,

$$f_\nu \cdot \theta = \langle f_\nu, \xi\theta \rangle \to \langle f, \xi\theta \rangle = f \cdot \theta \qquad \text{Q.E.D.}$$

As usual, we shall say that an infinite series $\Sigma_{\nu=1}^\infty f_\nu$ converges in $\mathcal{P}'_T$ if the sequence of its partial sums,

$$\left\{ \sum_{\nu=1}^n f_\nu \right\}_{n=1}^\infty$$

converges in $\mathcal{P}'_T$.

We now list some operations under which the space $\mathcal{P}'_T$ is closed. Since the elements of $\mathcal{P}'_T$ are by definition elements of $\mathfrak{D}'$, each of these operations has already been defined in Sec. 1.7. However, entirely equivalent definitions can be given in terms of the dot products of the elements in $\mathcal{P}'_T$ and those of $\mathcal{P}_T$. These are the definitions that are stated below. It is understood here that f and g are arbitrary distributions in $\mathcal{P}'_T$ and that each equation holds as θ traverses $\mathcal{P}_T$. We leave it to the reader to show that the definitions listed below are consistent with those stated in Sec. 1.7 and that $\mathcal{P}'_T$ is closed under these operations.

1. *The addition of periodic distributions in $\mathcal{P}'_T$:*

$$(f + g) \cdot \theta \triangleq f \cdot \theta + g \cdot \theta \tag{9}$$

2. *The multiplication of a periodic distribution by a constant α:*

$$(\alpha f) \cdot \theta \triangleq \alpha(f \cdot \theta) \tag{10}$$

(Since $\mathcal{P}'_T$ is closed under these two operations, it can now be shown that $\mathcal{P}'_T$ is a linear space.)

3. *The shifting of a periodic distribution* (let τ be a real number):

$$f(t - \tau) \cdot \theta(t) \triangleq f(t) \cdot \theta(t + \tau) \tag{11}$$

4. *The transposition of a periodic distribution:*

$$f(-t) \cdot \theta(t) \triangleq f(t) \cdot \theta(-t) \tag{12}$$

5. *The differentiation of a periodic distribution:*

$$f^{(1)} \cdot \theta \triangleq -f \cdot \theta^{(1)} \tag{13}$$

By repetition, we get

$$f^{(k)} \cdot \theta = (-1)^k f \cdot \theta^{(k)} \qquad k = 1, 2, 3, \ldots \tag{14}$$

6. *The multiplication of a distribution f in $\mathcal{P}'_T$ by a periodic testing function ρ in $\mathcal{P}_T$:*

$$(\rho f) \cdot \theta \triangleq f \cdot (\rho \theta) \tag{15}$$

Note that *differentiation is a continuous linear operation in the space* $\mathcal{P}'_T$ in the following sense. For linearity, if f and g are in $\mathcal{P}'_T$ and α and β are fixed numbers, then

$$\frac{d}{dt}(\alpha f + \beta g) = \alpha \frac{df}{dt} + \beta \frac{dg}{dt}$$

For continuity, if $\{f_\nu\}_{\nu=1}^\infty$ converges in $\mathcal{P}'_T$ to f, then $\{f_\nu^{(1)}\}_{\nu=1}^\infty$ also converges in $\mathcal{P}'_T$ to $f^{(1)}$. This also means that any infinite series that converges in $\mathcal{P}'_T$ can be differentiated term by term. Similar statements may be made about the other operations listed above.

PROBLEMS

1 If f is in $\mathfrak{D}'$, can we always say that

$$\sum_{n=-\infty}^{\infty} f(t - nT)$$

is in $\mathcal{P}'_T$?

2 In the proofs of Theorems 11.3-2 and 11.3-4 we asserted that, if $\{f_\nu\}_{\nu=1}^\infty$ converges in $\mathfrak{D}'$ to f and if every f_ν is in $\mathcal{P}'_T$, then f is also in $\mathcal{P}'_T$. Prove this.

3 Show that the space $\mathcal{P}'_T$ is closed under the six operations listed at the end of this section. Also, show that the definitions of these operations are consistent with the definitions given in Sec. 1.7.

4 Verify that $\mathcal{P}'_T$ satisfies the axioms for a linear space.

5 Prove that differentiation is a continuous linear operation in the space $\mathcal{P}'_T$.

6 If $f(t)$ is in $\mathcal{P}'_T$, what can we say about $f(at)$, where a is a real constant?

11.4 *T* - CONVOLUTION

We have seen the importance of convolution in the theory of distributions and in the solving of various types of differential, difference, and integral equations. However, when the distributions under consideration are periodic, their convolution does not exist (unless one of them is the zero distribution). For example, if $f(t) = g(t) = 1$ for all t, then $f(t) \times g(\tau) = 1$ over the (t, τ) plane, so that

$$\langle f * g, \phi \rangle = \langle 1, \phi(t + \tau) \rangle$$

and this fails to exist for every nonzero ϕ in $\mathfrak{D}$. On the other hand, there is a different kind of convolution, which we shall call *T-convolution*, that plays an analogous role for periodic distributions. T-convolution is also called *finite convolution* for a reason that will become apparent later on.

Let f and g be arbitrary distributions in $\mathcal{P}'_T$, ξ and κ arbitrary unitary functions in $\mathcal{U}_T$, and θ any testing function in $\mathcal{P}_T$. Then, the T-convolution of f with g, which we shall denote by $f \triangle g$, is the functional on $\mathcal{P}_T$ defined by

$$(f \triangle g) \cdot \theta \triangleq f(t) \cdot [g(\tau) \cdot \theta(t + \tau)] \\ = f(t) \cdot \langle g(\tau), \kappa(\tau)\theta(t + \tau)\rangle \tag{1}$$

As a consequence of Theorem 2.7-2,

$$\langle g(\tau), \kappa(\tau)\theta(t + \tau)\rangle$$

is a function of t that is infinitely smooth. Also, it is clearly periodic with period T. Hence, it is in $\mathcal{P}_T$. Thus, definition (1) may be replaced by the equivalent definition

$$(f \triangle g) \cdot \theta \triangleq \langle f(t), \langle g(\tau), \xi(t)\kappa(\tau)\theta(t + \tau)\rangle\rangle \\ = \langle f(t) \times g(\tau), \xi(t)\kappa(\tau)\theta(t + \tau)\rangle \tag{2}$$

Note that $\xi(t)\kappa(\tau)\theta(t + \tau)$ is in $\mathcal{D}_{t,\tau}$ so that the right-hand side of (2) always exists for every choice of f and g from $\mathcal{P}'_T$. This is in contrast to the usual convolution of distributions in $\mathcal{D}'$, where additional restrictions on the distributions had to be imposed in order to ensure the existence of the convolution.

Theorem 11.4-1

The T-convolution of two distributions in $\mathcal{P}'_T$ yields a distribution in $\mathcal{P}'_T$. In other words, the space $\mathcal{P}'_T$ is closed under T-convolution.

In view of Theorem 11.3-2, we can prove this theorem by showing that (2) defines a continuous linear functional on the space $\mathcal{P}_T$. We leave the details to the reader.

Example 11.4-1 Let f and g be locally integrable periodic functions of period T. Then, for each θ in $\mathcal{P}_T$,

$$(f \triangle g) \cdot \theta = f(\tau) \cdot [g(t) \cdot \theta(t + \tau)]$$

As was shown in Example 11.3-1,

$$g(t) \cdot \theta(t + \tau) = \int_b^{b+T} g(t)\theta(t + \tau)\, dt \tag{3}$$

where b is any real number. This is clearly a periodic function of τ of period T. It is also infinitely smooth because we may differentiate the right-hand side of (3) under the integral sign any number of times

(Kestelman [1], p. 170). Thus, by Example 11.3-1 again,

$$(f \Delta g) \cdot \theta = \int_a^{a+T} f(\tau) \int_b^{b+T} g(t)\theta(t + \tau) \, dt \, d\tau$$
$$= \int_a^{a+T} f(\tau) \int_b^{b+T} g(t - \tau)\theta(t) \, dt \, d\tau$$

where a is also any real number. By Fubini's theorem, we may interchange the order of integration to obtain

$$(f \Delta g) \cdot \theta = \int_b^{b+T} \theta(t) \int_a^{a+T} f(\tau)g(t - \tau) \, d\tau \, dt$$
$$= \left[\int_a^{a+T} f(\tau)g(t - \tau) \, d\tau \right] \cdot \theta(t)$$

Thus, we have shown that in this special case

$$f \Delta g = \int_a^{a+T} f(\tau)g(t - \tau) \, d\tau$$

the right-hand side again being a locally integrable periodic function of period T (Titchmarsh [1], p. 391). This is just like an ordinary convolution except that now the integration is over a *finite* interval of length T. This is the reason why T-convolution is also called "finite convolution."

Theorem 11.4-2

T-convolution is commutative and associative. That is, if f, g, and h are all in $\mathcal{O}'_T$, then

$$f \Delta g = g \Delta f \qquad \text{commutativity} \tag{4}$$

and

$$f \Delta (g \Delta h) = (f \Delta g) \Delta h \qquad \text{associativity} \tag{5}$$

Proof: Since the direct product is commutative and associative (see Theorems 5.3-2 and 5.3-3), this theorem follows directly from (2).

Theorem 11.4-3

T-convolution commutes with the shifting operator and with differentiation. That is, if f and g are in $\mathcal{O}'_T$, if $h = f \Delta g$, and if a is a real constant, then

$$h(t - a) = f(t - a) \Delta g(t) = f(t) \Delta g(t - a) \tag{6}$$

and

$$h^{(1)} = f^{(1)} \Delta g = f \Delta g^{(1)} \tag{7}$$

Proof: By using definition (1), we may write for each θ in $\mathcal{O}_T$

$$h(t - a) \cdot \theta(t) = h(t) \cdot \theta(t + a) = f(t) \cdot [g(\tau) \cdot \theta(t + \tau + a)]$$
$$= f(t) \cdot [g(\tau - a) \cdot \theta(t + \tau)] = [f(t) \Delta g(t - a)] \cdot \theta(t)$$

Thus, $h(t - a) = f(t) \Delta g(t - a)$. The commutativity of T-convolution establishes the rest of (6). Equation (7) is proved in a similar way.

Example 11.4-2 Consider

$$\delta_T(t) \triangleq \sum_{n=-\infty}^{\infty} \delta(t - nT)$$

the distribution in $\mathcal{P}'_T$ that was discussed in Example 11.3-2. If f is also in $\mathcal{P}'_T$ and θ is any element in $\mathcal{P}_T$, then

$$(f \Delta \delta_T) \cdot \theta = f(t) \cdot [\delta_T(\tau) \cdot \theta(t + \tau)] = f(t) \cdot \theta(t)$$

and therefore

$$f \Delta \delta_T = f \tag{8}$$

Thus, δ_T is the unit element in the T-convolution algebra, which we shall describe later on. In accordance with the last theorem, we may also write

$$f^{(1)} = \frac{d}{dt}(f \Delta \delta_T) = f \Delta \delta_T^{(1)}$$

and, more generally,

$$f^{(k)} = f \Delta \delta_T^{(k)} \qquad k = 0, 1, 2, \ldots \tag{9}$$

T-convolution is clearly a linear operation in the sense that, if f, g, and h are in $\mathcal{P}'_T$ and if α and β are real numbers, then

$$f \Delta (\alpha g + \beta h) = \alpha(f \Delta g) + \beta(f \Delta h)$$

By virtue of this and (9), any linear differential expression with constant coefficients can be represented as a T-convolution so long as we restrict ourselves to the distributions in $\mathcal{P}'_T$:

$$a_n f^{(n)} + a_{n-1} f^{(n-1)} + \cdots + a_0 f$$
$$= (a_n \delta_T^{(n)} + a_{n-1} \delta_T^{(n-1)} + \cdots + a_0 \delta_T) \Delta f \qquad f \in \mathcal{P}'_T \tag{10}$$

Similarly, if a is a real number, then

$$f(t - a) = f(t) \Delta \delta_T(t - a) \tag{11}$$

and thus a linear difference expression with constant coefficients can also be represented as a T-convolution:

$$a_n f(t + nT) + a_{n-1} f(t + nT - T) + \cdots + a_0 f(t)$$
$$= [a_n \delta_T(t + nT) + a_{n-1} \delta_T(t + nT - T) + \cdots + a_0 \delta_T(t)] \Delta f(t)$$
$$f \in \mathcal{P}'_T \tag{12}$$

Theorem 11.4-4

T-convolution is a continuous operation in the following sense. If $\{f_\nu\}_{\nu=1}^{\infty}$ converges in $\mathcal{O}'_T$ to f and if g is in $\mathcal{O}'_T$, then $\{f_\nu \Delta g\}_{\nu=1}^{\infty}$ converges in $\mathcal{O}'_T$ to $f \Delta g$.

Proof: First note that f is also in $\mathcal{O}'_T$, since $\mathcal{O}'_T$ is closed under convergence. Now, if θ is in $\mathcal{O}_T$ and ξ and κ are in $\mathcal{U}_T$, then

$$(f_\nu \Delta g) \cdot \theta = \langle f_\nu(t), \langle g(\tau), \xi(t)\kappa(\tau)\theta(t+\tau) \rangle \rangle \tag{13}$$

But $\langle g(\tau), \xi(t)\kappa(\tau)\theta(t+\tau) \rangle$ is a testing function in $\mathcal{D}_t$, according to Corollary 2.7-2a. By Theorem 11.3-4, $\{f_\nu\}_{\nu=1}^{\infty}$ converges in $\mathcal{D}'_t$ to f, so that the right-hand side of (13) converges to

$$\langle f(t), \langle g(\tau), \xi(t)\kappa(\tau)\theta(t+\tau) \rangle \rangle = (f \Delta g) \cdot \theta \qquad \text{Q.E.D.}$$

It now follows that, if $\sum_{\nu=1}^{\infty} f_\nu$ converges in $\mathcal{O}'_T$ and if g is in $\mathcal{O}'_T$, then

$$g \Delta \sum_{\nu=1}^{\infty} f_\nu = \sum_{\nu=1}^{\infty} g \Delta f_\nu \tag{14}$$

We end this section by pointing out a simple identity that we shall employ when discussing the Fourier series of a periodic function. Let $\omega = 2\pi/T$, so that $e^{i\omega t}$ is in $\mathcal{O}_T$. Then, for f in $\mathcal{O}'_T$

$$f(t) \Delta e^{i\omega t} = f(\tau) \cdot e^{i\omega(t-\tau)} \tag{15}$$

Indeed, if θ is in $\mathcal{O}_T$, then

$$\begin{aligned}
[f(t) \Delta e^{i\omega t}] \cdot \theta(t) &= f(\tau) \cdot [e^{i\omega t} \cdot \theta(t+\tau)] \\
&= f(\tau) \cdot [e^{i\omega(t-\tau)} \cdot \theta(t)] \\
&= [f(\tau) \cdot e^{-i\omega \tau}][e^{i\omega t} \cdot \theta(t)] \\
&= [f(\tau) \cdot e^{i\omega(t-\tau)}] \cdot \theta(t)
\end{aligned}$$

PROBLEMS

1 Prove Theorem 11.4-1.

2 Use the definition (1) of T-convolution to show that $f \Delta \delta_T^{(1)} = f^{(1)}$. (Do not use Theorem 11.4-3, which was used in Example 11.4-2.) Then, use this result and the associativity and commutativity of T-convolution to establish (7).

3 Show that, if f is in $\mathcal{O}'_T$ and θ is in $\mathcal{O}_T$, then

$$f(t) \Delta \theta(t) = f(\tau) \cdot \theta(t-\tau)$$

and that this is a testing function in $\mathcal{O}_T$. This is the analogue in the space $\mathcal{O}'_T$ of the regularization operation in the space $\mathcal{D}'$ (see Sec. 5.5).

4 Compute the following T-convolutions. In each case, $T = 2\pi$.

a. $(\sin t) \triangle (\cos t)$

b. $1 \triangle \text{Pf} \dfrac{1}{\sin t}$

c. $f(t) \triangle \text{Pf} \dfrac{1}{\sin t}$

$$f(t) = \begin{cases} 1 & 0 \le t < \pi \\ -1 & \pi \le t < 2\pi \end{cases}$$
$$f(t) = f(t - 2\pi) \qquad -\infty < t < \infty$$

HINT: You may leave your answer in terms of the convergent integral

$$\int_0^{2\pi} \left(\frac{1}{\sin t} - \frac{1}{t} \right) dt$$

d. $f(t) \triangle e^{i\nu t} \qquad \nu = 0, \pm 1, \pm 2, \ldots$

$$f(t) = \sum_{n=-\infty}^{\infty} g(t - 2\pi n)$$
$$g(t) = \text{Pf} \frac{1_+(t) 1_+(2\pi - t)}{t}$$

HINT:

$$\int_0^{2\pi} \frac{e^{-i\nu\tau} - 1}{\tau} d\tau = \text{Ei}^*(-2\pi i \nu) - \gamma - \log|2\pi\nu| + i\frac{\pi}{2}$$

where $\text{Ei}^*(x)$ is the exponential integral function (see Example 8.3-2 for its definition) and γ is Euler's constant.

11.5 THE T - CONVOLUTION ALGEBRA $\mathcal{O}'_T$

In Sec. 6.2 we described the axioms that characterize a space of distributions as a convolution algebra. If we take addition and T-convolution as our additive and multiplicative operations, respectively, it turns out that the space $\mathcal{O}'_T$ is also an algebra. (We shall refer to it as the T-convolution algebra.) Indeed, the following axioms are satisfied:

1. $\mathcal{O}'_T$ is a linear space.
2. $\mathcal{O}'_T$ is closed under T-convolution.
3. T-convolution is associative for all distributions in $\mathcal{O}'_T$.

Moreover, $\mathcal{O}'_T$ is a commutative algebra because $f \triangle g = g \triangle f$. It possesses the unit element δ_T, since $f \triangle \delta_T = f$.

A *T-convolution equation* is an equation of the form

$$f \triangle u = g \qquad (1)$$

where f and g are given elements of $\mathcal{O}'_T$ and u is an unknown element that is also required to be in $\mathcal{O}'_T$. The technique that was described in Sec. 6-2 for solving a convolution equation in some convolution algebra can be applied here to solve (1) in $\mathcal{O}'_T$. Indeed, the theorems of Sec. 6.2 and their proofs can be applied directly to the present case if we merely replace $\mathfrak{D}'_R$ by $\mathcal{O}'_T$, the symbol $*$ by the symbol Δ, and δ by δ_T. For the sake of easy reference, we shall restate these theorems here, incorporating these changes without repeating any proofs.

As before, the *inverse in* $\mathcal{O}'_T$ of an element f in $\mathcal{O}'_T$ is an element $f^{\Delta-1}$ of $\mathcal{O}'_T$ such that

$$f^{\Delta-1} \Delta f = \delta_T \tag{2}$$

Not all distributions in $\mathcal{O}'_T$ possess such inverses. For example, a testing function in $\mathcal{O}_T$ cannot have an inverse in $\mathcal{O}'_T$ because its T-convolution with each element of $\mathcal{O}'_T$ yields a testing function in $\mathcal{O}_T$, as was asserted in Sec. 11.4, Prob. 3.

Theorem 11.5-1

Let f be a given distribution in $\mathcal{O}'_T$. A necessary and sufficient condition for (1) to have at least one solution in $\mathcal{O}'_T$ for every g in $\mathcal{O}'_T$ is that f possess an inverse $f^{\Delta-1}$ in $\mathcal{O}'_T$. When f does have an inverse in $\mathcal{O}'_T$, this inverse is unique and (1) possesses a unique solution in $\mathcal{O}'_T$ given by

$$u = f^{\Delta-1} \Delta g \tag{3}$$

Theorem 11.5-2

If h and j are in $\mathcal{O}'_T$ and possess inverses in $\mathcal{O}'_T$, then

$$(h \Delta j)^{\Delta-1} = h^{\Delta-1} \Delta j^{\Delta-1} \tag{4}$$

Simultaneous T-convolution equations can be written in the matrix form

$$\mathbf{f} \Delta \mathbf{u} = \mathbf{g} \tag{5}$$

where $\mathbf{f}$ is a given $n \times n$ matrix, $\mathbf{g}$ is a given $n \times 1$ matrix, $\mathbf{u}$ is an unknown $n \times 1$ matrix, and all elements of $\mathbf{f}$, $\mathbf{g}$, and $\mathbf{u}$ are in $\mathcal{O}'_T$. Here it is understood that the manipulations one would perform with matrices by using ordinary multiplication are now performed by using T-convolution instead. Thus, the determinant det $\mathbf{f}$, adjoint $\mathbf{a}$, and inverse $\mathbf{f}^{\Delta-1}$ of the matrix $\mathbf{f}$ are defined in the usual way.

In accordance with our customary notation, $\mathcal{O}'_{T,n \times m}$ shall denote the space of all $n \times m$ matrices whose elements are in $\mathcal{O}'_T$. Thus, we say that the inverse in $\mathcal{O}'_{T,n \times n}$ of a matrix $\mathbf{f}$ in $\mathcal{O}'_{T,n \times n}$ is a matrix $\mathbf{f}^{\Delta-1}$ in $\mathcal{O}'_{T,n \times n}$ satisfying

$$\mathbf{f} \Delta \mathbf{f}^{\Delta-1} = \boldsymbol{\delta}_{T,n \times n} \tag{6}$$

where $\delta_{T,n\times n}$ is the $n \times n$ matrix whose main diagonal elements are all δ_T and whose elements off the main diagonal are the zero distribution. Indeed $\delta_{T,n\times n}$ *is the $n \times n$ unit matrix in the T-convolution algebra $\mathcal{O}'_T$.* As usual, a given $\mathbf{f}$ in $\mathcal{O}'_{T,n\times n}$ may not possess any inverse in $\mathcal{O}'_{T,n\times n}$.

By repeating the proof of Theorem 6.2-3 with appropriate changes, we obtain

Theorem 11.5-3

Let $\mathbf{f}$ be a given matrix in $\mathcal{O}'_{T,n\times n}$. A necessary and sufficient condition for (5) to have at least one solution in $\mathcal{O}'_{T,n\times 1}$ for each $\mathbf{g}$ in $\mathcal{O}'_{T,n\times 1}$ is that $\det \mathbf{f}$ *possess an inverse in $\mathcal{O}'_T$. If this is the case, the inverse in $\mathcal{O}'_{T,n\times n}$ of $\mathbf{f}$ is unique and there is precisely one solution in $\mathcal{O}'_{T,n\times 1}$ for $\mathbf{u}$. This solution is*

$$\mathbf{u} = \mathbf{f}^{\Delta-1} \Delta \mathbf{g}$$

and $\mathbf{f}^{\Delta-1}$ is given by

$$\mathbf{f}^{\Delta-1} = (\det \mathbf{f})^{\Delta-1} \Delta \mathbf{a}$$

where $\mathbf{a}$ is the adjoint of $\mathbf{f}$.

Thus, when solving the T-convolution equation (1) or (5), the problem is essentially that of finding an inverse for f or $\mathbf{f}$, respectively. It turns out that the Fourier series is a convenient tool for doing this. In fact, this series can be used in the same way as the Laplace transformation was for solving convolution equations in $\mathcal{D}'_R$. The Fourier series of periodic distributions is discussed, therefore, in the next section.

Before leaving this section, one difference between the convolution algebra $\mathcal{D}'_R$ and the T-convolution algebra $\mathcal{O}'_T$ will be pointed out. In Sec. 6.2 we mentioned (without proving) that the convolution algebra $\mathcal{D}'_R$ does not have divisors of zero. That is to say, if f and g are both in $\mathcal{D}'_R$ and if $f * g = 0$, then $f = 0$ or $g = 0$ (or both equal zero). This is not so for the T-convolution algebra $\mathcal{O}'_T$; *it does possess divisors of zero*. An example will suffice to show this.

Example 11.5-1 Let $f = 1$ and $g = e^{i\omega t}$, where $\omega = 2\pi/T$. Then,

$$f \Delta g = \int_a^{a+T} f(t-\tau)g(\tau)\, d\tau$$
$$= \int_a^{a+T} e^{i\omega\tau}\, d\tau = 0$$

In fact, the same result will hold for $g = e^{i\nu\omega t}$ ($\nu = \pm 1, \pm 2, \pm 3, \ldots$).

A consequence of this distinction between $\mathcal{D}'_R$ and $\mathcal{O}'_T$ is the following. Assume that f is in $\mathcal{D}'_R$, $f \neq 0$, and f does not have an inverse in $\mathcal{D}'_R$. Then the fact that there are no divisors of zero in $\mathfrak{D}'_R$ ensures that the equation $f * u = g$ possesses either no solution in $\mathfrak{D}'_R$ or precisely one solution in $\mathfrak{D}'_R$. It cannot possess more than one solution in $\mathfrak{D}'_R$. (We asked the reader to prove this in Sec. 6.2, Prob. 4.)

On the other hand, when f is in $\mathcal{O}'_T$, $f \neq 0$, and f does not have an inverse in $\mathcal{O}'_T$, the equation $f \Delta u = g$ may have an infinity of solutions in $\mathcal{O}'_T$. Example 11.5-1 indicates one such f for which this occurs. Indeed, to any particular solution u of the equation $1 \Delta u = g$ one may add any solution $e^{i\nu\omega t}$ ($\nu = \pm 1, \pm 2, \pm 3, \ldots$) of the homogeneous equation $1 \Delta u = 0$ to obtain another solution of $1 \Delta u = g$. Thus, we have an additional complication when dealing with T-convolution equations.

PROBLEMS

1 Convince yourself that the theorems of this section are valid.

2 Show that the inverse in $\mathcal{O}'_T$ of

$$f \triangleq \delta_T{}^{(1)} - \alpha \delta_T$$

where α is a constant not equal to $i\nu 2\pi/T$ ($\nu = 0, \pm 1, \pm 2, \ldots$), is the periodic function given by

$$f^{\Delta-1}(t) = \frac{e^{\alpha T}}{1 - e^{\alpha T}} \qquad 0 \leq t < T$$
$$f^{\Delta-1}(t) = f^{\Delta-1}(t - T) \qquad -\infty < t < \infty$$

By using this result, solve in $\mathcal{O}'_T$ the differential equation

$$u^{(1)}(t) - u(t) = \sin \frac{2\pi t}{T}$$

3 By using Theorem 11.5-2 and the assertions of the preceding problem, solve in $\mathcal{O}'_T$ the differential equation

$$u^{(2)}(t) - 3u^{(1)}(t) + 2u(t) = \cos \frac{2\pi t}{T}$$

11.6 THE FOURIER SERIES

Our purpose in this section is to show that a distribution f is periodic and of period T if and only if it has the series expansion

$$f(t) = \sum_{\nu = -\infty}^{\infty} F_\nu e^{i\nu\omega t} \qquad \omega = \frac{2\pi}{T} \qquad (1)$$

where the constant coefficients F_ν are given by

$$F_\nu = \frac{1}{T} f(t) \cdot e^{-i\nu\omega t} \tag{2}$$

and have the property of being of *slow growth*. This latter property is defined as follows. A sequence of constants $\{F_\nu\}_{\nu=-\infty}^{\infty}$ is said to be of slow growth if there exist a constant M and an integer k such that $|F_\nu| \leq M|\nu|^k$ for all nonzero ν. The series (1) is called the *Fourier series* of $f(t)$, and the constants F_ν are called the *Fourier coefficients* of $f(t)$.

We have here a generalization to periodic distributions of the classical Fourier series. As has happened in so many other applications of distribution theory, this generalized Fourier series has a simpler structure than the classical Fourier series. Unlike the classical case, a trigonometric series whose coefficients are the Fourier coefficients (2) of some periodic distribution f always converges (in the distributional sense) to f. Moreover, before distribution theory was invented, a variety of methods had been devised for summing trigonometric series that diverge in the classical sense. (See, for example, the Cesaro summation described in Titchmarsh [1], pp. 411–416.) Now we can rely on a distributional interpretation, since every trigonometric series whose coefficients are of slow growth always converges in $\mathcal{S}'$, as we shall show. Furthermore, it can be shown that, if the coefficients are not of slow growth, the trigonometric series will not converge even in $\mathcal{D}'$.

First, we shall demonstrate that every trigonometric series having coefficients of slow growth is a Fourier series.

Theorem 11.6-1

Any series

$$\sum_{\nu=-\infty}^{\infty} b_\nu e^{i\nu\omega t} \qquad \omega = \frac{2\pi}{T} > 0 \tag{3}$$

whose constant coefficients b_ν comprise a sequence of slow growth converges in $\mathcal{S}'$ to a periodic distribution $f(t)$ of period T. Moreover, if the constants F_ν are defined by (2) (i.e., the F_ν are the Fourier coefficients of the limit distribution f), then $F_\nu = b_\nu$.

Proof: Corollary 2.4-3b asserts that (3) converges in $\mathcal{D}'$. Moreover, every partial sum of (3) is in $\mathcal{O}'_T$. By Theorem 11.3-4 and the fact that $\mathcal{O}'_T$ is closed under convergence, (3) converges in $\mathcal{O}'_T$ and its limit f is also in $\mathcal{O}'_T$. The proof of Corollary 2.4-3b also shows that, except for the constant term b_0, (3) is a finite-order derivative of a trigonometric series that converges uniformly for all t to a continuous function. Consequently, by Theorem 4.3-1, (3) converges in $\mathcal{S}'$.

Finally, the fact that (3) converges in $\mathcal{P}'_T$ allows us to form the "dot product" of (3) with any element of $\mathcal{P}_T$ term by term. Hence,

$$F_\nu = \frac{1}{T} f(t) \cdot e^{-i\nu\omega t} = \frac{1}{T} \sum_{k=-\infty}^{\infty} b_k \int_a^{a+T} e^{i\omega t(k-\nu)}\, dt \qquad (4)$$

But

$$\int_a^{a+T} e^{i\omega t(k-\nu)}\, dt = \begin{cases} T & k = \nu \\ 0 & k \neq \nu \end{cases}$$

Therefore, $F_\nu = b_\nu$, which completes the proof.

Conversely, we shall now show that every periodic distribution possesses a Fourier series.

Theorem 11.6-2

Let f be any periodic distribution with a period T and let the constants F_ν be given by (2). Then, $\{F_\nu\}_{\nu=1}^{\infty}$ is a sequence of slow growth and the series

$$\sum_{\nu=-\infty}^{\infty} F_\nu e^{i\nu\omega t} \qquad (5)$$

converges in $\mathcal{S}'$ to $f(t)$.

Proof: By using a fixed ξ in $\mathcal{U}_T$, we may rewrite (2) as

$$F_\nu = \frac{1}{T} \langle f(t), \xi(t) e^{-i\nu\omega t} \rangle$$

By Theorem 3.3-1, there exist a constant C and a nonnegative integer r such that

$$|F_\nu| \leq \frac{C}{T} \sup_t \left| \frac{d^r}{dt^r} [\xi(t) e^{-i\nu\omega t}] \right|$$
$$\leq \frac{C}{T} \sum_{\mu=0}^{r} (\nu\omega)^\mu \binom{r}{\mu} \sup_t |\xi^{(r-\mu)}(t)|$$
$$\leq M|\nu|^r \qquad \nu \neq 0$$

where M is a sufficiently large constant. Thus, the F_ν comprise a sequence of slow growth.

Again by Corollary 2.4-3b and Theorem 4.3-1, the series (5) converges in $\mathcal{S}'$. That its limit is f is all that remains to be shown.

By using the change of variable $x = tT/2\pi$ and the fact that $\delta(at) = \delta(t)/a$ $(a > 0)$, we can alter (11) of Sec. 2.4 into the following series expansion for $\delta_T(t)$:

$$\delta_T(x) = \frac{1}{T} \sum_{\nu=-\infty}^{\infty} e^{i\nu\omega x} \qquad (6)$$

Then,

$$f(t) = f(t) \,\Delta\, \delta_T(t)$$
$$= f(t) \,\Delta\, \frac{1}{T} \sum_{\nu} e^{i\nu\omega t}$$

By the continuity of T-convolution (see Theorem 11.4-4) and the fact that (6) converges in $\mathcal{O}'_T$ (see Theorem 11.3-4), we may interchange the summation and T-convolution. By invoking (15) of Sec. 11.4, we can then write

$$f(t) = \frac{1}{T} \sum_{\nu} f(t) \,\Delta\, e^{i\nu\omega t} = \frac{1}{T} \sum_{\nu} e^{i\nu\omega t}[f(\tau) \cdot e^{-i\nu\omega t}]$$
$$= \sum_{\nu} F_\nu e^{i\nu\omega t} \qquad \text{Q.E.D.}$$

Because the space $\mathcal{S}'$ is closed under convergence and because (5) converges in $\mathcal{S}'$, Theorem 11.6-2 implies that *every periodic distribution is a distribution of slow growth*. It also implies the uniqueness of the Fourier series: *if two distributions f and g in $\mathcal{O}'_T$ have the same Fourier coefficients, then $f = g$.*

Example 11.6-1 The Fourier coefficients of $\delta_T^{(k)}(t)$ ($k = 1, 2, 3, \ldots$) are

$$F_\nu = \frac{1}{T} \delta_T^{(k)}(t) \cdot e^{-i\nu\omega t} = \frac{(-1)^k}{T} \delta_T(t) \cdot \frac{d^k}{dt^k} e^{-i\nu\omega t}$$
$$= \frac{1}{T} (i\nu\omega)^k$$

Thus, the Fourier series of $\delta_T^{(k)}(t)$ is

$$\delta_T^{(k)}(t) = \frac{1}{T} \sum_{\nu=-\infty}^{\infty} (i\nu\omega)^k e^{i\nu\omega t} \qquad (7)$$

an expression that can also be obtained by differentiating (6) under the summation sign.

An essential property of the Fourier series that makes it so useful as a tool for solving T-convolution equations is that it converts T-convolution into multiplication.

Theorem 11.6-3

If f and g are in $\mathcal{O}'_T$ and if F_ν and G_ν ($\nu = 0, \pm 1, \pm 2, \ldots$) are their respective Fourier coefficients, then the Fourier coefficients of $f \,\Delta\, g$ are $TF_\nu G_\nu$.

Proof

$$\frac{1}{T}[f(t) \,\Delta\, g(t)] \cdot e^{-i\nu\omega t} = \frac{1}{T} f(t) \cdot [g(\tau) \cdot e^{-i\nu\omega(t+\tau)}]$$

$$= \frac{1}{T}[f(t) \cdot e^{-i\nu\omega t}][g(\tau) \cdot e^{-i\nu\omega\tau}]$$

$$= TF_\nu G_\nu \quad \text{Q.E.D.}$$

Example 11.6-2 We have seen that for f in $\mathcal{O}'_T$

$$f^{(k)} = \delta_T^{(k)} \,\Delta\, f \qquad k = 1, 2, 3, \ldots \tag{8}$$

and that the Fourier coefficients of $\delta_T^{(k)}$ are $(i\nu\omega)^k/T$. By applying the previous theorem to (8) and denoting the Fourier coefficients of f by F_ν, we find the Fourier coefficients of $f^{(k)}$ to be $(i\nu\omega)^k F_\nu$.

We have also seen that, if τ is a real number, then

$$f(t - \tau) = \delta_T(t - \tau) \,\Delta\, f(t) \tag{9}$$

Since the Fourier coefficients of $\delta_T(t - \tau)$ are clearly $e^{-i\nu\omega\tau}/T$, we also find that those of $f(t - \tau)$ are $e^{-i\nu\omega\tau} F_\nu$.

PROBLEMS

1 Compute the Fourier series of the following periodic distributions. You may leave your answers for the Fourier coefficients in terms of convergent definite integrals.

   a. $f(t) = \begin{cases} 1 & 0 \leq t < 1 \\ 0 & 1 \leq t < 2 \end{cases}$

   $f(t) = f(t - 2) \qquad -\infty < t < \infty$

   b. $f(t) = \sum_{n=-\infty}^{\infty} g(t - 2\pi n)$

where

$$g(t) = \text{Pf} \, \frac{1_+(t) 1_+(2\pi - t)}{t}$$

   c. $\delta_T(t - a) + \delta_T(t + a) \qquad T = 1,\, 0 < a < \tfrac{1}{2}$

   d. $\text{Pf} \, \dfrac{1}{\sin t}$

   e. $\text{Pf} \, \dfrac{1}{|\sin t|}$

   f. $\text{Pf} \tan t$

2 Show that, if the coefficients b_ν are not of slow growth, then the series (3) will not converge even in $\mathfrak{D}'$.

3 Determine the Fourier series of the periodic function

$$f(t) = e^t \qquad 0 \le t < 2\pi$$
$$f(t) = f(t - 2\pi) \qquad -\infty < t < \infty$$

first by a direct computation and then in the following alternative way. Differentiate $f(t)$ and its Fourier series $\Sigma_\nu F_\nu e^{i\nu t}$ and, by using the Fourier series for $\delta_{2\pi}(t)$, show that

$$\sum_\nu (1 - i\nu) F_\nu e^{i\nu t} = \frac{e^{2\pi} - 1}{2\pi} \sum_\nu e^{i\nu t}$$

A term-by-term comparison determines the F_ν.

4 Show that, if the distribution $f(t)$ satisfies

$$(e^{-i\omega t} - 1) f(t) = 0 \qquad \omega = \frac{2\pi}{T}$$

for all t, then $f = C\delta_T$, where C is an arbitrary constant. HINT: See Sec. 3.2.

5 Use the assertion of the preceding problem to show that

$$\frac{1}{T} \sum_{\nu = -\infty}^{\infty} e^{i\nu\omega t} = \delta_T(t)$$

6 Find a necessary and sufficient condition on a periodic distribution f that ensures that its primitives $f^{(-1)}$ are also periodic.

7 Prove that the terms of a Fourier series can be grouped in any fashion before summing. HINT: This can be done to any absolutely convergent series of complex numbers.

8 Show that, if a periodic distribution f is even, then its Fourier coefficients are given by

$$F_\nu = \frac{1}{T} f(t) \cdot \cos \nu\omega t$$

and its Fourier series becomes

$$f(t) = F_0 + 2 \sum_{\nu = 1}^{\infty} F_\nu \cos \nu\omega t$$

What can be said if f is odd?

11.7 THE FINITE FOURIER TRANSFORMATION AND THE SOLUTION OF T - CONVOLUTION EQUATIONS

A somewhat different way to view the Fourier series makes its analogy to the Fourier or Laplace transformations still more evident. The Fourier transformation converts any distribution in $\mathfrak{D}'$ into an ultradistribution, and the Laplace transformation converts certain distributions into functions that are analytic in at least a strip. On the other hand, one can apply the Fourier-coefficients formula (2) of Sec. 11.6 to any distribution in $\mathcal{O}'_T$ to convert it into a sequence of complex numbers. This defines the *direct finite Fourier transformation* $\mathfrak{F}_T$, which is, therefore, a one-to-one mapping of the space $\mathcal{O}'_T$ onto the space of sequences of numbers of slow growth. The *inverse finite Fourier transformation* $\mathfrak{F}_T^{-1}$ is defined in turn by the Fourier series representation (1) of Sec. 11.6. The symbolism for these transformations is

$$\mathfrak{F}_T f = \{F_\nu\}_{\nu=-\infty}^{\infty}$$

and

$$\mathfrak{F}_T^{-1}\{F_\nu\}_{\nu=-\infty}^{\infty} = f$$

where f is an element of $\mathcal{O}'_T$ and $\{F_\nu\}_{\nu=-\infty}^{\infty}$ is the corresponding sequence of Fourier coefficients.

An important application of the finite Fourier transformation is in the resolution of the T-convolution equation

$$f \, \Delta \, u = g \tag{1}$$

where f and g are given elements of $\mathcal{O}'_T$ and the solution u is also required to be in $\mathcal{O}'_T$. As was discussed in Sec. 11.5, this problem can be solved quite simply if f has a T-convolution inverse $f^{\Delta-1}$, which is by definition an element of $\mathcal{O}'_T$ that satisfies the equation

$$f \, \Delta \, f^{\Delta-1} = \delta_T \tag{2}$$

By applying the direct finite Fourier transformation to (2) and invoking Theorem 11.6-3, we obtain the following infinite set of equations, wherein the X_ν denote the Fourier coefficients of $f^{\Delta-1}$:

$$T F_\nu X_\nu = \frac{1}{T} \qquad \nu = 0, \pm 1, \pm 2, \ldots \tag{3}$$

All these equations can be satisfied only if every F_ν is different from zero. In this case, we can divide both sides of (3) by TF_ν and then apply the inverse finite Fourier transformation to obtain

$$f^{\Delta-1} = \mathfrak{F}_T^{-1}\{X_\nu\}_{\nu=-\infty}^{\infty} = \frac{1}{T^2} \mathfrak{F}_T^{-1} \left\{ \frac{1}{F_\nu} \right\}_{\nu=-\infty}^{\infty}$$

$$= \frac{1}{T^2} \sum_{\nu=-\infty}^{\infty} \frac{1}{F_\nu} e^{i\nu\omega t} \tag{4}$$

However, the right-hand side will have a sense if (and only if, according to Sec. 11.6, Prob. 2) $\{1/F_\nu\}_{\nu=-\infty}^{\infty}$ is a sequence of slow growth.

Under these conditions on the F_ν, we can employ Theorems 11.5-1 and 11.6-3 to obtain a solution to (1) for any g in $\mathcal{O}'_T$ by using the finite Fourier transformation. Indeed, with $\{G_\nu\}_{\nu=-\infty}^{\infty} = \mathfrak{F}_T g$, the solution to (1) is

$$u(t) = \mathfrak{F}_T^{-1}\left\{\frac{G_\nu}{TF_\nu}\right\}_{\nu=-\infty}^{\infty} = \frac{1}{T}\sum_{\nu=-\infty}^{\infty}\frac{G_\nu}{F_\nu}e^{i\nu\omega t} \tag{5}$$

Thus, we have arrived at

Theorem 11.7-1

The T-convolution equation (1) has a unique solution in $\mathcal{O}'_T$ for every g in $\mathcal{O}'_T$ if none of the Fourier coefficients F_ν of f are zero and if $\{1/F_\nu\}_{\nu=-\infty}^{\infty}$ is a sequence of slow growth. In this case, the solution is given by (5).

On the other hand, if some of the F_ν are zero, then $f^{\Delta-1}$ does not exist, since there is no set of X_ν for which all the equations (3) will be satisfied. This means that (1) will not have a solution for every g in $\mathcal{O}'_T$. However, for certain g, (1) will have a solution and, in fact, an infinite number of solutions.

Indeed, let us assume that $F_\nu = 0$ for $\nu = \nu_k$ ($k = 0, \pm 1, \pm 2, \ldots$) and that $F_\nu \neq 0$ for the rest of the ν, say for $\nu = \mu_j$ ($j = 0, \pm 1, \pm 2, \ldots$). Then, (1) will have at least one solution in $\mathcal{O}'_T$ if the $G_{\nu_k} = 0$ and if the G_{μ_j} are such that the G_{μ_j}/TF_{μ_j} either are finite in number or form a sequence of slow growth. Any such solution can be constructed as follows. Set $U_{\mu_j} = G_{\mu_j}/TF_{\mu_j}$; choose the U_{ν_k} arbitrarily, provided that the resulting sequence $\{U_\nu\}_{\nu=-\infty}^{\infty}$ is of slow growth. Then

$$u(t) = \mathfrak{F}_T^{-1}\{U_\nu\}_{\nu=-\infty}^{\infty} = \sum_{\nu=-\infty}^{\infty} U_\nu e^{i\nu\omega t} \tag{6}$$

There are clearly an infinity of such solutions because of our freedom of choice for the U_{ν_k}. [We have left it to the reader to show that, even with the U_{ν_k} chosen in this arbitrary fashion, (6) still constitutes a solution to (1).]

We turn now to simultaneous T-convolution equations. In matrix notation we have

$$\mathbf{f} \Delta \mathbf{u} = \mathbf{g} \qquad \mathbf{f} \in \mathcal{O}'_{T,n\times n}; \mathbf{u}, \mathbf{g} \in \mathcal{O}'_{T,n\times 1} \tag{7}$$

Let $\mathbf{F}_\nu$ be the $n \times n$ matrices of Fourier coefficients obtained by applying the finite Fourier transformation to $\mathbf{f}$, and let $\mathbf{G}_\nu$ be the $n \times 1$ matrices corresponding to $\mathbf{g}$. Thus, (7) may be transformed into

$$T\mathbf{F}_\nu \mathbf{U}_\nu = \mathbf{G}_\nu \qquad \nu = 0, \pm 1, \pm 2, \ldots \tag{8}$$

where the $\mathbf{U}_\nu$ are unknown $n \times 1$ matrices of complex numbers. If det $\mathbf{F}_\nu$ is not zero for every ν and if $\{1/\det \mathbf{F}_\nu\}_{\nu=-\infty}^{\infty}$ is a sequence of slow growth, then det $\mathbf{f}$ does possess an inverse in $\mathcal{O}'_T$ and we can invoke Theorem 11.5-3 to conclude that (7) possesses a unique solution in $\mathcal{O}'_{T,n\times 1}$. It can be found as follows. Upon solving (8) we have

$$\mathbf{U}_\nu = \frac{1}{T} \mathbf{F}_\nu^{-1} \mathbf{G}_\nu \qquad \nu = 0, \pm 1, \pm 2, \ldots$$

Note that the corresponding elements in the sequence of matrices $\{\mathbf{U}_\nu\}_{\nu=-\infty}^{\infty}$ will comprise sequences of slow growth because these elements are polynomials of $1/\det \mathbf{F}_\nu$ and of the elements of $\mathbf{F}_\nu$ and $\mathbf{G}_\nu$. Thus, we may apply the inverse finite Fourier transformation to get

$$\mathbf{u} = \mathfrak{F}_T^{-1}\{\mathbf{U}_\nu\}_{\nu=-\infty}^{\infty} = \mathfrak{F}_T^{-1}\left\{\frac{\mathbf{F}_\nu^{-1}\mathbf{G}_\nu}{T}\right\}_{\nu=-\infty}^{\infty} \tag{9}$$

In Fourier series form, the element in the pth row of $\mathbf{u}$ is

$$u_p = \sum_{\nu=-\infty}^{\infty} U_{\nu p} e^{i\nu\omega t}$$

where $U_{\nu p}$ is the element in the pth row of $\mathbf{U}_\nu$.

By employing the same technique for the equation

$$\mathbf{f} \triangle \mathbf{f}^{\triangle-1} = \delta_{T,n\times n}$$

we find that $\mathbf{f}^{\triangle-1}$ is given by

$$\mathbf{f}^{\triangle-1} = \frac{1}{T^2} \mathfrak{F}_T^{-1}\{\mathbf{F}_\nu^{-1}\}_{\nu=-\infty}^{\infty} \tag{10}$$

As a summary of these statements, we have

Theorem 11.7-2

The simultaneous T-convolution equations (7) have a unique solution $\mathbf{u}$ in $\mathcal{O}'_{T,n\times 1}$ for every $\mathbf{g}$ in $\mathcal{O}'_{T,n\times 1}$ if det $\mathbf{F}_\nu$ is different from zero for each ν and if $\{1/\det \mathbf{F}_\nu\}_{\nu=-\infty}^{\infty}$ is a sequence of slow growth. In this case, the solution is given by (9) and $\mathbf{f}^{\triangle-1}$ is given by (10).

When det $\mathbf{F}_\nu = 0$ for some ν (say $\nu = \nu_k$), the situation becomes considerably more complicated. Rather than attempt a complete discussion, we shall merely point out some of the difficulties that arise. First of all, the simultaneous equations, represented by

$$T\mathbf{F}_{\nu_k}\mathbf{U}_{\nu_k} = \mathbf{G}_{\nu_k} \tag{11}$$

either may be inconsistent or may be linearly dependent (Mirsky [1], chap. V). When they are inconsistent, they and, therefore, (7) will have no solutions. When the simultaneous equations are consistent but linearly dependent, they will have an infinity of solutions. (In this latter case, the $\mathbf{G}_{\nu_k}$ may not have to be zero for any such solution to exist.) Moreover, the solutions $\mathbf{U}_{\nu_k}$ cannot be chosen arbitrarily but must be so selected that they satisfy (11). In addition, the chosen $\mathbf{U}_{\nu_k}$ and the rest of the $\mathbf{U}_\nu$ must be such that sequences of corresponding elements in $\{\mathbf{U}_\nu\}_{\nu=-\infty}^{\infty}$ are of slow growth. The upshot of all this is that the simultaneous equations (7) will have either an infinity of solutions or none at all and that the Fourier coefficients of any such solution must still satisfy certain restrictions even for those ν for which det $\mathbf{F}_\nu = 0$.

PROBLEMS

1 Demonstrate that the finite Fourier transformation is a continuous operation in the following sense. If $\{f_\mu\}_{\mu=1}^\infty$ converges in $\mathcal{O}'_T$ to f and if

$$\mathfrak{F}_T f_\mu = \{F_{\mu\nu}\}_{\nu=1}^\infty \qquad \mathfrak{F}_T f = \{F_\nu\}_{\nu=1}^\infty$$

then, for each fixed ν, $F_{\mu\nu} \to F_\nu$ as $\mu \to \infty$.

2 Show that (6) is a solution to (1) even when the U_{ν_k} are chosen in the described arbitrary fashion.

3 Find a solution in $\mathcal{O}'_T$ for the following two T-convolution equations. In both cases, $T = 1$.

a. $\left[\delta_T\left(t - \dfrac{1}{\sqrt{5}}\right) + \delta_T\left(t + \dfrac{1}{\sqrt{5}}\right) \right] \Delta u(t) = \delta_T^{(1)}(t)$

b. $[\delta_T(t - \tfrac{1}{8}) + \delta_T(t + \tfrac{1}{8})] \Delta u(t) = \delta_T(t - \tfrac{1}{4}) - \delta_T(t + \tfrac{1}{4})$

11.8 SOME APPLICATIONS TO DIFFERENTIAL AND DIFFERENCE EQUATIONS

In this last section, the theory of periodic distributions will be applied to solve in $\mathcal{O}'_T$ ordinary linear differential and difference equations with constant coefficients.

Ordinary linear differential equations with constant coefficients Consider the differential equation

$$a_n u^{(n)} + a_{n-1} u^{(n-1)} + \cdots + a_0 u = g \qquad a_n \neq 0, n > 0 \tag{1}$$

or, equivalently,

$$(a_n \delta_T^{(n)} + a_{n-1} \delta_T^{(n-1)} + \cdots + a_0 \delta_T) \Delta u = g$$

where the a_k ($k = 0, 1, \ldots, n$) are constants, g is a given element of $\mathcal{O}'_T$,

340 *Distribution theory and transform analysis* *Chap. 11*

and u is an unknown element in $\mathcal{O}'_T$ that we seek. By virtue of Example 11.6-2, the finite Fourier transformation converts (1) into

$$[a_n(i\nu\omega)^n + a_{n-1}(i\nu\omega)^{n-1} + \cdots + a_0]U_\nu = G_\nu,$$
$$\omega = \frac{2\pi}{T}; \nu = 0, \pm 1, \pm 2, \ldots \quad (2)$$

If it turns out that none of the roots of the polynomial

$$P(\zeta) = a_n\zeta^n + a_{n-1}\zeta^{n-1} + \cdots + a_0 \quad (3)$$

coincide with any of the values $i\nu\omega = i\nu 2\pi/T$ ($\nu = 0, \pm 1, \pm 2, \ldots$), then each U_ν is uniquely determined by (2). In this case $\{1/P(i\nu\omega)\}_{\nu=-\infty}^{\infty}$ is clearly a sequence of slow growth. Thus, according to Theorem 11.7-1, the unique solution to (1) is given by the following Fourier series:

$$u(t) = \sum_{\nu=-\infty}^{\infty} \frac{G_\nu}{P(i\nu\omega)} e^{i\nu\omega t} \quad (4)$$

Equation (1) can also be solved directly in the T-convolution algebra $\mathcal{O}'_T$ by using the T-convolution inverse of

$$f \triangleq a_n \delta_T{}^{(n)} + a_{n-1}\delta_T{}^{(n-1)} + \cdots + a_0 \delta_T \quad (5)$$

instead of using the finite Fourier transformation. Let us first factor the polynomial (3) into the form

$$P(\zeta) = a_n(\zeta - \gamma_1)(\zeta - \gamma_2) \cdots (\zeta - \gamma_n)$$

where some or all of the roots γ_k may be equal to each other. Certainly, f can be written in the form

$$f = a_n(\delta_T{}^{(1)} - \gamma_1 \delta_T) \Delta (\delta_T{}^{(1)} - \gamma_2 \delta_T) \Delta \cdots \Delta (\delta_T{}^{(1)} - \gamma_n \delta_T) \quad (6)$$

Moreover, still assuming that none of the γ_k equal $i\nu\omega$ ($\nu = 0, \pm 1, \pm 2, \ldots$), one can easily verify that the T-convolution inverse of $\delta_T{}^{(1)} - \gamma_k \delta_T$ is the periodic function h_k defined by the equations

$$\begin{aligned} h_k(t) &\triangleq \frac{e^{\gamma_k t}}{1 - e^{\gamma_k T}} & 0 \leq t < T \\ h_k(t) &\triangleq h_k(t - T) & -\infty < t < \infty \end{aligned} \quad (7)$$

By virtue of Theorem 11.5-2, the T-convolution inverse of (5) is

$$f^{\Delta-1} = \frac{1}{a_n} h_1 \Delta h_2 \Delta \cdots \Delta h_n$$

and, consequently, the solution to (1) is found to be

$$u = \frac{1}{a_n} h_1 \Delta h_2 \Delta \cdots \Delta h_n \Delta g \quad (8)$$

Let us compare these two techniques for solving (1). The first method is computationally easy to perform except that the evaluation of the Fourier coefficients G_ν is a possible stumbling block. Moreover, the solution is rendered as a Fourier series (4). The second method yields the solution in a closed form (8), but it requires the determination of the roots of the polynomial (3) and the evaluation of n T-convolutions.

In both cases, one must check to make sure that none of the values $i\nu\omega$ ($\nu = 0, \pm 1, \pm 2, \ldots$) is a root of the polynomial (3) in order to validate the solution. In many instances, this check may be made without factoring (3). If none of the roots are purely imaginary, then they cannot coincide with any of the $i\nu\omega$. There are a variety of tests, such as the Routh-Hurwitz criteria, by which one can ascertain whether the roots lie in certain regions of the complex plane that do not contain the imaginary axis. (See, for example, Kaplan [2], chap. 7.) If (3) satisfies these tests, the need to factor it can be avoided in the first method.

Example 11.8-1 Let us solve in $\mathcal{P}'_T$ the differential equation

$$u^{(2)}(t) - u(t) = e^{i\omega t} \qquad \omega = \frac{2\pi}{T} \tag{9}$$

by these two methods. By applying the finite Fourier transformation to (9), we obtain

$$(-\omega^2 \nu - 1) U_\nu = \begin{cases} 1 & \nu = 1 \\ 0 & \nu \neq 1 \end{cases}$$

Hence,

$$u(t) = -\frac{e^{i\omega t}}{1 + \omega^2} \tag{10}$$

Our first method is utterly simple for this example, because it is perfectly obvious what the Fourier coefficients of the right-hand side of (9) are. In general, this will not be the case. Moreover, for the sake of comparing the relative difficulty of the two methods, this example is unfair, since such a complete simplification does not arise in the second method, as we shall now see. However, it does illustrate the techniques involved.

We may rewrite (9) as

$$f \Delta u = (\delta_T{}^{(1)} - \delta_T) \Delta (\delta_T{}^{(1)} + \delta_T) \Delta u(t) = e^{i\omega t}$$

In accordance with (7), we set

$$h_1(t) = \frac{e^t}{1 - e^T} \qquad h_2(t) = \frac{e^{-t}}{1 - e^{-T}} \qquad 0 \leq t < T$$

and

$$h_1(t) = h_1(t - T) \qquad h_2(t) = h_2(t - T) \qquad -\infty < t < \infty$$

Hence,
$$f^{\Delta-1}(t) = \int_0^T h_1(t-\tau)h_2(\tau)\,d\tau$$
$$= \frac{e^t}{2(1-e^T)} - \frac{e^{-t}}{2(1-e^{-T})} \qquad 0 \le t < T \tag{11}$$
and
$$f^{\Delta-1}(t) = f^{\Delta-1}(t-T) \qquad -\infty < t < \infty$$
Thus, we now have
$$u(t) = f^{\Delta-1} \Delta\, e^{i\omega t} = -\frac{e^{i\omega t}}{1+\omega^2} \tag{12}$$
which is the same solution as (10).

Finally, let us note that (9) has other solutions which we can obtain by adding to (12) the general solution $Ae^t + Be^{-t}$ $(-\infty < t < \infty)$ of the homogeneous equation $u^{(2)} - u = 0$. However, $Ae^t + Be^{-t}$ is not in $\mathcal{O}'_T$, and this is the reason why it was not generated as part of the solution. In other words, the techniques developed above merely pick out the solutions of a differential equation that are in the space $\mathcal{O}'_T$ and ignore all other solutions.

Simultaneous ordinary linear differential equations with constant coefficients Simultaneous equations of this type can be represented in the form
$$(\mathbf{L}\delta_T) \Delta\, \mathbf{u} = \mathbf{g} \tag{13}$$
where the given $\mathbf{g}$ and the unknown $\mathbf{u}$ are in $\mathcal{O}'_{T,n\times 1}$ and $\mathbf{L}\delta_T$ is an $n \times n$ matrix each element of which has the form
$$a_m \delta_T{}^{(m)} + a_{m-1}\delta_T{}^{(m-1)} + \cdots + a_0 \delta_T \tag{14}$$
the a_k being constants. The finite Fourier transformation converts (13) into
$$T\mathbf{F}_\nu \mathbf{U}_\nu = \mathbf{G}_\nu \qquad \nu = 0, \pm 1, \pm 2, \ldots \tag{15}$$
where the $n \times n$ matrices $\mathbf{F}_\nu$ have elements of the form
$$\frac{1}{T}[a_m(i\nu\omega)^m + a_{m-1}(i\nu\omega)^{m-1} + \cdots + a_0] \qquad \omega = \frac{2\pi}{T}$$

It follows then that $\det \mathbf{F}_\nu$ is a polynomial $P(\zeta)$ in $\zeta = i\nu\omega$. If $P(\zeta)$ is not identically zero and if its roots do not coincide with any of the values of $i\nu\omega$, then $\{1/\det \mathbf{F}_\nu\}_{\nu=-\infty}^{\infty}$ is a sequence of slow growth and we may apply Theorem 11.7-2. Thus, (15) may be solved to determine the $\mathbf{U}_\nu$ and thereby $\mathbf{u}(t)$. Here again, tests such as the Routh-Hurwitz criteria may suffice for checking the root location of $P(\zeta)$.

Example 11.8-2 Let $g_1(t)$ be the periodic "pulse" function of period $T = 2$ defined by
$$g_1(t) = \begin{cases} 1 & 0 \leq t < 1 \\ 0 & 1 \leq t < 2 \end{cases}$$
$$g_1(t) = g_1(t-2) \quad -\infty < t < \infty$$
and consider the simultaneous differential equations
$$\begin{aligned} u_1^{(1)} + u_1 + u_2 &= g_1 \\ u_1^{(1)} + u_2^{(1)} - u_2 &= 0 \end{aligned} \tag{16}$$
Since $\omega = 2\pi/T = \pi$, the Fourier coefficients $G_{1\nu}$ of g_1 are
$$G_{1\nu} = \tfrac{1}{2} \int_0^1 e^{-i\nu\pi t}\, dt = \begin{cases} \dfrac{1}{i\nu\pi} & \nu \text{ odd} \\ 0 & \nu \text{ even} \end{cases}$$

It follows from (16) that the Fourier coefficients $U_{1\nu}$ and $U_{2\nu}$ of $u_1(t)$ and $u_2(t)$ are related to $G_{1\nu}$ by
$$\begin{aligned} (i\nu\pi + 1)U_{1\nu} + U_{2\nu} &= G_{1\nu} \\ i\nu\pi U_{1\nu} + (i\nu\pi - 1)U_{2\nu} &= 0 \end{aligned}$$
which may be solved for $U_{1\nu}$ and $U_{2\nu}$. This yields the following solution for (16), wherein the summations are over all odd integral values of ν,
$$u_1(t) = \sum_{\nu \text{ odd}} \frac{1 - i\nu\pi}{-\nu^2\pi^2 + i\nu\pi(\nu^2\pi^2 + 1)} e^{i\nu\pi t}$$
$$u_2(t) = \sum_{\nu \text{ odd}} \frac{1}{\nu^2\pi^2 + 1 + i\nu\pi} e^{i\nu\pi t}$$

Linear difference equations with constant coefficients We restrict our discussion to difference equations of the form
$$a_n u(t + n\tau) + a_{n-1} u(t + n\tau - \tau) + \cdots + a_0 u(t) = g(t) \tag{17}$$
where τ is a real constant and the a_ν are complex constants with $a_n \neq 0$, $a_0 \neq 0, n \geq 1$. Again by Example 11.6-2, the Fourier coefficients of u and g are related by
$$(a_n e^{i\nu\omega n\tau} + a_{n-1} e^{i\nu\omega(n-1)\tau} + \cdots + a_0) U_\nu = G_\nu \tag{18}$$
In order to solve (18) for the U_ν (assuming that the $G_\nu \neq 0$), the polynomial
$$P(\zeta) = a_n \zeta^n + a_{n-1} \zeta^{n-1} + \cdots + a_0 \tag{19}$$
must have no roots equal to the $e^{i\nu\omega\tau}$ ($\nu = 0, \pm 1, \pm 2, \ldots$). This will certainly be the case if every root ζ_k of $P(\zeta)$ has a magnitude differing from unity. (In many cases, this can again be ascertained by using tests that are analogous to the Routh-Hurwitz criteria.) Assuming that this is

true, it is again obvious that $\{1/P(e^{i\nu\omega\tau})\}_{\nu=-\infty}^{\infty}$ is a bounded sequence and therefore a sequence of slow growth. The solution of (17) is, therefore,

$$u(t) = \sum_{\nu=-\infty}^{\infty} \frac{G_\nu}{P(e^{i\nu\omega\tau})} e^{i\nu\omega t}$$

The discussion of simultaneous linear difference equations proceeds in a conventional way. Now it is the determinant of a matrix whose elements have the form (19) that must be checked for possible roots with unit magnitudes.

PROBLEMS

1 Verify (11) and (12).

2 Solve the following differential equations in $\mathcal{P}'_T (T=1)$:
 a. $u^{(2)}(t) + 2u^{(1)}(t) + u(t) = \sin 2\pi t$
 b. $u^{(2)}(t) - 4u(t) = g(t)$
  $g(t) = t \qquad 0 \le t < 1$
  $g(t) = g(t-1) \qquad -\infty < t < \infty$
 c. $u^{(2)}(t) + u(t) = \delta_1^{(1)}(t) \triangleq \sum_{\nu=-\infty}^{\infty} \delta^{(1)}(t-\nu)$

3 Find all possible solutions in $\mathcal{P}'_T$ ($T = 2\pi$) of the differential equation
$u^{(2)}(t) + 4u(t) = e^{it}$

4 Solve the following difference equations in $\mathcal{P}'_T$:
 a. $u(t+2) + 5u(t+1) + 6u(t) = \cos t \qquad T = 2\pi$
 b. $u(t+3) + 2u(t+2) + u(t+1) + 2u(t) = \sin \pi t \qquad T = 2$
 c. $u(t+4) + u(t) = \delta_1(t) \triangleq \sum_{\nu=-\infty}^{\infty} \delta(t-\nu) \qquad T = 1$

5 Find all possible solutions in $\mathcal{P}'_T$ ($T = 2\pi$) of the following difference-differential equations:
 a. $u^{(1)}(t+1) + u(t) = e^{it}$
 b. $u^{(1)}(t+\pi) + iu(t) = e^{it}$
 c. $u^{(2)}(t) + u^{(1)}(t) + u(t+1) = \delta_{2\pi}(t) \triangleq \sum_{\nu=-\infty}^{\infty} \delta(t-2\pi\nu)$

6 Solve the following simultaneous equations in $\mathcal{P}'_T$ ($T = 2\pi$):
 a. $u_1^{(1)}(t) + u_1(t) + 2u_2^{(1)}(t) - u_2(t) = \sin t$
  $u_1^{(1)}(t) - 2u_1(t) + u_2(t) = \delta_\pi^{(1)}(t) \triangleq \sum_{\nu=-\infty}^{\infty} \delta^{(1)}(t-\nu\pi)$
 b. $u_1(t+1) - u_1(t) + u_2(t+1) - 3u_2(t) = \operatorname{Pf} \dfrac{1}{\sin t}$
  $2u_1(t+1) + u_2(t+1) + 2u_2(t) = 0$
 c. $u_1^{(1)}(t+1) + u_2(t+1) + u_2^{(1)}(t) = \delta_{2\pi}(t)$
  $u_1^{(1)}(t) - u_1(t) + u_2(t) = 0$

Appendix A The Axioms for a Linear Space

A space $\mathfrak{B}$ of elements $f, g, h, \ldots$ is said to be a *linear space* if the following conditions are satisfied. There is an operation $+$, called "addition," such that any pair of elements f and g can be combined to yield a unique element $f + g$, which is also in $\mathfrak{B}$. Similarly, there is another operation, called "multiplication by a complex number," such that any complex number α and any element f can be combined to yield a unique element αf, which is again in $\mathfrak{B}$. Moreover, the following axioms are fulfilled for every choice of the elements f, g, and h and the complex numbers α and β:

1. $f + g = g + f$
2. $(f + g) + h = f + (g + h)$
3. There exists a unique element $\emptyset$ in $\mathfrak{B}$ such that $f + \emptyset = f$. (Instead of the symbol $\emptyset$, we have always used 0 to denote this element in our various spaces of functions and distributions.)
4. For each f in $\mathfrak{B}$ there exists a unique element $-f$ such that $f + (-f) = \emptyset$.
5. $\alpha(\beta f) = (\alpha\beta)f$
6. $1f = f$
7. $\alpha(f + g) = \alpha f + \alpha g$
8. $(\alpha + \beta)f = \alpha f + \beta f$

Appendix B Tables of Formulas for the Right-sided Laplace Transformation

This appendix is divided into two tables. The first one presents a number of operational formulas for the right-sided Laplace transformation, and the second one lists a variety of transform formulas for this transformation. Throughout these two tables the following definitions for some of our symbols will be understood. The meanings of other symbols are stated in Appendix C.

a, τ	real numbers
b	a real positive number
k	a nonnegative integer
α, β, η	complex numbers
γ	= Euler's constant = $0.5772 \cdots$
C	$= e^\gamma$
$\psi(k)$	$= -\gamma + \sum_{\nu=1}^{k-1} \frac{1}{\nu} \quad k = 2, 3, 4, \ldots$

More extensive tables of ordinary Laplace transformation formulas can be found in Erdélyi (ed.) [1] and Van der Pol and Bremmer [1]. The latter reference presents formulas for the ordinary two-sided Laplace transformation as well as for the ordinary right-sided one. A long list of distributional transform formulas is given by Lavoine [1].

Table B.1 Operational formulas for the right-sided Laplace transformation

In this table we assume that $f(t)$ and $g(t)$ are right-sided Laplace-transformable distributions and that

$$\mathfrak{L}f(t) = F(s) \quad \mathrm{Re}\, s > \sigma_f$$
$$\mathfrak{L}g(t) = G(s) \quad \mathrm{Re}\, s > \sigma_g$$

No.	$h(t)$	$\mathfrak{L}h(t)$
1	$\alpha f(t)$	$\alpha F(s) \quad \mathrm{Re}\, s > \sigma_f$
2	$f(t) + g(t)$	$F(s) + G(s) \quad \mathrm{Re}\, s > \sigma_f + \sigma_g$, at least
3	$f(t - \tau)$	$e^{-s\tau}F(s) \quad \mathrm{Re}\, s > \sigma_f$
4	$e^{-\alpha t}f(t)$	$F(s + \alpha) \quad \mathrm{Re}\, s > \sigma_f - \mathrm{Re}\,\alpha$
5	$f(bt)$	$\dfrac{1}{b}F\left(\dfrac{s}{b}\right) \quad \mathrm{Re}\, s > b\sigma_f$
6	$\dfrac{1}{b}e^{\alpha t/b}f\left(\dfrac{t}{b}\right)$	$F(bs - \alpha) \quad \mathrm{Re}\, s > \dfrac{\sigma_f - \mathrm{Re}\,\alpha}{b}$
7	$t^k f(t)$	$(-1)^k F^{(k)}(s) \quad \mathrm{Re}\, s > \sigma_f$
8	$f^{(k)}(t)$	$s^k F(s) \quad \mathrm{Re}\, s > \sigma_f$
9	$f(t) * g(t)$	$F(s)G(s) \quad \mathrm{Re}\, s > \sup(\sigma_f, \sigma_g)$
10	$1_+(t) * f(t)$	$\dfrac{F(s)}{s} \quad \mathrm{Re}\, s > \sup(\sigma_f, 0)$

[In this case,

$$1_+(t) * f(t) = 1_+(t - \tau)\int_\tau^t f(x)\, dx$$

whenever f is a locally integrable function whose support is bounded on the left at $t = \tau$]

11	Let f be a distribution of bounded support:	
	$\displaystyle\sum_{\nu=0}^{\infty} f(t - \nu b)$	$\dfrac{F(s)}{1 - e^{-bs}} \quad \mathrm{Re}\, s > 0$

Table B.2 Transform formulas for the right-sided Laplace transformation

In the following table a given pseudofunction may be a regular distribution for certain values of its parameters; in this case the Pf notation may be dropped. Also, when a Laplace transform is a multivalued function, it is the principal-branch values that are understood.

By our convention, the region of convergence is always taken as an open set (Re $s > \sigma_1$), even though in certain cases the Laplace transform definition has a sense on the boundary (Re $s = \sigma_1$).

No.	$f(t)$	$\mathfrak{L}f(t)$
1	$1_+(t)$	$\dfrac{1}{s}$ Re $s > 0$
2	Pf $t^\eta 1_+(t)$	$\dfrac{\Gamma(\eta+1)}{s^{\eta+1}}$ Re $s > 0$; $\eta \neq -1, -2, -3, \ldots$
3	Pf $(t-b)^\eta 1_+(t-b)$	$e^{-bs} \dfrac{\Gamma(\eta+1)}{s^{\eta+1}}$ Re $s > 0$; $\eta \neq -1, -2, -3, \ldots$
4	Pf $\dfrac{1_+(t)}{t}$	$-\log Cs$ Re $s > 0$
5	Pf $\dfrac{1_+(t)}{t^2}$	$s(\log Cs - 1)$ Re $s > 0$
6	Pf $\dfrac{1_+(t)}{t^3}$	$-\dfrac{s^2}{2}(\log Cs - \tfrac{3}{2})$ Re $s > 0$
7	Pf $\dfrac{1_+(t)}{t^k}$	$-\dfrac{(-s)^{k-1}}{(k-1)!}[\log s - \psi(k)]$ Re $s > 0$; $k = 2, 3, 4, \ldots$
8	Pf $\dfrac{1_+(t)b}{t(b+t)}$	$e^{bs}\,\mathrm{Ei}(-bs) - \log Cs$ Re $s > 0$
9	Pf $\dfrac{1_+(t)b^2}{t^2(b+t)}$	$bs(\log Cs - 1) - e^{bs}\,\mathrm{Ei}(-bs) + \log Cs$ Re $s > 0$
10	Pf $\dfrac{1_+(t)b}{t(t-b)}$	$\log Cs - e^{-bs}\,\mathrm{Ei}^*(bs)$ Re $s > 0$
11	Pf $\dfrac{1_+(t)b^2}{t^2(t-b)}$	$\log Cs - sb(\log Cs - 1) - e^{-bs}\,\mathrm{Ei}^*(bs)$ Re $s > 0$
12	Pf $\dfrac{1_+(t)1_+(b-t)}{t}$	$\mathrm{Ei}(-bs) - \log Cs$ $-\infty < \mathrm{Re}\, s < \infty$
13	Pf $\dfrac{1_+(t)}{t-b}$	$-e^{-bs}\,\mathrm{Ei}^*(bs)$ Re $s > 0$

Table B.2 *Transform formulas for the right-sided Laplace transformation (continued)*

No.	$f(t)$	$\mathfrak{L}f(t)$
14	$\dfrac{1_+(t)}{t+b}$	$-e^{-bs}\,\mathrm{Ei}(-bs) \qquad \mathrm{Re}\,s > 0$
15	$\mathrm{Pf}\,\dfrac{1_+(t)1_+(b-t)}{b-t}$	$e^{-bs}[\mathrm{Ei}^*(bs) - \log Cs] \qquad -\infty < \mathrm{Re}\,s < \infty$
16	$\mathrm{Pf}\,\dfrac{1_+(t)1_+(b-t)}{(b-t)^2}$	$se^{-bs}[\mathrm{Ei}^*(bs) - \log Cs + 1] - \dfrac{1}{b}$ $\qquad -\infty < \mathrm{Re}\,s < \infty$
17	$1_+(t)\log t$	$-\dfrac{\log Cs}{s} \qquad \mathrm{Re}\,s > 0$
18	$1_+(t)1_+(b-t)\log t$	$\dfrac{1}{s}[\mathrm{Ei}(-bs) - \log Cs - e^{-bs}\log b]$ $\qquad -\infty < \mathrm{Re}\,s < \infty$
19	$1_+(t)\log(t+b)$	$-\dfrac{1}{s}[e^{bs}\,\mathrm{Ei}(-bs) - \log b] \qquad \mathrm{Re}\,s > 0$
20	$1_+(t)\log|t-b|$	$-\dfrac{1}{s}[e^{-bs}\,\mathrm{Ei}^*(bs) - \log b] \qquad \mathrm{Re}\,s > 0$
21	$1_+(t)\log|t^2 - b^2|$	$\dfrac{1}{s}[2\log b - e^{bs}\,\mathrm{Ei}(-bs) - e^{-bs}\,\mathrm{Ei}^*(bs)]$ $\qquad \mathrm{Re}\,s > 0$
22	$1_+(t)(\log t)^2$	$\dfrac{1}{s}\left[(\log Cs)^2 + \dfrac{\pi^2}{6}\right] \qquad \mathrm{Re}\,s > 0$
23	$\mathrm{Pf}\,\dfrac{1_+(t)\log t}{t}$	$\tfrac{1}{2}\left[(\log Cs)^2 + \dfrac{\pi^2}{6}\right] \qquad \mathrm{Re}\,s > 0$
24	$1_+(t)e^{-\alpha t}$	$\dfrac{1}{s+\alpha} \qquad \mathrm{Re}\,s > -\mathrm{Re}\,\alpha$
25	$1_+(t)\sin\alpha t$	$\dfrac{\alpha}{s^2+\alpha^2} \qquad \mathrm{Re}\,s > |\mathrm{Im}\,\alpha|$
26	$1_+(t)\cos\alpha t$	$\dfrac{s}{s^2+\alpha^2} \qquad \mathrm{Re}\,s > |\mathrm{Im}\,\alpha|$
27	$1_+(t)\sinh\alpha t$	$\dfrac{\alpha}{s^2-\alpha^2} \qquad \mathrm{Re}\,s > |\mathrm{Re}\,\alpha|$
28	$1_+(t)\cosh\alpha t$	$\dfrac{s}{s^2-\alpha^2} \qquad \mathrm{Re}\,s > |\mathrm{Re}\,\alpha|$
29	$\dfrac{e^{\alpha t} - e^{\beta t}}{\alpha - \beta}1_+(t)$	$\dfrac{1}{(s-\alpha)(s-\beta)} \qquad \mathrm{Re}\,s > \sup(\mathrm{Re}\,\alpha, \mathrm{Re}\,\beta)$

Table B.2 *Transform formulas for the right-sided Laplace transformation (continued)*

No.	$f(t)$	$\mathscr{L}f(t)$
30	$\dfrac{\alpha e^{\alpha t} - \beta e^{\beta t}}{\alpha - \beta} 1_+(t)$	$\dfrac{s}{(s-\alpha)(s-\beta)}$ $\operatorname{Re} s > \sup(\operatorname{Re}\alpha, \operatorname{Re}\beta)$
31	$\operatorname{Pf} 1_+(t) t^\eta e^{\alpha t}$	$\dfrac{\Gamma(\eta+1)}{(s-\alpha)^{\eta+1}}$ $\operatorname{Re} s > \operatorname{Re}\alpha;\ \eta \neq -1, -2, -3, \ldots$
32	$\operatorname{Pf} \dfrac{1_+(t) e^{\alpha t}}{t}$	$-\log[C(s-\alpha)]$ $\operatorname{Re} s > \operatorname{Re}\alpha$
33	$\operatorname{Pf} \dfrac{1_+(t) e^{\alpha t}}{t^k}$	$-\dfrac{(-s+\alpha)^{k-1}}{(k-1)!}[\log(s-\alpha) - \psi(k)]$ $\operatorname{Re} s > \operatorname{Re}\alpha;\ k = 2, 3, 4, \ldots$
34	$e^{-bt^2} 1_+(t)$	$\dfrac{1}{2}\sqrt{\dfrac{\pi}{b}} e^{s^2/4b} \operatorname{erfc}\dfrac{s}{2\sqrt{b}}$ $-\infty < \operatorname{Re} s < \infty$
35	$\operatorname{Pf} 1_+(t) t^\eta \sin \alpha t$	$\dfrac{\Gamma(\eta+1)}{2i}\left[\dfrac{1}{(s-i\alpha)^{\eta+1}} - \dfrac{1}{(s+i\alpha)^{\eta+1}}\right]$ $\operatorname{Re} s > |\operatorname{Im}\alpha|;\ \eta \neq -1, -2, -3, \ldots$
36	$\dfrac{1_+(t) \sin \alpha t}{t}$	$\dfrac{1}{2i}\log\dfrac{s+i\alpha}{s-i\alpha}$ $\operatorname{Re} s > |\operatorname{Im}\alpha|$
37	$\operatorname{Pf}\dfrac{1_+(t)\sin \alpha t}{t^k}$	$\dfrac{(-1)^{k-1}}{(k-1)!2i}\{(s+i\alpha)^{k-1}[\log(s+i\alpha) - \psi(k)]$ $- (s-i\alpha)^{k-1}[\log(s-i\alpha) - \psi(k)]\}$ $\operatorname{Re} s > |\operatorname{Im}\alpha|;\ k = 2, 3, 4, \ldots$
38	$\operatorname{Pf} 1_+(t) t^\eta \cos \alpha t$	$\dfrac{\Gamma(\eta+1)}{2}\left[\dfrac{1}{(s-i\alpha)^{\eta+1}} + \dfrac{1}{(s+i\alpha)^{\eta+1}}\right]$ $\operatorname{Re} s > |\operatorname{Im}\alpha|;\ \eta \neq -1, -2, -3, \ldots$
39	$\operatorname{Pf}\dfrac{1_+(t)\cos \alpha t}{t}$	$-\gamma - \log\sqrt{s^2+\alpha^2}$ $\operatorname{Re} s > |\operatorname{Im}\alpha|$
40	$\operatorname{Pf}\dfrac{1_+(t)\cos \alpha t}{t^k}$	$\dfrac{(-1)^k}{(k-1)!2}\{(s-i\alpha)^{k-1}[\log(s-i\alpha) - \psi(k)]$ $+ (s+i\alpha)^{k-1}[\log(s+i\alpha) - \psi(k)]\}$ $\operatorname{Re} s > |\operatorname{Im}\alpha|;\ k = 2, 3, 4, \ldots$
41	$\dfrac{1_+(t)(\sin \alpha t)^2}{t}$	$\tfrac{1}{4}\log\left(1 + \dfrac{4\alpha^2}{s^2}\right)$ $\operatorname{Re} s > 2|\operatorname{Im}\alpha|$
42	$\operatorname{Pf}\dfrac{1_+(t)(\cos \alpha t)^2}{t}$	$-\tfrac{1}{4}\log[C^4 s^2(s^2 + 4\alpha^2)]$ $\operatorname{Re} s > 2|\operatorname{Im}\alpha|$
43	$\dfrac{1_+(t) \sin\sqrt{bt}}{t}$	$\pi \operatorname{erf}\sqrt{\dfrac{b}{4s}}$ $\operatorname{Re} s > 0$

Table B.2 *Transform formulas for the right-sided Laplace transformation (continued)*

No.	$f(t)$	$\mathfrak{L}f(t)$
44	$\dfrac{1_+(t)\sin\sqrt{bt}}{\sqrt{t}}$	$\sqrt{\dfrac{\pi}{s}}\,e^{-b/4s}\,\mathrm{erg}\sqrt{\dfrac{b}{4s}}\quad \mathrm{Re}\,s>0$
45	$\dfrac{1_+(t)\cos\sqrt{bt}}{\sqrt{t}}$	$\sqrt{\dfrac{\pi}{s}}\,e^{-b/4s}\quad \mathrm{Re}\,s>0$
46	$\mathrm{Pf}\,\dfrac{1_+(t)\sin\sqrt{bt}}{t^2}$	$-\pi\left(s+\dfrac{b}{2}\right)\mathrm{erf}\sqrt{\dfrac{b}{4s}}-\sqrt{b\pi s}\,e^{-b/4s}$ $\mathrm{Re}\,s>0$
47	$\mathrm{Pf}\,\dfrac{1_+(t)\cos\sqrt{bt}}{t^{\frac{3}{2}}}$	$-\pi\sqrt{b}\,\mathrm{erf}\sqrt{\dfrac{b}{4s}}-2\sqrt{\pi s}\,e^{-b/4s}\quad \mathrm{Re}\,s>0$
48	$\mathrm{Pf}\,\dfrac{1_+(t)\cos\sqrt{bt}}{t^{\frac{5}{2}}}$	$\pi\sqrt{b}\left(s+\dfrac{b}{6}\right)\mathrm{erf}\sqrt{\dfrac{b}{4s}}+\dfrac{1}{3}\sqrt{\pi s}\,(4s+b)e^{-b/4s}$ $\mathrm{Re}\,s>0$
49	$\mathrm{Pf}\,1_+(t)t^\eta\sinh\alpha t$	$\dfrac{\Gamma(\eta+1)}{2}\left[\dfrac{1}{(s-\alpha)^{\eta+1}}-\dfrac{1}{(s+\alpha)^{\eta+1}}\right]$ $\mathrm{Re}\,s>|\mathrm{Re}\,\alpha|;\ \eta\neq -1,-2,-3,\ldots$
50	$\dfrac{1_+(t)\sinh\alpha t}{t}$	$\tfrac{1}{2}\log\dfrac{s+\alpha}{s-\alpha}\quad \mathrm{Re}\,s>|\mathrm{Re}\,\alpha|$
51	$\mathrm{Pf}\,\dfrac{1_+(t)\sinh\alpha t}{t^k}$	$\dfrac{(-1)^k}{(k-1)!2}\{(s-\alpha)^{k-1}[\log(s-\alpha)-\psi(k)]$ $-(s+\alpha)^{k-1}[\log(s+\alpha)-\psi(k)]\}$ $\mathrm{Re}\,s>|\mathrm{Re}\,\alpha|;\ k=2,3,4,\ldots$
52	$\mathrm{Pf}\,1_+(t)t^\eta\cosh\alpha t$	$\dfrac{\Gamma(\eta+1)}{2}\left[\dfrac{1}{(s-\alpha)^{\eta+1}}+\dfrac{1}{(s+\alpha)^{\eta+1}}\right]$ $\mathrm{Re}\,s>|\mathrm{Re}\,\alpha|;\ \eta\neq -1,-2,-3,\ldots$
53	$\mathrm{Pf}\,\dfrac{1_+(t)\cosh\alpha t}{t}$	$-\log\sqrt{s^2-\alpha^2}-\gamma\quad \mathrm{Re}\,s>|\mathrm{Re}\,\alpha|$
54	$\mathrm{Pf}\,\dfrac{1_+(t)\cosh\alpha t}{t^k}$	$\dfrac{(-1)^k}{(k-1)!2}\{(s-\alpha)^{k-1}[\log(s-\alpha)-\psi(k)]$ $+(s+\alpha)^{k-1}[\log(s+\alpha)-\psi(k)]\}$ $\mathrm{Re}\,s>|\mathrm{Re}\,\alpha|;\ k=2,3,4,\ldots$
55	$\dfrac{1_+(t)\cosh\sqrt{bt}}{\sqrt{t}}$	$\sqrt{\dfrac{\pi}{s}}\,e^{b/4s}\quad \mathrm{Re}\,s>0$
56	$\dfrac{1_+(t)\sinh\sqrt{bt}}{\sqrt{t}}$	$\sqrt{\dfrac{\pi}{s}}\,e^{b/4s}\,\mathrm{erf}\sqrt{\dfrac{b}{4s}}\quad \mathrm{Re}\,s>0$

Table B.2 *Transform formulas for the right-sided Laplace transformation* (continued)

No.	$f(t)$	$\mathfrak{L}f(t)$
57	$\dfrac{1_+(t)\sinh\sqrt{bt}}{t}$	$\pi\,\mathrm{erg}\sqrt{\dfrac{b}{4s}}\qquad \mathrm{Re}\,s>0$
58	$\mathrm{Pf}\,\dfrac{1_+(t)\cosh\sqrt{bt}}{t^{3/2}}$	$\pi\sqrt{b}\,\mathrm{erg}\sqrt{\dfrac{b}{4s}}-2\sqrt{\pi s}\,e^{b/4s}\qquad \mathrm{Re}\,s>0$
59	$\mathrm{Pf}\,\dfrac{1_+(t)\cosh\sqrt{bt}}{t^{5/2}}$	$-\pi\sqrt{b}\left(s-\dfrac{b}{6}\right)\mathrm{erg}\sqrt{\dfrac{b}{4s}}+\dfrac{1}{3}(4s-b)\sqrt{\pi s}\,e^{b/4s}$ $\mathrm{Re}\,s>0$
60	$\dfrac{1_+(t)\sinh at\sin at}{t}$	$\dfrac{1}{2}\tan^{-1}\left(\dfrac{2a^2}{s^2}\right)\qquad \mathrm{Re}\,s>|a|$
61	$\dfrac{1_+(t)\sinh at\cos at}{t}$	$\dfrac{1}{4}\log\dfrac{s^2+2as+2a^2}{s^2-2as+2a^2}\qquad \mathrm{Re}\,s>|a|$
62	$\dfrac{1_+(t)\cosh at\sin at}{t}$	$\dfrac{1}{2}\tan^{-1}\dfrac{2as}{s^2-2a^2}\qquad \mathrm{Re}\,s>|a|$
63	$\mathrm{Pf}\,\dfrac{1_+(t)\cosh at\cos at}{t}$	$-\dfrac{1}{4}\log(s^4+4a^4)-\gamma\qquad \mathrm{Re}\,s>|a|$
64	$\delta(t)$	$1\qquad -\infty<\mathrm{Re}\,s<\infty$
65	$\delta(t-\tau)$	$e^{-s\tau}\qquad -\infty<\mathrm{Re}\,s<\infty$
66	$\delta^{(k)}(t)$	$s^k\qquad -\infty<\mathrm{Re}\,s<\infty$
67	$\delta^{(k)}(t-\tau)$	$s^k e^{-s\tau}\qquad -\infty<\mathrm{Re}\,s<\infty$
68	$\displaystyle\sum_{\nu=0}^{\infty}\delta(t-b\nu)$	$\dfrac{1}{1-e^{bs}}\qquad \mathrm{Re}\,s>0$
69	$\delta(t)+2\displaystyle\sum_{\nu=1}^{\infty}(-1)^\nu\delta(t-2b\nu)$	$\tanh bs\qquad \mathrm{Re}\,s>0$
70	$\delta(t)+2\displaystyle\sum_{\nu=1}^{\infty}\delta(t-2b\nu)$	$\coth bs\qquad \mathrm{Re}\,s>0$
71	$\delta(t+a)-\delta(t-a)$	$2\sinh as\qquad -\infty<\mathrm{Re}\,s<\infty$
72	$\delta(t+a)+\delta(t-a)$	$2\cosh as\qquad -\infty<\mathrm{Re}\,s<\infty$
73	$\displaystyle\sum_{\nu=0}^{\infty}1_+(t-b\nu)$	$\dfrac{1}{s(1-e^{-bs})}\qquad \mathrm{Re}\,s>0$
74	$\displaystyle\sum_{\nu=0}^{\infty}(-1)^\nu 1_+(t-b\nu)$	$\dfrac{1}{s(1+e^{-bs})}\qquad \mathrm{Re}\,s>0$

Table B.2 *Transform formulas for the right-sided Laplace transformation* (continued)

No.	$f(t)$	$\mathfrak{L}f(t)$
75	$J_k(\alpha t)1_+(t)$	$\dfrac{(\sqrt{s^2+\alpha^2}-s)^k}{\alpha^k\sqrt{s^2+\alpha^2}}$ $\quad$ Re $s >$ |Im α|
76	$t^k J_k(\alpha t)1_+(t)$	$\dfrac{1\times 3\times 5\times\cdots\times(2k-1)\alpha^k}{(s^2+\alpha^2)^{k+\frac{1}{2}}}$ Re $s >$ |Im α|; $k = 1, 2, 3, \ldots$
77	Pf $\dfrac{J_0(\alpha t)1_+(t)}{t}$	$-\log\left[\dfrac{C}{2}(s+\sqrt{s^2+\alpha^2})\right]$ $\quad$ Re $s >$ |Im α|
78	$\dfrac{J_k(\alpha t)1_+(t)}{t}$	$\dfrac{1}{k\alpha^k}(\sqrt{s^2+\alpha^2}-s)^k$ Re $s >$ |Im α|; $k = 1, 2, 3, \ldots$
79	Pf $\dfrac{J_0(\alpha t)1_+(t)}{t^2}$	$s\log\left[\dfrac{C}{2}(s+\sqrt{s^2+\alpha^2})\right]-\sqrt{s^2+\alpha^2}$ Re $s >$ |Im α|
80	Pf $\dfrac{J_1(\alpha t)1_+(t)}{t^2}$	$-\dfrac{\alpha}{2}\left\{\log\left[\dfrac{C}{2}(\sqrt{s^2+\alpha^2}+s)\right]\right.$ $\left.+\dfrac{s}{\sqrt{s^2+\alpha^2}+s}-\dfrac{1}{2}\right\}$ Re $s >$ |Im α|
81	$\dfrac{J_k(\alpha t)1_+(t)}{t^2}$	$\dfrac{(\sqrt{s^2+\alpha^2}-s)^k(k\sqrt{s^2+\alpha^2}+s)}{\alpha^k k(k-1)(k+1)}$ Re $s >$ |Im α|; $k = 2, 3, 4, \ldots$
82	Pf $\dfrac{J_0(2\sqrt{\alpha t})1_+(t)}{t}$	$\mathrm{Ei}\left(-\dfrac{\alpha}{s}\right)-\log(C^2\alpha)$ $\quad$ Re $\alpha > 0$; Re $s > 0$
83	Pf $\dfrac{J_0(2\sqrt{\alpha t})1_+(t)}{t^2}$	$-(s+\alpha)\left[\mathrm{Ei}\left(-\dfrac{\alpha}{s}\right)-\log(C^2\alpha)+2\right]$ $-s(e^{-\alpha/s}-2)$ Re $\alpha > 0$; Re $s > 0$
84	$\dfrac{J_1(2\sqrt{\alpha t})1_+(t)}{\sqrt{\alpha t}}$	$\dfrac{1}{\alpha}(1-e^{-\alpha/s})$ $\quad$ Re $\alpha > 0$; Re $s > 0$
85	$J_2(2\sqrt{\alpha t})1_+(t)$	$\dfrac{1}{\alpha}-\left(\dfrac{1}{s}+\dfrac{1}{\alpha}\right)e^{-\alpha/s}$ $\quad$ Re $\alpha > 0$; Re $s > 0$
86	$\dfrac{J_2(2\sqrt{\alpha t})1_+(t)}{t}$	$1+\dfrac{s}{\alpha}(e^{-\alpha/s}-1)$ $\quad$ Re $\alpha > 0$; Re $s > 0$
87	$I_k(\alpha t)1_+(t)$	$\dfrac{(s-\sqrt{s^2-\alpha^2})^k}{\alpha^k\sqrt{s^2-\alpha^2}}$ $\quad$ Re $s >$ |Re α|

Table B.2 *Transform formulas for the right-sided Laplace transformation* (continued)

No.	$f(t)$	$\mathfrak{L}f(t)$
88	$t^k I_k(\alpha t) 1_+(t)$	$\dfrac{1 \times 3 \times 5 \times \cdots \times (2k-1)\alpha^k}{(s^2 - \alpha^2)^{k+\frac{1}{2}}}$ $\operatorname{Re} s > \lvert \operatorname{Re} \alpha \rvert; \ k = 1, 2, 3, \ldots$
89	$\operatorname{Pf} \dfrac{I_0(\alpha t) 1_+(t)}{t}$	$-\log\left[\dfrac{C}{2}(s + \sqrt{s^2 - \alpha^2})\right]$ $\operatorname{Re} s > \lvert \operatorname{Re} \alpha \rvert$
90	$\dfrac{I_k(\alpha t) 1_+(t)}{t}$	$\dfrac{1}{k\alpha^k}(s - \sqrt{s^2 - \alpha^2})^k$ $\operatorname{Re} s > \lvert \operatorname{Re} \alpha \rvert; \ k = 1, 2, 3, \ldots$
91	$\operatorname{Pf} \dfrac{I_0(\alpha t) 1_+(t)}{t^2}$	$s \log\left[\dfrac{C}{2}(s + \sqrt{s^2 - \alpha^2})\right] - \sqrt{s^2 - \alpha^2}$ $\operatorname{Re} s > \lvert \operatorname{Re} \alpha \rvert$
92	$\operatorname{Pf} \dfrac{I_1(\alpha t) 1_+(t)}{t^2}$	$\dfrac{\alpha}{2}\left\{\dfrac{1}{2} - \dfrac{s}{\alpha^2}(s - \sqrt{s^2 - \alpha^2}) - \log\left[\dfrac{C}{2}(s + \sqrt{s^2 - \alpha^2})\right]\right\}$ $\operatorname{Re} s > \lvert \operatorname{Re} \alpha \rvert$
93	$\dfrac{I_k(\alpha t) 1_+(t)}{t^2}$	$\dfrac{(s - \sqrt{s^2 - \alpha^2})^k (s + k\sqrt{s^2 - \alpha^2})}{\alpha^k k(k-1)(k+1)}$ $\operatorname{Re} s > \lvert \operatorname{Re} \alpha \rvert; \ k = 2, 3, 4, \ldots$
94	$\operatorname{Pf} \dfrac{I_0(2\sqrt{\alpha t}) 1_+(t)}{t}$	$\operatorname{Ei}^*\left(\dfrac{\alpha}{s}\right) - \log(C^2 \alpha)$ $\operatorname{Re} \alpha > 0; \ \operatorname{Re} s > 0$
95	$\operatorname{Pf} \dfrac{I_0(2\sqrt{\alpha t}) 1_+(t)}{t^2}$	$-(s - \alpha)\left[\operatorname{Ei}^*\left(\dfrac{\alpha}{s}\right) - \log(C^2 \alpha) + 2\right] - s(e^{\alpha/s} - 2)$ $\operatorname{Re} \alpha > 0; \ \operatorname{Re} s > 0$
96	$\dfrac{I_1(2\sqrt{\alpha t}) 1_+(t)}{\sqrt{\alpha t}}$	$\dfrac{1}{\alpha}(e^{\alpha/s} - 1)$ $\operatorname{Re} \alpha > 0; \ \operatorname{Re} s > 0$
97	$I_2(2\sqrt{\alpha t}) 1_+(t)$	$\dfrac{1}{\alpha} + \left(\dfrac{1}{s} - \dfrac{1}{\alpha}\right)e^{\alpha/s}$ $\operatorname{Re} \alpha > 0; \ \operatorname{Re} s > 0$
98	$\dfrac{I_2(2\sqrt{\alpha t}) 1_+(t)}{t}$	$\dfrac{s}{\alpha}(e^{\alpha/s} - 1) - 1$ $\operatorname{Re} \alpha > 0; \ \operatorname{Re} s > 0$
99	$1_+(t) \operatorname{erfc} \dfrac{b}{2\sqrt{t}}$	$\dfrac{e^{-b\sqrt{s}}}{s}$ $\operatorname{Re} s > 0$
100	$\sqrt{\dfrac{t}{\pi}} e^{-b^2/4t} 1_+(t)$	$\dfrac{e^{-b\sqrt{s}}}{2s}\left(\dfrac{1}{\sqrt{s}} + b\right)$ $\operatorname{Re} s > 0$

Table B.2 *Transform formulas for the right-sided Laplace transformation (continued)*

No.	$f(t)$	$\mathfrak{L}f(t)$	
101	$\dfrac{1_+(t)e^{-b^2/4t}}{\sqrt{\pi t}}$	$\dfrac{e^{-b\sqrt{s}}}{\sqrt{s}}$	$\operatorname{Re} s > 0$
102	$\dfrac{1_+(t)e^{-b^2/4t}}{t^{\frac{3}{2}}}$	$\dfrac{2\sqrt{\pi}}{b}e^{-b\sqrt{s}}$	$\operatorname{Re} s > 0$
103	$\operatorname{Ei}(-bt)1_+(t)$	$-\dfrac{1}{s}\log\dfrac{s+b}{b}$	$\operatorname{Re} s > 0$
104	$\operatorname{Ei}^*(bt)1_+(t)$	$-\dfrac{1}{s}\log\dfrac{s-b}{b}$	$\operatorname{Re} s > b$
105	$\operatorname{si}(bt)1_+(t)$	$-\dfrac{\tan^{-1}(s/b)}{s}$	$\operatorname{Re} s > 0$
106	$\operatorname{ci}(bt)1_+(t)$	$-\dfrac{1}{2s}\log\left(\dfrac{s^2}{b^2}+1\right)$	$\operatorname{Re} s > 0$

Appendix C Glossary of Symbols

Symbol	Meaning and page where discussed
a	The adjoint of the matrix distribution **f** in the convolution algebra $\mathcal{D}'_R$ or in the T-convolution algebra $\mathcal{O}'_T$ 154
A(s)	The adjoint of the matrix **F**(s) 264
$\mathcal{B}'$	The space of bounded distributions 15
$\mathbf{c}^T$	The matrix transpose of the matrix distribution **c** 154
ci(t)	$= - \int_t^\infty \dfrac{\cos x}{x}\, dx$ 355
$\mathcal{C}_R$	The space of continuous functions whose supports are bounded on the left 157
$\mathcal{C}_+$	The space of functions continuous over $0 \leq t < \infty$ and with supports contained in this interval 168
det **f**	The determinant of the matrix distribution **f** in the convolution algebra $\mathcal{D}'_R$ or in the T-convolution algebra $\mathcal{O}'_T$ 153, 328
det **F**(s)	The determinant of the matrix **F**(s) 264
$D(\mathfrak{N})$	The domain of the operator $\mathfrak{N}$ 143

Glossary of symbols

Symbol	Meaning and page where discussed
$D(w*)$	The domain of the convolution operator $w*$ 143
$D^k f$	A partial derivative of f of order $\hat{k} = k_1 + k_2 + \cdots + k_n$ 21
$D_t^k f$	The derivative $D^k f$ with respect to t 72
$\mathfrak{D}$	The space of testing functions of bounded support 2, 21
$\mathfrak{D}'$	The space of distributions 7, 23
$\mathfrak{D}_I$	The space of elements of $\mathfrak{D}$ whose supports are contained in the interval I 83
$\mathfrak{D}'_L$	The space of distributions whose supports are bounded on the right 135
$\mathfrak{D}_{L_1}$	The space of testing functions that are absolutely integrable over $-\infty < t < \infty$ as well as all their derivatives 15
$\mathfrak{D}_m$	The space of functions having continuous derivatives up to the order m and a bounded support 14
$\mathfrak{D}'_m$	The space of continuous linear functionals on $\mathfrak{D}_m$ 15
$\mathfrak{D}'_R$	The space of distributions whose supports are bounded on the left 135
$\mathfrak{D}_{0I}$	The space of all continuous functions whose supports are contained in the interval I 87
$\mathfrak{D}'_{R, n \times m}$	The space of $n \times m$ matrices whose elements are in $\mathfrak{D}'_R$ 154
$\mathfrak{D}'_+$	The space of distributions whose supports are contained in $0 \leq t < \infty$ 168
erf s	$= \dfrac{2}{\sqrt{\pi}} \int_0^s e^{-z^2} dz = \int_0^{s^2} \dfrac{e^{-z}}{\sqrt{\pi z}} dz$; the error function 130
erfc s	$= 1 - \text{erf } s = \dfrac{2}{\sqrt{\pi}} \int_s^\infty e^{-z^2} dz$; the complementary error function; here, the path of integration has a bounded imaginary part 350
erg s	$= \dfrac{2}{\sqrt{\pi}} \int_0^s e^{z^2} dz$ 351
Ei(s)	$= -\int_s^\infty \dfrac{e^z}{z} dz$; as $z \to \infty$, $\arg z \to \beta$, where $\dfrac{\pi}{2} \leq \beta \leq \dfrac{3\pi}{2}$ 348
Ei($-t$)	$= \int_{-\infty}^{-t} \dfrac{e^x}{x} dx \quad (t > 0)$ 348
Ei*(s)	$= i\pi + \text{Ei}(s)$ 228
$\mathcal{E}'$	The space of distributions that have bounded supports 135

358 *Distribution theory and transform analysis* App. C

Symbol	Meaning and page where discussed
$\mathfrak{e}(x)$	The energy function for a one-port 302
$\bar{f}$	The Fourier transform of f 2, 184
$\check{f}$	The transpose of f (that is, $\check{f}(t) = f(-t)$) 55
$\tilde{f}$	The complex conjugate of f 29
$f^{(k)}$	The kth-order derivative of a distribution f 47
$f^{(-k)}$	A kth-order primitive of a distribution f 67
f^{*-1}	The inverse of the distribution f in the convolution algebra $\mathfrak{D}'_R$ 150
$\mathbf{f}^{*-1}$	The inverse of the matrix distribution $\mathbf{f}$ in the convolution algebra $\mathfrak{D}'_R$ 154
$f^{*\nu}$	The convolution of f with itself $\nu - 1$ times 157
$f^{\Delta-1}$	The inverse of the periodic distribution f in the T-convolution algebra $\mathcal{O}'_T$ 328
$\mathbf{f}^{\Delta-1}$	The inverse of the matrix $\mathbf{f}$ in the T-convolution algebra $\mathcal{O}'_T$ 328
$f(t) \times g(\tau)$	The direct product of the distributions $f(t)$ and $g(\tau)$ 115
$f * g$	The convolution of the distributions f and g 122
$\text{Fp} \int \cdots dt$	Hadamard's finite part of a divergent integral 16
F_ν	The Fourier coefficients of the periodic distribution f 330
$F(s)$	The Laplace transform of $f(t)$ 213, 222
$\mathbf{F}^{-1}(s)$	The inverse of the matrix $\mathbf{F}(s)$ 263
$\mathfrak{F}$	The direct Fourier transformation 172, 184
$\mathfrak{F}^{-1}$	The inverse Fourier transformation 173, 185
$\mathfrak{F}_c^{-1}\bar{\phi}$	The analytic extension of the inverse Fourier transform of $\bar{\phi}$ 195
$\mathfrak{F}_T$	The direct finite Fourier transformation 336
$\mathfrak{F}_T^{-1}$	The inverse finite Fourier transformation 336
$\mathcal{K}$	The space of all derivatives of elements of $\mathfrak{D}$ 67
$\mathcal{K}_I$	The space of all derivatives of elements of $\mathfrak{D}_I$ 92
$\inf \phi(t)$	$= \inf_t \phi(t)$; the greatest lower bound on $\phi(t)$ for $-\infty < t < \infty$
$\inf (a, b, \ldots, q)$	The smallest number in the finite set of real numbers $\{a, b, \ldots, q\}$
$I_k(t)$	$= -i^k J_k(it)$; the kth-order modified Bessel function of first kind 353
J	The linear difference operator with constant coefficients 279
$\mathbf{J}$	The matrix linear difference operator with constant coefficients 285
$J_0(t)$	The zero-order Bessel function of first kind 135, 219

App. C *Glossary of symbols* **359**

Symbol	Meaning and page where discussed
$J_p(t)$	The pth-order Bessel function of first kind 135, 233
$\hat{k}$	The sum of the components of $k = \{k_1, k_2, \ldots, k_n\}$ 21
L	The linear differential or integrodifferential operator with constant coefficients 158, 273
L	The matrix linear differential or integrodifferential operator with constant coefficients 269, 276
$\mathfrak{L}$	The direct Laplace transformation 213, 222
$\mathfrak{L}_+$	The direct Laplace transformation as applied to a function whose support is contained in $0 \leq t < \infty$ 214
$\mathfrak{L}^{-1}$	The inverse Laplace transformation 216
$\mathfrak{M}$	The Mellin transformation 235
$n!$	$= n(n-1)(n-2) \cdots 1$; the factorial function for a positive integer n
$\mathfrak{N}$	An operator in $\mathfrak{D}'$ 142
$\mathcal{O}_M$	The space of functions having ordinary derivatives of all orders, which are all of slow growth 106
Pf	The symbol denoting a pseudofunction 17
Pv $\int \cdots dt$	Cauchy's principal value of a divergent integral 18
Pv $\cdots$	A symbolic prefix denoting a distribution generated by Cauchy's principal value 18
$\mathcal{P}_T$	The space of periodic testing functions with a period T 314
$\mathcal{P}'_T$	The space of periodic distributions with a period T 316
$\mathcal{P}'_{T,n\times m}$	The space of $n \times m$ matrices whose elements are in $\mathcal{P}'_T$ 328
$\mathcal{Q}$	The space of Mikusiński convolution quotients 169
$R(\mathfrak{N})$	The range of the operator $\mathfrak{N}$ 143
$\mathfrak{R}^1$	The one-dimensional real euclidean space 21
$\mathfrak{R}^2$	The two-dimensional real euclidean space 114
$\mathfrak{R}^n$	The n-dimensional real euclidean space 21
sgn x	$= 1_+(x) - 1_+(-x)$ 190
si(t)	$= -\int_t^\infty \dfrac{\sin x}{x} dx$ 355
sup $\phi(t)$	$= \sup_t \phi(t)$; the least upper bound on $\phi(t)$ for $-\infty < t < \infty$
$\sup_{t \in \Omega} \phi(t)$	The least upper bound on $\phi(t)$ for t in the set Ω
sup $(a, b, \ldots, q)$	The largest number in the finite set of real numbers $\{a, b, \ldots, q\}$
$\mathcal{S}$	The space of testing functions of rapid descent 100

Symbol	Meaning and page where discussed
S'	The space of distributions of slow growth 103
$\mathcal{U}_T$	The space of unitary functions with respect to T 315
$w(t)$	The unit impulse response of a one-port 296
$w *$	A convolution operator 143
$W(s)$	The immittance function of a one-port 305
$\mathcal{Z}$	The space of testing functions whose Fourier transforms are in $\mathcal{D}$ 195
$\mathcal{Z}'$	The space of ultradistributions 198
$\Gamma(z)$	$= \int_0^\infty e^{-x} x^{z-1}\, dx$; the gamma function 232
δ	The delta functional 10, 23
δ_T	$= \sum_{n=-\infty}^{\infty} \delta(t - nT)$; the periodic delta functional with a period T 319
$\delta_{n \times n}$	The $n \times n$ unit matrix in the convolution algebra $\mathcal{D}'_R$ 154
$\delta_{T, n \times n}$	The $n \times n$ unit matrix in the T-convolution algebra $\mathcal{S}'_T$ 328
$\psi(k)$	$= -\gamma + \sum_{\nu=1}^{k-1} \frac{1}{\nu}$ $(k = 2, 3, 4, \ldots)$
$\Omega_a \times \Omega_b$	The cartesian product of the set Ω_a in $\mathcal{R}_t$ with the set Ω_b in $\mathcal{R}_r$ 118
$1_+(t)$	The unit step function 12
$\mathbf{1}_{n \times n}$	The $n \times n$ unit matrix 263

Some additional special symbols

$\int_\Omega$	An integral extended over the set Ω
$\int \cdots dt < \infty$	The notation for a convergent integral of a positive function
$\{f_\nu\}$	A set with elements f_ν
$\{f_\nu\}_{\nu=1}^\infty$	The sequence $f_1, f_2, f_3, \ldots$
$t \to a+$	t tends to the real number a through real values that are greater than a
$t \to a-$	t tends to the real number a through real values that are less than a
$f(a+)$	$= \lim_{t \to a+} f(t)$
$f(a-)$	$= \lim_{t \to a-} f(t)$
$f(t) = O[g(t)]$ as $t \to a$	$\frac{f(t)}{g(t)}$ remains bounded as $t \to a$

$f(t) = o[g(t)]$ as $t \to a$	$\lim_{t \to a} \dfrac{f(t)}{g(t)} = 0$
$f(t) \sim g(t)$ as $t \to a$	$\lim_{t \to a} \dfrac{f(t)}{g(t)} = 1$
$*$	The convolution symbol
Δ	The T-convolution symbol
$\triangleq$	Equality by definition
$\equiv$	Identically equal
$\not\equiv$	Not identically equal
$\in$	Is a member of
$\binom{m}{n}$	$= \dfrac{m!}{n!(m-n)!}$; the binomial coefficient
$\subset$	Is contained in
$\supset$	Contains
$\cup$	Union symbol for sets
$\overline{\Lambda}$	The closure of a set Λ

Appendix D Bibliography

Beltrami, E. J.
  [1] Some Alternative Approaches to Distributions, *SIAM Rev.*, vol. 5, pp. 351–357, October, 1963.
Bochner, S.
  [1] "Lectures on Fourier Integrals," Annals of Mathematical Studies, no. 42, Princeton University Press, Princeton, N.J., 1959.
Borel, E.
  [1] "Methodes et problemes de theorie des fonctions," Gauthier-Villars, Paris, 1922.
Bremermann, H. J., and L. Durand
  [1] On Analytic Continuation, Multiplication, and Fourier Transformations of Schwartz Distributions, *J. Math. Phys.*, vol. 2, pp. 240–258, 1961.
Carslaw, H. S.
  [1] "Introduction to the Theory of Fourier's Series and Integrals," 3d ed., Dover Publications, Inc., New York, 1930.
Cauer, W.
  [1] The Poisson Integral for Functions with Positive Real Part, *Bull. Am. Math. Soc.*, vol. 38, pp. 713–717, October, 1932.
Courant, R.
  [1] "Methods of Mathematical Physics," vol. II, "Partial Differential Equations," Interscience Publishers, Inc., New York, 1962.

Doetsch, G.
  [1] "Handbuch der Laplace-transformation," Verlag Birkhauser, Basel, Switzerland, 1950.
Dolezal, V.
  [1] "Dynamics of Linear Systems," Publishing House of the Czechoslovak Academy of Sciences, Prague, 1964.
Dolezal, V., and Z. Vorel
  [1] Theory of Kirchhoff's Networks, *Casopis Pestovani Mat.*, vol. 87, pp. 440–476, 1962.
Ehrenpreis, L.
  [1] Analytic Functions and the Fourier Transform of Distributions, I, *Ann. Math.*, vol. 63, pp. 129–159, January, 1956.
Erdélyi, A.
  [1] "Operational Calculus and Generalized Functions," Holt, Rinehart and Winston, Inc., New York, 1962.
Erdélyi, A. (ed.)
  [1] "Tables of Integral Transforms," vols. I and II, McGraw-Hill Book Company, New York, 1954.
Frazer, R. A., W. J. Duncan, and A. R. Collar
  [1] "Elementary Matrices," Cambridge University Press, New York, 1957.
Friedman, A.
  [1] "Generalized Functions and Partial Differential Equations," Prentice-Hall, Inc., Englewood Cliffs, N.J., 1963.
Friedman, B.
  [1] "Principles and Techniques of Applied Mathematics," John Wiley & Sons, Inc., New York, 1956.
Gelfand, I. M., and G. E. Shilov
  [1] "Generalized Functions," vols. 1–3, Academic Press Inc., New York, 1964.
Hadamard, J.
  [1] "Lectures on Cauchy's Problem in Linear Partial Differential Equations," Dover Publications, Inc., New York, 1952.
Halperin, I.
  [1] "Introduction to the Theory of Distributions," University of Toronto Press, Toronto, Canada, 1952.
Hobson, E. W.
  [1] "The Theory of Functions of a Real Variable," vols. 1 and 2, Dover Publications, Inc., New York, 1957.
Ishihara, T.
  [1] On Generalized Laplace Transforms, *Proc. Japan Acad.*, vol. 37, pp. 556–561, 1961.
Jahnke, E., F. Emde, and F. Losch
  [1] "Tables of Higher Functions," 6th ed., McGraw-Hill Book Company, New York, 1960.
Kaplan, W.
  [1] "Ordinary Differential Equations," Addison-Wesley Publishing Company, Inc., Reading, Mass., 1958.
  [2] "Operational Methods for Linear Systems," Addison-Wesley Publishing Company, Inc., Reading, Mass., 1962.
Kestelman, H.
  [1] "Modern Theories of Integration," 2d ed., Dover Publications, Inc., New York, 1960.

König, H.
[1] Neue Begründung der Theorie der Distributionen, *Math. Nachr.*, vol. 9, pp. 129–148, 1953.

König, H., and J. Meixner
[1] Lineare Systeme und lineare Transformationen, *Math. Nachr.*, vol. 19, pp. 265–322, 1958.

Korevaar, J.
[1] Distributions Defined from the Point of View of Applied Mathematics, *Koninkl. Ned. Akad. Wetenschap.*, ser. A, vol. 58, pp. 368–389, 483–503, 663–674, 1955.

Lavoine, J.
[1] "Calcul symbolique, distributions et pseudo-fonctions," Centre National de la Recherche Scientifique, Paris, 1959.

Lighthill, M. J.
[1] "Fourier Analysis and Generalized Functions," Cambridge University Press, New York, 1958.

Liverman, T. P. G.
[1] "Generalized Functions and Direct Operational Methods," vol. 1, Prentice-Hall, Inc., Englewood Cliffs, N.J., 1964.

Loomis, L. H., and D. V. Widder
[1] The Poisson Integral Representation of Functions Which Are Positive and Harmonic in a Half-plane, *Duke Math. J.*, vol. 9, pp. 643–645, 1942.

Mikusiński, J.
[1] Sur la méthode de généralisation de Laurent Schwartz et sur la convergence faible, *Fundamenta Math.*, vol. 35, pp. 235–239, 1948.
[2] "Operational Calculus," Pergamon Press, New York, 1959.

Mirsky, L.
[1] "An Introduction to Linear Algebra," Oxford University Press, London, 1955.

Pietsch, A.
[1] Ein elementarer Beweis des Darstellungssatzes für Distributionen, *Math. Nachr.*, vol. 22, pp. 47–50, 1960.

Raisbeck, G.
[1] A Definition of Passive Linear Networks in Terms of Time and Energy, *J. Appl. Phys.*, vol. 25, pp. 1510–1514, December, 1954.

Rehberg, C. F.
[1] The Theory of Generalized Functions for Electrical Engineers, *NYU Dept. Elec. Eng. Tech. Rept.* 400-42, August, 1961.

Schwartz, L.
[1] "Théorie des distributions," vols. I and II, Actualités Scientifiques et Industrielles, Hermann & Cie, Paris, 1957, 1959.
[2] "Méthodes mathématiques pour les sciences physiques," Hermann & Cie, Paris, 1961.
[3] Transformation de Laplace des distributions, *Seminaire Math. Univ. Lund*, tome suppl. dedié à M. Riesz, 1952, pp. 196–206.

Sebastião e Silva, J.
[1] Les Fonctions analytiques comme ultra-distributions dans le calcul operationnel, *Math. Ann.*, vol. 136, pp. 58–96, 1958.

Soboleff, S. L.
[1] Méthode nouvelle à résoudre le problème de Cauchy pour les équations linéaires hyperboliques normales, *Mat. Sb.*, vol. 1, pp. 39–72, 1936.

Taylor, A. E.
[1] "Introduction to Functional Analysis," John Wiley & Sons, Inc., New York, 1958.
Temple, G.
[1] The Theory of Generalized Functions, *Proc. Roy. Soc. (London)*, ser. A, vol. 228, pp. 175–190, 1955.
Titchmarsh, E. C.
[1] "The Theory of Functions," 2d ed., Oxford University Press, Fair Lawn, N.J., 1939.
[2] "Introduction to the Theory of Fourier Integrals," 2d ed., Oxford University Press, Fair Lawn, N.J., 1948.
Van der Pol, B., and H. Bremmer
[1] "Operational Calculus Based on the Two-sided Laplace Transform," 2d ed., Cambridge University Press, New York, 1955.
Vasilach, S.
[1] Calcul operationnel algébrique des distributions à support dans R_+^n, $n \geq 1$, *Rev. Math. Pures Appl., Acad. Rep. Populaire Roumaine*, vol. 4, pp. 185–219, 1959; vol. 5, pp. 495–531, 1960; vol. 6, pp. 69–100, 1961.
Widder, D. V.
[1] "The Laplace Transform," Princeton University Press, Princeton, N.J., 1946.
[2] "Advanced Calculus," 2d ed., Prentice-Hall, Inc., Englewood Cliffs, N.J., 1961.
Wohlers, M. R., and E. J. Beltrami
[1] Distribution Theory as the Basis of Generalized Passive-network Analysis, *Grumman Aircraft Eng. Corp. Res. Doc.* 126, March, 1964. See also *IEEE Trans. Circuit Theory*, vol. CT-12, June, 1965.
Youla, D. C., L. J. Castriota, and H. J. Carlin
[1] Bounded Real Scattering Matrices and the Foundations of Linear Passive Network Theory, *IRE Trans. Circuit Theory*, vol. CT-6, pp. 102–124, March, 1959.
[2] Scattering Matrices and the Foundations of Linear Passive Network Theory, *Rept.* R-594-57, PIB-522, Polytechnic Institute of Brooklyn, Sept. 10, 1957.
Zemanian, A. H.
[1] An Approximate Means of Evaluating Integral Transforms, *J. Appl. Phys.*, vol. 25, pp. 262–266, February, 1954.
[2] Further Properties of Certain Classes of Transfer Functions, *Quart. Appl. Math.*, vol. 18, pp. 223–228, October, 1960; vol. 19, pp. 158–159, July, 1961.
[3] An Upper Bound on Nonnegative Transient Responses, *Quart. Appl. Math.*, vol. 20, pp. 88–89, April, 1962.
[4] An n-port Realizability Theory Based on the Theory of Distributions, *IEEE Trans. Circuit Theory*, vol. CT-10, pp. 265–274, June, 1963.
[5] Distributional Convolution and an Application to Network Realizability Theory, *Proc. Sixth Midwest Circuits Conf.*, University of Wisconsin, May, 1963, pp. K1–K9.

Index

Abelian theorems, 243, 249
Abscissa, of absolute convergence, 214
　of convergence, 223, 252, 255
Addition, of distributions, 26, 321
　of ultradistributions, 199
Adjoint, 154, 264
Associativity, of convolution, 128–129, 142
　of direct products, 122
　of T-convolution, 324

Beltrami, E. J., iv, 294
Bessel's differential equation, 56
Bessel's function, 56, 220, 233
Bochner, S., iii, iv
Borel, E., 119
Bounded distribution, 15
Boundedness property, 83, 109
Bremermann, H. J., iv
Bremmer, H., iii, 11, 45, 346
Brodskii, M. S., 37

Carlin, H. J., 294, 299, 302, 304
Carslaw, H. S., 5, 39, 72
Cartesian product, 118
Castriota, L. J., 294, 299, 302, 304
Cauchy principal value, 18

Cauer, W., 307
Causality, 300
Change of independent variable, 30
Characteristic equation, 158
Closed under convergence, 5
Cofactor, 154, 264
Collar, A. R., 271
Commutativity, of convolution, 125, 142
　of direct products, 119
　with shifting operator, 143, 299
　of T-convolution, 324
Complementary solution, 159, 268, 272
Completeness, 5
Conjugate space, 7
Continuity, of convolution, 136, 141
　of differentiation, 50
　of Fourier transformation, 187, 203
　of functionals, 6–7, 23
　of Laplace transformation, 231, 254, 257
　of one-ports, 299
　of operators in $\mathfrak{D}'$, 144, 146
　with respect to, distribution's parameter, 75
　　testing function's parameter, 72–73
Convergence, in $\mathfrak{D}$, 5, 22
　in $\mathfrak{D}'$, 37
　in $\mathfrak{D}_I$, 83–84
　in $\mathfrak{D}'_L$, 135

Convergence, in $\mathfrak{D}_{L_1}$, 15
  in $\mathfrak{D}_m$, 14
  in $\mathfrak{D}'_R$, 135
  in $\mathfrak{D}_{0I}$, 87
  in $\mathcal{E}'$, 135
  in $\mathcal{O}_T$, 314
  in $\mathcal{O}'_T$, 320
  in $\mathcal{S}$, 100
  in $\mathcal{S}'$, 104, 111
  in $\mathcal{Z}$, 195
  in $\mathcal{Z}'$, 201
Convolution, 122, 139, 142
  associativity of, 128–129, 142
  commutativity of, 125, 142
  continuity of, 136, 141
  differentiation of, 132, 140
  Fourier transformation of, 191, 206
  Laplace transformation of, 240
  linearity of, 125, 142
  shifting of, 131
  support of, 125
Convolution algebra, 149
Convolution operator, 143, 296
Convolution quotients, 169
Courant, R., iv

Definite integrals of distributions, 71, 96
Delta functional, 10, 23
  characterization of, 81, 98, 210
Determinant in convolution algebra, 153
Difference equation, linear, with constant coefficients, 279, 286, 343
Difference-differential equation, linear, with constant coefficients, 286
Differential equation, linear, with constant coefficients, 158, 266, 339
  with polynomial coefficients, 290
Differentiation, of convolutions, 132, 140
  of distributions, 47
  under integral sign, 77
  of periodic distributions, 321
  of pseudofunctions, 59, 61, 64
  with respect to testing function's parameter, 73, 111–113
  of ultradistributions, 200, 202
Direct product, 115
  associativity of, 122
  commutativity of, 119
  support of, 118

Distribution, 7, 23
  in $\mathcal{B}'$, 15
  in $\mathfrak{D}'$, 7, 23
  in $\mathfrak{D}'_L$, 135
  in $\mathfrak{D}'_m$, 15
  in $\mathfrak{D}'_R$, 135
  in $\mathfrak{D}'_+$, 168
  in $\mathcal{E}'$, 135
  in $\mathcal{O}'_T$, 316
  in $\mathcal{S}'$, 103
  bounded, 15
  boundedness property of, 83, 109
  depending on parameter, 75
  even, 27
  as finite-order derivative of continuous function, 86, 92
  imaginary part of, 29
  left-sided, 135
  nondecreasing, 74
  nonnegative, 74
  nonnegative-definite, 304
  odd, 27
  order of, 94–95
  periodic, 316
  real, 295
  real part of, 29
  right-sided, 135
  of slow growth, 102–103
Distributional convergence, 37
Divisors of zero, 150, 329
Doetsch, G., 243
Dolezal, V., 35, 293
Dominated convergence, 41
Dual space, 7
Duncan, W. J., 271
Durand, L., iv

Ehrenpreis, L., 193
Elementary solutions, 152
Emde, F., 130, 135, 228, 248
Energy function, 302
Equality, of distributions, 24–25
  of ultradistributions, 199
Erdélyi, A., iv, 169, 346
Error function, 130
Essential point, 25
Euler's constant, 222
Even distribution, 27
Exponential integral, 228, 232

Final-value theorems, 243, 249
Finite convolution, 323
Finite Fourier transformation, direct, 336
  inverse, 336
Finite part, 16, 58
Fourier coefficients, 331
Fourier integral, 78, 172
Fourier series, 331
Fourier transformation, distributional, 78, 184, 203
  continuity of, 187, 203
  of convolutions, 191, 206
  inverse, 185, 203
  operation-transform formulas, 186–187, 204
  uniqueness of, 185, 203
  finite, 336
  ordinary, 172
    inverse, 173
    linearity of, 179
    uniqueness of, 177, 179
Frazer, R. A., 271
Friedman, A., 172
Friedman, B., 10
Function, in $\mathcal{C}_+$, 168
  in $\mathcal{O}_M$, 106–107
  of slow growth, 104
Functional, 6, 23

Gelfand, I. M., 37, 172, 193, 299
Generalized functions, 172
Green's function, 159

Hadamard, J., 16
Hadamard's finite part, 16, 58
Hahn-Banach theorem, 96
Half-plane, of absolute convergence, 214
  of convergence, 223
Halperin, I., 56, 94
Heaviside unit step function, 12
Hobson, E. W., 31, 74
Homogeneous equation, 158

Immittance function, 305
Indefinite integral, 67
  (*See also* Primitive)
Infinitely smooth function, 2, 21
Initial conditions, 159, 275, 277, 283, 288

Initial-value theorems, 243
Integration, by parts, 70
  with respect to distribution's parameter, 75–76
Integrodifferential equation, linear, with constant coefficients, 165, 273
Inverse, in $\mathfrak{D}'_R$, 150, 154
  in $\mathcal{O}'_T$, 328
Ishihara, T., 213

Janke, E., 130, 135, 228, 248

Kaplan, W., 159, 341
Kestelman, H., 8, 25, 32, 33, 41, 117, 126, 174, 175, 324
König, H., v, 294, 310
Korevaar, J., v

Laguerre polynomials, 291
Laguerre's differential equation, 291
Laplace transformation, distributional, 223, 252, 254
  analyticity of, 225, 253, 257
  continuity of, 231, 254, 257
  of convolution, 240
  inversion criteria, 236, 258
  linearity of, 224, 253, 256
  of matrix, 263
  operation-transform formulas, 228, 254, 257, 347
  uniqueness of, 225, 253, 256
  ordinary, 213, 222, 230
    analyticity of, 215
    inverse, 216
    linearity of, 215
    uniqueness of, 217
  ultradistributional, 234
Lavoine, J., 65, 346
Left-sided Laplace transformation, 220, 252
Lighthill, M. J., v, 315
Linear space, 345
Linearity, of convolution, 125, 142
  of differentiation, 49
  of Fourier transformation, 179, 187, 203
  of functionals, 6, 23

Linearity, of Laplace transformation, 215, 224, 253, 256
 of one-ports, 298
 of operators in $\mathfrak{D}'$, 143
Liverman, T. P. G., v
Local behavior of distributions, 34
Locally finite covering, 32
Locally integrable function, 7
Loomis, L. H., v, 308
Losch, F., 130, 135, 228, 248

Meixner, J., 294, 310
Mellin transformation, 235
Mikusiński, J., iv, v, 157, 169, 170
Mikusiński operators, 169
Mikusiński's operational calculus, 168
Mirsky, L., 156, 304, 338
Multiplication, of distribution, by constant, 26, 321
 by infinitely smooth function, 28, 321
 of independent variable by constant, 27, 29, 200
 of ultradistribution, by constant, 199
 by multiplier in Z, 200
Multipliers in space Z, 198

Neighborhood, 30
Nondecreasing distribution, 74
Nonnegative distribution, 74
Nonnegative-definite distribution, 304
Norm for $\mathfrak{D}_{0I}$, 88
Null set, 25

Odd distribution, 27
One-port, 295
Operational calculus, Mikusiński's, 168
Operator in $\mathfrak{D}'$, 142–143
 convolution, 143
 real, 295
Order of distribution, 94–95, 162–163

Parseval's equation, 179–180
Particular solution, 159
Partitioning of unity, 32
Passivity, 302
Period, 314

Periodic distribution, 316
Periodic testing function, 314
Pietsch, A., 86
Poisson's summation formula, 189, 190
Positive-reality, 305
Primitive, 67, 108, 135
Pseudofunction, 17, 57, 63

Raisbeck, G., 294
Region, of absolute convergence, 214
 of convergence, 223, 252, 255
Regular distribution, 7, 23
Regular ultradistribution, 199
Regularization, 132
Rehberg, C. F., v
Riemann-Lebesgue lemma, 174
Right-sided Laplace transformation, 214, 223

Schwartz, L., iii, 17, 70, 94, 108, 125, 138, 144, 149, 150, 172, 191, 256, 258, 299, 304
Sebastião e Silva, J., 193
Seminorms, for $\mathfrak{D}_I$, 84
 for $\mathcal{S}$, 111
Sequence of slow growth, 331
Shifting, of convolutions, 131
 of distributions, 26, 321
 of ultradistributions, 200
Shilov, G. E., 37, 172, 193, 299
Sifting property, 11
Single-valuedness, 143, 297
Singular distribution, 10, 23
Soboleff, S. L., iii
Steiglitz, K., 45
Strip of convergence, 255
Subalgebra, 152
Support, of convolution, 125
 of direct product, 118
 of distribution, 25
 of testing function, 6, 22
Symbolic function, 10

T-convolution, 323
 associativity of, 324
 commutativity of, 324
 continuity of, 326
 differentiation of, 324
 linearity of, 325
 shifting of, 324

T-convolution algebra, 327
Taylor, A., 96, 150
Temple, G., v
Tensor product (*see* Direct product)
Testing function, in $\mathfrak{D}$, 2, 21
   in $\mathfrak{D}_I$, 83
   in $\mathfrak{D}_{L_1}$, 15
   in $\mathfrak{D}_m$, 14
   in $\mathfrak{D}_{0I}$, 87
   in $\mathcal{O}_T$, 314
   in $\mathcal{S}$, 100
   in $\mathcal{Z}$, 195
   of rapid descent, 100
Time-invariance, 299
Titchmarsh, E. C., 3, 46, 77, 88, 96, 127, 182, 189, 193–196, 215, 324, 331
Translation (*see* Shifting)
Transposition, of distribution, 27, 321
   of ultradistribution, 200
Trigonometric series, 51
   (*See also* Fourier series)
Two-sided Laplace transformation, 220, 254

Ultradistribution, 193, 198
   regular, 199
   Taylor's series for, 201

Unit impulse, 297
Unit impulse response, 297
Unit matrix, in $\mathfrak{D}'_R$, 154
   in $\mathcal{O}'_T$, 329
Unit step function, 12
Unitary function, 315

Van der Pol, B., iii, 11, 45, 346
Vasilach, S., 149
Volterra's integral equation, 157
Vorel, Z., 293

Weighting function, 159
Widder, D. V., v, 4, 16, 58, 73, 74, 214, 243, 308
Wohlers, M. R., 294

Youla, D. C., 294, 299, 302, 304

Zemanian, A. H., 181, 242, 294, 304

A CATALOG OF SELECTED
DOVER BOOKS
IN SCIENCE AND MATHEMATICS

A CATALOG OF SELECTED
DOVER BOOKS
IN SCIENCE AND MATHEMATICS

QUALITATIVE THEORY OF DIFFERENTIAL EQUATIONS, V.V. Nemytskii and V.V. Stepanov. Classic graduate-level text by two prominent Soviet mathematicians covers classical differential equations as well as topological dynamics and ergodic theory. Bibliographies. 523pp. 5⅜ x 8½. 65954-2 Pa. $14.95

MATRICES AND LINEAR ALGEBRA, Hans Schneider and George Phillip Barker. Basic textbook covers theory of matrices and its applications to systems of linear equations and related topics such as determinants, eigenvalues and differential equations. Numerous exercises. 432pp. 5⅜ x 8½. 66014-1 Pa. $10.95

QUANTUM THEORY, David Bohm. This advanced undergraduate-level text presents the quantum theory in terms of qualitative and imaginative concepts, followed by specific applications worked out in mathematical detail. Preface. Index. 655pp. 5⅜ x 8½. 65969-0 Pa. $14.95

ATOMIC PHYSICS (8th edition), Max Born. Nobel laureate's lucid treatment of kinetic theory of gases, elementary particles, nuclear atom, wave-corpuscles, atomic structure and spectral lines, much more. Over 40 appendices, bibliography. 495pp. 5⅜ x 8½. 65984-4 Pa. $13.95

ELECTRONIC STRUCTURE AND THE PROPERTIES OF SOLIDS: The Physics of the Chemical Bond, Walter A. Harrison. Innovative text offers basic understanding of the electronic structure of covalent and ionic solids, simple metals, transition metals and their compounds. Problems. 1980 edition. 582pp. 6⅛ x 9¼. 66021-4 Pa. $16.95

BOUNDARY VALUE PROBLEMS OF HEAT CONDUCTION, M. Necati Özisik. Systematic, comprehensive treatment of modern mathematical methods of solving problems in heat conduction and diffusion. Numerous examples and problems. Selected references. Appendices. 505pp. 5⅜ x 8½. 65990-9 Pa. $12.95

A SHORT HISTORY OF CHEMISTRY (3rd edition), J.R. Partington. Classic exposition explores origins of chemistry, alchemy, early medical chemistry, nature of atmosphere, theory of valency, laws and structure of atomic theory, much more. 428pp. 5⅜ x 8½. (Available in U.S. only) 65977-1 Pa. $11.95

A HISTORY OF ASTRONOMY, A. Pannekoek. Well-balanced, carefully reasoned study covers such topics as Ptolemaic theory, work of Copernicus, Kepler, Newton, Eddington's work on stars, much more. Illustrated. References. 521pp. 5⅜ x 8½. 65994-1 Pa. $12.95

PRINCIPLES OF METEOROLOGICAL ANALYSIS, Walter J. Saucier. Highly respected, abundantly illustrated classic reviews atmospheric variables, hydrostatics, static stability, various analyses (scalar, cross-section, isobaric, isentropic, more). For intermediate meteorology students. 454pp. 6⅛ x 9¼. 65979-8 Pa. $14.95

CATALOG OF DOVER BOOKS

RELATIVITY, THERMODYNAMICS AND COSMOLOGY, Richard C. Tolman. Landmark study extends thermodynamics to special, general relativity; also applications of relativistic mechanics, thermodynamics to cosmological models. 501pp. 5⅜ x 8½. 65383-8 Pa. $13.95

APPLIED ANALYSIS, Cornelius Lanczos. Classic work on analysis and design of finite processes for approximating solution of analytical problems. Algebraic equations, matrices, harmonic analysis, quadrature methods, much more. 559pp. 5⅜ x 8½. 65656-X Pa. $13.95

INTRODUCTION TO ANALYSIS, Maxwell Rosenlicht. Unusually clear, accessible coverage of set theory, real number system, metric spaces, continuous functions, Riemann integration, multiple integrals, more. Wide range of problems. Undergraduate level. Bibliography. 254pp. 5⅜ x 8½. 65038-3 Pa. $8.95

INTRODUCTION TO QUANTUM MECHANICS With Applications to Chemistry, Linus Pauling & E. Bright Wilson, Jr. Classic undergraduate text by Nobel Prize winner applies quantum mechanics to chemical and physical problems. Numerous tables and figures enhance the text. Chapter bibliographies. Appendices. Index. 468pp. 5⅜ x 8½. 64871-0 Pa. $12.95

ASYMPTOTIC EXPANSIONS OF INTEGRALS, Norman Bleistein & Richard A. Handelsman. Best introduction to important field with applications in a variety of scientific disciplines. New preface. Problems. Diagrams. Tables. Bibliography. Index. 448pp. 5⅜ x 8½. 65082-0 Pa. $12.95

MATHEMATICS APPLIED TO CONTINUUM MECHANICS, Lee A. Segel. Analyzes models of fluid flow and solid deformation. For upper-level math, science and engineering students. 608pp. 5⅜ x 8½. 65369-2 Pa. $14.95

ELEMENTS OF REAL ANALYSIS, David A. Sprecher. Classic text covers fundamental concepts, real number system, point sets, functions of a real variable, Fourier series, much more. Over 500 exercises. 352pp. 5⅜ x 8½. 65385-4 Pa. $11.95

PHYSICAL PRINCIPLES OF THE QUANTUM THEORY, Werner Heisenberg. Nobel Laureate discusses quantum theory, uncertainty, wave mechanics, work of Dirac, Schroedinger, Compton, Wilson, Einstein, etc. 184pp. 5⅜ x 8½. 60113-7 Pa. $6.95

INTRODUCTORY REAL ANALYSIS, A.N. Kolmogorov, S.V. Fomin. Translated by Richard A. Silverman. Self-contained, evenly paced introduction to real and functional analysis. Some 350 problems. 403pp. 5⅜ x 8½. 61226-0 Pa. $10.95

PROBLEMS AND SOLUTIONS IN QUANTUM CHEMISTRY AND PHYSICS, Charles S. Johnson, Jr. and Lee G. Pedersen. Unusually varied problems, detailed solutions in coverage of quantum mechanics, wave mechanics, angular momentum, molecular spectroscopy, scattering theory, more. 280 problems plus 139 supplementary exercises. 430pp. 6½ x 9¼. 65236-X Pa. $13.95

CATALOG OF DOVER BOOKS

ASYMPTOTIC METHODS IN ANALYSIS, N.G. de Bruijn. An inexpensive, comprehensive guide to asymptotic methods—the pioneering work that teaches by explaining worked examples in detail. Index. 224pp. 5⅜ x 8½. 64221-6 Pa. $7.95

OPTICAL RESONANCE AND TWO-LEVEL ATOMS, L. Allen and J. H. Eberly. Clear, comprehensive introduction to basic principles behind all quantum optical resonance phenomena. 53 illustrations. Preface. Index. 256pp. 5⅜ x 8½. 65533-4 Pa. $8.95

COMPLEX VARIABLES, Francis J. Flanigan. Unusual approach, delaying complex algebra till harmonic functions have been analyzed from real variable viewpoint. Includes problems with answers. 364pp. 5⅜ x 8½. 61388-7 Pa. $9.95

ATOMIC SPECTRA AND ATOMIC STRUCTURE, Gerhard Herzberg. One of best introductions; especially for specialist in other fields. Treatment is physical rather than mathematical. 80 illustrations. 257pp. 5⅜ x 8½. 60115-3 Pa. $7.95

APPLIED COMPLEX VARIABLES, John W. Dettman. Step-by-step coverage of fundamentals of analytic function theory—plus lucid exposition of five important applications: Potential Theory; Ordinary Differential Equations; Fourier Transforms; Laplace Transforms; Asymptotic Expansions. 66 figures. Exercises at chapter ends. 512pp. 5⅜ x 8½. 64670-X Pa. $12.95

ULTRASONIC ABSORPTION: An Introduction to the Theory of Sound Absorption and Dispersion in Gases, Liquids and Solids, A.B. Bhatia. Standard reference in the field provides a clear, systematically organized introductory review of fundamental concepts for advanced graduate students, research workers. Numerous diagrams. Bibliography. 440pp. 5⅜ x 8½. 64917-2 Pa. $11.95

UNBOUNDED LINEAR OPERATORS: Theory and Applications, Seymour Goldberg. Classic presents systematic treatment of the theory of unbounded linear operators in normed linear spaces with applications to differential equations. Bibliography. I99pp. 5⅜ x 8½. 64830-3 Pa. $7.95

LIGHT SCATTERING BY SMALL PARTICLES, H.C. van de Hulst. Comprehensive treatment including full range of useful approximation methods for researchers in chemistry, meteorology and astronomy. 44 illustrations. 470pp. 5⅜ x 8½. 64228-3 Pa. $12.95

CONFORMAL MAPPING ON RIEMANN SURFACES, Harvey Cohn. Lucid, insightful book presents ideal coverage of subject. 334 exercises make book perfect for self-study. 55 figures. 352pp. 5⅜ x 8¼. 64025-6 Pa. $11.95

OPTICKS, Sir Isaac Newton. Newton's own experiments with spectroscopy, colors, lenses, reflection, refraction, etc., in language the layman can follow. Foreword by Albert Einstein. 532pp. 5⅜ x 8½. 60205-2 Pa. $12.95

GENERALIZED INTEGRAL TRANSFORMATIONS, A.H. Zemanian. Graduate-level study of recent generalizations of the Laplace, Mellin, Hankel, K. Weierstrass, convolution and other simple transformations. Bibliography. 320pp. 5⅜ x 8½. 65375-7 Pa. $8.95

THE ELECTROMAGNETIC FIELD, Albert Shadowitz. Comprehensive undergraduate text covers basics of electric and magnetic fields, builds up to electromagnetic theory. Also related topics, including relativity. Over 900 problems. 768pp. 5⅝ x 8¼. 65660-8 Pa. $18.95

FOURIER SERIES, Georgi P. Tolstov. Translated by Richard A. Silverman. A valuable addition to the literature on the subject, moving clearly from subject to subject and theorem to theorem. 107 problems, answers. 336pp. 5⅜ x 8½. 63317-9 Pa. $9.95

THEORY OF ELECTROMAGNETIC WAVE PROPAGATION, Charles Herach Papas. Graduate-level study discusses the Maxwell field equations, radiation from wire antennas, the Doppler effect and more. xiii + 244pp. 5⅜ x 8½. 65678-0 Pa. $6.95

DISTRIBUTION THEORY AND TRANSFORM ANALYSIS: An Introduction to Generalized Functions, with Applications, A.H. Zemanian. Provides basics of distribution theory, describes generalized Fourier and Laplace transformations. Numerous problems. 384pp. 5⅜ x 8½. 65479-6 Pa. $11.95

THE PHYSICS OF WAVES, William C. Elmore and Mark A. Heald. Unique overview of classical wave theory. Acoustics, optics, electromagnetic radiation, more. Ideal as classroom text or for self-study. Problems. 477pp. 5⅜ x 8½. 64926-1 Pa. $13.95

CALCULUS OF VARIATIONS WITH APPLICATIONS, George M. Ewing. Applications-oriented introduction to variational theory develops insight and promotes understanding of specialized books, research papers. Suitable for advanced undergraduate/graduate students as primary, supplementary text. 352pp. 5⅜ x 8½. 64856-7 Pa. $9.95

A TREATISE ON ELECTRICITY AND MAGNETISM, James Clerk Maxwell. Important foundation work of modern physics. Brings to final form Maxwell's theory of electromagnetism and rigorously derives his general equations of field theory. 1,084pp. 5⅜ x 8½. 60636-8, 60637-6 Pa., Two-vol. set $25.90

AN INTRODUCTION TO THE CALCULUS OF VARIATIONS, Charles Fox. Graduate-level text covers variations of an integral, isoperimetrical problems, least action, special relativity, approximations, more. References. 279pp. 5⅜ x 8½. 65499-0 Pa. $8.95

HYDRODYNAMIC AND HYDROMAGNETIC STABILITY, S. Chandrasekhar. Lucid examination of the Rayleigh-Benard problem; clear coverage of the theory of instabilities causing convection. 704pp. 5⅜ x 8¼. 64071-X Pa. $14.95

CALCULUS OF VARIATIONS, Robert Weinstock. Basic introduction covering isoperimetric problems, theory of elasticity, quantum mechanics, electrostatics, etc. Exercises throughout. 326pp. 5⅜ x 8½. 63069-2 Pa. $9.95

DYNAMICS OF FLUIDS IN POROUS MEDIA, Jacob Bear. For advanced students of ground water hydrology, soil mechanics and physics, drainage and irrigation engineering and more. 335 illustrations. Exercises, with answers. 784pp. 6⅛ x 9¼. 65675-6 Pa. $19.95

CATALOG OF DOVER BOOKS

NUMERICAL METHODS FOR SCIENTISTS AND ENGINEERS, Richard Hamming. Classic text stresses frequency approach in coverage of algorithms, polynomial approximation, Fourier approximation, exponential approximation, other topics. Revised and enlarged 2nd edition. 721pp. 5⅜ x 8½. 65241-6 Pa. $15.95

THEORETICAL SOLID STATE PHYSICS, Vol. 1: Perfect Lattices in Equilibrium; Vol. II: Non-Equilibrium and Disorder, William Jones and Norman H. March. Monumental reference work covers fundamental theory of equilibrium properties of perfect crystalline solids, non-equilibrium properties, defects and disordered systems. Appendices. Problems. Preface. Diagrams. Index. Bibliography. Total of 1,301pp. 5⅜ x 8½. Two volumes. Vol. I: 65015-4 Pa. $16.95
Vol. II: 65016-2 Pa. $16.95

OPTIMIZATION THEORY WITH APPLICATIONS, Donald A. Pierre. Broad spectrum approach to important topic. Classical theory of minima and maxima, calculus of variations, simplex technique and linear programming, more. Many problems, examples. 640pp. 5⅜ x 8½. 65205-X Pa. $16.95

THE CONTINUUM: A Critical Examination of the Foundation of Analysis, Hermann Weyl. Classic of 20th-century foundational research deals with the conceptual problem posed by the continuum. 156pp. 5⅜ x 8½. 67982-9 Pa. $6.95

ESSAYS ON THE THEORY OF NUMBERS, Richard Dedekind. Two classic essays by great German mathematician: on the theory of irrational numbers; and on transfinite numbers and properties of natural numbers. 115pp. 5⅜ x 8½. 21010-3 Pa. $5.95

THE FUNCTIONS OF MATHEMATICAL PHYSICS, Harry Hochstadt. Comprehensive treatment of orthogonal polynomials, hypergeometric functions, Hill's equation, much more. Bibliography. Index. 322pp. 5⅜ x 8½. 65214-9 Pa. $9.95

NUMBER THEORY AND ITS HISTORY, Oystein Ore. Unusually clear, accessible introduction covers counting, properties of numbers, prime numbers, much more. Bibliography. 380pp. 5⅜ x 8½. 65620-9 Pa. $10.95

THE VARIATIONAL PRINCIPLES OF MECHANICS, Cornelius Lanczos. Graduate level coverage of calculus of variations, equations of motion, relativistic mechanics, more. First inexpensive paperbound edition of classic treatise. Index. Bibliography. 418pp. 5⅜ x 8½. 65067-7 Pa. $12.95

MATHEMATICAL TABLES AND FORMULAS, Robert D. Carmichael and Edwin R. Smith. Logarithms, sines, tangents, trig functions, powers, roots, reciprocals, exponential and hyperbolic functions, formulas and theorems. 269pp. 5⅜ x 8½. 60111-0 Pa. $6.95

THEORETICAL PHYSICS, Georg Joos, with Ira M. Freeman. Classic overview covers essential math, mechanics, electromagnetic theory, thermodynamics, quantum mechanics, nuclear physics, other topics. First paperback edition. xxiii + 885pp. 5⅜ x 8½. 65227-0 Pa. $21.95

CATALOG OF DOVER BOOKS

HANDBOOK OF MATHEMATICAL FUNCTIONS WITH FORMULAS, GRAPHS, AND MATHEMATICAL TABLES, edited by Milton Abramowitz and Irene A. Stegun. Vast compendium: 29 sets of tables, some to as high as 20 places. 1,046pp. 8 x 10½. 61272-4 Pa. $26.95

MATHEMATICAL METHODS IN PHYSICS AND ENGINEERING, John W. Dettman. Algebraically based approach to vectors, mapping, diffraction, other topics in applied math. Also generalized functions, analytic function theory, more. Exercises. 448pp. 5⅜ x 8¼. 65649-7 Pa. $10.95

A SURVEY OF NUMERICAL MATHEMATICS, David M. Young and Robert Todd Gregory. Broad self-contained coverage of computer-oriented numerical algorithms for solving various types of mathematical problems in linear algebra, ordinary and partial, differential equations, much more. Exercises. Total of 1,248pp. 5⅜ x 8½.
Two volumes. Vol. I: 65691-8 Pa. $16.95
Vol. II: 65692-6 Pa. $16.95

TENSOR ANALYSIS FOR PHYSICISTS, J.A. Schouten. Concise exposition of the mathematical basis of tensor analysis, integrated with well-chosen physical examples of the theory. Exercises. Index. Bibliography. 289pp. 5⅜ x 8½. 65582-2 Pa. $8.95

INTRODUCTION TO NUMERICAL ANALYSIS (2nd Edition), F.B. Hildebrand. Classic, fundamental treatment covers computation, approximation, interpolation, numerical differentiation and integration, other topics. 150 new problems. 669pp. 5⅜ x 8½. 65363-3 Pa. $16.95

INVESTIGATIONS ON THE THEORY OF THE BROWNIAN MOVEMENT, Albert Einstein. Five papers (1905–8) investigating dynamics of Brownian motion and evolving elementary theory. Notes by R. Fürth. 122pp. 5⅜ x 8½.
60304-0 Pa. $5.95

CATASTROPHE THEORY FOR SCIENTISTS AND ENGINEERS, Robert Gilmore. Advanced-level treatment describes mathematics of theory grounded in the work of Poincaré, R. Thom, other mathematicians. Also important applications to problems in mathematics, physics, chemistry and engineering. 1981 edition. References. 28 tables. 397 black-and-white illustrations. xvii + 666pp. 6⅛ x 9¼.
67539-4 Pa. $17.95

AN INTRODUCTION TO STATISTICAL THERMODYNAMICS, Terrell L. Hill. Excellent basic text offers wide-ranging coverage of quantum statistical mechanics, systems of interacting molecules, quantum statistics, more. 523pp. 5⅜ x 8½.
65242-4 Pa. $12.95

STATISTICAL PHYSICS, Gregory H. Wannier. Classic text combines thermodynamics, statistical mechanics and kinetic theory in one unified presentation of thermal physics. Problems with solutions. Bibliography. 532pp. 5⅜ x 8½.
65401-X Pa. $12.95

CATALOG OF DOVER BOOKS

ORDINARY DIFFERENTIAL EQUATIONS, Morris Tenenbaum and Harry Pollard. Exhaustive survey of ordinary differential equations for undergraduates in mathematics, engineering, science. Thorough analysis of theorems. Diagrams. Bibliography. Index. 818pp. 5⅜ x 8½. 64940-7 Pa. $18.95

STATISTICAL MECHANICS: Principles and Applications, Terrell L. Hill. Standard text covers fundamentals of statistical mechanics, applications to fluctuation theory, imperfect gases, distribution functions, more. 448pp. 5⅜ x 8½. 65390-0 Pa. $11.95

ORDINARY DIFFERENTIAL EQUATIONS AND STABILITY THEORY: An Introduction, David A. Sánchez. Brief, modern treatment. Linear equation, stability theory for autonomous and nonautonomous systems, etc. 164pp. 5⅜ x 8¼. 63828-6 Pa. $6.95

THIRTY YEARS THAT SHOOK PHYSICS: The Story of Quantum Theory, George Gamow. Lucid, accessible introduction to influential theory of energy and matter. Careful explanations of Dirac's anti-particles, Bohr's model of the atom, much more. 12 plates. Numerous drawings. 240pp. 5⅜ x 8½. 24895-X Pa. $7.95

THEORY OF MATRICES, Sam Perlis. Outstanding text covering rank, nonsingularity and inverses in connection with the development of canonical matrices under the relation of equivalence, and without the intervention of determinants. Includes exercises. 237pp. 5⅜ x 8½. 66810-X Pa. $8.95

GREAT EXPERIMENTS IN PHYSICS: Firsthand Accounts from Galileo to Einstein, edited by Morris H. Shamos. 25 crucial discoveries: Newton's laws of motion, Chadwick's study of the neutron, Hertz on electromagnetic waves, more. Original accounts clearly annotated. 370pp. 5⅜ x 8½. 25346-5 Pa. $10.95

INTRODUCTION TO PARTIAL DIFFERENTIAL EQUATIONS WITH APPLICATIONS, E.C. Zachmanoglou and Dale W. Thoe. Essentials of partial differential equations applied to common problems in engineering and the physical sciences. Problems and answers. 416pp. 5⅜ x 8½. 65251-3 Pa. $11.95

BURNHAM'S CELESTIAL HANDBOOK, Robert Burnham, Jr. Thorough guide to the stars beyond our solar system. Exhaustive treatment. Alphabetical by constellation: Andromeda to Cetus in Vol. 1; Chamaeleon to Orion in Vol. 2; and Pavo to Vulpecula in Vol. 3. Hundreds of illustrations. Index in Vol. 3. 2,000pp. 6⅛ x 9¼. 23567-X, 23568-8, 23673-0 Pa., Three-vol. set $44.85

CHEMICAL MAGIC, Leonard A. Ford. Second Edition, Revised by E. Winston Grundmeier. Over 100 unusual stunts demonstrating cold fire, dust explosions, much more. Text explains scientific principles and stresses safety precautions. 128pp. 5⅜ x 8½. 67628-5 Pa. $5.95

AMATEUR ASTRONOMER'S HANDBOOK, J.B. Sidgwick. Timeless, comprehensive coverage of telescopes, mirrors, lenses, mountings, telescope drives, micrometers, spectroscopes, more. 189 illustrations. 576pp. 5⅜ x 8¼. (Available in U.S. only) 24034-7 Pa. $11.95

CATALOG OF DOVER BOOKS

SPECIAL FUNCTIONS, N.N. Lebedev. Translated by Richard Silverman. Famous Russian work treating more important special functions, with applications to specific problems of physics and engineering. 38 figures. 308pp. 5⅜ x 8½. 60624-4 Pa. $9.95

OBSERVATIONAL ASTRONOMY FOR AMATEURS, J.B. Sidgwick. Mine of useful data for observation of sun, moon, planets, asteroids, aurorae, meteors, comets, variables, binaries, etc. 39 illustrations. 384pp. 5⅜ x 8¼. (Available in U.S. only) 24033-9 Pa. $8.95

INTEGRAL EQUATIONS, F.G. Tricomi. Authoritative, well-written treatment of extremely useful mathematical tool with wide applications. Volterra Equations, Fredholm Equations, much more. Advanced undergraduate to graduate level. Exercises. Bibliography. 238pp. 5⅜ x 8½. 64828-1 Pa. $8.95

POPULAR LECTURES ON MATHEMATICAL LOGIC, Hao Wang. Noted logician's lucid treatment of historical developments, set theory, model theory, recursion theory and constructivism, proof theory, more. 3 appendixes. Bibliography. 1981 edition. ix + 283pp. 5⅜ x 8½. 67632-3 Pa. $8.95

MODERN NONLINEAR EQUATIONS, Thomas L. Saaty. Emphasizes practical solution of problems; covers seven types of equations. ". . . a welcome contribution to the existing literature...."–*Math Reviews*. 490pp. 5⅜ x 8½. 64232-1 Pa. $13.95

FUNDAMENTALS OF ASTRODYNAMICS, Roger Bate et al. Modern approach developed by U.S. Air Force Academy. Designed as a first course. Problems, exercises. Numerous illustrations. 455pp. 5⅜ x 8½. 60061-0 Pa. $10.95

INTRODUCTION TO LINEAR ALGEBRA AND DIFFERENTIAL EQUATIONS, John W. Dettman. Excellent text covers complex numbers, determinants, orthonormal bases, Laplace transforms, much more. Exercises with solutions. Undergraduate level. 416pp. 5⅜ x 8½. 65191-6 Pa. $11.95

INCOMPRESSIBLE AERODYNAMICS, edited by Bryan Thwaites. Covers theoretical and experimental treatment of the uniform flow of air and viscous fluids past two-dimensional aerofoils and three-dimensional wings; many other topics. 654pp. 5⅜ x 8½. 65465-6 Pa. $16.95

INTRODUCTION TO DIFFERENCE EQUATIONS, Samuel Goldberg. Exceptionally clear exposition of important discipline with applications to sociology, psychology, economics. Many illustrative examples; over 250 problems. 260pp. 5⅜ x 8½. 65084-7 Pa. $8.95

LAMINAR BOUNDARY LAYERS, edited by L. Rosenhead. Engineering classic covers steady boundary layers in two- and three- dimensional flow, unsteady boundary layers, stability, observational techniques, much more. 708pp. 5⅜ x 8½. 65646-2 Pa. $18 95

LECTURES ON CLASSICAL DIFFERENTIAL GEOMETRY, Second Edition, Dirk J. Struik. Excellent brief introduction covers curves, theory of surfaces, fundamental equations, geometry on a surface, conformal mapping, other topics. Problems. 240pp. 5⅜ x 8½. 65609-8 Pa. $8.95

CATALOG OF DOVER BOOKS

ROTARY-WING AERODYNAMICS, W.Z. Stepniewski. Clear, concise text covers aerodynamic phenomena of the rotor and offers guidelines for helicopter performance evaluation. Originally prepared for NASA. 537 figures. 640pp. 6⅛ x 9¼.
64647-5 Pa. $16.95

DIFFERENTIAL GEOMETRY, Heinrich W. Guggenheimer. Local differential geometry as an application of advanced calculus and linear algebra. Curvature, transformation groups, surfaces, more. Exercises. 62 figures. 378pp. 5⅜ x 8½.
63433-7 Pa. $9.95

INTRODUCTION TO SPACE DYNAMICS, William Tyrrell Thomson. Comprehensive, classic introduction to space-flight engineering for advanced undergraduate and graduate students. Includes vector algebra, kinematics, transformation of coordinates. Bibliography. Index. 352pp. 5⅜ x 8½. 65113-4 Pa. $9.95

A SURVEY OF MINIMAL SURFACES, Robert Osserman. Up-to-date, in-depth discussion of the field for advanced students. Corrected and enlarged edition covers new developments. Includes numerous problems. 192pp. 5⅜ x 8½. 64998-9 Pa. $8.95

ANALYTICAL MECHANICS OF GEARS, Earle Buckingham. Indispensable reference for modern gear manufacture covers conjugate gear-tooth action, gear-tooth profiles of various gears, many other topics. 263 figures. 102 tables. 546pp. 5⅜ x 8½.
65712-4 Pa. $14.95

SET THEORY AND LOGIC, Robert R. Stoll. Lucid introduction to unified theory of mathematical concepts. Set theory and logic seen as tools for conceptual understanding of real number system. 496pp. 5⅜ x 8¼. 63829-4 Pa. $12.95

A HISTORY OF MECHANICS, René Dugas. Monumental study of mechanical principles from antiquity to quantum mechanics. Contributions of ancient Greeks, Galileo, Leonardo, Kepler, Lagrange, many others. 671pp. 5⅜ x 8½.
65632-2 Pa. $14.95

FAMOUS PROBLEMS OF GEOMETRY AND HOW TO SOLVE THEM, Benjamin Bold. Squaring the circle, trisecting the angle, duplicating the cube: learn their history, why they are impossible to solve, then solve them yourself. 128pp. 5⅜ x 8½. 24297-8 Pa. $4.95

MECHANICAL VIBRATIONS, J.P. Den Hartog. Classic textbook offers lucid explanations and illustrative models, applying theories of vibrations to a variety of practical industrial engineering problems. Numerous figures. 233 problems, solutions. Appendix. Index. Preface. 436pp. 5⅜ x 8½. 64785-4 Pa. $11.95

CURVATURE AND HOMOLOGY, Samuel I. Goldberg. Thorough treatment of specialized branch of differential geometry. Covers Riemannian manifolds, topology of differentiable manifolds, compact Lie groups, other topics. Exercises. 315pp. 5⅜ x 8½. 64314-X Pa. $9.95

HISTORY OF STRENGTH OF MATERIALS, Stephen P. Timoshenko. Excellent historical survey of the strength of materials with many references to the theories of elasticity and structure. 245 figures. 452pp. 5⅜ x 8½. 61187-6 Pa. $12.95

CATALOG OF DOVER BOOKS

GEOMETRY OF COMPLEX NUMBERS, Hans Schwerdtfeger. Illuminating, widely praised book on analytic geometry of circles, the Moebius transformation, and two-dimensional non-Euclidean geometries. 200pp. 5⅜ x 8¼. 63830-8 Pa. $8.95

MECHANICS, J.P. Den Hartog. A classic introductory text or refresher. Hundreds of applications and design problems illuminate fundamentals of trusses, loaded beams and cables, etc. 334 answered problems. 462pp. 5⅜ x 8½. 60754-2 Pa. $11.95

TOPOLOGY, John G. Hocking and Gail S. Young. Superb one-year course in classical topology. Topological spaces and functions, point-set topology, much more. Examples and problems. Bibliography. Index. 384pp. 5⅜ x 8¼. 65676-4 Pa. $10.95

STRENGTH OF MATERIALS, J.P. Den Hartog. Full, clear treatment of basic material (tension, torsion, bending, etc.) plus advanced material on engineering methods, applications. 350 answered problems. 323pp. 5⅜ x 8½. 60755-0 Pa. $9.95

ELEMENTARY CONCEPTS OF TOPOLOGY, Paul Alexandroff. Elegant, intuitive approach to topology from set-theoretic topology to Betti groups; how concepts of topology are useful in math and physics. 25 figures. 57pp. 5⅜ x 8½.
60747-X Pa. $3.95

ADVANCED STRENGTH OF MATERIALS, J.P. Den Hartog. Superbly written advanced text covers torsion, rotating disks, membrane stresses in shells, much more. Many problems and answers. 388pp. 5⅜ x 8½. 65407-9 Pa. $10.95

COMPUTABILITY AND UNSOLVABILITY, Martin Davis. Classic graduate-level introduction to theory of computability, usually referred to as theory of recurrent functions. New preface and appendix. 288pp. 5⅜ x 8½. 61471-9 Pa. $8.95

GENERAL CHEMISTRY, Linus Pauling. Revised 3rd edition of classic first-year text by Nobel laureate. Atomic and molecular structure, quantum mechanics, statistical mechanics, thermodynamics correlated with descriptive chemistry. Problems. 992pp. 5⅜ x 8½. 65622-5 Pa. $19.95

AN INTRODUCTION TO MATRICES, SETS AND GROUPS FOR SCIENCE STUDENTS, G. Stephenson. Concise, readable text introduces sets, groups, and most importantly, matrices to undergraduate students of physics, chemistry, and engineering. Problems. 164pp. 5⅜ x 8½. 65077-4 Pa. $7.95

THE HISTORICAL BACKGROUND OF CHEMISTRY, Henry M. Leicester. Evolution of ideas, not individual biography. Concentrates on formulation of a coherent set of chemical laws. 260pp. 5⅜ x 8½. 61053-5 Pa. $8.95

THE PHILOSOPHY OF MATHEMATICS: An Introductory Essay, Stephan Körner. Surveys the views of Plato, Aristotle, Leibniz & Kant concerning propositions and theories of applied and pure mathematics. Introduction. Two appendices. Index. 198pp. 5⅜ x 8½. 25048-2 Pa. $8.95

THE DEVELOPMENT OF MODERN CHEMISTRY, Aaron J. Ihde. Authoritative history of chemistry from ancient Greek theory to 20th-century innovation. Covers major chemists and their discoveries. 209 illustrations. 14 tables. Bibliographies. Indices. Appendices. 851pp. 5⅜ x 8½. 64235-6 Pa. $18.95

CATALOG OF DOVER BOOKS

DE RE METALLICA, Georgius Agricola. The famous Hoover translation of greatest treatise on technological chemistry, engineering, geology, mining of early modern times (1556). All 289 original woodcuts. 638pp. 6¾ x 11. 60006-8 Pa. $21.95

SOME THEORY OF SAMPLING, William Edwards Deming. Analysis of the problems, theory and design of sampling techniques for social scientists, industrial managers and others who find statistics increasingly important in their work. 61 tables. 90 figures. xvii + 602pp. 5⅜ x 8½. 64684-X Pa. $16.95

THE VARIOUS AND INGENIOUS MACHINES OF AGOSTINO RAMELLI: A Classic Sixteenth-Century Illustrated Treatise on Technology, Agostino Ramelli. One of the most widely known and copied works on machinery in the 16th century. 194 detailed plates of water pumps, grain mills, cranes, more. 608pp. 9 x 12.
28180-9 Pa. $24.95

LINEAR PROGRAMMING AND ECONOMIC ANALYSIS, Robert Dorfman, Paul A. Samuelson and Robert M. Solow. First comprehensive treatment of linear programming in standard economic analysis. Game theory, modern welfare economics, Leontief input-output, more. 525pp. 5⅜ x 8½. 65491-5 Pa. $14.95

ELEMENTARY DECISION THEORY, Herman Chernoff and Lincoln E. Moses. Clear introduction to statistics and statistical theory covers data processing, probability and random variables, testing hypotheses, much more. Exercises. 364pp. 5⅜ x 8½. 65218-1 Pa. $10.95

THE COMPLEAT STRATEGYST: Being a Primer on the Theory of Games of Strategy, J.D. Williams. Highly entertaining classic describes, with many illustrated examples, how to select best strategies in conflict situations. Prefaces. Appendices. 268pp. 5⅜ x 8½. 25101-2 Pa. $7.95

CONSTRUCTIONS AND COMBINATORIAL PROBLEMS IN DESIGN OF EXPERIMENTS, Damaraju Raghavarao. In-depth reference work examines orthogonal Latin squares, incomplete block designs, tactical configuration, partial geometry, much more. Abundant explanations, examples. 416pp. 5⅜ x 8¼.
65685-3 Pa. $10.95

THE ABSOLUTE DIFFERENTIAL CALCULUS (CALCULUS OF TENSORS), Tullio Levi-Civita. Great 20th-century mathematician's classic work on material necessary for mathematical grasp of theory of relativity. 452pp. 5⅜ x 8½. 63401-9 Pa. $11.95

VECTOR AND TENSOR ANALYSIS WITH APPLICATIONS, A.I. Borisenko and I.E. Tarapov. Concise introduction. Worked-out problems, solutions, exercises. 257pp. 5⅜ x 8¼. 63833-2 Pa. $8.95

THE FOUR-COLOR PROBLEM: Assaults and Conquest, Thomas L. Saaty and Paul G. Kainen. Engrossing, comprehensive account of the century-old combinatorial topological problem, its history and solution. Bibliographies. Index. 110 figures. 228pp. 5⅜ x 8½. 65092-8 Pa. $7.95

CATALOG OF DOVER BOOKS

CATALYSIS IN CHEMISTRY AND ENZYMOLOGY, William P. Jencks. Exceptionally clear coverage of mechanisms for catalysis, forces in aqueous solution, carbonyl- and acyl-group reactions, practical kinetics, more. 864pp. 5⅜ x 8½. 65460-5 Pa. $19.95

PROBABILITY: An Introduction, Samuel Goldberg. Excellent basic text covers set theory, probability theory for finite sample spaces, binomial theorem, much more. 360 problems. Bibliographies. 322pp. 5⅜ x 8½. 65252-1 Pa. $10.95

LIGHTNING, Martin A. Uman. Revised, updated edition of classic work on the physics of lightning. Phenomena, terminology, measurement, photography, spectroscopy, thunder, more. Reviews recent research. Bibliography. Indices. 320pp. 5⅜ x 8¼. 64575-4 Pa. $8.95

PROBABILITY THEORY: A Concise Course, Y.A. Rozanov. Highly readable, self-contained introduction covers combination of events, dependent events, Bernoulli trials, etc. Translation by Richard Silverman. 148pp. 5⅜ x 8¼. 63544-9 Pa. $7.95

AN INTRODUCTION TO HAMILTONIAN OPTICS, H. A. Buchdahl. Detailed account of the Hamiltonian treatment of aberration theory in geometrical optics. Many classes of optical systems defined in terms of the symmetries they possess. Problems with detailed solutions. 1970 edition. xv + 360pp. 5⅜ x 8½. 67597-1 Pa. $10.95

STATISTICS MANUAL, Edwin L. Crow, et al. Comprehensive, practical collection of classical and modern methods prepared by U.S. Naval Ordnance Test Station. Stress on use. Basics of statistics assumed. 288pp. 5⅜ x 8½. 60599-X Pa. $7.95

DICTIONARY/OUTLINE OF BASIC STATISTICS, John E. Freund and Frank J. Williams. A clear concise dictionary of over 1,000 statistical terms and an outline of statistical formulas covering probability, nonparametric tests, much more. 208pp. 5⅜ x 8½. 66796-0 Pa. $7.95

STATISTICAL METHOD FROM THE VIEWPOINT OF QUALITY CONTROL, Walter A. Shewhart. Important text explains regulation of variables, uses of statistical control to achieve quality control in industry, agriculture, other areas. 192pp. 5⅜ x 8½. 65232-7 Pa. $7.95

METHODS OF THERMODYNAMICS, Howard Reiss. Outstanding text focuses on physical technique of thermodynamics, typical problem areas of understanding, and significance and use of thermodynamic potential. 1965 edition. 238pp. 5⅜ x 8½. 69445-3 Pa. $8.95

STATISTICAL ADJUSTMENT OF DATA, W. Edwards Deming. Introduction to basic concepts of statistics, curve fitting, least squares solution, conditions without parameter, conditions containing parameters. 26 exercises worked out. 271pp. 5⅜ x 8½. 64685-8 Pa. $9.95

TENSOR CALCULUS, J.L. Synge and A. Schild. Widely used introductory text covers spaces and tensors, basic operations in Riemannian space, non-Riemannian spaces, etc. 324pp. 5⅜ x 8¼. 63612-7 Pa. $9.95

CATALOG OF DOVER BOOKS

A CONCISE HISTORY OF MATHEMATICS, Dirk J. Struik. The best brief history of mathematics. Stresses origins and covers every major figure from ancient Near East to 19th century. 41 illustrations. 195pp. 5⅜ x 8½. 60255-9 Pa. $8.95

A SHORT ACCOUNT OF THE HISTORY OF MATHEMATICS, W.W. Rouse Ball. One of clearest, most authoritative surveys from the Egyptians and Phoenicians through 19th-century figures such as Grassman, Galois, Riemann. Fourth edition. 522pp. 5⅜ x 8½. 20630-0 Pa. $11.95

HISTORY OF MATHEMATICS, David E. Smith. Nontechnical survey from ancient Greece and Orient to late 19th century; evolution of arithmetic, geometry, trigonometry, calculating devices, algebra, the calculus. 362 illustrations. 1,355pp. 5⅜ x 8½. 20429-4, 20430-8 Pa., Two-vol. set $26.90

THE GEOMETRY OF RENÉ DESCARTES, René Descartes. The great work founded analytical geometry. Original French text, Descartes' own diagrams, together with definitive Smith-Latham translation. 244pp. 5⅜ x 8½. 60068-8 Pa. $8.95

THE ORIGINS OF THE INFINITESIMAL CALCULUS, Margaret E. Baron. Only fully detailed and documented account of crucial discipline: origins; development by Galileo, Kepler, Cavalieri; contributions of Newton, Leibniz, more. 304pp. 5⅜ x 8½. (Available in U.S. and Canada only) 65371-4 Pa. $9.95

THE HISTORY OF THE CALCULUS AND ITS CONCEPTUAL DEVELOPMENT, Carl B. Boyer. Origins in antiquity, medieval contributions, work of Newton, Leibniz, rigorous formulation. Treatment is verbal. 346pp. 5⅜ x 8½. 60509-4 Pa. $9.95

THE THIRTEEN BOOKS OF EUCLID'S ELEMENTS, translated with introduction and commentary by Sir Thomas L. Heath. Definitive edition. Textual and linguistic notes, mathematical analysis. 2,500 years of critical commentary. Not abridged. 1,414pp. 5⅜ x 8½. 60088-2, 60089-0, 60090-4 Pa., Three-vol. set $32.85

GAMES AND DECISIONS: Introduction and Critical Survey, R. Duncan Luce and Howard Raiffa. Superb nontechnical introduction to game theory, primarily applied to social sciences. Utility theory, zero-sum games, n-person games, decision-making, much more. Bibliography. 509pp. 5⅜ x 8½. 65943-7 Pa. $13.95

THE HISTORICAL ROOTS OF ELEMENTARY MATHEMATICS, Lucas N.H. Bunt, Phillip S. Jones, and Jack D. Bedient. Fundamental underpinnings of modern arithmetic, algebra, geometry and number systems derived from ancient civilizations. 320pp. 5⅜ x 8½. 25563-8 Pa. $8.95

CALCULUS REFRESHER FOR TECHNICAL PEOPLE, A. Albert Klaf. Covers important aspects of integral and differential calculus via 756 questions. 566 problems, most answered. 431pp. 5⅜ x 8½. 20370-0 Pa. $8.95

CATALOG OF DOVER BOOKS

CHALLENGING MATHEMATICAL PROBLEMS WITH ELEMENTARY SOLUTIONS, A.M. Yaglom and I.M. Yaglom. Over 170 challenging problems on probability theory, combinatorial analysis, points and lines, topology, convex polygons, many other topics. Solutions. Total of 445pp. 5⅜ x 8½. Two-vol. set.
Vol. I: 65536-9 Pa. $7.95
Vol. II: 65537-7 Pa. $7.95

FIFTY CHALLENGING PROBLEMS IN PROBABILITY WITH SOLUTIONS, Frederick Mosteller. Remarkable puzzlers, graded in difficulty, illustrate elementary and advanced aspects of probability. Detailed solutions. 88pp. 5⅜ x 8½.
65355-2 Pa. $4.95

EXPERIMENTS IN TOPOLOGY, Stephen Barr. Classic, lively explanation of one of the byways of mathematics. Klein bottles, Moebius strips, projective planes, map coloring, problem of the Koenigsberg bridges, much more, described with clarity and wit. 43 figures. 210pp. 5⅜ x 8½. 25933-1 Pa. $6.95

RELATIVITY IN ILLUSTRATIONS, Jacob T. Schwartz. Clear nontechnical treatment makes relativity more accessible than ever before. Over 60 drawings illustrate concepts more clearly than text alone. Only high school geometry needed. Bibliography. 128pp. 6⅛ x 9¼. 25965-X Pa. $7.95

AN INTRODUCTION TO ORDINARY DIFFERENTIAL EQUATIONS, Earl A. Coddington. A thorough and systematic first course in elementary differential equations for undergraduates in mathematics and science, with many exercises and problems (with answers). Index. 304pp. 5⅜ x 8½. 65942-9 Pa. $8.95

FOURIER SERIES AND ORTHOGONAL FUNCTIONS, Harry F. Davis. An incisive text combining theory and practical example to introduce Fourier series, orthogonal functions and applications of the Fourier method to boundary-value problems. 570 exercises. Answers and notes. 416pp. 5⅜ x 8½. 65973-9 Pa. $11.95

AN INTRODUCTION TO ALGEBRAIC STRUCTURES, Joseph Landin. Superb self-contained text covers "abstract algebra": sets and numbers, theory of groups, theory of rings, much more. Numerous well-chosen examples, exercises. 247pp. 5⅜ x 8½.
65940-2 Pa. $8.95

STARS AND RELATIVITY, Ya. B. Zel'dovich and I. D. Novikov. Vol. 1 of *Relativistic Astrophysics* by famed Russian scientists. General relativity, properties of matter under astrophysical conditions, stars and stellar systems. Deep physical insights, clear presentation. 1971 edition. References. 544pp. 5⅜ x 8½.
69424-0 Pa. $14.95

Prices subject to change without notice.
Available at your book dealer or write for free Mathematics and Science Catalog to Dept. GI, Dover Publications, Inc., 31 East 2nd St., Mineola, N.Y. 11501. Dover publishes more than 250 books each year on science, elementary and advanced mathematics, biology, music, art, literature, history, social sciences and other areas.